Techniques and Concepts of High-Energy Physics V

NATO ASI Series

Advanced Science Institutes Series

A series presenting the results of activities sponsored by the NATO Science Committee, which aims at the dissemination of advanced scientific and technological knowledge, with a view to strengthening links between scientific communities.

The series is published by an international board of publishers in conjunction with the NATO Scientific Affairs Division

A	**Life Sciences**	Plenum Publishing Corporation
B	**Physics**	New York and London
C	**Mathematical and Physical Sciences**	Kluwer Academic Publishers Dordrecht, Boston, and London
D	**Behavioral and Social Sciences**	
E	**Applied Sciences**	
F	**Computer and Systems Sciences**	Springer-Verlag
G	**Ecological Sciences**	Berlin, Heidelberg, New York, London,
H	**Cell Biology**	Paris, and Tokyo

Recent Volumes in this Series

Series B: Physics

Techniques and Concepts of High-Energy Physics V

Edited by

Thomas Ferbel

University of Rochester
Rochester, New York

Springer Science+Business Media, LLC

Proceedings of the Fifth NATO Advanced Study Institute on
Techniques and Concepts of High-Energy Physics,
held July 14–25, 1988,
in St. Croix, U.S. Virgin Islands

ISBN 978-1-4615-8003-4   ISBN 978-1-4615-8001-0 (eBook)
DOI 10.1007/978-1-4615-8001-0

Library of Congress Cataloging in Publication Data

NATO Advanced Study Institute on Techniques and Concepts of High-Energy
Physics (5th: 1988: Saint Croix, V.I.)
    Techniques and concepts of high-energy physics V / edited by Thomas
Ferbel.
        p.   cm—(NATO ASI series. Series B, Physics; v. 204)
    "Proceedings of the Fifth NATO Advanced Study Institute on Techniques and
Concepts of High-Energy Physics, held July 14–25, 1988, in St. Croix, U.S. Virgin
Islands"—Verso of t.p.
    "Published in cooperation with NATO Scientific Affairs Division."
    Includes bibliographical references.

    1. Particles (Nuclear physics)—Congresses. I. Ferbel, Thomas. II. North Atlan-
tic Treaty Organization. Scientific Affairs Division. III. Title. IV. Series.
QC793.N38   1988   89-22990
539.7′2-dc20   CIP

© 1990 Springer Science+Business Media New York
Originally published by Plenum Press, New York in 1990
Softcover reprint of the hardcover 1st edition 1990

PREFACE

The fifth Advanced Study Institute (ASI) on Techniques and Concepts of High Energy Physics was held again at the Hotel on the Cay, in the scenic harbor of Christiansted, St. Croix, U. S. Virgin Islands. The ASI brought together a total of 71 participants, from 17 different countries. It was another great success, due to the dedication of the inspiring lecturers, the exceptional study body, and, of course, the beautiful setting.

The primary support for the meeting was again provided by the Scientific Affairs Division of NATO. The ASI was cosponsored by the U.S. Department of Energy, by Fermilab, by the National Science Foundation, and by the University of Rochester. A special contribution from the Oliver S. and Jennie R. Donaldson Charitable Trust provided an important degree of flexibility, as well as support for worthy students from developing nations.

As in the case of the previous ASI's, the scientific program was designed for advanced graduate students and recent PhD recipients in experimental particle physics. The present volume of lectures should complement the material published in the first four ASI's, and prove to be of value to a wider audience of physicists.

It is a pleasure to acknowledge the encouragement and support that I have continued to receive from colleagues and friends in organizing this meeting. I am indebted to the members of my Advisory Committee for their infinite patience and excellent advice. I am grateful to my distinguished lecturers for their enthusiasm and participation in the ASI, and, of course for their hard work in preparing the lectures and providing the superb manuscripts for the proceedings. I thank Scott Ogg of the West Indies Lab for his fascinating description of the geology and marine life of St. Croix, and Albert Lang for talking him into this. I thank Frederic Perrier for organizing the student presentations. I also thank Earle Fowler, Bernard Hildebrand and Bill Wallenmeyer for support from the Department of Energy, David Berley for assistance from the National Science Foundation, and Leon Lederman for providing me with access to the talents of Angela Gonzales at Fermilab. At Rochester, I am indebted to Judy Mack, Sal Spinnichia, and especially Connie Jones, for organizational assistance and typing. I owe thanks to Andrew Pappas and the managers of the facilities at the Hotel on the Cay, for their and their staff's hospitality and to Margi Levy and her colleagues at Southerland Tours in Christiansted who have been helping me for years with local arrangements. I wish to acknowledge the generosity of Chris Lirakis and Mrs. Marjorie Atwood of the Donaldson Trust, and support of George Blanar of LeCroy Research Systems Corp. Finally, I thank Luis da Cunha of NATO for his cooperation and confidence.

T. Ferbel
Rochester, New York
April 1989

LECTURERS

U. Amaldi	CERN
J. Dorfan	Stanford Linear Accelerator Center
G. Giacomelli	University of Bologna
J. Harvey	Rutherford Appleton Laboratory
L. Ibanez	University of Madrid
G. Martinelli	University of Rome
S. Peggs	Central Design Group of the SSC
M. Regler	Austrian Academy of Sciences
B. Winstein	University of Chicago

SCIENTIFIC ADVISORY COMMITTTEE

M. Jacob	CERN
R. Palmer	Brookhaven National Laboratory
R. Peccei	DESY
D. Perkins	Oxford University
C. Quigg	Central Design Group of the SSC and FNAL
P. Soding	DESY
R. Taylor	Stanford Linear Accelerator Center
M. Tigner	Central Design Group of the SSC and Cornell University

SCIENTIFIC DIRECTOR

T. Ferbel   University of Rochester

CONTENTS

ASPECTS OF THE PHYSICS OF THE STANDARD MODEL

Guido MARTINELLI

INFN - Istituto Nazionale Fisica Nucleare and
Dipartimento di Fisica, Università di Roma "La Sapienza"
P.le A.Moro 2, 00185 Roma, Italy

ABSTRACT

Some selected aspects of the physics of the Standard Model are reviewed. In particular the present status of the measurement of the basic parameters of the Standard Model and their comparison with theoretical predictions, including radiative corrections, are discussed, together with W and Z^0 physics at collider energies and hadron production of heavy flavours.

1. INTRODUCTION

In this series of lectures I have selected some aspects of the physics of the Standard Model[1] which seemed to me of particular interest at the time of the St.Croix School.

The Standard SU(2)xU(1) model has been so far spectacularly successfull. It has provided a consistent picture of QED and of the Fermi theory of weak charged current interactions (including parity violation, quark mixing, CP violation etc.) and predicted the weak neutral current interaction and the existence and properties of the W and Z^0 bosons. However all physicists believe that the Standard Model cannot be the ultimate theory of nature since it leaves too many questions unanswered. Including QCD the model has too many free parameters (~ 20) and give no explanation for the existence of three distinct gauge couplings, the equality of the proton and electron electromagnetic charge, the quark and lepton families, mixing angles and masses. Furthermore it requires the fine tuning of some parameters (as the QCD θ-parameter) to avoid conflict with experiments.

An important sector of the present and future experimental and theoretical work is to subject the Standard Model to diverse and stringent tests to find indications on new physics and map out the excluded domains. In sect.2 and 3 I will review the present test of the Standard Model, including the effects of radiative corrections. The physics of the W and Z^0 at collider energies, where the Standard Model has found its more dramatic confirmation, will be discussed in sec.4.

In sect.5 I will review another subject, related to the physics of the Standard Model, i.e. the theoretical and experimental aspects of the production of heavy flavours in hadronic collisions. This subject is particularly relevant for the search of the top quark at CERN and FNAL and important theoretical advances have been done in the last year.

In the course of the discussion other related topics, as upper limits on new quark/lepton doublets, the production of extra W and Z heavy vector bosons or extra light neutrinos species will be briefly discussed as complementary results to the tests of the Standard Model.

A more complete review of the physics beyond the Standard Model can be found in the lectures of L.Ibanez. The low energy phenomenology of weak interactions in kaon-decays, with particular attention to CP violation, is discussed in the parallel lectures by B.Winstein.

2. THE STANDARD MODEL

The standard model is a gauge theory which describes strong, weak and electro-magnetic interactions. The model is based on a SU(3) group of colour, a SU(2) group of weak isospin and a U(1) group of weak hypercharge. The SU(2)xU(1) symmetry is spontaneously broken with the help of scalar mesons called Higgs scalars: in the resulting theory we find two charged and one neutral massive bosons, $W^\pm$ and Z^o, and one massless neutral vector boson, the photon[1]. The Lagrangian which describes the interaction of these vector bosons with fermion fields is given by

$$
(1) \qquad \mathcal{L} = \frac{g}{\sqrt{2}} \; (J_\mu^- W_\mu^+ + J_\mu^+ W_\mu^-) \; \overset{\text{weak neutral currents}}{+ \frac{g}{\cos \theta_w} \; (J_\mu^3 - \sin^2 \theta_w \, J_\mu^{e.m.}) \, Z_\mu +}
$$

$$
\overset{\text{electromagnetic currents}}{+ \, g \, \sin \, \theta_w \, J_\mu^{e.m} \, A_\mu} \quad ,
$$

where $J_\mu^\pm$ are the weak-isospin currents, $J_\mu^{e.m.}$ is the electromagnetic current, $tg\theta_w = g'/g$ and g and g' are the couplings associated with the gauged SU(2) and U(1) group, respectively.

From eq.(1), it follows that

$$
(2) \qquad e = g \sin\theta_w \; ,
$$

where e is the electric charge. The currents in eq.(1) are given once the isospin and hypercharge assignment of fermions is known. To describe present phenomenology, we put the left-handed fermions into isospin doublets:

$$\begin{pmatrix} \nu_e \\ e^- \end{pmatrix}_L , \begin{pmatrix} \nu_\mu \\ \mu^- \end{pmatrix}_L , \begin{pmatrix} \nu_\tau \\ \tau^- \end{pmatrix}_L , \qquad \text{leptons,}$$

(3)
$$\begin{pmatrix} u \\ d' \end{pmatrix}_L , \begin{pmatrix} c \\ s' \end{pmatrix}_L , \begin{pmatrix} t \\ b' \end{pmatrix}_L , \qquad \text{quarks,}$$

In eq.(3) u, c, t are the charge $+\frac{2}{3}$ quarks and d', s', b' are related by a unitary transformation to the mass eigenstate charge $-\frac{1}{3}$ d, s and b quarks. Right-handed fermions are weak-isospin singlets.

W exchanges give rise to charged-current processes like, for example, the muon decay into electron plus neutrinos:

(4) $\mu^- \rightarrow e^- + \bar\nu_e + \nu_\mu$.

This process is illustrated in fig. 1. Since the momentum trasfer $|q^2| \ll M_W^2$, the amplitude for this process can be obtained by an effective Hamiltonian of the form

(5) $H^{eff} = \dfrac{g^2}{8M_W^2} \{ [\bar\nu_\mu \gamma^\mu(1 - \gamma_5) \mu] [\bar e\gamma_\mu(1 - \gamma_5)\nu_e] + h.c. \}.$

We can then relate the coupling g and the W mass to the Fermi constant G_F:

(6) $\dfrac{G_F}{\sqrt 2} = \dfrac{g^2}{8M_W^2}$, $M_W^2 = \dfrac{\pi\alpha}{\sqrt 2\, G_F \sin^2\theta_w}$,

where α is the fine-structure constant.

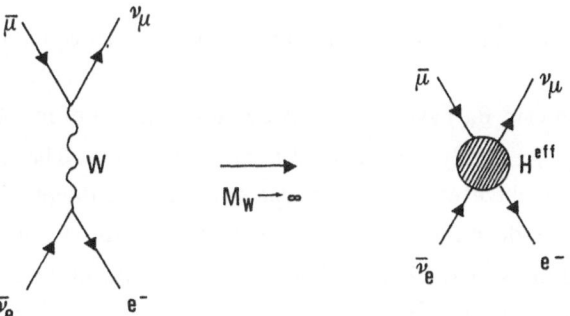

Fig.1. - Relevant diagram for the muon decay. For $M_W \rightarrow \infty$, the diagram is described by an effective four-fermion Hamiltonian.

In analogy with this particular case, the low-energy effective Hamiltonian for weak interactions takes the form

$$(7) \qquad H^{\text{eff}} = \frac{4G_F}{\sqrt{2}} \, [J_\mu^+ J_\mu^- + \rho(J_\mu^3 - \sin^2\theta_w \, f_\mu^{\cdot m \cdot})^2] \, .$$

In eq.(7), we have introduced the parameter ρ [2]:

$$(8) \qquad \rho = \frac{M_w^2}{M_z^2 \cos^2\theta_w} = \frac{\displaystyle\sum_i [I^{(i)}(I^{(i)} + 1) - I_3^{(i)2}] \, \lambda_{(i)}^2}{2\displaystyle\sum_i I_3^{(i)2} \, \lambda_{(i)}^2} \, ,$$

where the sums are over the representations of scalar fields with weak isospin $I^{(i)}$; $I_3^{(i)}$ and $\lambda_{(i)}$ are the third component of the weak isospin and the vacuum expectation value of the neutral member of the (i) Higgs multiplet. In the GWS model with only doublets of Higgs scalars, $\rho=1$.

For a complete determination of the weak-current couplings (excluding fermion and Higgs masses and mixing angles) we have to fix four (three if ρ is chosen to be 1) constants. Two of them are usually taken to be

$$(9) \qquad \begin{cases} \alpha^{-1} \sim 137.036 & \text{from Josephson effect or } g_e - 2 \text{ experiments} \\ G_F \sim 1.166\ 34 \cdot 10^{-5} \ (\text{GeV})^{-2} & \text{from the } \mu \text{ lifetime .} \end{cases}$$

Including radiative corrections, the μ lifetime is given by

$$(10) \qquad \tau_\mu^{-1} = \frac{G_F^2}{192\pi^3} \, m_\mu^5 \left(1 - \frac{8m_e^2}{m_\mu^2}\right) \left[1 - \frac{\alpha}{2\pi}\left(\pi^2 - \frac{25}{4}\right) + O\left(\alpha^2 \ln \frac{m_\mu}{m_e}\right)\right] .$$

The other parameters to be fixed could be chosen to be $\sin^2\theta_w$ and ρ taken from low-energy neutral and charged leptonic processes or, alternatively, the masses of the intermediate vector bosons M_w and M_z.

The parameters ρ and $\sin^2\theta_w$ have been measured for a variety of probles and reactions at momentum scales ranging roughly from 1 MeV to about 100 GeV. The experiments include purely leptonic processes like $e^+e^- \rightarrow \mu^+\mu^-$ or $\nu_\mu e$ scattering, semileptonic reactions like νp elastic scattering and νN deep inelastic processes, atomic parity violation experiments etc. Moreover the masses of the W and Z^0 vector bosons have been directly measured at the CERN Sp$\bar{\text{p}}$S Collider and at the FNAL Tevatron.

At low energy the cleanest way to determine $\sin^2\theta_w$ and ρ is through the study of pure leptonic processes, since for these interactions theoretical predictions are completely unambigous. The problem is, generally speaking, a problem of statistical errors, since for these processes the rates are rather low.

On the other hand, an impressive amount of experimental data have been collected in lepton-hadron reactions, expecially in deep inelastic scattering, to investigate the neutral current structure of the neutrino-quark interaction. These processes are described in the framewok of the quark parton model, assuming a QCD parametrization of the nucleon structure functions and the theoretical analysis has a large number of parameters which must be measured in order to control the uncertainty typical of all processes involving hadrons.

In the following I will discuss in turn two typical examples, νe elastic scattering and $\nu_\mu \, \mathcal{N}$ deep inelastic scattering.

$\nu_\mu e$ reactions were observed in bubble chamber experiments, where the rates are very low but the signal is very clear, and in counter experiments using high energy neutrino beams. $\bar{\nu}_e e$ processes have been studied using reactor produced neutrino beams. The effective Hamiltonian which governs neutrino initiated leptonic processes can be derived from the diagrams in fig.(2)

(11) $\quad H = \dfrac{G_F}{\sqrt{2}} \, [\bar{\nu} \gamma^\mu (1 - \gamma_5) \, \nu] \, J^e_\mu$

where

(12) $\quad J^e_\mu = \bar{e} \, \gamma_\mu \, (g^e_V - g^e_A \, \gamma_5) \, e$.

The expressions of $g^e_{V,A}$ in the standard model are reported in Table I for different processes. From the Hamiltonian in eq.(11) one easy derives the cross-section of the reaction $\nu(\bar{\nu})e \rightarrow \nu(\bar{\nu})e$:

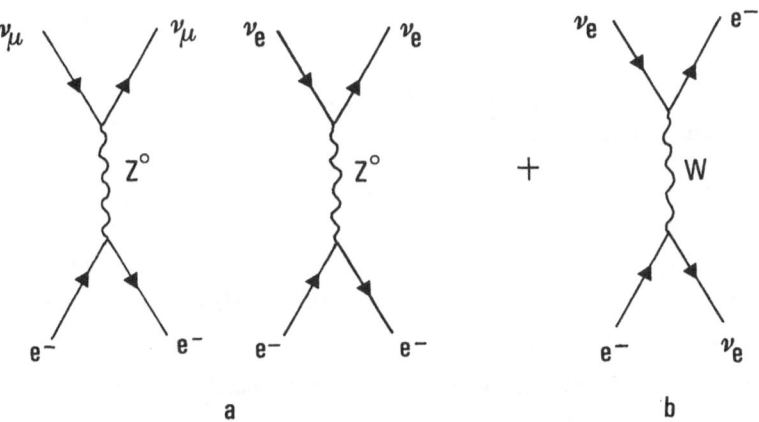

Fig.2a,b. Relevant diagrams for: a) $\nu_\mu e$ scattering; b) $\nu_e e$ scattering.

TABLE I
The standard model expressions for $g_{V,A}^e$, defined in eq.(12).

	g_V^e	g_A^e
$\nu_\mu e$	$-\frac{1}{2} + 2 \sin^2\theta_w$	$-\frac{1}{2}$
$\bar{\nu}_\mu e$	$-\frac{1}{2} + 2 \sin^2\theta_w$	$+\frac{1}{2}$
$\nu_e e$	$+\frac{1}{2} + 2 \sin^2\theta_w$	$+\frac{1}{2}$
$\bar{\nu}_e e$	$+\frac{1}{2} + 2 \sin^2\theta_w$	$-\frac{1}{2}$

$$(13) \quad \sigma \simeq \frac{G_F^2 \, M_e \, E_\nu}{2\pi} (A + \frac{B}{3}) \quad ; \quad G_F^2 \, M_e \simeq 10^{-41} \, cm^2 \, GeV^{-1}$$

where $A = (g_V^e + g_A^e)^2$ and $B = (g_V^e - g_A^e)^2$ for νe reactions and $A = (g_V^e - g_A^e)^2$ and $B = (g_V^e + g_A^e)^2$ for $(\bar{\nu})e$ processes.

Notice that a precise determination of the $\nu_e(\bar{\nu}_e)$ cross-section would also give the sign of the axial coupling of the electron to the Z^0, as a consequence of the interference between neutral and charged currents.

The most precise measurements involve $R_{\nu_\mu e} = \dfrac{\sigma_{\nu_\mu e}}{\sigma_{\bar{\nu}_\mu e}}$, in which ρ cancels:

$$(14) \quad R_{\nu_\mu e} = \frac{\sigma_{\nu_\mu e}}{\sigma_{\bar{\nu}_\mu e}} = 3 \frac{1 - 4 \sin^2 \theta_w + \frac{16}{3} \sin^4 \theta_w}{1 - 4 \sin^2 \theta_w + 16 \sin^4 \theta_w}$$

$R_{\nu_\mu e}$ in particularly suitable for the determination of $\sin^2 \theta_w$ also because many experimental systematic effects cancel in the ratio.

A collection of results for the measured νe and $\bar{\nu} e$ cross-sections is reported in Table II[3]. From $\nu_\mu e$ data, one gets:

TABLE II

Results for $\overset{(-)}{\nu}\, e \rightarrow \overset{(-)}{\nu}\, e$ elastic scattering. Except for the UCI reactor experiment all σ/E are in units of 10^{-42} cm²/GeV. The two uncertainties for the E734 and CHARM experiments are statistical and systematic, respectively, with the systematic uncertainties strongly correlated between $\sigma_{\nu_\mu e}$ and $\sigma_{\bar\nu_\mu e}$. The numbers in parantheses are the predictions of the standard model for $\sin^2\theta_w = 0.230$.

Group	Experimental Value		Reference
E734 (BNL)	$\dfrac{\sigma_{\nu_\mu e}}{E_\nu} = 1.60 \pm 0.29 \pm 0.27$ (1.55)	$\dfrac{\sigma_{\bar\nu_\mu e}}{E_{\bar\nu}} = 1.16 \pm 0.20 \pm 0.14$ (1.33)	L. A. Ahrens et al., Phys. Rev. Lett. **51**. 1514 (1983); **54**, 18 (1985).
CHARM (CERN)	$\dfrac{\sigma_{\nu_\mu e}}{E_\nu} = 1.9 \pm 0.4 \pm 0.4$ (1.6)	$\dfrac{\sigma_{\bar\nu_\mu e}}{E_{\bar\nu}} = 1.5 \pm 0.3 \pm 0.4$ (1.3)	F. Bergsma et al., . Phys. Lett. **117B**, 272 (1982), Phys. Lett. **147B**, 481 (1984), C. Santoni, private communication.
CB (FNAL 15') (preliminary)	$\dfrac{\sigma_{\nu_\mu e}}{E_\nu} = 1.52 \pm 0.39$ (1.59)		A. M. Cnops et al., Phys. Rev. Lett. **41**, 357 (1978); M. F. Bregman, Thesis, Nevis 248 (1984); M. Murtagh, private communication.
VMWOF (FNAL)	$\dfrac{\sigma_{\nu_\mu e}}{E_\nu} = 1.40 \pm 0.50$ (1.59)		R. H. Heisterberg et al., Phys. Rev. Lett. **44**. 635 (1980).
GGM (CERN)	$\dfrac{\sigma_{\nu_\mu e}}{E_\nu} = 2.4^{+1.2}_{-0.9}$ (1.6)		N. Armenise et al., Phys. Lett. **B86**. 225 (1979).
AP (CERN)	$\dfrac{\sigma_{\nu_\mu e}}{E_\nu} = 1.1 \pm 0.6$ (1.6)	$\dfrac{\sigma_{\bar\nu_\mu e}}{E_{\bar\nu}} = 2.2 \pm 1.0$ (1.3)	H. Faissner et al., Phys. Rev. Lett. **41**, 213 (1978).
GGM (CERN)		$\dfrac{\sigma_{\bar\nu_\mu e}}{E_{\bar\nu}} = 1.0^{+1.3}_{-0.6}$ (1.3)	J. Blietschau et al., Nucl. Phys. **B114**, 189 (1976).
UCI (Savannah River Reactor) $1.5 < E_e < 3.0\ MeV$	$\sigma_{\bar\nu_e e} = (4.6 \pm 1.3) \times 10^{-45}\ cm^2/fission$ (3.33)		F. Reines et al., Phys. Rev. Lett **37**, 315 (1976).
$3.0 < E_e < 4.5\ MeV$	$\sigma_{\bar\nu_e e} = (1.12 \pm 0.29) \times 10^{-45}\ cm^2/fission$ (0.55)		
ILM (LANL)	$\dfrac{\sigma_{\nu_e e}}{E_\nu} = 8.9 \pm 3.5$ (9.4)		R. C. Allen et al., Phys. Rev. Lett. **55**, 2401 (1985).

(15) $\sin^2 \theta_w = 0.221 \pm 0.021 \pm 0.003$

where the uncertainty on $\sin^2 \theta_w$ is dominated by the statistical error. The measurement of $R_{\nu_\mu e}$ will become more precise in the future. The CHARM II experiment at CERN is a high statistics experiment which is intended to measure $R_{\nu_\mu e}$ with such an accuracy as to obtain $\sin^2 \theta_w$ with a final error of 0.005 (statistical + systematic)[4].

In $\nu \mathbf{N}$ reactions the effective Hamiltonian responsible for charged current interactions ($\nu \mathbf{N} \to \mu X$) is given by:

(16) $H^{c.c.} = \dfrac{4G_F}{\sqrt{2}} \, J_\mu^+ \, J_\mu$,

where:

$$J_\mu = [\,\bar{u},\bar{c},\bar{t}\,] \, U_{CKM} \, \gamma_\mu \begin{bmatrix} d \\ s \\ b \end{bmatrix}_L + [\bar{\nu}_e, \bar{\nu}_\mu, \bar{\nu}_\tau] \, \gamma_\mu \begin{bmatrix} e^- \\ \mu^- \\ \tau \end{bmatrix}_L .$$

U_{CKM} is the Cabibbo-Kobayashi-Maskawa mixing matrix (assuming $m_\nu = 0$, $U_{CKM}^{lept} = \hat{I}$). In the valence approximation (i.e. the nucleon made only by valence u and d quarks), the active part of $H^{c.c.}$ is:

(17) $H^{c.c.} = \dfrac{G_F}{\sqrt{2}} [\bar{\mu}\gamma^\mu (1 - \gamma_5)\nu_\mu] \, [\bar{u}\gamma_\mu (1 - \gamma_5)d] + h.c.$ $(\cos \theta_c = 1)$.

From eq.(17) in the naive parton model approximation, one finds for ν scattering on isoscalar targets fig.(3):

$$\frac{d\sigma^{\nu_\mu I=0}_{c.c.}}{dxdy} = \frac{2G_F^2 \, M_P E_\nu}{\pi} \, [xu(x) + xd(x)] \ ,$$

(18)
$$\frac{d\sigma^{\bar{\nu}_\mu I=0}_{c.c.}}{dxdy} = \frac{2G_F^2 \, M_P E_{\bar{\nu}}}{\pi} \, [xu(x) + xd(x)] \, (1-y)^2 .$$

In eqs.(18) $E_{\nu,\bar{\nu}}$ is the incident-neutrino (anti-neutrino) energy; M_P is the nucleon mass, and

$x = \dfrac{Q^2}{2M_P\nu}$, $y = E_h/E_{\nu,\bar{\nu}}$; E_h = final hadronic energy, $0 < x, y < 1$,

where $Q^2 = |(p_\nu - p'_\nu)^2|$ and $\nu = E_\nu - E'_\nu$.

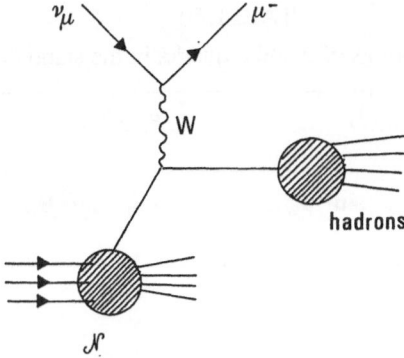

Fig. 3. The neutrino scatters a quark off the nucleon. The final state is constituted by hadronic fragments of the nucleon and of the scattered quark+the muon.

On the other hand, for neutral-current processes

$$(19) \quad H^{n.c.} = \frac{G_F}{\sqrt{2}} \, [\bar{\nu}_\mu \, \gamma^\mu \, (1 - \gamma_5) \, \nu_\mu] \, J_\mu^H \ ,$$

where

$$(20) \quad J_\mu^H = \sum_f \bar{q}_f \, \gamma_\mu \, (g_V^f - g_A^f \, \gamma_5) \, q_f = \sum_f \bar{q}_f \, [\epsilon_L^f \, \gamma_\mu \, (1 - \gamma_5) + \epsilon_R^f \, \gamma_\mu \, (1 + \gamma_5)] \, q_f$$

The expression for $\epsilon_{L,R}(u,d)$ in the standard model are reported in Table III. From eqs.(19) and (20), one computes

$$\frac{d\sigma_{n.c.}^{\nu_\mu I=0}}{dxdy} = \frac{2G_F^2 \, M_P E_\nu}{\pi} \, [(|\epsilon_L(u)|^2 + |\epsilon_L(d)|^2) +$$

$$+ (|\epsilon_R(u)|^2 + |\epsilon_R(d)|^2) \, (1 - y)^2] \, [xu(x) + xd(x)],$$

$$(21) \quad \frac{d\sigma_{n.c.}^{\bar{\nu}_\mu I=0}}{dxdy} = \frac{2G_F^2 \, M_P E_{\bar{\nu}}}{\pi} \, [(|\epsilon_L(u)|^2 + |\epsilon_L(d)|^2) \, (1 - y)^2 +$$

$$+ (|\epsilon_R(u)|^2 + |\epsilon_R(d)|^2)] \, [xu(x) + xd(x)].$$

9

TABLE III
Neutral-current couplings of u and d quarks in the standard model ($\rho=1$).

$\varepsilon_L(u)$	$\varepsilon_L(d)$	$\varepsilon_R(u)$	$\varepsilon_L(d)$
$\frac{1}{2} - \frac{2}{3} \sin^2 \theta_w$	$-\frac{1}{2} + \frac{1}{3} \sin^2 \theta_w$	$-\frac{2}{3} \sin^2 \theta_w$	$\frac{1}{3} \sin^2 \theta_w$

From the expressions given in eqs.(21), we observe that measurements of deep inelastic c.c. and n.c. reactions on isoscalar targets will only allow the determination of the left-right handed couplings $g_{L,R}$ defined as

$$g_L = [|\varepsilon_L(u)|^2 + |\varepsilon_L(d)|^2]^{1/2} , \quad g_R = [|\varepsilon_R(u)|^2 + |\varepsilon_R(d)|^2]^{1/2} .$$

The allowed regions of coupling constants may be represented as annuli in the $\varepsilon_L(u) - \varepsilon_L(d)$ and $\varepsilon_R(u) - \varepsilon_R(u)$ planes. Combining eqs.(18) and (21), one easily derives

$$R^\nu = \frac{\sigma^{\nu_\mu I=0}_{n.c.}}{\sigma^{\nu_\mu I=0}_{c.c.}} = |\varepsilon_L(u)|^2 + |\varepsilon_L(d)|^2 + \frac{1}{3} \left(|\varepsilon_R(u)|^2 + |\varepsilon_R(d)|^2 \right) ,$$

(22)

$$R^{\bar\nu} = \frac{\sigma^{\bar\nu_\mu I=0}_{n.c.}}{\sigma^{\bar\nu_\mu I=0}_{c.c.}} = \left(|\varepsilon_L(u)|^2 + |\varepsilon_L(d)|^2 \right) + 3 |\varepsilon_R(u)|^2 + |\varepsilon_R(d)|^2 .$$

In the GWS model, eqs.(22) can be rewritten as (cf. table III)

$$(23) \quad R^\nu = \frac{1}{2} - \sin^2 \theta_w + \frac{20}{27} \sin^4 \theta_w , \quad R^{\bar\nu} = \frac{1}{2} - \sin^2 \theta_w + \frac{20}{9} \sin^4 \theta_w .$$

$R^{\nu,\bar\nu}$ have been measured in several high energy experiments on deep inelastic $\nu_\mu(\bar\nu_\mu)$ scattering from approximatively isoscalar targets. The most precise results are from two CERN (CDHS and CHARM) and two FNAL (CCFRR and FMM) experiments[5-8]. The results are reported in Table IV and V. The theoretical uncertainties in the determination of $\sin^2 \theta_w$ from deep inelastic scattering are usually extimated by varying the structure functions over reasonable ranges, by using independent extimate of higher twist effects, by including some extimate of the longitudinal structure function etc. The major uncertainty turns out to be the charm threshold which is mainly important in the changed current denominator of R^ν. The threshold is treated by assuming slow rescaling with a charm mass which is allowed to vary in the range $m_c = 1.2 \rightarrow 1.8$ GeV. Note that for $\sin^2 \theta_w \simeq 0.23$:

TABLE IV

Results for deep inelastic scattering from approximately isoscalar targets. $R_\nu \equiv \sigma_{\nu N}^{NC}/\sigma_{\nu N}^{CC}$ and $R_{\bar\nu} \equiv \sigma_{\bar\nu N}^{NC}/\sigma_{\bar\nu N}^{CC}$ where NC and CC represent neutral current and charged current, respectively, while $g_i^2 \equiv \in_i(u)^2 + \in_i(d)^2$ for i = L or R. The numbers in parentheses below the experimental values are the standard model predictions for $\sin^2\theta_w = 0.230$.

Group	Experimental Value		Reference
CDHS (CERN) (iron)	$R_\nu = 0.3072 \pm 0.0032$ (.3035)		H. Abramowicz et al. Phys. Rev. Lett. **57**. 298 (1986)
CHARM (CERN) (marble-isoscalar corrected)	$R_\nu = 0.3093 \pm 0.0031$ (.3127)		J. V. Allaby et al. Phys. Lett. **177B**, 446 (1986).
CCFRR (FNAL) (iron)	$g_L^2 = 0.292 \pm 0.009$ (.301)	$g_R^2 = 0.030 \pm 0.009$ (.029)	P. G. Reutens et al. Phys. Lett. **152B**, 404 (1985) F. Merritt, private communication
FMM (FNAL) (sand-steel)	$g_L^2 = 0.282 \pm 0.014$ (.301)	$g_R^2 = 0.044 \pm .014$ (.029)	D. Bogert et al. Phys. Rev. Lett. **55**, 1969 (1985) F. Taylor, private communication
CDHS (CERN) (iron)	$R_\nu = 0.301 \pm 0.007$ (.303)	$R_{\bar\nu} = 0.363 \pm 0.015$ (.375)	H. Abramowicz et al. Z. Phys. **C28**, 51 (1985)
SKAT (Serpukhov) (freon)	$R_\nu = 0.33 \pm 0.02$ (.32)	$R_{\bar\nu} = 0.44 \pm 0.11$ (.40)	V. V. Ammosov et al. Z. Phys. **C30**. 569 (1986)
FIIM (FNAL 15') (neon-hydrogen)		$R_{\bar\nu} = 0.406 \pm 0.028$ (.367)	P. A. Gorichev et al. Yad Fiz **39**. 626 (1984) (SJNP **39**. 396 (1984)).
ABCDILOS (CERN-BEBC) (neon-hydrogen)	$R_\nu = 0.345 \pm 0.018$ (.313)	$R_{\bar\nu} = 0.364 \pm 0.030$ (.364)	P. Bosetti et al. Nucl. Phys. **B217**, 1 (1983)
ABBPPST (CERN-BEBC) (deuterium)	$R_\nu = 0.33 \pm 0.03$ (.31)	$R_{\bar\nu} = 0.35 \pm 0.05$ (.37)	D. Allasia et al. Phys. Lett. **133B**. 129 (1983)
FNAL 15' (neon-hydrogen)	$R_\nu = 0.32 \pm 0.03$ (.33)		J. Marriner et al., Phys. Rev. **D27**. 2569 (1983)
CHARM (CERN) (marble)	$R_\nu = 0.320 \pm 0.010$ (.314)	$R_{\bar\nu} = 0.377 \pm 0.020$ (.362)	M. Jonker et al. Phys. Lett. **99B**. 265 (1981) Phys. Lett. **103B**. 469(E) (1981)
CITF (FNAL) (iron)	$R_\nu = 0.28 \pm 0.03$ (.30)	$R_{\bar\nu} = 0.35 \pm 0.11$ (.39)	F. S. Merritt et al. Phys. Rev. **D17**, 2199 (1978)
HPWF (FNAL) (CH_2)	$R_\nu = 0.30 \pm 0.04$ (.33)	$R_{\bar\nu} = 0.33 \pm 0.09$ (.35)	P. Wanderer et al. Phys. Rev. **D17**. 1679 (1978)

TABLE V

Deep inelastic scattering from non-isoscalar targets. $R_\nu^p \equiv \sigma_{\nu p}^{NC}/\sigma_{\nu p}^{CC}$, $R_\nu^n \equiv \sigma_{\nu n}^{NC}/\sigma_{\nu n}^{CC}$ and $R_{\bar\nu}^p \equiv \sigma_{\bar\nu p}^{NC}/\sigma_{\bar\nu p}^{CC}$, while $R_\nu^{n/p} \equiv \sigma_{\nu n}^{NC}/\sigma_{\nu p}^{CC}$ and $R_{\bar\nu}^{n/p} \equiv \sigma_{\bar\nu n}^{NC}/\sigma_{\bar\nu p}^{NC}$. r_ν is defined as $(\sigma_{\nu n}^{NC}/\sigma_{\nu p}^{NC})/(\sigma_{\nu n}^{CC}/\sigma_{\nu p}^{CC})$, with an analogous definition for $r_{\bar\nu}$. The numbers in parentheses are the standard model predictions for $\sin^2\theta_W = 0.230$.

Group	Experimental Value		Reference
BBCIMOU (CERN-BEBC) (hydrogen)	$R_\nu^p = 0.384 \pm 0.028$ (.403)	$R_{\bar\nu}^p = 0.338 \pm 0.021$ (.333)	G. T. Jones *et al.* Phys. Lett. **178B**, 329 (1986); R. W. L. Jones and S. Burke, private communication. S. J. Towers, Oxford Thesis RAL T013 (1985).
BEBC-TST (CERN) (hydrogen)	$R_\nu^p = 0.47 \pm 0.04$ (.45)	$R_{\bar\nu}^p = 0.33 \pm 0.04$ (.28)	N. Armenise *et al.* Phys. Lett. **122B**, 448 (1983); J. Moreels *et al.* Phys. Lett. **138B**, 230 (1984).
SIMTT (FNAL 15') (deuterium)	$R_\nu^p = 0.49 \pm 0.06$ (.44) $R_\nu^n = 0.22 \pm 0.03$ (.23)		T. Kafka *et al.* Phys. Rev. Lett. **48**, 910 (1982)
FNAL 15' (hydrogen)		$R_{\bar\nu}^p = 0.36 \pm 0.06$ (.28)	D. D. Carmony *et al.* Phys. Rev. **D26**, 2965 (1982)
ABCMO (CERN-BEBC) (hydrogen)	$R_\nu^p = 0.51 \pm 0.04$ (.52)		J. Blietschau *et al.* Phys. Lett. **88B**, 381 (1979).
FNAL 15' (hydrogen)	$R_\nu^p = 0.48 \pm 0.17$ (.46)		F. A. Harris *et al.*, Phys. Rev. Lett. **39**, 437 (1977).
FIIM (FNAL 15') (neon-hydrogen)		$R_{\bar\nu}^{n/p} = 0.88 \pm 0.17$ (1.03)	P. A. Gorichev *et al.* Yad. Fiz. **39**, 626 (1984) (SJNP **39**, 396 (1984)).
FNAL 15' (neon-hydrogen)	$R_\nu^{n/p} = 1.08 \pm 0.19$ (1.13)		J. Marriner *et al.* Phys. Rev. **D27**, 2569 (1983).
ABBPPST (CERN-BEBC) (deuterium)	$r_\nu = 0.06 \pm 0.06$ (.06)	$r_{\bar\nu} = 0.02 \pm 0.09$ (−.01)	D. Allasia *et al.* Phys. Lett. **133B**, 129 (1983).

$$(24) \qquad \frac{\delta R^{\nu}}{\delta \sin^2 \theta_w} \simeq -0.67 , \qquad \frac{\delta R^{\bar{\nu}}}{\delta \sin^2 \theta_w} \simeq -0.02 .$$

Eqs.(24) imply that R^{ν} is particularly sensitive to the precise value of $\sin^2 \theta_w$. The very simple expressions gives in eq.(23) are modified by taking into account

    i) the neutron excess in the target,

    ii) the sea ($q\bar{q}$ pairs and gluons) content of the nucleon,

    iii) the effects of scaling violations on parton densities induced by strong interactions,

    iv) $O(\alpha_s)$ (α_s = strong-interaction coupling constant) corrections,

    v) the fact that experimentally the longitudinal structure function defined as $F_L = F_2 - 2 \times F_1$ is different from zero unlike the naive parton model.

Other effects, coming from electroweak radiative corrections must also be included in the analysis of the data (see also below).

In refs.(9) and (10) a comprehensive analysis of the existing data on weak neutral currents and W and Z masses has been presented. All the published data on ν-hadron, νe, eN, μN, e^+e^- reactions, atomic parity violation and W/Z masses were incorporated in these analyses. Radiative electro-weak corrections, which allow us to relate different processes occurring at different energies (and which will be briefly discussed in sect. 3) were also taken into account. The main conclusion of refs.(9) and (10) is that there is no experimental evidence for deviation from the standard model. The results for $\sin^2 \theta_w$ are summarized in table VI for $\rho = 1$. Notice that the accuracy on $\sin^2 \theta_w$ obtained from the measurement of the W and Z^0 masses is already comparable to the accuracy of the best determinations given by deep inelastic scattering experiments. Allowing ρ to vary as a free parameter one finds:

$$\rho = 0.998 \pm 0.0086$$

(25)

$$\sin^2 \theta_w = 0.229 \pm 0.0064$$

The results reported in Table VI and in eqs.(25) have been obtained by assuming three generations, a mass for the top $m_{top} \lesssim 100$ GeV and a mass for the Higgs (in the minimal Standard Model with only one Higgs doublet) $m_H \lesssim 1$ TeV.

TABLE VI

Determination of $\sin^2\theta_w$ from various reactions. Where two errors are shown the first is experimental and the second is theoretical. In the other cases the theoretical and experimental errors are combined.

Reaction	$\sin^2\theta_w$
Deep inelastic (isoscalar)	$0.233 \pm .003 \pm [.005]$
$\overset{(-)}{\nu}_\mu p \rightarrow \overset{(-)}{\nu}_\mu p$	$0.210 \pm .033$
$\overset{(-)}{\nu}_\mu e \rightarrow \overset{(-)}{\nu}_\mu e$	$0.233 \pm .018 \pm [.002]$
W, Z	$0.228 \pm .007 \pm [.002]$
Atomic parity violation	$0.209 \pm .018 \pm [.014]$
SLAC eD	$0.221 \pm .015 \pm [.013]$
μC	$0.25 \pm .08$
All data	0.230 ± 0.0048

3. RADIATIVE CORRECTIONS TO THE STANDARD MODEL PARAMETERS[11,12,13]

The relations between the weak couplings occurring in neutral current reactions, as for example in tables I and III, and the fundamental parameters of the theory (α, G_F, $\sin^2\theta_w$ and ρ), are valid at the tree level in the Standard Model. Higher order electro-weak radiative corrections modify these relations. Moreover a precise determination of the weak couplings, at the level of radiative corrections, allows to extract as accurate a value of $\sin^2\theta_w$ as possible for comparison with extensions of the Standard Model (grand unified theories, supersymmetry etc.) and to search for indications of (or set limits on) such new physics as additional Z bosons, heavy fermions or non- standard Higgs representations. In this sections I will briefly review the subject of radiative corrections.

The inclusion of higher order terms modify the Born relations given in eq.(6) of sect. 2 and the neutral current couplings given in tables I and III:

$$\sin^2\theta_w|_{exp} = \sin^2\theta_w(\lambda) + \delta_{exp}$$

$\sin^2 \theta_w$ (λ) is a theoretical universal quantity defined at a scale λ (in the following, we will assume for simplicity $\lambda \simeq M_{Z,W}$) in a given renormalization scheme. δ_{exp} is a process (experiment)-dependent correction.

A *priori*, we expect for δ_{exp} the following hierarchy (relative to the size) of corrections:

$$\alpha \ln (M^2/\mu^2) \simeq 10\alpha \ ,$$

$$\frac{\alpha}{\sin^2 \theta_w} \simeq 5\alpha \ ,$$

(26)

$$[\alpha \ln (M^2/\mu^2)]^2 \simeq 0.8\alpha \ ,$$

$$\alpha \ \ ......$$

μ is the energy scale of the process under consideration ($\mu^2 \simeq 0.3$ (GeV)^2), $M \simeq M_{W,Z} \simeq 80$ GeV. Further contributions of order $O(\alpha \ln (m_t^2/\mu^2))$, where m_t is a typical fermion mass, must also be considered. From eq.(26), we see that the most important corrections arise from leading-logarithm ($\sim \alpha \ln (M^2/\mu^2)$) contributions; a precise control of $O(\alpha^2 \ln^2 (M^2/\mu^2))$ terms is also needed since they are comparable to $O(\alpha)$ terms.

A very simple and universal treatment of the leading-logarithm corrections (and subsequently of the $O(\alpha^2 \ln^2 (M^2/\mu^2))$ corrections) can be obtained in the framework of the renormalization group. This approach will be discussed in subsect. 3.1. In 3.2, we will consider in detail the radiative corrections which enter into the determination of $R^{v,\bar{v}}$. The renormalization group methods exposed for this particular case can be easily extended to any other process. Details of the renormalization group approach can be found in ref.(13).

3.1. *Renormalization group applied to the effective weak Hamiltonian*

As we have already seen in sect. 2, in the limit $M_{W,Z} \to \infty$ ($\mu \ll M_{W,Z}$), the bare effective Hamiltonian for weak interactions can be written as

(27) $$H^{(0)} = \sum_i c_i \ O_i \ ,$$

where O_i are four fermion local operators. Examples of $H^{(0)}$ have been given in eqs.(5), (7), (11) and (17). We choose for O_i a basis of operators which are multiplicately renormalized under e.m. corrections. As an explicit example, I report in fig. 4 the one-loop diagrams which renormalize the weak Hamiltonian given in eq.(17). The diagrams are computed with an ultraviolet momentum cut-off Λ. The result has the form

(28) $$(O_i)^{renormalized} = (1 + \alpha \gamma_i \ln \Lambda^2)(O_i)^{bare} \ .$$

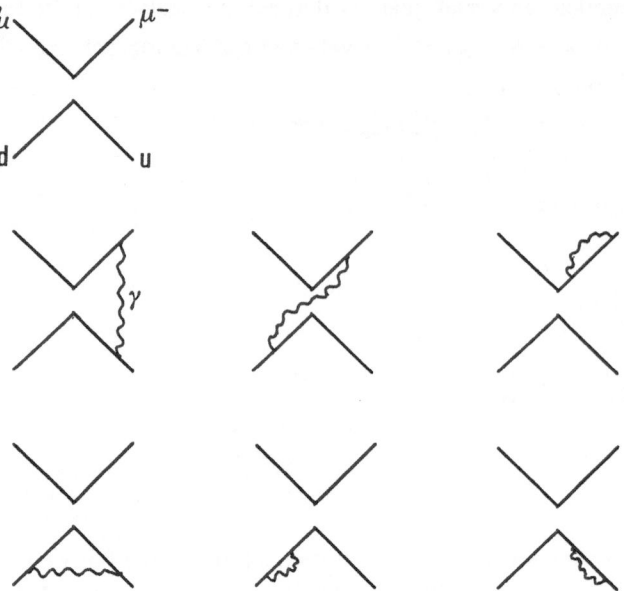

Fig.4. Lowest order and one-loop diagrams which enter into the renormalization of the Hamiltonian in eq.(17).

The scale of the ultraviolet cut-off corresponds to the mass of the intermediate vector bosons. The renormalization group states that the effective Hamiltonian at a scale $\mu << M$ is expressed in the following way:

(29) $H = \sum_i c_i(\mu)\, O_i(\mu)$.

where the coefficients $c_i(\mu)$ in eq.(29) satisfy the renormalization group equations

(30) $\left[-\dfrac{\partial}{\partial t} + \beta(\alpha)\, \dfrac{\partial}{\partial \alpha} + \Gamma_i(\alpha) \right] c_i(\mu) = 0, \quad t = \ln (M^2/\mu^2)$.

$O_i(\mu)$ are the four fermion operators renormalized at the scale μ. $\beta(\alpha)$ and $\Gamma_i(\alpha)$ are the coupling constant β-function and the anomalous dimensions of the operators O_i:

(31) $\beta(\alpha) = \dfrac{\partial}{\partial \ln \mu^2}\, \alpha(\mu) = \beta_0 \alpha^2 + ..., \quad \Gamma_i(\alpha) = \alpha \gamma_i + ...$

and the γ_i's have been introduced in eq.(28).

The solution of eq.(30) for the coefficients c_i, has the form

(32) $\quad c_i(t) = c_i(t = 0, \alpha(t)) \exp\left[\int_0^t \Gamma[\alpha(t')]\ dt'\right] \simeq c_i^{(0)} \left[\dfrac{\alpha(M)}{\alpha(\mu)}\right]^{\gamma_i/\beta_0} (1 + ...).$

As an example, we now apply the renormalization group equations given above to the computation of the perturbative corrections to the determination of $R^{\nu,\bar\nu}$.

3.2. Corrections to the evaluation of $R^{\nu,\bar\nu}$

We start by considering the corrections to the Hamiltonian responsible for neutrino charged-current reactions:

(33) $\quad H^{(o)} = \dfrac{G_F^{(o)}}{\sqrt{2}}\ [\bar\mu\gamma^\mu(1 - \gamma_5)\ \nu_\mu]\ [\bar u\gamma_\mu(1 - \gamma_5)\ d] + h.c. = c^o\ O^o$

$G_F^{(o)}$ is the bare Fermi constant (called simply G_F in the previous section). One can show that $G_F^{(o)} = G_F$ in the leading logarithmic approximation if one uses the μ-life time to define the Fermi constant.

After a straightforward evaluation of the diagrams in fig. 4, one finds for the anomalous dimension of the operator in eq.(33)

(34) $\quad \gamma = \dfrac{1}{2\pi}\ .$

Then

(35) $\quad H = \dfrac{G_F^{(o)}}{\sqrt{2}}\ \left[\dfrac{\alpha(M)}{\alpha(\mu)}\right]^{\gamma/b_Q} [\bar\mu\gamma^\mu(1 - \gamma_5)\ \nu_\mu]\ [\bar u\gamma_\mu(1 - \gamma_5)\ d]\ ,$

where $b_Q = \sum\limits_f\ (Q_f^2/3\pi)$ and

(36) $\quad \dfrac{\alpha(M)}{\alpha(\mu)} = \dfrac{1}{1 - \alpha(\mu)\ b_Q\ \ln\ (M^2/\mu^2)}\ .$

Thus leading-logarithmic corrections renormalize the strength of the charged-current amplitudes.

For neutral currents, the relevant bare Hamiltonian is given by:

(37) $\quad H^{(o)} = \dfrac{G_F^{(o)}}{\sqrt{2}}\ \rho\ [\bar\nu_\mu\gamma^\mu(1 - \gamma_5)\ \nu_\mu]\left[J_\mu^Z + \dfrac{2}{\rho}\ \bar\mu\gamma_\mu(1 - \gamma_5)\ \mu\right],$

17

where

$$J_\mu^Z = \sum_f \bar{f} \gamma_\mu [\tau_3^f - 4 \sin^2 \theta_w(M) Q^f - \tau_3^f \gamma_5] f \ ,$$

$\sin^2 \theta_w(M) = \sin^2 \theta_w(\lambda=M)$, $\tau_3^f = \pm \frac{1}{2}$ and Q^f is the electric charge for the fermion of flavour f. Only the last three diagrams of fig. 4 are present in this case. However, they cannot give leading-logarithmic contributions because both the vector and axial vector currents are conserved in the massless limit. The leading-logarithmic corrections come in this case from a new type of diagram, the so-called 'penguin' diagram, as illustrated in fig. 5. The fermions which loop in the diagram of fig. 5 give a factor

(38) $\sim \ln \Lambda^2 (q_\mu q_\nu - g_{\mu\nu} q^2)$.

The $q_\mu q_\nu$ terms give zero when contracted with the conserved e.m. current; the term proportional to
$g_{\mu\nu} q^2$ kills the photon propagator ($\sim 1/q^2$), giving a local four-fermion operator. Thus, for each looping flavour, neglecting isospin and charge factors, one obtains

(39)   'penguin' $= - \frac{\alpha}{3\pi} \ln \Lambda^2 [\bar{v}_\mu \gamma^\mu (1 - \gamma_5) \, v_\mu][\bar{f} \gamma_\mu Q^f f]$.

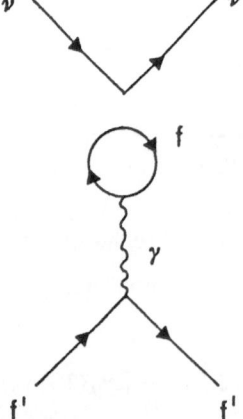

Fig. 5. 'Penguin' diagram which renormalizes the neutral-current Hamiltonian of eq.(37)

It is clear from fig.5 and eq.(39) that 'penguin' diagrams cannot renormalize the fermion axial vector part of the Hamiltonian because they contain only the vector current piece $\sim \bar{f}\gamma_\mu f$. Let us rewrite eq.(37) in the following way:

$$(40) \qquad H^{(o)} = \frac{G_F^{(o)}}{\sqrt{2}}\ \rho \left[\frac{4b_z}{b_Q} - \frac{2}{\rho}\ \frac{1}{3\pi b_Q} - 4\ \sin^2\theta_w(M)\right] \cdot$$

$$\cdot\ [\bar{v}_\mu\gamma^\mu(1-\gamma_5)\ v_\mu]\left[\sum_f(\bar{f}\gamma_\mu Q^f f)\right] + H_\perp = c_Q^o\ O_Q + H_\perp$$

where

$$b_z = \frac{1}{4}\sum_f \frac{tr(\tau_3^f Q^f)}{3\pi}\ ,\quad b_Q = \sum_f \frac{tr(Q^f)^2}{3\pi}$$

$H_\perp$ is the axial piece of the Hamiltonian not renormalized by 'penguin' diagrams. In fact,

$$\tau_3^f - \frac{\displaystyle\sum_f tr(\tau_3^f Q^f)}{\displaystyle\sum_f tr(Q^f)^2}\ Q^f$$

is orthogonal to Q_f. The operator $O_Q = [\bar{v}_\mu\gamma^\mu(1-\gamma_5)\ v_\mu]\left[\sum_f(\bar{f}\gamma_\mu Q^f f)\right]$ is multiplicatively

renormalized by the 'penguin' diagram as it can be read in eq.(39). From eq.(39), one finds

$$(41) \qquad (O_Q)^{renormalized} = [1 - \alpha b_Q\ \ln\Lambda^2]\ (O_Q)^{bare}$$

which gives $C_{O_Q} = \left[\frac{\alpha(M)}{\alpha(\mu)}\right]^{-1}$; then

$$(42) \qquad H = \frac{G_F^{(o)}}{\sqrt{2}}\ \rho\ \left[\frac{4b_z}{b_Q} - \frac{2}{\rho}\ \frac{1}{3\pi b_Q} - 4\ \sin^2\theta_w(M)\right]\frac{\alpha(\mu)}{\alpha(M)}O_Q + H_\perp\ .$$

It is possible to write the renormalized Hamiltonian of eq.(42) in the same form as the bare one, eq.(37), provided we redefine the value of $\sin^2\theta_w$:

(43) $\sin^2 \theta_w(\mu) = \sin^2 \theta_w(M) \dfrac{\alpha(\mu)}{\alpha(M)} - \dfrac{\alpha(\mu) - \alpha(M)}{\alpha(M)} \dfrac{1}{b_Q}\left(b_z - \dfrac{1}{6\pi\rho}\right).$

The effect of the leading-logarithm corrections is to make the value $\sin^2 \theta_w$ to vary by changing the scale from M to μ. This implies that the measured value of $\sin^2 \theta_w$ depends on the scale at which a given physical process occurs. From eq.(43), we may derive the following renormalization group equation for the weak coupling $\alpha_w = \alpha/\sin^2 \theta_w$:

(44) $\dfrac{\partial}{\partial \ln \mu^2} \alpha_w^{-1}(\mu^2) = -\left(b_z - \dfrac{1}{6\pi\rho}\right).$

The term $\sim b_z$ is just the expected contribution to the renormalization of the SU(2) coupling constant coming from fermion loops. The extra $1/\rho$ term is usually absent in the ordinary evolution equation for this coupling.

Besides the leading-logarithm correction considered above, we have to take into account other potentially large corrections of order $\alpha_w = \alpha/\sin^2 \theta_w$. These corrections are more easily estimated in the limit[13]:

(45) $\alpha \to 0, \qquad \sin^2 \theta_w \to 0, \qquad \alpha_w$ fixed .

In this limit, the photon coincides with the U(1) gauge boson, Z^0 coincides with W^3 and the bare weak Hamiltonian has the form

(46) $H^{(0)} = \dfrac{G_F^{(0)}}{\sqrt{2}} J_\mu J_\mu,$

where

$$J_\mu = \sum_f \left[\bar{f}\gamma_\mu \dfrac{1 - \gamma_5}{2} \tau f \right].$$

Fig.6. Diagrams contributing to the $O(\alpha_w)$ corrections.

The corrected Hamiltonian at order α_w is obtained by computing the contribution coming from the double-W-exchange diagrams of fig.6. Because of the global SU(2) symmetry of the Hamiltonian of eq.(46), the corrected Hamiltonian must have the form

(47) $$H = \frac{G_F^{(o)}}{\sqrt{2}} \; [(1 + \varepsilon_1) \, J_\mu \, J_\mu + \varepsilon_2 \, J_\mu^o \, J_\mu^o] \; ,$$

where

$$J_\mu^o = \sum_f \left[\bar{f} \gamma_\mu \; \frac{1 - \gamma_5}{2} \, f \right]$$

is the SU(2) singlet current. ε_1 simply redefines the Fermi constant: $G_F = G_F^{(o)}(1 + \varepsilon_1)$. ε_2 has been evaluated in refs.(12) and (13):

(48) $$\varepsilon_2 = -\frac{9}{16} \frac{\alpha_w}{\pi} \; .$$

and modify the value of neutral-current couplings to fermions.

Let us consider now the effects of the radiative corrections discussed above on the determination of $\sin^2 \theta_w$ in deep inelastic scattering.

In the previous section, we have reported the expression for R^ν in the Born approximation:

$$R^\nu = \frac{1}{2} - \sin^2 \theta_w + \frac{20}{27} \sin^4 \theta_w \; .$$

The radiative corrections to the effective charged and neutral current Hamiltonian, previously considered, modify this equation to:

(49) $$R^\nu = \frac{1}{F^2} \left(\frac{1}{2} - \sin^2 \theta_w(\mu) + \frac{20}{27} \sin^4 \theta_w(\mu) - \frac{1}{3} \sin^2 \theta_w(\mu) \, \varepsilon_2 \right) \; ,$$

where the factor $F = [\alpha(M)/\alpha(\mu)]^{\gamma/bQ}$ comes from charged-current corrections, eq.(35), $\sin^2 \theta_w \rightarrow \sin^2 \theta_w(\mu)$ from neutral currents, eq.(43), and ε_2 comes from the $O(\alpha_w)$ corrections of eq.(48). Eq.(49) implies a shift in $\sin^2 \theta_w$:

(50) $$\delta s^2 \simeq \frac{R^\nu \, \delta F^2 + (1/3) \sin^2 \theta_w \, \varepsilon_2}{(40/27) \sin^2 \theta_w - 1} \; .$$

Using $R_v \simeq 0.31$ and $\sin^2 \theta_w \sim 0.23$ in eq.(50), one finds a correction $\delta s^2 \simeq 0.004$ at $\mu \simeq 1.4$ GeV.

In summary, the above discussion has shown that the radiative corrections modify the Born relations and that the effective values of the parameters of the standard model, such as $\sin^2 \theta_w$ and ρ, depend on the scale at which they are measured.

We now come to the effect of radiative corrections on the W and Z^0 masses and widths and their relations with $\sin^2 \theta_w$ and ρ. Beyond the tree level a very convenient (and almost universality used) definition of $\sin^2 \theta_w$ is through its relation to $M_{W,Z}$[11]:

$$(51) \quad \sin^2 \theta_w(M) = \sqrt{1 - \frac{M_W^2}{M_Z^2}}$$

The advantage of the definition of $\sin^2 \theta_w$ in eq.(51) is that one does not have to extrapolate $\sin^2 \theta_w$ from low-energy processes up to a large scale; this would require the evolution of the vacuum polarization diagram of fig.7 for the IVB. Such evolution is affected by large theoretical incertitudes mainly due to light quark masses. On the contrary, the extrapolation of the e.m. coupling constant α (photon vacuum polarization) can be done on a more solid basis by using as input the experimentally measured cross-section for e^+e^- into hadrons. The disadvantage of the definition in eq.(51) is its sensitivity to small errors in $M_{Z,W}$.

In the leading-logarithmic approximation the bare relations given in eqs.(6) are modified as follows:

$$M_W^2 = \frac{\pi\alpha(M)}{\sqrt{2}\, G_F \sin^2 \theta_w(M)} \quad,$$

$$(52) \quad M_Z^2 = \frac{M_W^2}{\cos \theta_w(M)} \quad,$$

$$\Gamma_{Z,W}/M_{Z,W} = \text{unrenormalized} \quad,$$

where $\Gamma_{Z,W}$ is the Z(W) decay width.

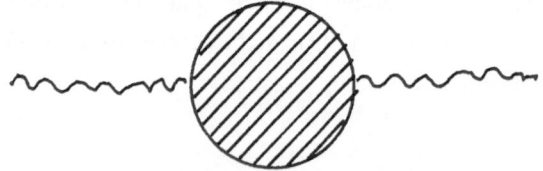

Fig.7. Vacuum polarization diagram.

Next-to-leading logarithmic corrections modify eqs.(52) by terms of order of α_W and α. Using for $\sin^2 \theta_w$ the definition of eq.(51), the following relation can be written at all orders in α

$$(53) \quad M_W = \cos \theta_w(M) \, M_Z = \left[\frac{\pi \alpha(M)}{\sqrt{2} \, G_F} \right]^{1/2} \frac{1}{\sin \theta_w (M)} \, (1+\delta_M) \,.$$

δ_M includes the effect of all the next-to-leading corrections and it has been evaluated at order $O(\alpha)$[11,12]. A detailed analysis shows that leading-logarithm corrections order of 4%, while finite corrections are of opposite sign and $\sim 0.4\%$, confirming that the most important contribution comes from the leading piece. It is convenient to write eq.(53) as follows:

$$(54) \quad M_W = \left[\frac{\pi \alpha(0)}{\sqrt{2} \, G_F} \right]^{1/2} \frac{1}{\sin \theta_w (M)} \left[\frac{\alpha(M)}{\alpha(0)} \right]^{1/2} (1+\delta_M) =$$

$$= \left[\frac{\pi \alpha(0)}{\sqrt{2} \, G_F} \right]^{1/2} \frac{1}{\sin \theta_w (M)} \, (1+\Delta) \,.$$

$$= \frac{A_o}{\sin \theta_w (M)(1-\Delta r)^{1/2}}$$

where

$$(55) \quad A_o = (\pi \alpha(0)/\sqrt{2} \, G_F)^{1/2} = 37.281 \text{ GeV}.$$

Δr takes into account the running of the electromatic coupling α up to ~ 100 GeV, $(\f(\alpha(100 \text{ GeV}),\alpha(\text{Thomson})) \sim 1.073$ or $\alpha^{-1} (100 \text{ GeV}) \simeq 128)$, as well as the other non-leading logarithmic corrections. Δr is predicted to be $\Delta r = 0.0713 \pm 0.0013$ for $m_{top} = 45$ GeV and $m_H = 100$ GeV.

Eqs.(53) and (54) allow us to predict the W and Z^o masses from the measurement of the Fermi constant, α and $\sin^2 \theta_w$. A collection of measured/predicted values of the W and Z masses is reported in Table VII.

The data are now precise enough to begin testing the Standard Model at the radiative correction level, at least for the leading logarithmic corrections.

A simultaneous fit of deep inelastic data and W and Z masses leads to $\Delta r = 0.077 \pm 0.037$ in excellent agreement with the predicted value given above.

TABLE VII

Measured and predicted values for the masses of the intermediate vector bosons.

	M_W(GeV)	M_Z(GeV)
UA1[14]	$83.5^{+1.1}_{-1.0}\pm2.7$	$93\pm1.4\pm3.0$
UA2[15]	$80.2\pm0.8\pm1.3$	$91.5\pm1.2\pm1.7$
UA1+UA2 COMBINED	80.9 ± 1.4	91.9 ± 1.8
THEORY $s^2_w=0.230$ $\Delta r=0.0713$	80.7	91.9
FROM R_ν $s^2_w=0.233\pm0.006$ $\Delta r=0.0713$	80.2 ± 1.1	91.6 ± 0.9
NO RADIATIVE CORRECTIONS $\Delta r=0$ $s^2_0=0.242\pm0.006$	75.9 ± 1.0	87.1 ± 0.7

It is quite difficult to reconciliate the W and Z^o masses predicted without radiative corrections (last row of Table VII) with the masses measured by the UA1 and UA2 collaborations[14,15]. Notice that most of the radiative correction effects come from the running of α from low energy up to ~ 100 GeV. The major uncertainties in the predictions of the Standard Model with three fermion families come from the unknown masses of the top quark and of the Higgs particle which affect Δr and ρ. The Higgs mass dependence is small, although not completely negligible as long as $m_H \lesssim 1$ TeV. It typically introduces an uncertainty of about 0.002 on $\sin^2\theta_w$. For larger values of m_H per perturbative calculation of the radiative corrections becomes suspect because the Higgs becomes a strongly interacting particle.

The sensitivity to m_{top} is much larger. For example, for large top masses, ρ is quadratically divergent with m_{top} (but only logarithmically in m_H). This dependence can be summarized by the formula:

$$(56) \quad \delta\rho = \frac{3G_F \, (250 \text{ GeV})^2}{8\sqrt{2} \, \pi^2} \left\{ \frac{m_{top}^2 - (45 \text{ GeV})^2}{(250 \text{ GeV})^2} \right\}$$

for $m_{top} \gg m_{bot}$ and $m_H \gg M_W$. Similarly Δr exhibit a large sensitivity to m_{top}. The data from the measurement of the W/Z masses and of ρ are clearly inconsistent (for all values of m_H < 1 TeV) with a mass of the top larger than ~ 180 ÷ 200 GeV. Similar limits can be found on the mass splitting between quarks or leptons in a fourth fermion family: $|m_{top'} - m_{b'}| <$ 180 GeV and $|m_L - m_{vL}| <$ 310 GeV[9,10].

I finally mention that the analysis of the parameters of the Standard Model has allowed to put limits on the possible existence of extra Z bosons, which are expected in some popular models, and to compare the measured value of $\sin^2 \theta_w$ with the predictions of GUT's. From the analysis of several models, which predicts extra Z's which can mix with the usual Z^0, and by using the expected coupling of the extra Z's to light fermions, it has been found that the neutral currents constraints are usually more stringent than the direct production limits from the Sp$\bar{\text{p}}$S Collider. Nevertheless masses as low as 120-300 GeV, depending on the model are still typically allowed.

The central value of $\sin^2 \theta_w$ is $\geq$ 2.5 standard deviations above the prediction of minimal SU(5) or similar models for all the allowed values of m_{top}. On the other side the experimental number is somehow low with respect to the predictions of simple supersymmetric GUT's.

4. STUDIES OF W AND Z PRODUCTION, MASSES AND DECAY PROPERTIES AT THE COLLIDER

A very important chapter in the physics of the Standard Model in the '80s has been written by the collider experiments which studied the properties of the intermediate vector bosons.

The UA1 and UA2 experiments found direct evidence for intermediate vector boson production in p - $\bar{\text{p}}$ interaction at high energy via the processes:

$$p + \bar{p} \rightarrow W^{\pm} + X$$
$$\phantom{p + \bar{p} \rightarrow} \mathrel{\rlap{\kern1ex\lower1ex{\hbox{$\rightarrow$}}}{\vert}} e^{\pm} v$$

$$p + \bar{p} \rightarrow Z^0 + X$$
$$\phantom{p + \bar{p} \rightarrow} \mathrel{\rlap{\kern1ex\lower1ex{\hbox{$\rightarrow$}}}{\vert}} e^+ e^-$$

The measurement of the W and Z^0 masses yielded to a value of $\sin^2 \theta_w$ and ρ consistent with the results derived from low energy experiments; the W $\rightarrow$ ev decay angular distribution was found consistent with the predictions of the V-A Standard Model; the transverse and longitudinal momentum distribution of the vector bosons were shown in agreement with QCD expectation and the results for $W \to e\nu$, $W \to \mu\nu$ and $W \to \tau\nu$ gave a test of lepton universality at large momentum transfer ($Q^2 \sim M_W^2$). In this section I review some aspects of the W and Z^0 physics at the Collider.

4.1. *Production cross-sections.*

W and Z production at the Collider proceeds through the Drell-Yan mechanism, which was originally proposed as the basic mechanism for the reaction $p + \bar{p} \to e^+ + e^- + X$. The interpretation of this process in the framework of the parton model is that lepton pairs originate from quark-antiquark annihilation into a virtual photon of mass Q^2 (the invariant mass of the lepton pair) as illustrated in fig.8[16]. Drell-Yan processes are very important because they probe the parton content of the hadrons and, being electroweak processes, they can realistically be predicted in perturbation theory, once the parton densities has been measured in deep inelastic scattering. In the naïve parton model the cross-section is given by:

$$(57) \quad \frac{d\sigma}{dQ^2} = \frac{4\pi\alpha^2}{3N_cQ^2S} \sum_f e_f^2 \int \frac{dx_1}{x_1} \frac{dx_2}{x_2} \left[q_f^{h1}(x_1)\, \bar{q}_f^{h2}(x_2) + \bar{q}_f^{h1}(x_1)\, q_f^{h2}(x_2) \right] .$$

Notice the colour factor in the denominator of eq.(57).

In the QCD improved parton model, the formula in eq.(57) is modified by strong interactions. Leading logarithmic corrections change, according to the factorization theorem, the scaling densities $q(x)$ into $q(x,Q^2)$. Next-to-leading corrections, add terms of order $O(\alpha_s)$ to the simple formula (57). The relevant diagrams at order α_s are shown in figs. 9a,b,c. Eq.(57) becomes[17]:

$$\frac{d\sigma}{dQ^2} = \frac{4\pi\alpha^2}{9Q^2S} \sum_f e_f^2 \int dx_1\, dx_2 \left\{ [q_f^{h1}(x_1)\, \bar{q}_f^{h2}(x_2) + \bar{q}_f^{h1}(x_1)\, q_f^{h2}(x_2)] \right.$$

$$(58)$$

$$\left. [\delta(1-z) + \frac{\alpha_s}{\pi} f_q(z)] + ([q_f^{h1}(x_1,Q^2) + \bar{q}_f^{h1}(x_1,Q^2)]\, G^{h2}(x_2,Q^2) + 1 \leftrightarrow 2)\, \frac{\alpha_s}{\pi} f_G(z) \right\}$$

$$z = \frac{Q^2}{x_1 x_2 S} .$$

Unlike for deep inelstic scattering the next-to-leading corrections ($\alpha_s f_q$ and $\alpha_s f_G$) turn out to be very important (~100%) at fixed target and ISR energies, leading to a breakdown of perturbation theory. The correction are so large that they tend to cancel the $1/N_c$ colour factor of the cross-section.

Perturbative correction are large, mainly because of the presence in $\alpha_s f_q(z)$ of a term of the form:

(59) $\dfrac{\alpha_s}{2\pi} \dfrac{4}{3}\pi^2 \, \delta(1\text{-}z)$.

The origin of such a large term ($\pi^2 \sim 10$ and $\alpha_s \sim 0.2 \div 0.3$) can be traced back to the quark Sudakov form factor[18]. Consider the virtual correction shown in fig.10 to the quark vertex. At large momentum transfer, one finds that the (leading) double logarithmic correction has the form:

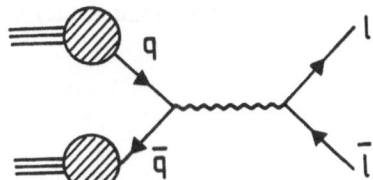

Fig.8. Quark-antiquark annihilation in a lepton pair.

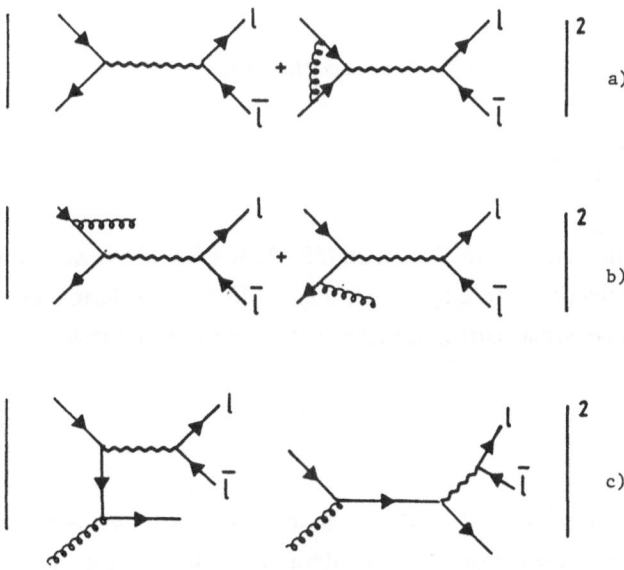

Fig.9a,b,c. O(α_s) diagrams for Drell-Yan processes.

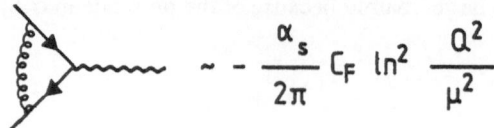

Fig.10. Diagram which gives a form factor to the quark at order $O(\alpha_s)$.

(60) $\quad - \dfrac{\alpha_s}{2\pi}\, C_F \ln^2(Q^2/\mu^2) \qquad C_F = \dfrac{4}{3}$.

This correction can be as large as the lowest order, thus giving a break-down of the perturbative expansion. However, one can show that double logarithms may be resummed to all orders in the double logarithmic approximation, with the following result for the quark form factor:

(61) $\quad \Gamma(q^2) \sim e^{-(\alpha_s/2\pi)\, C_F \ln^2 Q^2/\mu}$.

This result agree with our intuition that when a charged particle is accelerated it must radiate an infinite number of soft gluons and consequently the elastic form factor is damped to zero. Now one defines the quark densities from deep inelastic scattering, where the transferred momentum is negative and the Sudakov correction has the form:

(62) $\quad - \dfrac{\alpha_s}{2\pi}\, C_F \ln^2(-Q^2/\mu^2)$.

The analytic continuation to positive Q^2 gives a contribution:

(63) $\quad - \dfrac{\alpha_s}{2\pi}\, C_F\, (\ln^2 \dfrac{|Q^2|}{\mu^2} - \pi^2)$,

which gives rise to the large correction in eq.(59). However, since we know that the $\ln^2 Q^2$ Sudakoff terms exponentiate, we may expect that so do the "π^2" terms, associated with soft gluon emission and that we can extract from the correction an overall factor:

$$\exp(\dfrac{2\alpha_s}{3\pi}\, \pi^2) \ .$$

The remaining correction of order α_s is now small and this may bring perturbation theory under control. Arguments have been provide in the literature in favour of the exponentiation of the "π^2" terms[19], which has been confirmed by the recent calculation of ref.(20).

In any case, the correction becomes reasonable (~30-35%) in W and Z⁰ production at the collider. The Drell-Yan production of W and Z⁰ proceeds through the same mechanism illustrated in fig.8, by substituting the virtual photon with the intermediate vector bosons. In eq.(57), we have only to change the overall normalization and the combination of quark densities to take into account the different boson-quark-quark couplings. In fig. 11 and 12 I compare the theoretical predictions[21] with the experimental results from the UA1 and UA2 groups at $\sqrt{s}$ = 546 and 630 GeV. The error band in the theoretical predictions comes from several uncertainties

i] an uncertainty of about 30% due to the possible choices of the structure functions and of the scale Q^2 to be used in $\alpha_s(Q^2)$ in higher order terms.

ii] The uncertainty in the branching ratios $B(W \to e\nu)$ and $B(Z \to ll)$ due to the as yet unknown top mass.

iii] A possible increase of ~20% in the production cross-section due to the recent BCDMS data[22]. Within these uncertainties the agreement between theory and experiments is really good. We observe that the measurement of the W and Z⁰ production cross-section is one of the best places where we can check the colour factor. In fact not only the colour factor suppresses the production cross-section, but it also affects the branching ratios because it modifies the W and Z decay rates into quarks.

The measurement of the ratios of the σ-B partial production cross-sections for the various W and Z⁰ decays allows also a test of the lepton universality at large momentum transfer $Q^2 \sim M_{W,Z}^2$ independently from any theoretical calculation.

Defining the weak charged coupling constants g_i and the weak neutral coupling constant K_i, we have:

$$\frac{\sigma_W \; B(W \to l_1 \; \nu_1)}{\sigma_W \; B(W \to l_2 \; \nu_2)} = \left(\frac{g_1}{g_2}\right)^2$$

(64)

$$\frac{\sigma_W \; B(Z \to l_1 \; \bar{l}_1)}{\sigma_W \; B(Z \to l_2 \; \bar{l}_2)} = \left(\frac{K_1}{K_2}\right)^2$$

The results of the UA1[23] group are:

$$g_\mu/g_e = 1.00 \pm 0.07 \pm 0.04$$

(65) $$g_\tau/g_e = 1.01 \pm 0.10 \pm 0.06$$

$$K_\mu/K_e = 1.02 \pm 0.15 \pm 0.04$$

These results confirm the lepton universality at a better than 15% level.

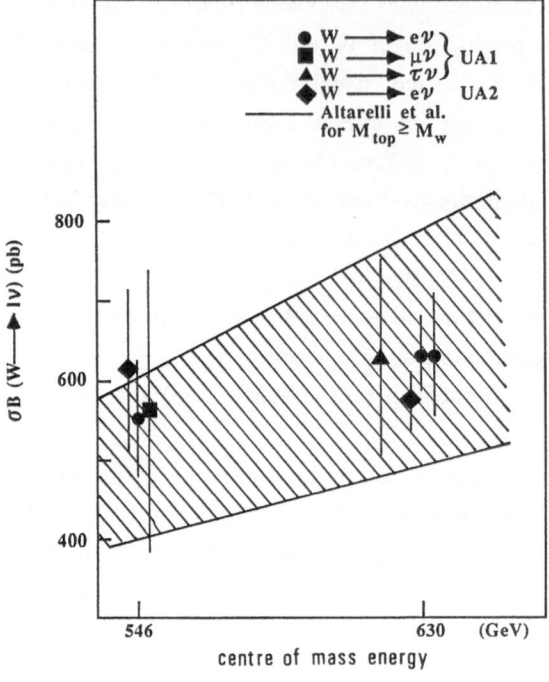

Fig. 11

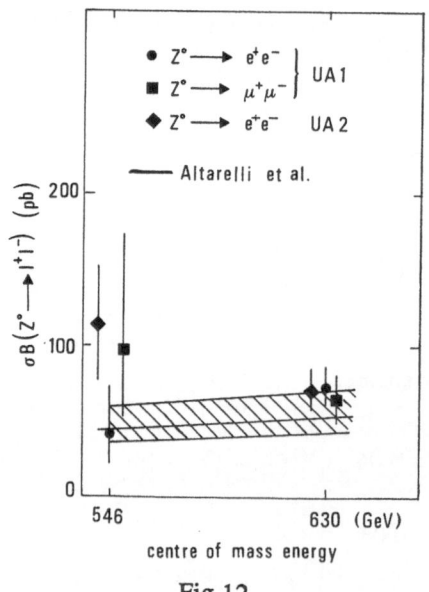

Fig.12

Figs. 11 and 12. Partial cross-sections versus the center of mass energy; 11] for W→ ev production, 12] for Z → ll production. The theoretical predictions and their error band are taken from ref.21 and the leptonic branching ratios W→ ev and Z → ll have been calculated under the assumption that there is no kinematically allowed W → tb̄ and Z → tt̄ decay.

From the measurement of σ_W B(W $\rightarrow$ ev) and σ_Z B(Z $\rightarrow$ e$^+$e$^-$), by taking the production cross-sections from the theoretical calculations one finds:

(66)
$$B(W \rightarrow ev) = 0.10 \pm 0.014^{+0.02}_{-0.03}$$

$$B(Z^o \rightarrow e^+e^-) = 0.046 \pm 0.009^{+0.008}_{-0.019}$$

where the second error comes from the theoretical uncertainties on the production cross-sections.

At lowest order the standard model predicts:

(67)
$$\Gamma(W \rightarrow ev_e) \sim \frac{G_F m_W^3}{6\pi\sqrt{2}} \approx 250 \text{ MeV}$$

$$\Gamma_W = 3\Gamma(W \rightarrow ev_e) + (2+\epsilon) \times 3 \; \Gamma(W \rightarrow ev_e)(1 + \frac{\alpha_s(M_W)}{\pi}) \approx 2.8 \text{ GeV}$$
$$\uparrow_{colour}$$

and

(68)
$$\Gamma(Z^o \rightarrow e^+e^-) = \frac{G_F m_Z^3}{24\pi\sqrt{2}}[1 + (1 - 4\sin^2\theta_w)] \approx 90 \text{ MeV}$$

$$\Gamma_Z \approx (2.8 + 0.18 \, \Delta N_v) \text{ GeV}$$

where ϵ is the correction due to the presence of the top quark and ΔN_v is the number of additional light neutrino species besides the three known ($v_{e,\mu,\tau}$) neutrinos. From eqs.(67) and (68) one derives:

(69)
$$B(W \rightarrow ev) \approx 0.089 \sim \frac{1}{12}$$

$$B(Z \rightarrow e^+e^-) \approx 0.032$$

in good agreement with the experimental results reported in eqs.(66). Moreover the mass distributions of the electronic channels can be used to extract 90% confidence level limits on the total widths:

$\Gamma_W < 5.4\,\text{GeV}$

$$\text{UA1}$$

$\Gamma_Z < 5.2\,\text{GeV}$

(70)

$\Gamma_W < 7.0\,\text{GeV}$

$$\text{UA2}$$

$\Gamma_Z < 5.6\,\text{GeV}$

The good theoretical extimates of the production cross-section allow us also to put limits on the masses of additional heavy vector bosons which arise naturally in many possible extensions of the Standard Model.

The mass dependence of $\sigma_{W'}\,B_{W'}$ and $\sigma_{Z'}\,B_{Z'}$, assuming standard couplings to fermions, is reported in fig.13. The theoretical predictions for the production cross-sections can be converted to lower mass limits for W' and Z'

$M_{W'} \geq 246\,\text{GeV}$ 90% CL

(71)

$M_{Z'} \geq 186\,\text{GeV}$ UA1+UA2 combined

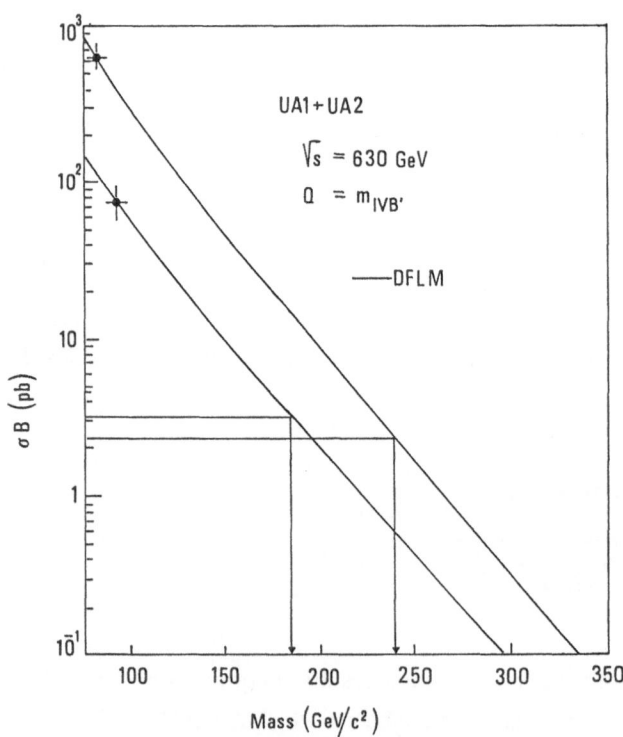

Fig.13. Limits on the partial production cross-sections of extra Z's and W's with standard couplings to fermions from the UA1 and UA2 collaborations.

On the other hand, most of the extensions of the Standard Model predict couplings which depend on the model under consideration. In the simplest cases, where there is one extra Z besides the usual one, the physical mass eigen-state Z bosons are mixed with each other:

$$Z_1 = Z_1^0 \cos\theta + Z_2^0 \sin\theta$$

(72)

$$Z_2 = - Z_1^0 \sin\theta + Z_2^0 \cos\theta$$

The most popular model are the $SU(2)_L \times SU(2)_R \times U(1)$ left-right symmetric model, $SO(10)$ which breaks to $SU(5) \times U(1)$, $E(6) \rightarrow SO(10) \times U(1)$, $E(6) \rightarrow SU(3) \times SU(2) \times U(1) \times U(1)$ ("strings" in fig. 14) or more generally $E(6) \rightarrow G \times U(1)$, where G contains the standard model. For all these cases the production cross-section can be computed and used to put lower limits on the masses of the extra Z bosons as reported in fig.(14).

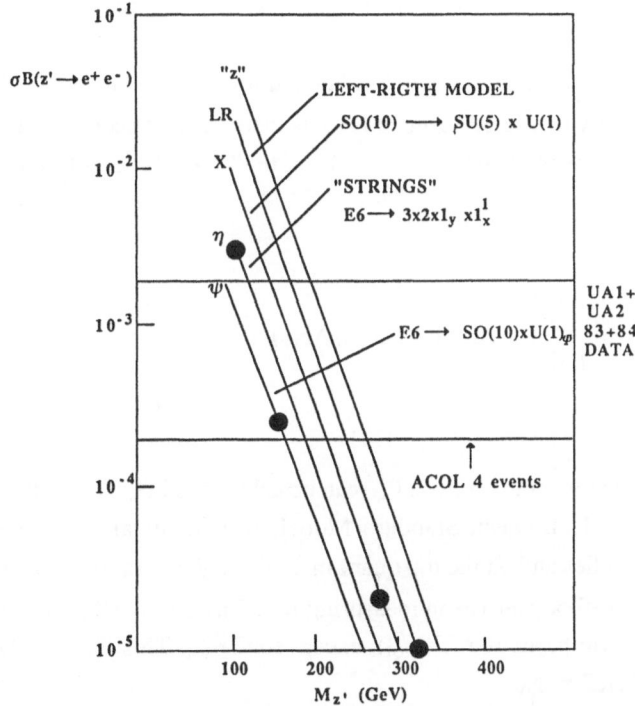

Fig.14. Production cross-section times branching ratio of exotic extra Z⁰'s predicted by several popular models. The limits from the UA1+UA2 combined data and those which will probably be obtained by ACOL are also shown. The actual limits are in general lower than the limits which can be derived from the analysis of the values of the parameters (ρ and Δr) of the Standard Model [9,10] (black circles).

33

4.2. Limits on the number of light neutrinos from the measurement of $R=\sigma(W \to l\nu)/\sigma(Z \to ll)$

The number of $W \to l\nu$ and $Z^0 \to ll$ events observed in experiments like UA1, UA2 or CDF is sensitive through the W, Z leptonic branching ratios to additional open channels, such as $W \to t\bar{b}$ and $Z^0 \to \nu_i \bar{\nu}_i$. The absolute production rates $\sigma(W \to l\nu)$ and $\sigma(Z \to ll)$ however are not suitable to deduce directly N_ν, as the uncertainties are too large. This is due on the one hand to the ~ 30% spread on the theoretical predictions on W, Z production rates[21], and on the other to the uncertainty on the experimental luminosity which is, for example, 8% for UA2 and 15% for UA1. What is considered instead, as initially suggested in ref.24, is the ratio:

(73) $R = \sigma(W \to l\nu)/\sigma(Z \to ll) = \sigma_W/\sigma_Z \, BR(W \to l\nu)/BR(Z \to ll)$

as in this ratio most of the experimental and theoretical uncertainties cancel. The number N_ν of neutrino species clearly affects the $Z^0 \to ll$ branching ratio. The ratio R also depends significantly on the top quark mass, if 45 GeV $< m_{top} <$ 75 GeV, as for $m_{top} < M_Z/2$ both the $Z^0 \to t, \bar{t}$ and $W \to t\bar{b}$ channels are open, while for $M_Z/2 < m_{top} < M_W$ only $W \to t\bar{b}$ remains.

The (top quark mass dependent) central value and upper limit on N_ν is obtained from the comparison of the directly measured value of R to its theoretical expectation. The latter one can be expressed in terms of the total and partial widths of the W and Z^0 as follows:

(74) $R_{th} = \left[\dfrac{\sigma_W}{\sigma_Z} \right] \left[\dfrac{\Gamma_{ev}^W \Gamma_{tot}^Z}{\Gamma_{ee}^Z \Gamma_{tot}^W} \right] = R_\sigma \cdot R_\Gamma (m_{top}, \Delta N_\nu) \ .$

The ratio of total production cross sections R_σ can be calculated in QCD, while the second term R_Γ is predicted by the electroweak Standard Model. It contains all the dependence on the number of neutrino families and on the top quark mass through the ratio $\Gamma^Z_{tot}/\Gamma^W_{tot}$.

In conclusion this method consists in measuring R and assuming all terms on the right hand side of eq.(74) known (in particular Γ^W_{tot}), except for Γ^Z_{tot}. This then allows to determine indirectly Γ^Z_{tot} and therefore N_ν.

The CERN collider experiments UA1 and UA2 have measured the ratio R[14,15], and evaluated the uncertainties originating from the relative W and Z^0 selection and detection efficiencies, the background subtraction, and the statistical error on the numbers of observed events. The results of the two experiments can been combined, leading to:

(75) $R = 8.4^{+1.2}_{-0.9}$ and $R < 10.1$ (90% C.L.)

$R < 10.5$ (95% C.L.)

The uncertainties on these measurements are predominantly of statistical origin, due to the limited number of Z^0 events. These results will certainly be improved in the next future by CDF and at ACOL. The possible additional neutrinos are assumed to be light with respect to the Z^0 which is here produced on-shell. This allows to neglect any phase space suppression in $Z^0 \rightarrow \nu_i$ $\bar{\nu}_i$ (for $M_\nu < 10$ GeV the suppression does not exceed 4%). It is also assumed that the charged lepton partners (L_i) of these additional neutrinos, and the quarks of the same generations (Q_i), are heavy enough so that their contribution to the W and Z^0 total width can be neglected, i.e. that $W \rightarrow L_i \bar{\nu}_i$ or $Z^0 \rightarrow Q_i \bar{Q}_i$ decays are kinematically forbidden.

The second theoretical input in this method is the ratio of total cross sections R_σ. It is not known with comparable precision to R_Γ. It suffers from uncertainties on $\sin^2\theta_w$ and, more important, on the structure functions relating the partonic and hadronic cross sections[21,25,26]. Simplifying, R_σ can be written as

$$(76) \quad R_\sigma = \frac{\sigma_W}{\sigma_Z} \approx \frac{2\, f_{u\bar{d}}(M_W^2/s)\, |V_{ud}|^2}{C_u f_{u\bar{u}}\,(M_Z^2/s) + C_d f_{d\bar{d}}\,(M_Z^2/s)}$$

where $f_{qq'}(\tau) = \tau(dL^{qq'}/d\tau)$ are the appropriate partonic luminosities i.e. constrained products of quark densities $\sim \int u(x_1)\, \bar{d}(x_2)\, \delta(x_1 x_2 - M^2/s)\, dx_1 dx_2$, C_u and C_d are twice the $Z^0 \rightarrow u\,\bar{u}$, $d\,\bar{d}$ couplings $(g_V^2 + g_A^2)$ and $|V_{ud}|^2$ is the first generation mixing term $(\cos^2\theta_C)$. This expression shows explicitly the dependence of R_σ on the structure functions in $\tau(dL^{qq'}/d\tau)$, and on $\sin^2\theta_w$, through $M_Z - M_W$ and the coefficients C_u and C_d.

Theoretical expectations for R_σ vary from about 2.95 to 3.5 according to the various possible choices of structure functions and the uncertainty on $\sin^2\theta_w$. It is possible however to reduce the uncertainty on R_σ using available experimental data on deep inelastic μ-N scattering at large Q^2. As visible from relation (76), R_σ is essentially determined by the ratio $d(x)/u(x)$ of quarks densities in the region around $x_W = M_W/\sqrt{s} = 0.13$ and $x_Z = M_Z/\sqrt{s} = 0.15$ (at $\sqrt{s} = 630$ GeV), as $C_u f_{u\bar{u}} \gg C_d f_{d\bar{d}}$. This ratio $d(x)/u(x)$ in turn is practically given by the measured ratio of the F_2 deep inelastic structure functions $F_2^n(x)/F_2^p(x)$. The data on F_2^n/F_2^p of the two most recent μ-Deuterium and μ-Hydrogen deep-inelastic experiments EMC and BCDMS[27] are shown in fig.15 where they are compared to theoretical expectations for various sets of structure functions: DFLM[25], EHLQ[28], DO1[29], GHR[30]. In the region of interest $(x < 0.4)$, both sets of data are between the EHLQ and DO1 curves, lying close to the GHR and DFLM expectations. This means a value of R_σ between 3.1 and 3.4.

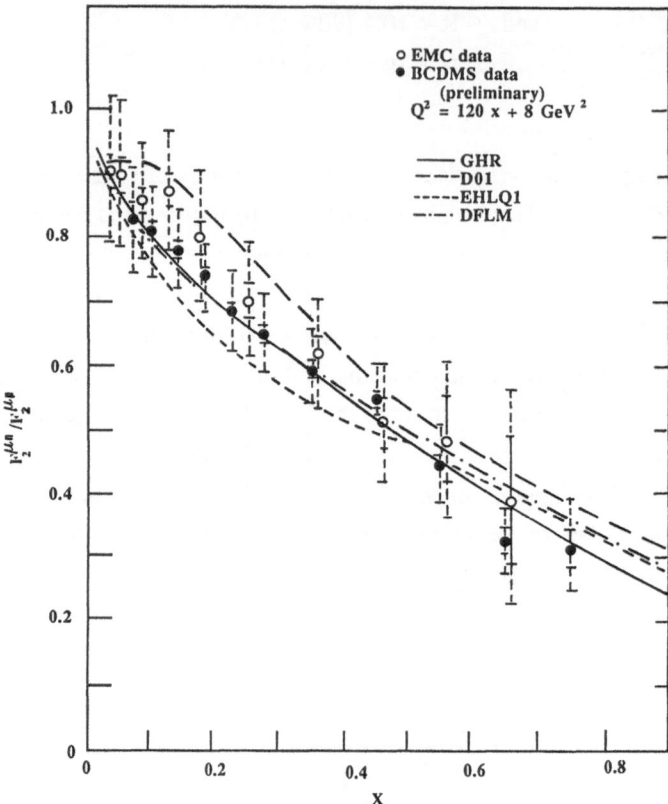

Fig.15. Comparison of the EMC and BCDMS deep inelastic muon scattering data[27] on the ratio of structure functions F_2^p/F_2^n with predictions from various sets of structure functions calculated at the appropriate Q^2[25-28-29-30].

The analysis of these muon scattering data[26] yields for R_σ :

(77) R_σ (muon data) = 3.25±0.10 ,

while a comprehensive analysis of neutrino deep-inelastic scattering data, from which the new set of structure functions (DFLM) is extracted[25], gives similarly:

(78) R_σ (neutrino data) = 3.28±0.15 .

As a central value we can take R_σ = 3.25, and R_σ = 3.15 as a reasonable lower limit, which determines the upper limit on N_ν. All the ingredients needed to obtain N_ν are now at hand.

The expected variation of R as a function of m_{top} for N_ν = 3 and 5 is shown in fig.16 for the central value R_σ = 3.25, the shaded band corresponding to the theoretical uncertainty δR_σ = ±0.10. The theoretical predictions are compared to the combined UA1 and UA2 experimental

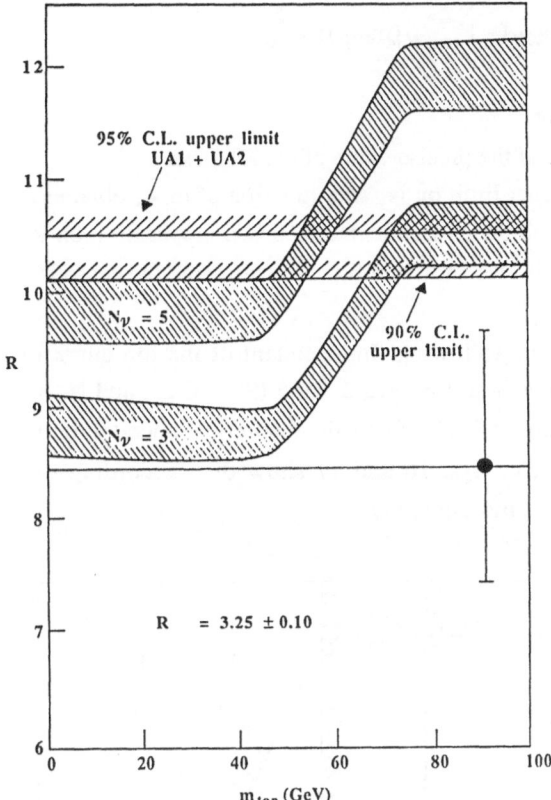

Fig. 16. Comparison of the theoretical predictions for the ratio R as a function of m_{top}, with the theoretical input $R_\sigma = 3.25 \pm 0.10$, with the experimental results of UA1/2. The horizontal continuous line represents the UA1 and UA2 combined measurement of R, and the hatched lines are the 90% and 95% C.L. upper limits implies by this measurement. The theoretical expectations are shown for 3 and 5 massless neutrinos, the error band corresponds to the theoretical uncertainty $\delta R_\sigma = \pm 0.10$.

central value and upper limits (90% and 95% C.L.) of R in the same figure. A lower value of R_σ is clearly less constraining for N_ν, as well as for the possible upper limit on m_{top} suggested by this comparison. Fig. 16 show clearly that N_ν is limited to < 6 for whatever top mass, and for large top masses the constraint is clearly stronger.

From relation (74), the (indirect) upper limit on the Z total width is given by:

$$(79) \quad \Gamma_{tot}^Z < \Gamma_{tot,up}^Z(m_{top}) = \frac{R_{up}}{R_\sigma} \frac{\Gamma_{ee}^Z}{\Gamma_{ev}^W} \Gamma_{tot}^W(m_{top}),$$

where R_{up} is the experimental upper limits on R. If the measured central value R is used instead in this expression, it yields an indirect measure of Γ_{tot}^Z itself. With this upper limit on the Z^0 width, if we assume that the excess over what is expected for 3 generations can only be due to new neutrino flavours, the upper limit on $\Delta N_\nu = N_\nu - 3$ as a function of m_{top} is given by:

(80) $\Delta N_\nu < (\Gamma^Z_{tot,up}(m_{top}) - \Gamma^Z_{tot,3G}(m_{top}))/\Gamma^Z_{\nu\nu}$

where 3G = 3 generations.

This limit is independent of the precise value of the Z^0 mass.

Fig. 17 shows the upper limit on N_ν as a function of m_{top}, obtained from the conservative lower value $R_\sigma = 3.15$ and the experimental 95% C.L. upper limit on R. The central values of N_ν as a function of m_{top} is also shown in fig. 17, using now the central value $R_\sigma = 3.25$ and the measured central value of R.

The conclusion are the following. Independent of the top quark mass, the limits on the number of neutrino flavours are: $N_\nu < 5.5 \pm 0.5$ (90% C.L.) and $N_\nu < 6.3 \pm 0.5$ (95% C.L.). The uncertainty on the upper limit reflects the theoretical uncertainty on R_σ. If the top mass is higher than about 75 GeV, figs. 16 and 17 show that, according to this method, a fourth generation light neutrino is most unlikely.

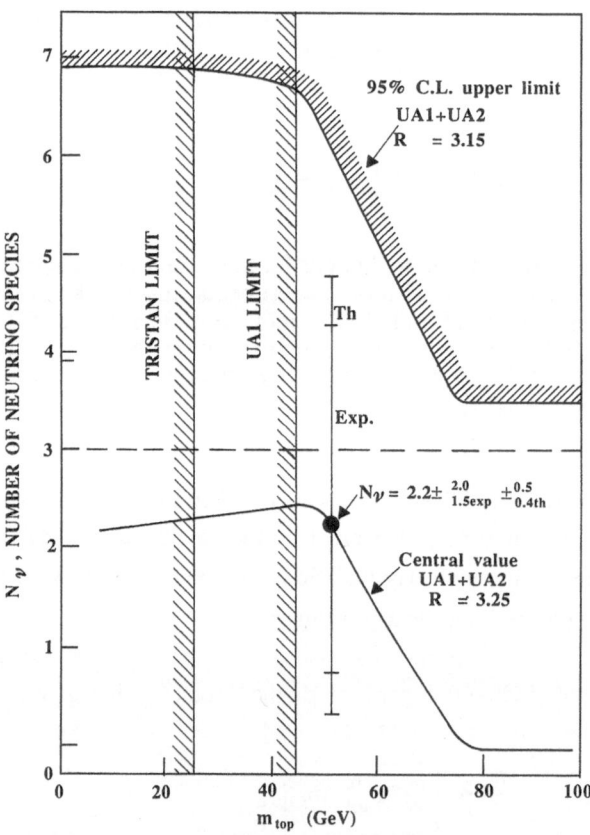

Fig.17. Total number of light neutrino species N_ν (solid line) as extracted from the combined UA1 and UA2 measurement of R, and the 95% C.L. upper limit on this number, as a function of m_{top}. The theoretical input is $R_\sigma = 3.25$ for the central value N_ν, while it is $R_\sigma = 3.15$ for the upper limit. The present TRISTAN and UA1 limits on m_{top} are shown.

4.3. Transverse momentum distribution of W/Z at the Collider

Hard scattering of gluons give rise to high q_T events in Drell-Yan processes through the diagrams of figs. 9b,c. It is easy to compute the high q_T distribution of lepton pairs in first order perturbation theory. However to predict the q_T distribution in a wider range of q_T, especially in W/Z production, all order effects must be taken into account[31] because of the presence in the perturbative series of large terms of the order:

$$(81) \qquad \frac{1}{q_T^2} \, \alpha_s^n \, (q_T^2) \, \ln^m \frac{Q^2}{q_T^2}$$

when $\Lambda_{QCD} \ll q_T \ll Q^2$, Q being $\sim M_{W,Z}$. The large logarithms in eq.(81) are in close relation with the Sudakov form factor discussed above.

Consider the quantity:

$$(82) \qquad \Sigma \, (q_T^2, Q^2) = \frac{1}{\sigma(Q^2)} \int^{q_T^2} \frac{d\sigma}{dp_T^2} \, dp_T^2 \;\;,$$

Σ can be interpreted as the probability of emitting a gluon with transverse momentum less or equal to q_T^2. To order α_s, we may write Σ as (1- {the probability of emitting a gluon with momentum larger than q_T^2}):

$$(83) \qquad \Sigma \, (q_T^2, Q^2) = 1 - \int_{q_T^2}^{Q^2} \frac{dp_T^2}{p_T^2} \, \frac{\alpha_s(p_T^2)}{2\pi} \, C_F \int_0^{1-p_T^2/Q^2} dx \, P_{qq}(x) \;\;,$$

where the upper limit on x is dictated by kinematical constraints. Remembering that $P_{qq} = (1+x^2)/(1 - x)$ we obtain:

$$(84) \qquad \Sigma \, (q_T^2, Q^2) \simeq 1 - \int_{q_T^2}^{Q^2} \frac{dp_T^2}{p_T^2} \, \frac{\alpha_s(p_T^2)}{2\pi} \, 2C_F \ln \frac{Q^2}{p_T^2} \;\;.$$

This is related to the lowest-order expansion of the Sudakov form factor. In fact, neglecting the dependence on the scale of α_s, one finds the double logarithm of eq.(60). Deriving eq.(84) with respect to q_T^2, one finds:

$$(85) \quad \frac{1}{\sigma} \frac{d\sigma}{dq_T^2} = \frac{\alpha_s(q_T^2)}{\pi} \frac{C_F}{q_T^2} \ln \left(\frac{Q^2}{q_T^2} \right) = v(q_T^2) \ .$$

In the region of transverse momenta much smaller than the relevant scale Q, the emission of n gluons factorizes into n independent emissions of a single gluon, whose probability is given in eq.(85):

$$(86) \quad \frac{1}{\sigma} \frac{d\sigma}{dq_T^2} = \sum_{n=0}^{\infty} \frac{1}{n!} \left(\prod_{i=1}^{n} \int d^2 k_{\perp}^{i} \frac{v(k_{\perp}^{i})}{\pi} \right) \delta^{(2)} (\vec{q}_T - \sum_i \vec{k}_{\perp}^{i}) \ .$$

Writing the two-dimensional δ-function in the familiar impact parameter space:

$$(87) \quad \delta^{(2)} (\vec{q}_T - \sum_i \vec{k}_{\perp}^{i}) = \frac{1}{(2\pi)^2} \int d^2\vec{b} \exp [- i\vec{b} \ (\vec{q}_T - \sum_i \vec{k}_{\perp}^{i})]$$

one obtains the eikonal result:

$$(88) \quad \frac{1}{\sigma} \frac{d\sigma}{dq_T^2} = \frac{1}{(2\pi)^2} \int d^2\vec{b} \exp [- i\vec{b} \ \vec{q}_T] \exp \{ \int d^2 p_T (v(p_T)/\pi) \ (e^{-i\vec{b} \ \vec{p}_T} -1) \} \ .$$

The leading contribution, obtained by exponentiating eq.(85), would have the form:

$$(89) \quad \frac{1}{\sigma} \frac{d\sigma}{dq_T^2} \sim \frac{\alpha_s}{q_T^2} \ln \left(\frac{Q^2}{q_T^2} \right) \exp [- \frac{\alpha_s}{2\pi} C_F \ln^2 \left(\frac{Q^2}{q_T^2} \right)] \ .$$

In this approximation, the cross-section is damped at $q_T = 0$, because of the Sudakov suppression, and goes to zero faster than any power in q_T (remember that it was diverging at lowest order in perturbation theory). The momentum conservation in eq.(86) changes this result. In fact it allows for the emission of many gluons with compensating momenta even at q_T very small. The Sudakov form factor cannot kill the cross-section since gluon emission is allowed.

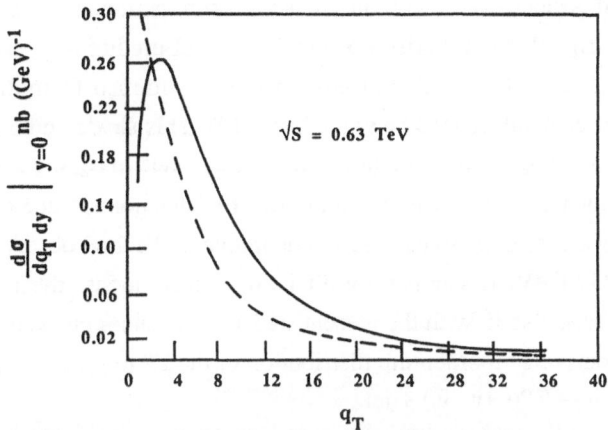

Fig.18. The lowest order result for $d\sigma^W/dq_T \, dy$ (dashed line) is compared with the result obtained by resumming all leading and next-to-leading contributions (solid line). The curves refer to W-production in p-p̄ collisions $\sqrt{s} = 630$ GeV.

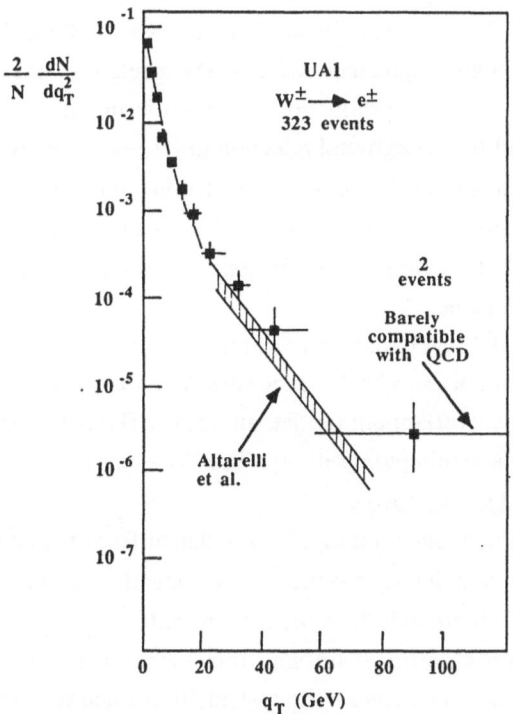

Fig.19. Comparison of the experimental q_T-distribution of W vector bosons as measured by the UA1 Group[23] with the theoretical predictions of ref.(21).

41

In fig. 18, the lowest-order q_T distribution is compared with the result obtained by resumming at all orders the leading (as well as the subleading) terms in $\ln(Q^2/q_T^2)$, for $d\sigma^W/dq_T$ dy at $\sqrt{s} = 630$ GeV. It is clear from this figure that the resummation is indeed important in the region $q_T \ll M_W$. In fig.19, a comparison of the theoretical predictions with the data of the UA1 group[23] is presented. Notice that at large q_T the resummed formula approaches the lowest-order perturbative result (shadowed area in fig.19). This shadowed area represents the theoretical uncertainty coming from the choice of Λ_{QCD}, the scale in α_s, the choice of the parton densities, etc. Notice the last experimental point in fig. 19. This point is due to two events with two observed jets with a very large transverse momentum and the total observed invariant mass of these events is ~250 GeV. It seems very difficult to accomodate these events as due to higher-order QCD effects, also if, with the present statistics, it cannot be excluded.

One can use the transverse momentum distribution of the Z^0 to set limits on the number of neutrino types from $p\,\bar{p} \rightarrow Z^0(\rightarrow\nu_i\,\bar{\nu}_i) + $ jet.

This method, is based on $p\,\bar{p} \rightarrow Z^0 + $ jet production followed by $Z \rightarrow \nu\,\bar{\nu}$, which is the QCD analogue of the QED process $e^+e^- \rightarrow \gamma + Z^0$. The simplest $Z^0 + $ jet QCD gluon-bresstrahlung production mechanism is represented by the diagrams in fig. 9b. A high transverse momentum Z^0 is produced recoiling against a (gluon) jet, with the Z^0 decaying (invisibly) into neutrino pairs. Topologically this is a large missing transverse energy monojet event. The Z (or W) transverse momentum distribution generated by these QCD radiative effects is well illustrated by the data from experiments UA1/2 at $\sqrt{s} = 630$ GeV and the experimental data on W production are in good agreement with QCD expectations as shown in fig. 19.

High transverse momentum Z^0 production is needed in this method to have an adequate experimental signature and for background rejection (a $Z^0 \rightarrow \nu\,\bar{\nu}$ decay from a Z^0 produced with a low transverse momentum p_t^Z cannot be detected in a $p\,\bar{p}$ collision, since a longitudinal missing energy cannot be measured in hadron colliders as numerous beam fragments always escape detection through the beam pipes). It appears as an apparent energy/momentum imbalance in the transverse plane.

The expected Standard Model sources of genuine large missing transverse energy jetty events representing the main physics backgrounds are: $W \rightarrow \tau\nu$ decays, W + jet events with W $\rightarrow$ e,μ,τ+ν decay products overlapping the jet, and heavy flavour production of c $\bar{c}$, b $\bar{b}$ or possibly t $\bar{t}$, followed by a semileptonic decay. The $W \rightarrow e\nu$ decays on the other hand are relatively easily recognised and removed.

In this method then, the upper limit on N_ν is obtained from the comparison between the observed event numbers and jet E_t spectra to theoretical expectations, once the known instrumental and physics backgrounds have been subtracted.

Monojet events with a significant missing transverse energy have been detected and analysed by UA1[32]. This analysis gives a limit on additional neutrino species of $\Delta N_\nu < 7$.

In its final analysis UA1 observes a total of 56 events containing one (or more) high transverse energy jets ($E_t > 12$ GeV), with an (isolated) missing transverse energy measured at a ≥ 4 σ significance level.

The net acceptance of UA1 to detect a missing E_t event from $Z^0 \to \nu \bar{\nu}$ is 1.8%, including the effect of the ≥ 4 σ E_t^{mis} cut. For the experimental sensitivity of ≈ 0.7 events/pb of UA1, this gives ≈ 2.0 expected events from $Z^0 \to \nu \bar{\nu}$ for each neutrino species. Apart from $W \to \tau\nu$ events, this is the largest source of high E_t monojets in the UA1 signal region. The τ decays can be characterised through a criterion (the τ-likelihood L_τ)[32] incorporating the jet collimation and the low charged track multiplicity characteristics of τ-jets. Once most of the τ decays have been eliminated by an appropriate cut ($L_\tau < 0$), all the known Standard Model sources, including $Z^0 \to \nu \bar{\nu}$ with $N_\nu = 3$, account for 21 ± 5 events, to be compared to the 24 events observed. The error on the expected number of events includes all absolute normalisation and Monte Carlo simulation uncertainties.

Due to limited statistics UA1 gives only an upper limit on N_ν. From the observed number of events at $E_t^{jet} < 40$ GeV compared to the predicted one, the limit on the number of neutrino species in $N_\nu < 10$ at a 90% C.L.. The expected contribution for seven extra neutrino species is compared to data in fig.20. The upper limit on N_ν determined by this method is comparable to the one from individual e^+e^- experiments (from $e^+e^- \to Z^0 + \gamma$).

Beyond the simplest extension of the Standard Model with additional neutrino families, other possible source of monojet events are the production of a fourth generation heavy lepton from $W \to L\nu$, or of supersymmetric particles. If present, these contributions would only tend to reduce the present limit on N_ν, as they compete for the same number of observed events. The absence of excess missing transverse energy events has been used by UA1, and recently by CDF, to put lower limits on lepton, squark and gluino masses.

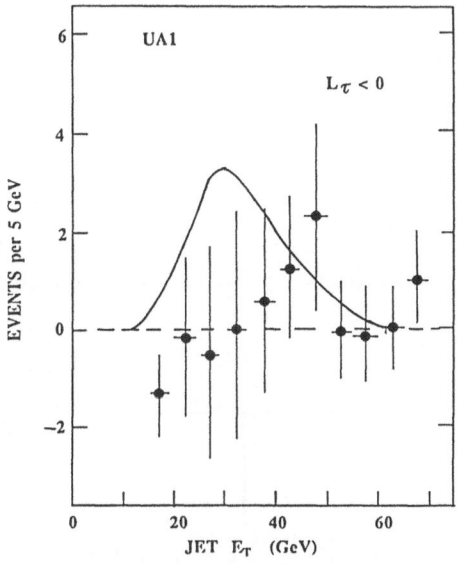

Fig.20. Jet transverse energy distribution for background-subtracted data (including the contribution for $N_\nu = 3$) passing the cut $L_\tau < 0$ (points with error bars), compared with the expected contribution for 7 extra massless neutrino species, solid line.

5. HEAVY FLAVOUR PRODUCTION AT THE COLLIDER

5.1. The production cross-section at order α_s^3

In the QCD improved parton model the production cross-section for the inclusive production of a heavy quark Q:

(90) $H_1(P_1) + H_2(P_2) \to Q(p_3) + X$

is given by:

(91) $$\frac{E_3 d^3\sigma}{d^3 p_3} = \sum_{ij} \int dx_1 \, dx_2 \left[\frac{E d^3\sigma_{ij}(x_1 P_1, x_2 P_2, p_3, \mu)}{d^3 p_3}\right] F_i^1(x_1, \mu^2) \, F_j^2(x_2, \mu^2)$$

where E_3 and p_3 are the heavy quark energy and momentum respectively; $F_{i,j}^{1,2}$ are the distribution functions of the light partons (gluons, light quarks and antiquarks) i,j inside the hadrons 1,2 evaluated at the scale μ. The scale μ is expected to be of the order of the heavy quark mass m. σ_{ij} is the short distance parton cross-section from which all mass singularities have been factored out. σ is calculated as a perturbative series in $\alpha_s(\mu^2)$. The lowest order which contributes in eq.(91) is $O(\alpha_s^2)$. At this order there are two relevant partonic processes:

(92) $q(p_1) + \bar{q}(p_2) \to Q(p_3) + \bar{Q}(p_4)$ *quark anti-quark annihilation*

(93) $g(p_1) + g(p_2) \to Q(p_3) + \bar{Q}(p_4)$ *gluon gluon fusion*

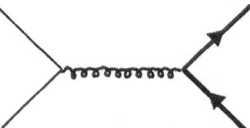

Fig.21. Lowest order diagrams for heavy flavour production: $q\bar{q}$ annihilation.

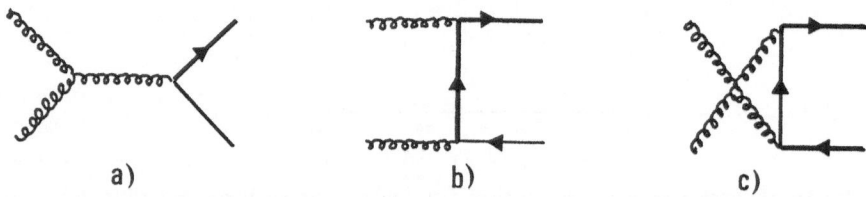

Fig.22. Lowest order diagrams for heavy flavour production: $gg \to Q\bar{Q}$.

The corresponding Feynman diagrams are shown in figs. 21 and 22a-c. As discussed in ref.(33) heavy flavour production, for $m \gg \Lambda_{QCD}$, is a short distance process and can be safely computed in perturbation theory. At order α_s^2 we can write eq.(91) in the form:

(94) $\quad \dfrac{E_3 d^3\sigma}{d^3 p_3} = \dfrac{\alpha_s^2(\mu^2)}{S^2} \sum_{ij} \int \dfrac{dx_1}{x_1^2} \dfrac{dx_2}{x_2^2} T_{ij}\,(\tau_1,\tau_2,\rho)\delta(1 - \tau_1 - \tau_2)\, F_i^1(x_1,\mu^2)\, F_j^2(x_2,\mu^2)$

where:

(95) $\quad \tau_1 = \dfrac{(P_1 \cdot p_3)}{(P_1 \cdot P_2)x_2}$

$\tau_2 = \dfrac{(P_2 \cdot p_3)}{(P_1 \cdot P_2)x_1}$

$\rho = \dfrac{4m^2}{x_1 x_2 S} \quad S = (P_1 + P_2)^2$

and the transition probabilities T_{ij} are given by[34]:

(96) $\quad T_{q\bar{q}} = \dfrac{N^2 - 1}{2N^2}\,(\tau_1^2 + \tau_2^2 + \tfrac{1}{2}\rho)$

(97) $\quad T_{gg} = \dfrac{1}{2(N^2 - 1)N}\,(\dfrac{N^2 - 1}{\tau_1 \tau_2} - 2N^2)\,(\tau_1^2 + \tau_2^2 + \rho - \dfrac{\rho^2}{4\tau_1\tau_2})$

At order α_s^3 the relevant parton subprocesses which contribute to the cross-section are:

$$q + \bar{q} \rightarrow Q + \bar{Q} \qquad\qquad \alpha_s^2,\ \alpha_s^3$$

$$g + g \rightarrow Q + \bar{Q} \qquad\qquad \alpha_s^2,\ \alpha_s^3$$

(98) $\quad q + \bar{q} \rightarrow Q + \bar{Q} + g \qquad\qquad \alpha_s^3$

$$g + q \rightarrow Q + \bar{Q} + g \qquad\qquad \alpha_s^3$$

$$g + \bar{q} \rightarrow Q + \bar{Q} + \bar{q} \qquad\qquad \alpha_s^3$$

Up to and including $O(\alpha_s^3)$ terms the results can be summarized as follows[35].

Let us write the total production cross-section, obtained by integrating eq.(91) over p_3 in the form:

(99) $\quad \sigma(S) = \displaystyle\sum_{ij} \int dx_1\, dx_2\, \hat{\sigma}_{ij}(x_1 x_2 S, m^2, \mu^2) F_i^1(x_1, \mu^2)\, F_j^2(x_2, \mu^2)$

where S is the square of the centre-of-mass energy of the colliding hadrons. The total short distance cross-section $\hat{\sigma}_{ij}$ can be written as:

(100) $\quad \hat{\sigma}_{ij}(x_1 x_2 S, m^2, \mu^2) = \dfrac{\alpha_s^2(\mu^2)}{m^2}\, f_{i,j}(\rho, \dfrac{\mu^2}{m^2})$

where the dimensionless functions $f_{i,j}$ have the following perturbative expansion:

(101) $\quad f_{i,j}(\rho, \dfrac{\mu^2}{m^2}) = f_{i,j}^{(0)}(\rho) + \alpha_s(\mu^2)\left[f_{i,j}^{(1)}(\rho) + \bar{f}_{i,j}^{(1)}(\rho)\ln\left(\dfrac{\mu^2}{m^2}\right)\right] + O(\alpha_s^2)$

and $\alpha_s(\mu^2)$ is the running coupling constant which obeys the renormalization group equation:

(102) $\quad \dfrac{d\alpha_s(\mu^2)}{d\ln(\mu^2)} = -\beta_0 \alpha_s^2 - \beta_1 \alpha_s^3 + O(\alpha_s^4)$

$$\beta_0 = \dfrac{33 - 2n_{l\,f}}{12\pi} \quad , \quad \beta_1 = \dfrac{153 - 19n_{l\,f}}{24\pi^2}$$

$n_{l\,f}$ being the number of light flavours. Renormalization and factorization of mass singularities are done at the scale μ.

The functions $f_{i,j}^{(0)}$ defined in eq.(100) are:

(103) $\quad f_{q\,q}^{(0)}(\rho) = \dfrac{\pi\beta\rho}{27}\, [2 + \rho]$

(104) $\quad f_{gg}^{(0)}(\rho) = \dfrac{\pi\beta\rho}{192}\, [\dfrac{1}{\beta}\,(\rho^2 + 16\rho + 16)\ln(\dfrac{1+\beta}{1-\beta}) - 28 - 31\rho]$

where $\beta = \sqrt{1-\rho}$ and:

$$f_{gq}^{(0)}(\rho) = f_{gq}^{(0)}(\rho) = 0$$

The $\bar{f}_{i,j}^{(1)}$ terms are determined by renormalization group arguments from the lowest order cross-sections:

$$(105) \quad \bar{\mathcal{F}}_{i,j}^{(1)}(\rho) = \frac{1}{2\pi} \left[4\pi \, \beta_0 \, \mathcal{F}_{i,j}^{(0)}(\rho) - \int_\rho^1 dz_1 \, \mathcal{F}_{k,j}^{(0)} \left(\frac{\rho}{z_1} \right) P_{k,i}(z_1) - \int_\rho^1 dz_2 \, \mathcal{F}_{i,k}^{(0)} \left(\frac{\rho}{z_2} \right) P_{k,j}(z_2) \right]$$

where $P_{i,j}$ are the Altarelli-Parisi kernels.

The quantities $\mathcal{F}_{i,j}^{(1)}$ depend on the scheme used for renormalization and factorization. Their explicit expression in an extension of the $\overline{MS}$ scheme has been given in ref.[35].

5.2. Top production

We discuss first the production of heavy flavours with a mass $m \geq 25$ GeV[36]. This case is important for the search for new heavy quarks at present and future accelerators and is a particularly favourable case for the relatively precise predictions that can be made at $Sp\bar{p}S$ - Collider and Tevatron energies.

First we consider the μ-dependence of the computed cross-section at a fixed value of Λ_{QCD} (Λ_{QCD} = 200 MeV for 5 flavours) and for a given set of parton densities[25]. In fig. 23 the cross-section for the production of a heavy quark of mass m = 40 GeV at leading and next-to-leading accuracy is reported as a function of μ at $\sqrt{S}$ = 630 GeV for $p\bar{p}$ collisions. The range of μ explored is $\frac{1}{5} m \leq \mu \leq 5m$. The variation of the result is much larger for the lowest order cross-section than for the corrected one. Thus there is an increase of the stability by increasing the order in α_s as it should be (of course the result at all orders is independent of μ). One also sees that the cross-section at the next-to-leading order shows a maximum for $\mu \simeq m/2$ and that this maximum approximately coincides with the point where the leading and next-to-leading results intersect. In other words the optimized scales defined by the condition $d\sigma/d\mu = 0$ or by the condition that the K-factor, $K = \sigma^{\alpha_s^3}/\sigma^{\alpha_s^2}$, $K(\mu) = 1$ coincide. The fact that the optimization scales are of order m and not for example $\simeq \Lambda_{QCD}$ or $\gg m$ is quite reassuring and shows that perturbation theory is quite all right in this case. From the figure it is apparent that it makes little sense to quote a value for the K-factor because it depends on μ (the same would be true for Υ decays or for the jet production cross-section). In fact, because the lowest order cross-section is of $O(\alpha_s^2)$, the K-factor is strongly dependent on the scale μ. So for example, for m = 40 GeV, if the K-factor is one at $\mu \simeq m/2$, it must be $\simeq 1.6$ at $\mu = 2m$ simply because this is the order of magnitude of the ratio $\alpha_s^2(m/2)/\alpha_s^2(2m)$ in the two cases. Thus the quantity that should be quoted is not the K-factor but the fact that for a given interval of variation of μ the next-to-leading cross-section exhibits a much smaller variation.

The present case has to be contrasted with cases where the lowest order cross-section is of order one (and not of order α_s to some power) as for example Drell-Yan processes. Also in that case the exact form of the next-to-leading corrections depends on the scale at which the parton densities and the coupling constant are defined. However the relative change, for reasonable choices of the factorization scale $\mu \simeq M_{l^+l^-}$, is small and the K-factor has a well defined

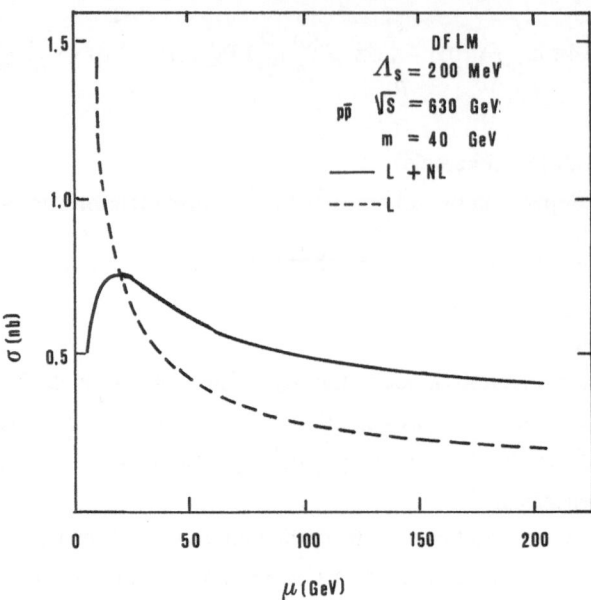

Fig.23. Cross-section for the production of a heavy quark pair of mass m = 40 GeV as a function of μ at $\sqrt{S}$ = 630 GeV in $p\bar{p}$ annihilation.

physical meaning. In the following we shall take the range $\frac{m}{2} \leq \mu \leq 2m$ as a reference for error estimates.

While μ represents an intrinsic theoretical error, additional ambiguities are due to our ignorance of the value of Λ_{QCD} and of the parton densities.

In fig.24 I report a certain number of determinations of α_s as a function of the scale Q as measured by deep inelastic experiments, e^+e^- annihilation and Υ decays[37]. In the figure the value of $\alpha_s(Q)$ at next-to-leading accuracy for three values of Λ_{QCD} (five flavours) are also reported. To estimate the error coming from the indetermination on Λ_{QCD}, given the available experimental information on the coupling constant, we let it vary between 90 and 250 MeV (i.e. Λ_{QCD} = (170±80) MeV). For a consistent evaluation of the error coming from our ignorance of Λ_{QCD} one has to derive the parton densities from D.I.S. and evolve them separately at each value of Λ_{QCD}. This is especially important for processes where the gluon density plays an important role: in fact it is well known that the gluon distribution is strongly correlated to the value of Λ_{QCD}. For these reasons the DFLM densities have been rederived and evolved separately for Λ_{QCD} = 90, 170 and 250 MeV[25,36].

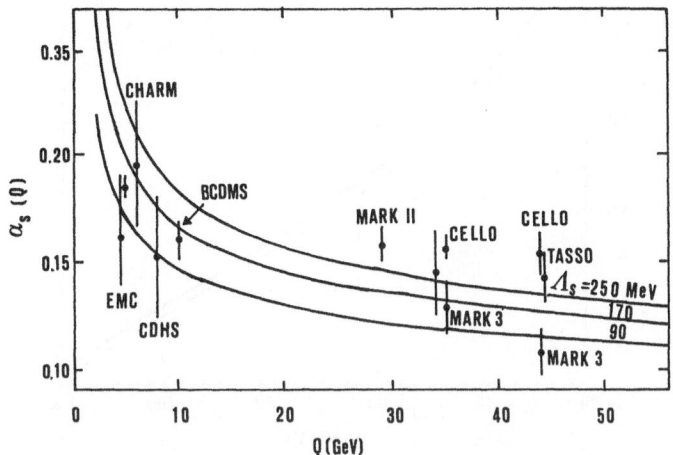

Fig.24. The two-loop QCD coupling $\alpha_s(Q)$ for different values of Λ_{QCD} in the MS scheme for five flavours is reported together with a sample of the most recent determinations from D.I.S., Υ decays and e^+e^- annihilation.

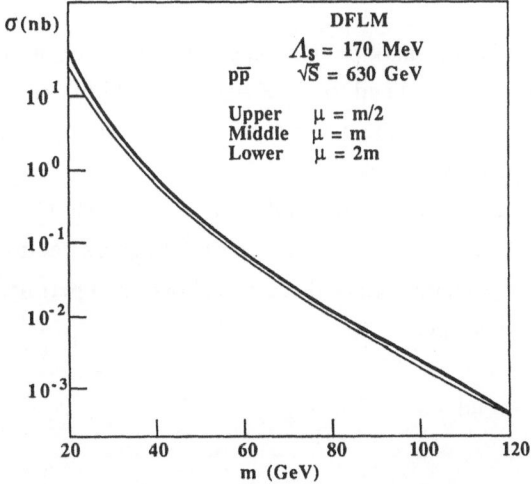

Fig.25. Cross-section at $\sqrt{S} = 630$ GeV as a function of m. The band represents the uncertainty coming from a variation of μ in the range m/2 - 2m.

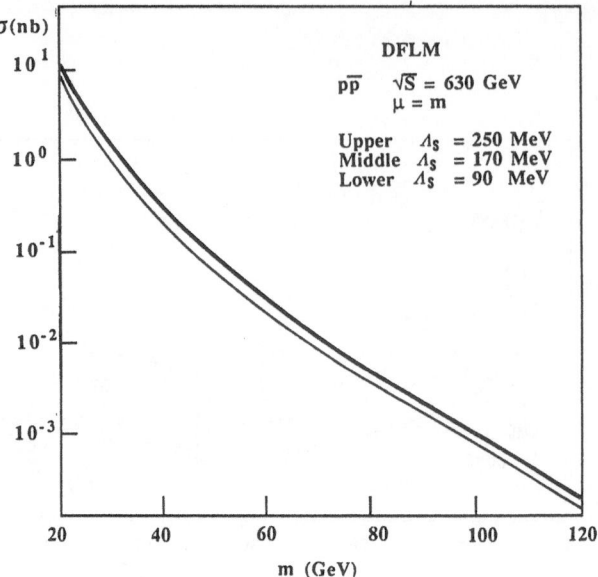

Fig.26. As in fig.25. The band represents the uncertainty coming from a variation of Λ_{QCD}, five flavours, between 90 and 250 MeV. The gluon densities chosen for the computation are consistent with the values of Λ_{QCD}.

The cross-section for heavy quark production at $\sqrt{S}$ = 630 GeV as a function of m with the renormalization scale $m/2 \leq \mu \leq m$ and for $90 \leq \Lambda_{QCD} \leq 250$ MeV is reported in figs. 25 and 26 respectively. Finally in Table VIII the μ and Λ dependences are presented for several values of $\sqrt{S}$ and of the mass of the heavy quark.

In fig. 27 the theoretical results in $p\bar{p}$ collisions at $\sqrt{S}$ = 630 GeV are compared with the UA1 95% confidence level limits on the $t\bar{t}$ and $b'\bar{b}'$ cross sections[38]. The theoretical prediction with its error band, derived as discussed above, is superimposed on the UA1 limits. From the theoretical band one gets:

(106) $m_t > 41$ GeV   and

   $m_{b'} > 34$ GeV .

These results essentially confirm the limits originally states in ref.(38), but the limits are now much better justified. Notice that by using other parton parametrizations, based mostly on harder gluon densities derived from the old CDHS results, as for example the EHLQ densities[28], one would get larger values of the lower limit for the t or b' masses.

TABLE VIII

Heavy quark (mass m) production cross-section in $p\bar{p}$ collisions. The cross sections (in nb) in the second column are expressed by a "central" value, obtained for $\mu = m$ and $\Lambda_S = 170$ MeV, and "errors" derived from varying μ and Λ_S in the range $\frac{1}{2}m \leq \mu \leq 2m$ and 90 MeV $\leq \Lambda_S \leq 250$ GeV (columns 3-6) by adding the corresponding uncertainties in quadrature.

		$\sqrt{S} = 0.63$ TeV			
m (GeV)	σ (nb)	$\mu = \frac{1}{2}m$ $\Lambda = 170$ MeV	$\mu = 2m$ $\Lambda = 170$ MeV	$\mu = m$ $\Lambda = 90$ MeV	$\mu = m$ $\Lambda = 250$ MeV
20	$25.6 \pm {}^{5.6}_{7.9}$	30.9	20.2	19.8	27.3
30	$3.04 \pm {}^{0.62}_{0.94}$	3.62	2.47	2.29	3.26
40	$0.643 \pm {}^{0.11}_{0.20}$	0.738	0.532	0.481	0.696
50	$0.188 \pm {}^{0.025}_{0.056}$	0.208	0.158	0.142	0.204
60	$\left(0.669 \pm {}^{0.067}_{0.190}\right)10^{-1}$	0.718×10^{-1}	0.569×10^{-1}	0.508×10^{-1}	0.716×10^{-1}
70	$\left(0.267 \pm {}^{0.022}_{0.074}\right)10^{-1}$	0.284×10^{-1}	0.229×10^{-1}	0.204×10^{-1}	0.281×10^{-1}
80	$\left(0.114 \pm {}^{8.009}_{0.031}\right)10^{-1}$	0.122×10^{-1}	0.989×10^{-2}	0.880×10^{-2}	0.118×10^{1}
90	$\left(0.511 \pm {}^{0.037}_{0.136}\right)10^{-2}$	0.548×10^{-2}	0.443×10^{-2}	0.394×10^{-2}	0.517×10^{-2}
100	$\left(0.234 \pm {}^{0.019}_{0.062}\right)10^{-2}$	0.254×10^{-2}	0.203×10^{-2}	0.181×10^{-2}	0.232×10^{-2}
110	$\left(0.110 \pm {}^{0.009}_{0.029}\right)10^{-2}$	0.119×10^{-2}	0.946×10^{-3}	0.847×10^{-3}	0.107×10^{-2}
120	$\left(0.517 \pm {}^{0.044}_{0.142}\right)10^{-3}$	0.562×10^{-3}	0.441×10^{-3} ·	0.398×10^{-3}	0.494×10^{-3}

		$\sqrt{S} = 1.8$ TeV			
20	$228 \pm {}^{55}_{60}$	277	190	182	252
40	$9.63 \pm {}^{1.7}_{2.5}$	11.4	7.97	7.74	9.97
60	$1.27 \pm {}^{0.19}_{0.35}$	1.46	1.06	1.01	1.31
80	$0.285 \pm {}^{0.037}_{0.077}$	0.322	0.241	0.222	0.296
100	$\left(0.873 \pm {}^{0.109}_{0.230}\right)10^{-1}$	0.974×10^{-1}	0.755×10^{-1}	0.675×10^{-1}	0.910×10^{-1}
120	$\left(0.331 \pm {}^{0.034}_{0.086}\right)10^{-1}$	0.362×10^{-1}	0.289×10^{-1}	0.257×10^{-1}	0.346×10^{-1}
140	$\left(0.144 \pm {}^{0.013}_{0.037}\right)10^{-1}$	0.155×10^{-1}	0.127×10^{-1}	0.112×10^{-1}	0.150×10^{-1}
160	$\left(0.691 \pm {}^{0.047}_{0.173}\right)10^{-2}$	0.732×10^{-2}	0.607×10^{-2}	0.540×10^{-2}	0.712×10^{-2}
180	$\left(0.352 \pm {}^{0.019}_{0.086}\right)10^{-2}$	0.369×10^{-2}	0.311×10^{-2}	0.276×10^{-2}	0.359×10^{-2}
200	$\left(0.187 \pm {}^{0.008}_{0.045}\right)10^{-2}$	0.195×10^{-2}	0.166×10^{-2}	0.147×10^{-2}	0.189×10^{-2}

TABLE VIII (continued)

m (GeV)	σ (nb)	$\mu = \tfrac{1}{2}m$ $\Lambda = 170$ MeV	$\mu = 2m$ $\Lambda = 170$ MeV	$\mu = m$ $\Lambda = 90$ MeV	$\mu = m$ $\Lambda = 250$ MeV
		$\sqrt{S} = 2.0$ TeV			
20	$275 \pm {}^{67}_{71}$	334	231	219	307
40	$12.2 \pm {}^{2.3}_{3.1}$	14.4	10.1	9.86	12.7
60	$1.66 \pm {}^{0.25}_{0.44}$	1.91	1.38	1.32	1.71
80	$0.378 \pm {}^{0.051}_{0.101}$	0.429	0.319	0.296	0.390
100	$0.117 \pm {}^{0.015}_{0.031}$	0.131	0.100	0.091	0.121
120	$\left(0.446 \pm {}^{0.048}_{0.116}\right)10^{-1}$	0.491×10^{-1}	0.387×10^{-1}	0.346×10^{-1}	0.464×10^{-1}
140	$\left(0.196 \pm {}^{0.018}_{0.051}\right)10^{-1}$	0.212×10^{-1}	0.172×10^{-1}	0.153×10^{-1}	0.204×10^{-1}
160	$\left(0.952 \pm {}^{0.071}_{0.240}\right)10^{-2}$	0.101×10^{-1}	0.836×10^{-2}	0.743×10^{-2}	0.986×10^{-2}
180	$\left(0.496 \pm {}^{0.029}_{0.122}\right)10^{-2}$	0.522×10^{-2}	0.438×10^{-2}	0.388×10^{-2}	0.509×10^{-2}
200	$\left(0.271 \pm {}^{0.013}_{0.065}\right)10^{-2}$	0.283×10^{-2}	0.240×10^{-2}	0.213×10^{-2}	0.275×10^{-2}

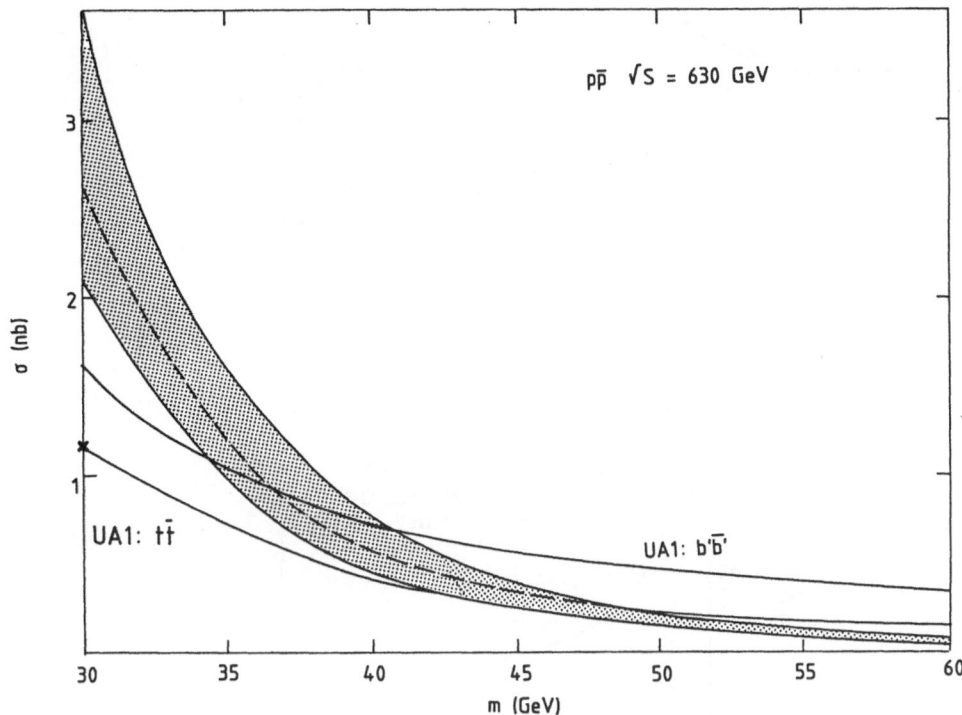

Fig.27. Comparison of the predicted cross-section with the UA1 95% upper bounds on $t\bar{t}$ and $b'\bar{b}'$ pair production cross-sections. The shaded area has been obtained by combining in quadrature the theoretical uncertainties. The dashed line corresponds to the lower edge of the band if the EHLQ densities are used.

5.3. Bottom production

For b production the corrections are expected to be larger since $m_b \ll m_t$ and consequently $\alpha_s(m_b) > \alpha_s(m_t)$. They were found to be particularly large[35,36] at Collider energies because in this case we have also $2m/\sqrt{S} \ll 1$. When $2m/\sqrt{S} \ll 1$ in fact the cross-section is essentially given by the diagrams of order α_s^3 shown in fig.28. These diagrams correspond to the exchange of a gluon (i.e. a particle of spin 1) in the t-channel, instead of a particle of spin 1/2 as in lowest order. At large value of $\sqrt{S}/m$ the asymptotic behavior of the cross-section is dominated by the fixed pole in the t-channel with the largest spin. This fact, together with the large gluon luminosity at small x, leads to large next-to-leading corrections. The diffraction-like production of $Q\bar{Q}$ pairs due to the diagrams in fig.28 has characteristic rapidity and p_T distributions so that the presence of anomalously large corrections is not to be necessarily interpreted as a failure of the perturbative expansion, in the sense that still higher order corrections can well be of moderate size.

Fig.28. Diagrams with gluon exchange in the t-channel.

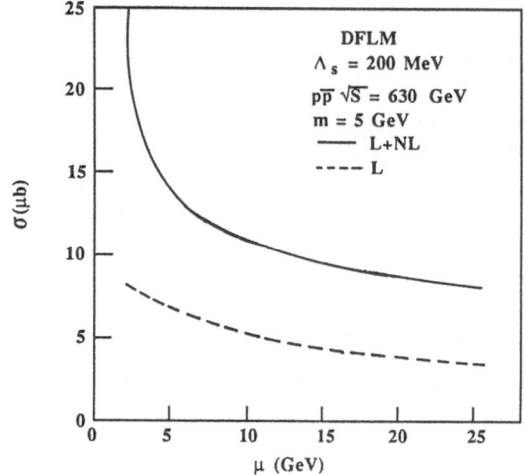

Fig.29. Variation of the cross-section as a function of μ, as in fig.23, for bottom production.

In fig.29 the analogue of fig.23 for top production is reported. The situation for the bottom is completely different: in the range $m/5 \leq \mu \leq 5m$ there is no maximum for the corrected cross-section and no scale for which $K(\mu) = 1$. Not only this, but the uncertainty associated to a variation of μ in a given range is even larger for the corrected cross section than for the lowest order one. The situation improves if one computes the total cross-section at ISR energies and below, where $2m/\sqrt{S}$ is not exceedingly small. Only in this range of energies can one reliably compute the cross-section. The results for b production at $\sqrt{S} = 41$ GeV and 62 GeV in pp collisions are reported in table IX together with the results at $\sqrt{S} = 630$ GeV in p$\bar{p}$ collisions, which are to be taken with the words of caution specified above. All the results in table IX include the uncertainties due to a variation of μ and of Λ_{QCD} as in the case of the top. The results are always reported for $m_b = 4.5$ and 5 GeV separately.

TABLE IX

Same as Table VII, for b production in pp, p$\bar{\text{p}}$ and π^-N collisions.

m_b (GeV)	σ	$\mu = \frac{1}{2}m$ $\Lambda = 170$ MeV	$\mu = 2m$ $\Lambda = 170$ MeV	$\Lambda = 90$ MeV $\mu = m$	$\Lambda = 250$ MeV $\mu = m$
	$\sqrt{S} = 41$ GeV			pp	
4.5	$23 \pm {}^{21}_{15}$ nb	40	12	13	34
5	$9.0 \pm {}^{8.4}_{5.9}$ nb	16	4.7	4.9	14
	$\sqrt{S} = 62$ GeV			pp	
4.5	$142 \pm {}^{98}_{80}$ nb	231	81	91	182
5	$66 \pm {}^{47}_{38}$ nb	109	37	41	86
	$\sqrt{S} = 630$ GeV			p$\bar{\text{p}}$	
4.5	$19 \pm {}^{10}_{8}$ μb	27	15	12.5	25
5	$12 \pm {}^{7}_{4}$ μb	18	10	9	16
	$\sqrt{S} = 24.5$ GeV			π N	
4.5	$7.6 \pm {}^{4.7}_{3.8}$ nb	12	4.6	5.2	10
5	3.1 ± 1.5 nb	4.4	1.9	2.2	3.9

The only existing study of b-production at the Sp$\bar{\text{p}}$S-Collider has been done by the UA1 collaboration by looking at the semileptonic decays of the b. They have obtained:

(107) $\sigma(\text{p}\bar{\text{p}} \to b\bar{b}X) = (10.2 \pm 3.3)\mu$b

It is impressive that in spite of the large theoretical uncertainties, a reasonable agreement with experiment is obtained (table IX). At ISR energies on the contrary the results of the table seem to be in severe disagreement with the result of ref.(39):

(108) $\sigma_{\text{partial}} B(\Lambda_b \to pD^0\pi^-) = (150 - 500)$nb

where σ_{partial} refers to $\sqrt{S} = 63$ GeV, $y[p(K^-\pi^+)\pi^-] > 1.4$ and $x_F(p) > 0.3$.

It is also interesting to compare the theoretical predictions from table IX with the results of the CERN experiment WA78 which has studied multimuon events by dumping a 320 GeV beam of π^- on uranium[40]. At $\sqrt{S} = 24.5$ GeV, WA78 found:

(109) $\sigma(\pi^-N \to b\bar{b}X) = (2.0 \pm 0.3 \pm 0.9)$nb

where N is an isoscalar nucleon. A large value had been previously reported by the NA10 collaboration at $\sqrt{S} = 23.0$ GeV[41]:

$$(110) \quad \sigma(\pi^- N \to b\bar{b}X) = (14^{+7}_{-6})\text{nb}$$

To obtain the results of table IX the pion structure function of ref.(42) has been used. The agreement with the result of WA78 is good for $m_b \simeq 5$ GeV, while the NA10 result prefers $m_b \simeq 4.5$ GeV. Notice that the experimental analysis needed to extract the cross-section depends on the assumptions done on the rapidity and p_T distribution of the heavy quark.

5.4. Charm production

It is clear that the calculations for charm production are affected by even larger errors because m_c is so small. The theoretical results for the production cross-section have to be taken more as a rough estimate rather than a solid prediction. In fig.30 the charm cross-section in pp collisions for energies in the range $\sqrt{S} = 10$ - 62 GeV is compared with the experimental data[43] by assuming $m_c = 1.5$ and 1.2 GeV. Not all the available data are mutually consistent within the quoted errors. The band of values corresponding to $m_c = 1.5$ GeV gives a reasonably good description of the data. Notice that a much smaller value of m_c is needed to reproduce the experimental results with the lowest order cross-section.

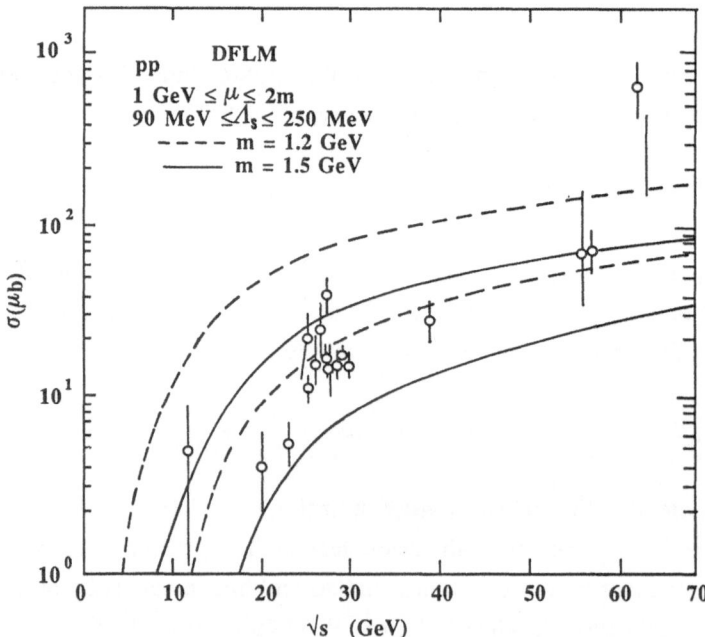

Fig.30. Total cross-section for charm production in pp collisions. The data are from ref.(43). The solid (dashed) curves determine the band due to the theoretical uncertainties, for $m_c = 1.5$ GeV ($m_c = 1.2$ GeV).

ACKNOWLEDGEMENTS

I wish to thanks Tom Ferbel for the perfect organization and the very pleasant atmosphere of the school, the students which patiently attended to my lectures and L.Ibanez, whose collaboration and help was essential to our random sailing at St.Croix.

REFERENCES

1) S.L.Glashow, Nucl.Phys. 22 (1961) 579; S.Weinberg, Phys.Rev.Lett. 19 (1967) 1264; A.Salam in "Elementary Particle Theory", ed. N.Svartholm (Almquist and Wiksells, Stockholm, 1969) p.367; S.L.Glashow, J.Iliopoulos and L.Maiani, Phys.Rev. D2 (1970) 1285.

2) B.W.Lee in "Proceedings of the XVI International Conference on High Energy Physics", edited by J.D.Jackson and A.Roberts, vol.4 (FNAL, Batavia, 1972), p.266.

3) See the references reported in Table II.

4) See for example V.Zacek, CHARM II Collaboration. Proceedings of the XXIV International Conference on High Energy Physics, München, August 4-10, 1988, R.Kothans and J.H.Kühn eds. p.917.

5) CDHS Collaboration, H.Abramowicz et al., Phys.Rev.Lett. 57 (1986) 298; Z.Phys. C28 (1985) 51.

6) CHARM Collaboration, J.V.Allaby et al., Phys.Lett. 177B (1986) 446; M.Jonker et al., Phys.Lett. 99B (1981) 265; Phys.Lett. 103B (1981) 469 (E).

7) CCFRR Collaboration, P.G.Reutens et al., Phys.Lett. 152B (1985) 404.

8) FMM Collaboration, D.Bogart et al., Phys.Rev.Lett. 55 (1985) 1969.

9) U.Amaldi et al., Phys.Rev. D36 (1987) 1385.

10) G.Costa et al., Nucl.Phys. B297 (1988) 244.

11) A.Sirlin, Phys.Rev. D22 (1980) 971.

12) W.Marciano and A.Sirlin, Phys.Rev. D22 (1980) 2695; F.Antonelli, M.Consoli and G.Corbò, Phys.Lett. B91 (1980) 90; M.Veltman, Phys.Lett. B91 (1980) 95; W.Marciano and A.Sirlin, Nucl.Phys. B189 (1981) 442; C.H.Llewellyn Smith and J.F.Wheather, Phys.Lett. B105 (1981) 486; Nucl.Phys. B208 (1982) 185; D.Yu.Bardin, P.Ch.Christova and O.M.Fedorenko, Nucl.Phys. B197 (1982) 1; G.G.Ross, C.H.Llewellyn Smith and J.F.Wheather, Nucl.Phys. B177 (1981) 263. For a complete list of references see also M.A.B.Bég and A.Sirlin, Phys.Rep. 88 (1982) 1; M.Consoli, S.Lo Presti and L.Maiani, Nucl.Phys. B223 (1983) 474.

13) F.Antonelli and L.Maiani, Nucl.Phys. B186 (1981) 269; S.Bellucci, M.Lusignoli and L.Maiani, Nucl.Phys. B169 (1981) 329.

14) UA1 Collaboration, J.Dowell, Proceedings of the XXIV International Conference on High Energy Physics, München, August 4-10, 1988, p.697; J.Müller CERN preprint EP/88-48; G.Arnison et al., Phys.Lett. 166B (1986) 484; C.Albajar et al., Phys.Lett. 198B (1987) 271.

15) UA2 Collaboration, R.Ansari et al., Phys.Lett. 186B (1987) 440; 194B (1987) 158.

16) S.D.Drell and T.M.Yan, Phys.Rev.Lett. 25 (1970) 316.

17) G.Altarelli, R.K.Ellis and G.Martinelli, Nucl.Phys. B143 (1978) 521; (E) B146 (1978) 544; J.Kubar André and F.Paige, Phys.Rev. D19 (1979) 221.

18) V.Sudakov, Sov.Phys. JETP 3 (1956) 65.

19) G.Parisi, Phys.Lett. 90B (1980) 295; G.Curci and M.Greco, Phys.Lett. 92B (1980) 175.

20) T.Matsunra, S.C. van der Marck and W.L. van Neerven, Leiden preprint (1988).

21) G.Altarelli et al., Nucl.Phys. B246 (1984) 12; G.Altarelli, R.K.Ellis and G.Martinelli, Z.f.Phys. C27 (1985) 617.

22) See for example R.Voss, Proceedings of the 1987 International Symposium on Lepton and Photon Interactions at High Energies, Hamburg (july, 1987).

23) UA1 Collaboration, C.Albajar et al., CERN-EP/88-168 (November 1988) and references therein.

24) N.Cabibbo, Proc. of the Third Topical Workshop on Photon-Antiproton Collisions, Rome (January '83), CERN 83-04 (1983) p.567.

25) M.Diemoz et al., Z.Phys. C39 (1988) 31.

26) P.Colas, D.Denegri and C.Stubenrauch, CERN-EP 88-16, to appear in Z.Phys.C.

27) EMC Collaboration, J.J. Aubert et al., Nucl.Phys. B293 (1987) 740; BCDMS Collaboration, A.Milsztajn, Ph.D. Thesis, Université Paris XI.

28) E.Eichten et al., Rev.Mod.Phys. 56 (1984) 579.

29) D.W.Dukes and J.F.Owens, Phys.Rev. D30 (1984) 49.

30) M.Gluck, E.Hoffmann and E.Reya, Z.Phys. C13 (1982) 119.

31) Yu.L.Dokshitzer, D.L.Dyakonov and S.I.Troyan, Phys.Lett. 78B (1978) 290; G.Parisi and R.Petronzio, Nucl.Phys. B154 (1979) 427.

32) UA1 Collaboration, C.Albajar et al., Phys.Lett. B185 (1987) 241.

33) J.C.Collins, D.E.Soper and G.Sterman, Nucl.Phys. B263 (1986) 37.

34) B.L.Combridge, Nucl.Phys. B151 (1979) 429.

35) S.Dawson, R.K.Ellis and P.Nason, Nucl.Phys. B303 (1988) 607.

36) G.Altarelli, M.Diemoz, G.Martinelli and P.Nason, Nucl.Phys. B308 (1988) 724.

37) F.Bergsma et al., Phys.Lett. 123B (1983) 269; 153B (1985) 111; J.V.Allaby et al., Phys.Lett. 197B (1987) 281; H.Abramowicz et al., Zeit.f.Phys. C25 (1984) 291; J.J.Aubert et al., Nucl.Phys. B272 (1986) 158; A.C.Benvenuti et al., Phys.Lett. B195 (1987) 91. See for example the summary by S.L.Wu, Proceedings of the 1987 International Symposium on Lepton and Photon Interactions at High Energies, Hamburg 1987; W.Kwong et al., E.Fermi Inst., preprint EFI-87-31, 1987.

38) C.Abajar et al., Zeit.f.Phys. C37 (1988) 505.

39) L.Cifarelli et al., (BCF collaboration), CERN-EP/87-189, 1987.

40) M.G.Catanesi et al., Phys.Lett. B202 (1988) 453.

41) P.Bordalo et al., CERN-EP/88-39, 1987 to appear in Zeit.f.Phys. C.

42) J.F.Owens, Phys.Rev. D30 (1984) 943.

43) S.P.Tavernier, Reports on Progress in Physics 50 (1987) 1439.

GRAND UNIFICATION, SUPERSYMMETRY, SUPERSTRINGS:

AN INTRODUCTION TO PHYSICS BEYOND THE STANDARD MODEL

L.E. Ibañez

Theory Division, CERN, 1211 Geneva 23, Switzerland

1. INTRODUCTION

During the last ten years the particle physics theorists have been bombing their experimental colleagues with new theories, symmetries, particles and promising experiments to be done. This includes grand unification, proton decay, neutrino oscillations, neutron oscillations, neutrino masses, supersymmetry, supergravity, gluinos, squarks, photinos, axions (visible, invisible and in between), fifth forces, heavy leptons, new quarks, extra Z^0's, strings, etc. None of these very bright ideas have, up to now, been checked experimentally, and there is a reasonable tendency towards discouragement. However, one should think that the present situation is quite difficult for theorists. The leitmotiv of all those new ideas is to go beyond the standard model and try to understand a considerable number of puzzles left unexplained by that theory. This seems to require information about energy scales for which the experimental data are very scarce (if at all existent). One must hope that present and future accelerators (and underground experiments) will eventually unravel some of the expected new physics. In these lectures, I give an elementary introduction to some of the most popular ideas about extensions of the standard model (SM). This includes grand unification, supersymmetry and strings but the emphasis is put on the second. The reason is that supersymmetry is a theory which can be checked experimentally at accelerators. Unfortunately, the same cannot be said for grand unification and strings which will have to be checked (if at all) indirectly. Since the audience of the lectures was of experimentalists, I tried to be as little technical as possible although this is sometimes unavoidable. In Section 2, I discuss some of the features of the SM which are not well understood. The grand unification idea is discussed in Section 3 in which a simplified introduction to the simplest GUTs is given. The longest part of the lectures, devoted to supersymmetry, is contained in Section 4. After describing the properties of supersymmetric interactions, I present in some detail the supersymmetric version of the SM. Plausible SUSY spectra and experimental signatures are also given. Finally, in Section 5 I give an oversimplified description of string theories, particularly as candidates for unified theories of all interactions. A few final comments are left for Section 6.

These lectures are not intended to be self-contained but just to give a flavour of the topics covered. The reader is referred to the original literature as well as to the good number of reviews existing about these topics[1-3]. Some of these are given in the list of references. The latter is not intended to be complete and I apologize for omissions.

2. PUZZLES IN THE STANDARD MODEL

It is really impressive how the SU(3)×SU(2)×U(1) Standard Model (SM) essentially describes all present high energy physics data (except for low p_T strong interactions in which non-perturbative effects play an important role). We are so much used to this model that we very often forget all the aspects of it we do not really understand. The ideal theory for a particle theorist is probably one in which there are essentially no free parameters (all measurable ones are predicted) and an overall beautiful symmetry (as simple as possible) governs all the aspects of the theory. The standard model is far from this situation. The number of parameters "external" to the model (not predicted by it) is large. We have three gauge coupling constants g_1, g_2, g_3 as well as the θ-parameter governing the CP-violating sector ($\theta F_{\mu\nu} \tilde{F}_{\mu\nu}$) of the QCD Lagrangian (the latter is experimentally close to zero, theorists do not know why). To these we have to add the nine masses of quarks and leptons plus the possible Majorana masses of neutrinos (experimentally very small). We further have the three Cabibbo-Kobayashi-Maskawa angles plus the KM CP violating phase. Finally, one has the couplings appearing in the Higgs sector of the model. Thus the standard model is a field theory with a number (~20) of free parameters that, when given appropriate (experimental) values, describe remarkably well all known high-energy physics data, no more, no less. It is obvious we should be ambitious and try to explain the values of the ~20 free parameters of the model. The efforts of physicists to understand them are often referred to as "beyond the standard model physics". All (but the gauge coupling constraints) are parameters in one way or another related to the Higgs-Yukawa sector of the theory, a sector poorly understood and which has not been tested experimentally. One must admit from the outset that the efforts made in the last 15 years have not been very successful from the point of view of explaining the values of the free parameters of the standard model. However, the particular values of those parameters are not the only points not understood in the SM, as we will discuss below. I believe that in the last 15 years, a lot of conceptual progress in understanding the origin of the SM and its connection with gravity has been achieved.

Let us review now different puzzling qualitative features of the SM. First, let us recall the quantum numbers of each of the three generations with respect to $SU(3)_C \times SU(2)_L \times U(1)_Y$:

$$
\begin{bmatrix}
Q_L \equiv (U,D)_L & \rightarrow & (3,2,1/6) \\
U_L^c & \rightarrow & (\bar{3},1,-2/3) \\
D_L^c & \rightarrow & (\bar{3},1,1/3) \\
L \equiv (\nu,E^-)_L & \rightarrow & (1,2,-1/2) \\
E_L^c & \rightarrow & (1,1,1)
\end{bmatrix}
\tag{1}
$$

Altogether each generation contains 3×2+3+3+2+1 = 15 Weyl spinors with the above quantum numbers. One of the most relevant facts of the SM is that quark-lepton generations are "chiral". From the mathematical point of view, this means that the group representation of each family [i.e., quantum numbers in (1)] is complex, e.g., there are no left-handed quarks with $(\bar{3},2,-1/6)$ quantum numbers. Another way of stating the chirality property is to say that left- and right-handed quarks and leptons couple differently to the gauge interactions [right-handed fields do not couple to $SU(2)_L$, the charged weak interactions].

An important consequence of chirality is that quarks and leptons cannot have mass terms in the SM Lagrangian. This is obvious because the mass terms of a fermion f is given by

$$m \; \bar{f}_R f_L + h.c. \tag{2}$$

where $f_{L(R)} = ((1\mp\gamma_5)/2)f$, and a term like (2) is not invariant under the $SU(2)\times U(1)$ symmetry (instead of being a singlet transforms like a doublet). In order that quarks and leptons get a mass, one must break $SU(2)\times U(1)$ spontaneously. If there is a scalar (Higgs) with Yukawa couplings to fermions like

$$h \; H \; \bar{f}_R f_L + h.c., \tag{3}$$

quarks and leptons get effective masses $\sim h\langle H\rangle$ once the Higgs gets a vacuum expectation value (v.e.v.) $\langle H\rangle \neq 0$.

Why did Nature choose fermions to be chiral? A very symmetric and reasonable (non-chiral) assignment for candidate quarks and leptons would have been:

$$Q \rightarrow (3,2,1/6) + (\bar{3},2,-1/6)$$
$$L \rightarrow (1,2,-1/2) + (1,2,1/2) \tag{4}$$

but Nature decided to choose the more asymmetric (chiral) assignments of (1). There is no answer to this question within the SM.

Let us look back again at the quantum numbers of (1). Why those funny hypercharge assignments? Since charge Q and the diagonal generator of $SU(2)$, T^3 are related to hypercharge by

$$Q = T^3 + Y, \tag{5}$$

an equivalent way of stating the same question is the following. Why is the charge, e.g., of the d-quark one-third of that of the electron (why are the charges of proton and positron the same)? This is called charge quantization. Apparently there is no reason why this should be so. In fact, one can convince oneself that charge quantization is naturally required by the quantum consistency of the theory, i.e., the absence of gauge "anomalies"[4]. Gauge theories with chiral fermions, like the SM, have potential dangers (breakdown of unitarity of the S-matrix) with the existence of linear divergences in "triangle graphs" of the type shown in Fig. 1. The dangerous graphs involving the gauge boson coupled to Y are shown in that figure. Let us suppose we know the $SU(3)\times SU(2)$ quantum numbers of the quarks and leptons in a family [as in (1)] but we did not know the hypercharge assignments. Consider giving them arbitrary hyper-charge values, e.g.

$$Y(Q_L) = a, \; Y(U_L^c) = b, \; Y(D_L^c) = c, \; Y(L) = d, \; Y(E_L^c) = e. \tag{6}$$

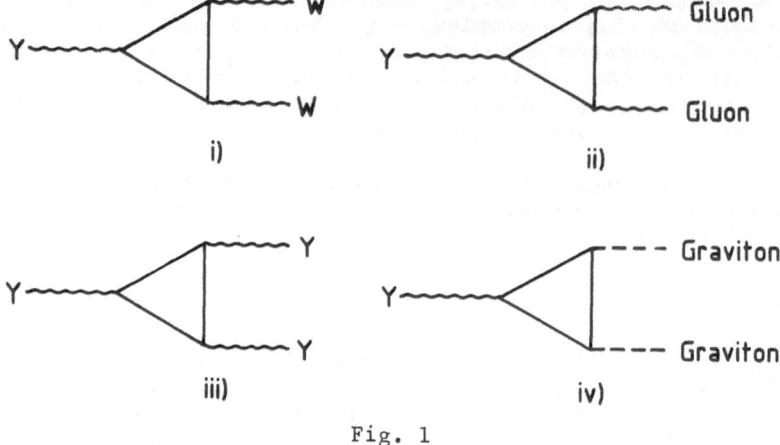

Fig. 1

Each of the graphs in Fig. 1 are proportional to

i) $3a(1/2) + d(1/2)$

ii) $2a(1/2) + b(1/2) + c(1/2)$

iii) $6a^3 + 3b^3 + 3c^3 + 2d^3 + e^3$

iv) $6a + 3b + 3c + 2d + e$

(7)

For example, the first term in i) is the contribution from Q_L to the triangle loop [factor 3 for three colours, a factor a from Y-boson coupling, a factor $1/2 = tr(T^3)^2$ from the coupling of two W's, etc.), whereas the second is the contribution from L.   Now, anomaly freedom dictates that all i)-iv) should vanish.   Since an overall normalization is irrelevant, we may choose, e.g., $Y(E_L^c) = e = 1$.   Then the reader can easily verify that the only solution allowed is

$$a = 1/6, \ b = -2/3, \ c = 1/3, \ d = -1/2, \ e = 1,$$
(8)

i.e., the assignments given in (1) are dictated by the absence of anomalies and so is charge quantization.   The reader has probably noticed that the fourth graph involves gravitons[5].   There are reasons to believe that one should consider gravitational anomalies as well as purely gauge anomalies and thus, Eq. (7.iv) has to be reinforced.   Anyway, it seems that charge quantization is not that arbitrary, even in the context of the SM, and is just a consequence of quantum consistency (freedom of anomalies).   Notice, however, that we imposed cancellations of anomalies family by family, which is not strictly necessary.   One could imagine a situation in which the contribution of each generation was anomalous but the sum of the contributions of the three families vanished.   Why did anomalies have to cancel family by family?   Again, this is something which is not understood within the scheme of the SM.

One can wonder about other remarkable qualitative features of the SM fermion content.   We have seen that the cancellation of anomalies in the SM occurs in a very intricate way, quark and lepton contributions cancel against each other.   A more simple world could be imagined in which quarks would feel only the strong force and leptons the weak force, e.g., families transforming like

Table 1

Lepton	$e \to 5 \times 10^{-4}$	$\mu \to 1 \times 10^{-1}$	$\tau \to 1.8$
D-quarks	$d \to 7 \times 10^{-3}$	$s \to 1.5 \times 10^{-1}$	$b \to 4.5$
U-quarks	$u \to 4 \times 10^{-3}$	$c \to 1.2$	$t \to \gtrsim 45$

$$(3,1,0) + (\bar{3},1,0) + (2,1,-1/2) + (2,1,1/2) + \text{etc.} \tag{9}$$

Again this kind of particle contents is not chiral but why did Nature have to choose the chiral one of (1)? Why a gauge group like SU(3)×SU(2)×U(1) and no other, to start with? One could think of more symmetric groups like SU(3)×SU(3) or SU(2)^4, etc. There is another very well-known puzzle which is the existence of three copies (generations) of quarks and leptons with identical gauge properties. Why three and not one, two or 13?

Eventually, we would like to understand more quantitative matters like the values of quark-lepton masses and mixing angles. The quark-lepton masses in GeV are shown in Table 1. One can observe three peculiar properties of the spectrum. i) The quark-lepton masses are, in general, very small compared with the natural scale of the masses in the SM. Since all masses are given by v.e.v.'s of the Higgs field, one would guess a natural mass scale to be $\sim M_W \sim g \langle H \rangle$. This is not the case and the first generation is even six orders of magnitude lighter. ii) There is a hierarchy of masses amongst the generations (3^d heavier than the second which is heavier than the first). iii) Quarks in each family are heavier than the corresponding lepton. To these three properties of the masses one has to add another important piece of information. The Cabibbo-Kobayashi-Maskawa mixing matrix is, to a good approximation, close to the unit matrix (the biggest off-diagonal element is the Cabibbo angle $\theta \sim 1/5$). Within the scheme of the SM both masses and mixings depend on the shape and size of the Yukawa coupling constants G_D, G_E, G_U appearing in the Lagrangian:

$$G_D^{ij} Q_L^i D_L^{cj} H + G_E^{ij} L^i E_L^{cj} + G_U^{ij} Q_L^i U_L^{cj} H^* + \text{h.c.}, \tag{10}$$

where $i,j = 1,2,3$ run over the three generations. The values of those couplings are "external" to the model so that we do not have any prediction for them within the SM. Of course, one would like to understand at least the above-mentioned four qualitative features of masses and mixing angles. To achieve that, we necessarily have to go beyond the SM.

All the aspects of the SM described above are probably sufficient to believe that the "theory of everything" cannot be just the standard model. There are further theoretical arguments that point towards difficulties associated with the SM field theory. To start with, the U(1) piece of the SM as well as the Yukawa and Higgs sectors are non-asymptotically free field theories. Then, one knows that at a very large scale the coupling constants will get too large (Landau singularities). Furthermore, there are reasons to believe[6] that the purely Higgs sector of the theory would be trivial (S matrix = 1). All this indicates that one cannot extrapolate the SM to very large energies. On the other hand, one knows that anyhow, when the energies considered are of the order of the Planck mass $M_P \simeq 10^{18}$ GeV, the effective gravitational constant $G \sim E^2/M_P^2 \sim 1$ and gravitational interactions cannot be neglected in any standard model computation.

There is another (aesthetical?) problem for the SM which appears even at scales much below the Planck scale and that is the "naturality problem". The point is that it is hard to understand how and why the scale of weak interactions is so small compared with the Planck scale, $M_W \ll M_P$. We know that the weak scale M_W is of the order of the Higgs mass parameter we put in our Lagrangian, $M_W \sim |\mu_{Higgs}|$. We can fix $|\mu_{Higgs}|$ to be ~100 GeV, as experimentally required. However, radiative loop corrections tend to destabilize that fixing and give large masses to the Higgs scalar. The point is that graphs like those in Fig. 2 give rise to quadratically divergent contributions to the Higgs mass. Of course, one can just choose a renormalization prescription, order by order, in which $|\mu_{Higgs}| \sim 100$ GeV. However, this will require a ridiculous fine-tuning of the counterterms in the Lagrangian in order to get the desired result[7]. This problem can be stated as the violation of a naturality criterium which goes as follows[8]. "A small parameter (like $|\mu_{Higgs}|$) is unnatural unless the symmetry is increased by setting it = 0". Setting $|\mu_{Higgs}| = 0$ does not increase the original symmetry of the SM. Notice that a similar naturality problem does not exist for fermions. The point is that one can see that all loop contributions to mass renormalization for a fermion are always proportional to the original bare mass. Thus, if that mass is set to zero, it will continue being zero even after radiative corrections. What happens is that when we set the mass of a fermion to zero, a "chiral symmetry" is preserved by which we can rotate f_R or/and f_L by arbitrary phases (a mass term would violate this). This symmetry is preserved by all the Lagrangian and hence remains true after radiative corrections. Thus, a vanishing fermion mass can be "natural", the symmetry of the Lagrangian increases. This is not the case for a scalar; a mass term like $m^2\phi^*\phi$ cannot be forbidden by any known consistent symmetry of a Lagrangian.

In the above sense, the SM is unnatural, the weak scale M_W is un-naturally low. In order to avoid this naturality problem, many different

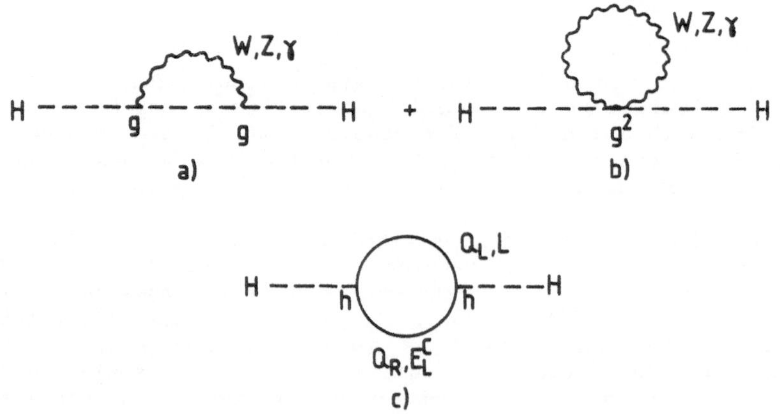

Fig. 2

possibilities have been considered which can be essentially classified into two: compositeness and supersymmetry. The most attractive idea put forward in the first is the possibility that Higgses were not elementary but composite of some new fermions ("techniquarks") bound together by a new strong force ("technicolour")[9]. Although the original idea is very nice, it gets very cumbersome when one tries to implement the details and, in particular, give masses to quarks and leptons. Unless one assumes very peculiar behaviour of the technicolour interaction, one gets wrong predictions for flavour changing neutral current processes (FCNC) (rates too high for reactions like $K^0 \rightarrow \mu^+\mu^-$, $\mu \rightarrow e\gamma$, etc.)[10]. Other ideas involve strongly interacting W bosons[11] or strongly interacting Higgses[12]. Finally, there are "preon" models in which all quarks and leptons are composite. In all these schemes no outstanding model has been found and one typically finds that the number of "preonic" particles one has to introduce, is often bigger than the observed number of quarks and leptons so that not much simplicity is achieved. Furthermore, all the dynamics of these models is unknown so that one cannot really give any model-independent prediction (apart from the expectation that if the confining scale of preons is reached at an accelerator, one would start producing many new particles).

From all the above, the clever reader has probably realized that I am not particularly fond of the compositeness scenarios considered up to now (except, maybe, technicolour). In these lectures, I will discuss the supersymmetry alternative in some detail. There is an additional motivation for that. I have been discussing above a number of puzzles of the SM, but the SM is not all physics. It is well known that eventually gravity has to be included in the scheme and a consistent quantum gravity has not yet been constructed. One would like to have a consistent quantum theory of all interactions including gravity. In this direction, it is known that supersymmetry (particularly superstrings) is quite promising. Furthermore, eventually one would like to answer seemingly more fundamental questions like: why does space-time exist? Why do we live in four dimensions? Why is one of the co-ordinates timelike? etc. To answer these questions, it seems one would need a consistent unified theory of all interactions, including gravity. The work done in the last 15 years (grand unification, supersymmetry, supergravity, superstrings) seems to be on the right track. In the next lecture, I will very briefly discuss these topics emphasizing the chances of an experimental verification. I must remark that none of these ideas has been checked experimentally and so must be considered at the moment just as attractive speculations.

3.   THE GRAND UNIFICATION IDEA

Before discussing the supersymmetry solution to the naturality problem, we are going to consider a beautiful generalization of the SM, grand unified theories[13,14]. In a remarkable paper[15], Georgi, Quinn and Weinberg pointed out the following. Consider the couplings of the SM model α_1, α_2, α_3 ($\alpha_i = g_i^2/4\pi$) as measured at accelerators. It is well known that the size of the coupling constants depend on the energies at which they are measured ("running coupling constants"). The value at any other scale μ is given by the one-loop renormalization group equations:

$$\frac{1}{\alpha_i(\mu)} = \frac{1}{\alpha_i(M_W)} - 8\pi b_i \, \log(\mu/M_W), \quad i = 1,2,3 \qquad (11)$$

where b_i are the coefficients of the corresponding "beta functions" and $\alpha_i(M_W)$ are the measured values at a scale, say, $\sim M_W$. Plugging the known values for $\alpha_i(M_W)$ and b_i in Eq. (11), one obtains a qualitative scale

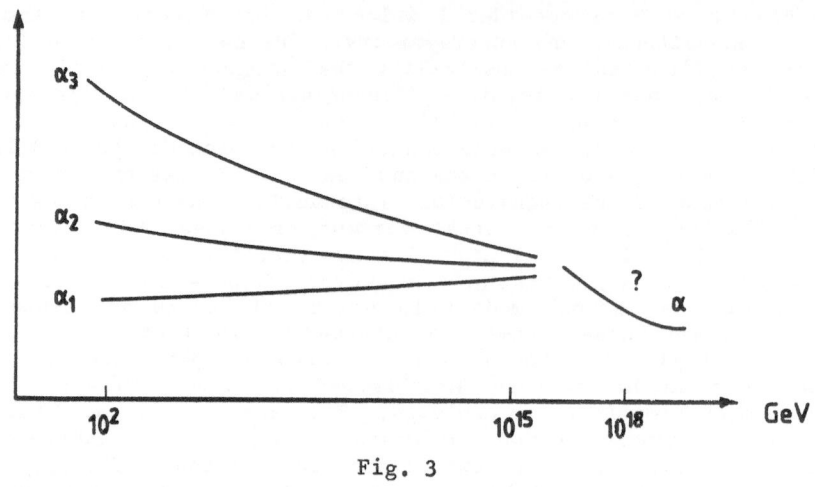

Fig. 3

dependence as shown in Fig. 3. The gauge coupling constants seem to get together at an extremely large mass scale $\sim 10^{15}$ GeV, not far away from the Planck mass ($\sim 10^{18}$ GeV). Thus, the following idea seems very plausible: the couplings join because beyond a scale $\sim 10^{15}$ GeV, there is a "unified theory" of the three SU(3)×SU(2)×U(1) interactions into a single one[13,14]. This will have a simple group G gauge invariance such that $G \supset SU(3)\times SU(2)\times U(1)$ and there will be a unique gauge coupling constant, g. The first task is then to look for a simple group G which contains the SM. The Lie algebras of groups were classified a long time ago by Cartan. The associated groups correspond to the SU(n), SO(n) (n odd), SO(n) (n even) and Sp(n) ("symplectic") classical groups. There is also a finite number (5) of Lie groups which are called "exceptional" and denoted G_2, F_4, E_6, E_7, E_8. The rank (number of commuting diagonal generators) of the SM model is four [two from SU(3) + one from SU(2) + one from U(1)] so that the candidate group needs to have a rank at least equal to four. Cartan's classification for simple algebras of rank $\geqslant 4$ is shown in Table 2. A further constraint one has to impose to get a chiral spectrum is that the group must admit "complex representations" (recall our statements about chirality in the previous section). As it turns out, only the SU(N), SO(2n+4) and E_6 groups have complex representations and then we are left only with the groups inside a square in Table 2. The group of minimal rank is SU(5) and this is the simplest candidate. For rank 5, we have SO(10) and SU(6) and for rank 6 we have E_6 and SU(7). The SU(6) and SU(7) groups do not have any particular advantage over their subgroup SU(5) so that it seems that SU(5), SO(10) and E_6 are the most attractive groups for "grand unification". Let us discuss them briefly in turn[1]:

i) SU(5). This is the simplest possible grand unification scheme[13]. It has $5^2 - 1 = 24$ gauge bosons which we can schematically put in a matrix:

$$\underline{24} = \begin{pmatrix} \text{gluons} & \begin{matrix} X_1^- Y_1^- \\ X_2^- Y_2^- \\ X_3^- Y_3^- \end{matrix} \\ \hline \begin{matrix} X_1^+ X_2^+ X_3^+ \\ Y_1^+ Y_2^+ Y_3^+ \end{matrix} & \begin{matrix} Z^0 W^- \\ W^+ \gamma \end{matrix} \end{pmatrix} \tag{12}$$

Table 2

RANK	CLASSICAL			EXCEPTIONAL	
4	SU(5)	SO(9)	Sp(8)	SO(8)	F$_4$
5	SU(6)	SO(11)	Sp(10)	SO(10)	
6	SU(7)	SO(13)	Sp(12)	SO(12)	E$_6$
7	SU(8)	SO(15)	Sp(14)	SO(14)	E$_7$
8	SU(9)	SO(17)	Sp(16)	SO(16)	E$_8$
⋮	⋮	⋮	⋮	⋮	
N	SU(N+1)	SO(2N+1)	Sp(2N)	SO(2N)	
⋮	⋮	⋮	⋮	⋮	
				SO(2N+4)	

Apart from the known[12] gauge bosons (8 gluons+$W^\pm+\gamma+Z^0$), there are 12 new ones $X_i^\pm$, $Y_i^\pm$ (i = 1,2,3 colours) which are triplets of colour and SU(2) doublets. They have fractional charges $\pm 4/3$, $\pm 1/3$ and couple both to quarks and leptons. Since the complete SU(5) symmetry is only realized above the unification scale $M_X \sim 10^{15}$ GeV, that will be the order of magnitude of the masses of $X^\pm$ and $Y^\pm$. Thus, below M_X the SU(5) symmetry will be spontaneously broken by some Higgs mechanism down to SU(3)×SU(2)× ×U(1), and there will be essentially no obvious trace of the SU(5) symmetry.

The quarks and leptons of each family in SU(5) sit in a five-plet and a ten-plet under SU(5):

$$
\underline{5} = \begin{bmatrix} d_1^c \\ d_2^c \\ d_3^c \\ e \\ \nu_e \end{bmatrix}_L
\quad ; \quad
\underline{10} = \begin{bmatrix} 0 & u_3^c & -u_2^c & u_1 & d_1 \\ & 0 & u_1^c & u_2 & d_2 \\ & & 0 & u_3 & d_3 \\ & & & 0 & e^c \\ & & & & 0 \end{bmatrix}_L
\tag{13}
$$

where we have represented the ten-plet by an antisymmetric 5×5 matrix (this multiplet results from the antisymmetric product of two five-plets and thus has 5×(5-1)/2 independent components). These assignments may

Fig. 4

look bizarre but one can convince oneself that a $(\underline{10+\bar{5}})$ representation is the simplest chiral, anomaly-free fermion content within SU(5). Notice that the $\underline{15}$ states of each SM generation just fit. There is no place for more quarks or leptons in (13). This is a remarkable fact.

Since the new gauge bosons $X^{\pm}$ and $Y^{\pm}$ have both colour and SU(2) quantum numbers, they can convert quarks into leptons and also quarks into antiquarks (i.e., they violate baryon and lepton numbers). Thus processes like the ones in Fig. 4 are possible.

They give rise to proton decay with a lifetime

$$\tau_p \sim \frac{M_X^4}{m^5} \sim 10^{30} \text{ years} \tag{14}$$

where m is a mass scale of order of the mass of the proton.

A model like SU(5) has two widely different symmetry breaking scales, $M_X \sim 10^{15}$ GeV and $M_W \sim 10^2$ GeV:

$$SU(5) \xrightarrow{M_X} SU(3)_C \times SU(2)_L \times U(1)_Y \xrightarrow{M_W} SU(3)_C \times U(1)_Q \tag{15}$$

The first breaking is assumed to be produced by a Higgs mechanism triggered by a v.e.v. of a scalar field transforming like the adjoint representation of SU(5). This will be a field with $5^2-1 = 24$ components whose v.e.v. can be represented in matrix form as:

$$\langle \phi^{a,b} \rangle = \begin{pmatrix} \nu & & & & \\ & \nu & & & 0 \\ & & \nu & & \\ & 0 & & -3\nu/2 & \\ & & & & -3\nu/2 \end{pmatrix}, \ \nu \sim 10^{15} \text{ GeV} \tag{16}$$

There is an explicit SU(3)×SU(2) symmetry left unbroken in (16). The second breaking at the weak scale will proceed as in the usual SM, with an $SU(2)_L$ doublet of Higgses. Within the original SU(5) structure that doublet is contained in a five-plet H^a (a = 1-5) of Higgses.

ii) $\underline{SO(10)}^{16}$. This gauge group contains SU(5):

$$SO(10) \supset SU(5) \times U(1) \tag{17}$$

and thus it is convenient to analyze it in terms of this subgroup. [Other interesting subgroups are $SU(4) \times SU(2)_L \times SU(2)_R$ and $SU(3)_C \times SU(2)_L \times SU(2)_R \times U(1)_{B-L}$.] SO(10) has $10 \times (10-1)/2 = 45$ gauge bosons which in terms of the SU(5) subgroup transform as

$$\underline{45} = \underline{24} + \underline{10} + \overline{\underline{10}} + \underline{1} \tag{18}$$

There are then 33 extra gauge bosons in this model. An attractive feature of SO(10) is that [unlike SU(5)] all particles within a family fit in the simplest SO(10) chiral multiple which has dimension 16:

$$16 = (\nu \; u_1 \; u_2 \; u_3; \; e \; d_1 \; d_2 \; d_3; \; \underbrace{}_{\text{LEFT}} \; d_3^c \; d_2^c \; d_1^c \; e^c; \; u_3^c, \; u_2^c, \; u_1^c, \; \nu^c)_L \qquad (19)$$

LEFT   RIGHT

and has an explicit left-right symmetry. This implies the existence in each generation of an extra particle (the 16th fermion which is not in the SM) which has the quantum numbers of a right-handed neutrino ν_L^c (it is a singlet under the standard model interactions but it has lepton number).

As happened with SU(5), the extra gauge bosons, in general, violate baryon and lepton numbers and hence proton is again unstable with a lifetime $\sim 10^{30}$ years.

Since the rank of SO(10) is five, there are more symmetry breaking possibilities. One in general needs at least two different Higgs fields to do the breaking of SO(10) down to the SM and this breaking may proceed in several stages. Some symmetry breaking possibilities are[17]:

$$SO(10) \xrightarrow{\phi, \phi' \neq 0} SU(3)_C \times SU(2)_L \times U(1)_Y$$

$$SO(10) \xrightarrow{\phi \neq 0} SU(3)_C \times SU(2)_L \times SU(2)_R \times U(1)_{B-L} \xrightarrow{\phi' \neq 0} SU(3)_C \times SU(2)_L \times U(1)_Y \qquad (20)$$

$$SO(10) \xrightarrow{\phi \neq 0} SU(4) \times SU(2)_L \times SU(2)_R \xrightarrow{\phi' \neq 0} SU(3)_C \times SU(2)_L \times U(1)_Y, \text{ etc.}$$

where the $U(1)_{B-L}$ in (20) corresponds to a gauge boson which couples to baryon-lepton number. The latter is gauged within SO(10). We must remark that the symmetry breaking in SO(10). We must remark that the symmetry breaking in SO(10) requires scalar fields with relatively high dimensionalities (ϕ has dimension 45 or 54 whereas ϕ' has dimension 16 or 126).

iii) $\underline{E_6}$. This gauge group[18] contains SO(10)

$$E_6 > SO(10) \times U(1) \qquad (21)$$

so that it is useful to analyze it in terms of that subgroup. [Other interesting subgroups include $SU(3)_C \times SU(3)_L \times SU(3)_R$.] E_6 has 78 gauge bosons which in terms of the SO(10) subgroup have the multiplet structure

$$\underline{78} = \underline{45} + \underline{16} + \overline{\underline{16}} + \underline{1} \qquad (22)$$

There are then plenty of extra gauge bosons which, as with SU(5) and SO(10), can mediate proton decay. A complete generation of quarks and leptons fit in the fundamental multiplet of E_6 which has dimension 27. However, in this case an E_6 generation contains many (12) extra yet unobserved quarks and leptons. In terms of SO(10), and E_6 generation contains:

$$\underline{27} = \underline{16} + \underline{10} + \underline{1} \qquad (23)$$

i.e., apart from an SM generation, one has two neutrino-like objects (ν_L^c and N) and new quarks ($g + \bar{g}$) and leptons ($\ell + \bar{\ell}$). Here, $g(\ell)$ has the SU(3)×SU(2)×U(1) quantum numbers of a right-handed d-quark (left-handed lepton).

Again since the rank of E_6 is large (6), there are plenty of possibilities for breaking the symmetry down to the standard model. All the large symmetry breaking may be done with a single scalar field of quite a large multiplicity, 351. This finishes our short review on SU(5), SO(10) and E_6 grand unified theories (GUTs).

There are a number of properties and predictions which appear in most GUTs. They include the following:

a) Charge quantization. Since the charge generator is $Q = Y + T^3$ and both $SU(2)_L \times U(1)_Y$ are included amongst the generators of the unifying simple group G, Q has to be traceless, like any other G generator. This implies then the same condition [(7.iv) = 0] that we found from imposing absence of the mixed hypercharge-gravitational anomaly (Fig. 1.iv). Thus charge quantization is an automatic property in GUTs.

b) Glashow-Weinberg angle prediction. Since the three coupling constants g_1, g_2, g_3 are assumed to join at the "grand unification mass" M_X, one can compute what is the latter by finding at what scale, e.g., g_2 and g_3 join. That is done by using the renormalization group equation in (11). Once we know the actual value of M_X, the value of the third coupling constant (e.g., g_1) may be computed by running it from M_X down to low energies where experiments are performed. Instead of using α_1, α_2, α_3, it is more convenient to use the fine structure constant α_e, $\sin^2\theta_W$ and α_3. The above computations yield (in leading log approximation)

$$\log\left(\frac{M_X}{M_W}\right) = \left(\frac{1}{\alpha_e(M_W)} - \frac{8}{3\alpha_3(M_W)}\right) \bigg/ 8\pi\left(\frac{5}{3}b_1 + b_2 - \frac{8}{3}b_3\right) \qquad (24a)$$

$$\sin^2\theta_W(M_W) = \frac{3}{8} - 5\pi\alpha_e(M_W)(b_1 - b_2)\log\left(\frac{M_X}{M_W}\right) . \qquad (24b)$$

Assuming that the GUT symmetry breaks directly to the SM (no intermediate symmetry breaking), one finds a value for $M_X \simeq 10^{14-15}$ GeV, implying a value of the Weinberg angle $\sin^2\theta_W \simeq 0.20$, which may be considered a successful prediction (we will further comment on this below). Notice that in models [like SO(10) or E_6] in which an intermediate scale of symmetry breaking is possible, Eqs. (24) get modified and there is more flexibility to accommodate different values of $\sin^2\theta_W$[17].

c) Quark-lepton mass relations. In GUTs, quarks and leptons live in the same multiplets and one expects to find relationships between the Yukawa couplings of quarks and leptons. Thus, e.g., in SU(5) one has the following couplings[19] between the fermions in $\underline{5} = (d_i^c, L)$ and $\underline{10} = (Q_L, u_i^c, e^c)$:

$$G_{D,L}^{ij}\bar{\psi}_{\underline{5}_i}\psi_{\underline{10}_j}H + G_u^{ij}\bar{\psi}_{\underline{10}_i}\psi_{\underline{10}_j}H^* + h.c. \qquad (25)$$

Notice that we only have two independent terms instead of the three in the standard model [Eq. (10)]. We then have equal Yukawa couplings for d-quarks and leptons ($G_L = G_D$) as long as SU(5) is a good symmetry. Once SU(5) is broken, there are radiative corrections which contribute differently to quarks and leptons (in particular gluon exchange). Computing these radiative effects and using the renormalization group, one gets[19]

$$\frac{m_D}{m_L} = \frac{h_D(m_D)}{h_L(m_L)} \simeq \frac{h_D(M_X)}{h_L(M_X)} \times \left(\frac{\alpha_3(M_D)}{\alpha_3(M_X)}\right)^{4/7} \simeq 3.0\text{-}4.0 \qquad (26)$$

which does compare well with the ratios m_b/m_τ and m_s/m_μ, but is definitely wrong for the first generation. However, one may argue that the first generation has some additional sources for its mass coming, e.g., from different type of Yukawas which makes it different. In SO(10) the Higgs Weinberg-Salam doublets are contained in the fundamental multiplet, a tenplet. Since all quarks and leptons are in the same multiplet, a 16-plet, there is only one Yukawa term:

$$G^{ij}\psi_{16i}\psi_{16j}H_{10} + h.c. \tag{27}$$

However, H_{10} contains two SU(2) doublets, one (H) coupling to d-quarks and charged leptons and the other ($\bar{H}$) coupling to U-quarks and neutrinos. Then, one has $m_{D,L} \simeq G\langle H\rangle$ but $m_{u,\nu} \simeq G\langle\bar{H}\rangle$. This gives us the same prediction for d-quarks and leptons as SU(5) but also an additional prediction $m_t/m_b \simeq m_c/m_s$ [of course, the renormalization effects, as it happened in Eq. (26), have to be taken into account]. Such a prediction is still not ruled out experimentally. Unfortunately, if the model contains one single ten-plet of Higgses, the mass matrices for U and D-quarks are going also to have identical shape and may be diagonalized by the same unitary matrices. U_U and U_D acting on the left-handed quarks ($U_U=U_D$). Since the Cabibbo-Kobayashi-Maskawa matrix $U = U_u^\dagger U_D = 1$, there would be no mixing at all in the simplest SO(10) model. On the other hand, this could be considered as a good feature of the model, since after all the CKM matrix is quite similar to the unit matrix. Again one could argue that some other effects (maybe loop contributions?) slightly modify the CKM matrix and gives non-vanishing off-diagonal entries. E_6 gives results similar to those of SO(10).

d) Neutrino masses. When discussing quark-lepton mass relationships in SO(10), we saw that one expects $m_\nu = m_u$ in the simplest model. This looks like a disaster since present limits for neutrino masses are in the range[20] $m_{\nu_e} \lesssim 15$ eV, $m_{\nu_\mu} \lesssim 150$ keV, $m_{\nu_\tau} \lesssim 150$ MeV. However, the above SO(10) prediction refers to the "Dirac ν-mass", which corresponds to a term in the Lagrangian:

$$m_\nu \bar{\nu}_R \nu_L + h.c. \tag{28}$$

and involves both ν_L and ν_R (or equivalently, ν_L^c). However, the right-handed neutrino ν_R may also have a "Majorana" mass, a term of the form

$$M \nu_L^c \nu_L^c + h.c. \tag{29}$$

which involves only the ν_R. In fact, the last term is allowed by the SU(3)×SU(2)×U(1) symmetry [unlike the one in Eq. (28)] and hence M may be as large as the unification mass (unlike m_ν which is at most $\sim M_W$). The natural value for M is $\sim M_X$ and that is usually the case in realistic models (anyway, one usually has $M \gg M_W$). We thus have a neutrino mass matrix for each generation[21]

$$M \sim \begin{array}{c} \\ \\ \nu_L \\ \\ \nu_L^c \end{array} \overset{\displaystyle \begin{array}{cc} \nu_L & \nu_L^c \end{array}}{\begin{bmatrix} 0 & m_\nu \\ m_\nu & M \end{bmatrix}} \tag{30}$$

which has eigenvalues $\sim m_\nu^2/M$ and $\sim M$. Thus there is an eigenstate (essentially ν_L^c) with a very large mass $\sim M$ and hence undetectable. The other eigenstate (mostly the ν_L) has a mass $\sim m_\nu^2/M$. Notice that for $m_\nu \sim m_u \sim 1$ Gev and $M \sim M_X \sim 10^{15}$ GeV, one expects neutrino masses extremely small, $\sim 10^{-6}$ eV. If M is smaller [as happens with SO(10) models with an intermediate $SU(2)_R$ symmetry], the values for neutrino masses may be close to the experimental limits.

A similar situation occurs for E_6. In the simplest version of SU(5), there is no room for a ν_L^c n the multiplet $10+\bar{5}$ so that there cannot be Dirac neutrino masses. One may of course add a singlet fermion to the SU(5) model to play the role of ν_L^c but that would be a bit artificial. An interesting effect which may appear if neutrinos have a mass is neutrino oscillations (e.g., $\nu_e \leftrightarrow \nu_\mu, \nu_\tau$ transitions). This may happen because, if the ν's have a mass, there will be in general a mixing matrix for them similar to the CKM matrix. Then, a neutrino produced in a weak interaction process will not be, in general, a mass eigenstate and the proportion of ν_e, ν_μ and ν_τ in a produced ν-state will oscillate with time as it propagates. Experimental searches for both neutrino masses and oscillations have been unsuccessful up to now[22].

e) Baryon number violation. As we mentioned before, all grand unified theories predict baryon number violating processes like proton decay due to the exchange of new heavy gauge bosons. The expected lifetime τ very much depends on details of each model. In particular, τ goes like the fourth power of M_X and e.g., a factor 1.8 in the value of M_X leads to a τ an order of magnitude larger. The minimal version of SU(5) (with just a $\underline{5}$ of Higgses and the usual three generations) has been in fact already ruled out by proton-decay experiments[23]. However, since the dependence of τ on M_X is so strong it is easy to slightly modify the model (by adding, e.g., some new scalar or fermion[24]) in such a way which is not ruled out by present experimental limits. Another baryon-number violating process which may appear in GUTs is neutron-antineutron oscillations ($n \leftrightarrow \bar{n}$) or n-n annihilation in nuclei. This is a process whose appearing or not is more model-dependent in relation with the particular Higgs particle content of the model under consideration[1].

The above are general features appearing in almost any model. Can one consider that the grand unification idea has been a success? Probably the most remarkable fact of GUTs [particularly SU(5) and SO(10)] is how well a quark-lepton generation fits into unified multiplets. Charge quantization is also automatic in GUT theories. Another attractive point is that GUTs give a (modest) reduction in the number of couplings since g_1, g_2, g_3 are related by their unification at M_X. This gives the prediction for $\sin^2\theta_W$ which we mentioned above and can be considered as a semi-quantitative success. In fact, detailed computations of $\sin^2\theta_W$ (including higher order effects) in the minimal SU(5) model give a result $\sin^2\theta_W = 0.214^{+0.004}_{-0.003}$ to be compared with the experimental world average $\sin^2\theta_W^{ex} = 0.228 \pm 0.0044$, which seems to be about three standard deviations away[25]. However, we repeat again that slight changes of the model can accommodate the experimental value. Thus, the prediction for the Weinberg angle may be considered as a semi-quantitative success. Concerning the predictions for quark-lepton mass ratios, the success (if any) is very limited since they certainly yield wrong results for the lightest generation. On the other hand, GUTs predict the observed experimental fact that quarks are heavier than their lepton partners within each family. This also points towards the existence of a "big desert" between the weak scale and a big scale sufficiently high so that the renormalization effects (logarithmic) have time to give a large contribution. Anyway, the grand unification idea, as sketched in the previous lines, is probably too naive. It completely ignores the gravitational interactions which are known to become important at a scale $\sim 10^{18}$ GeV, very close to the hypothetical grand unification mass. It sounds very unlikely that those two scales were not related and if they are, how can we ignore the effects of gravity? Since gravity is not an interaction like the others, no consistent quantum gravity field theory is known, one expects that new physics will govern the combined theory of the four interactions. Nevertheless, the above-mentioned attractive features (successes?) of GUTs seem to indicate that they may be partially right. It is plausible that GUTs may be just good "classification symmetries", in the same way that the SU(3) and SU(4) global flavour symmetries of the strong interactions are. Then GUT symmetries may be useful tools to study unified theories including gravity. This is in fact what happens in string theories in which, in general, one does not get a GUT structure of the type described above but grand unification tools are very useful in searching for potentially realistic strings. In the meantime GUTs have stimulated the search of baryon and lepton number violations, existence of monopoles, etc. whose experimental check is of the utmost importance.

4. SUPERSYMMETRY[2]

Let us come back now to the naturality problem we were discussing in Section 2. The problem is that there is no symmetry forbidding the appearance of a large mass term $m^2(H^*H)$ for the Higgs scalar, ruining the electroweak symmetry breaking. Supersymmetry[26] is an exception to this fact. It introduces a concept of "chirality" for scalars by assigning a fermionic bispinor of definite handedness to each complex scalar. This is the simplest version of supersymmetry ("N=1") in which any complex scalar has a fermionic partner. As we will discuss, this symmetry between bosons and fermions leads to cancellations which improve the ultraviolet behaviour of Green's functions. Thus, for example, the quadratic divergences appearing in the graphs of Fig. 2 are exactly cancelled by the diagrams in Fig. 5 in which "supersymmetric partners" (denoted by a hat) of the standard particles circulate in the loops. The cancellation comes about because the coupling constants in Figs. 2 and 5 are equal and because of the relative minus sign between fermionic and bosonic loops. We should nevertheless remark that supersymmetry ("SUSY") by itself does not solve the "hierarchy problem". An explanation why the Weinberg-Salam (WS) Higgs doublet is light compared to the Planck mass is still lacking. SUSY just tells us that if we find a reason why the WS doublet is light, its lightness will be protected from radiative corrections[27].

We will consider in what follows only N = 1 supersymmetric theories. Theories with N up to eight can be formulated. However, only N = 1 SUSY is chiral in the sense we discussed above, i.e., only N = 1 SUSY allows that left-handed and right-handed quarks and leptons transform differently under the gauge group. Let us now give a telegraphic introduction to the supersymmetry concept.

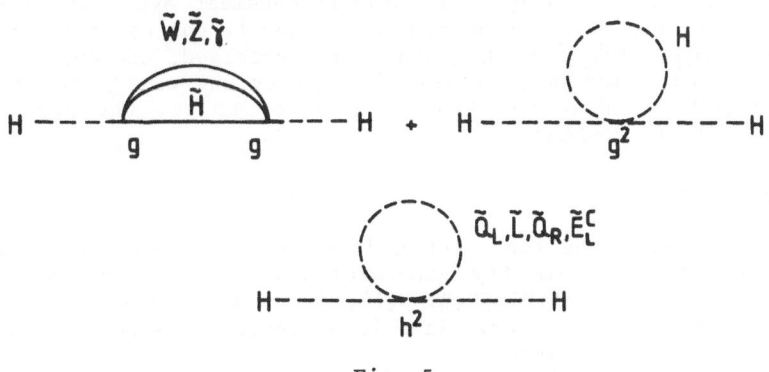

Fig. 5

41. Supersymmetric field theory

Supersymmetry is a symmetry between bosons and fermions. A dynamical system has supersymmetry if it is invariant under a "SUSY rotation" that changes fermions into bosons and vice versa. This requires having the same number of bosonic and fermionic degrees of freedom. The simplest such system one can think of in field theory is the set of fermion bispinor ψ_α ($\alpha = 1,2$) and a complex scalar ($\phi = A+iB$). The free massless Lagrangian of these two fields is itself supersymmetric

$$L_0 = \partial_\mu \phi^* \partial^\mu \phi + i\bar{\psi}\partial\psi \tag{31}$$

This system has a Noether current which is conserved once one applies the equations of motion. The conserved current is

$$J_\alpha^\mu = (\not{\partial}\phi\sigma^\mu\psi)_\alpha \;;\quad \partial_\mu J_\alpha^\mu = 0 \tag{32}$$

where $\sigma^\mu = (\mathbb{1}, \vec{\sigma})$ and $\alpha = 1,2$ is a bispinor index. The conservation of this current implies the conservation of a charge:

$$Q_\alpha \equiv \int dx^3 J_\alpha^0 \tag{33}$$

Something which is unusual about this charge operator, is that it is fermionic [see Eq. (32)] and it is not Lorentz scalar; it transforms as a bispinor.

Up to now the symmetry induced by the charge Q does not look very useful since it refers to a trivial Lagrangian like the one considered in Eq. (31). What makes SUSY interesting is that such a symmetry can be defined for interacting Lagrangians. Thus if one considers, e.g.,

$$L = L_0 - h^2 |\phi|^4 - h(\phi\bar{\psi}\psi + \text{h.c.}) \tag{34}$$

one can check again that a conserved current exists

$$J_\alpha^{\prime\mu} = J_\alpha^\mu + h\ \sigma_{\alpha\beta}^\mu (\phi^*)^2\ \bar{\psi}^\beta \tag{35}$$

The Lagrangian in Eq. (34) is the one of the "massless Wess–Zumino model". It involves a Yukawa as well as a $|\phi^4|$ interaction. In order to obtain a conserved current, it is crucial that the Yukawa coupling constant h is the "square root" of the $|\phi^4|$ coupling. This is an example of the relationships between coupling constants which appear in SUSY theories. We will show below how to write down Lagrangians which are automatically supersymmetric. Before doing that, let us consider again the conserved fermionic charge Q. If $\bar{Q}$ is the complex conjugate one can form the anti-commutator $\{Q,\bar{Q}\}$. Since both Q and $\bar{Q}$ are conserved, the anticommutator will be conserved. The only conserved quantity one can think of is the four-momentum P^μ contracted with σ^μ in order to get the right spinor structure. In fact, one has

$$\{Q_\alpha, \bar{Q}_\beta\} = 2P^\mu \sigma_{\alpha\beta}^\mu \tag{36}$$

Thus we see that the generator Q is peculiar in yet another aspect. It is not an internal symmetry generator like, e.g., the ones of the standard model symmetry $SU(3)\times SU(2)\times U(1)$, it is related to the generator of space-time translations, P^μ. In a loose sense, Q is the "square-root" of the generator of translations.

74

In order to do physics with this symmetry, we have to realize super-symmetry on fields. This will allow us to look for quantum field theories with actions invariant under supersymmetry. In the same way that one realizes Poincaré symmetry on fields with definite spin and mass [scalars $\phi(x)$, fermions $\psi_\alpha(x)$, gauge bosons $A^\mu(x)$, graviton $g^{\mu\nu}(x)$, etc.], one realizes supersymmetry on SUSY multiplets ("chiral multiplet"), "vector multiplet", etc.). In the same way that the scalar field is the simplest object in usual field theory, the "chiral multiplet" is the simplest supersymmetric object. A chiral multiplet has the following field content:

$$\phi \equiv (\phi, \psi_\alpha; F); \quad \begin{array}{l} \phi = \text{complex scalar} \\ \psi_\alpha = \text{bispinor} \\ F = \text{complex scalar (auxiliary)} \end{array} \qquad (37)$$

The field F is there for technical reasons but it disappears from the action once one uses the equations of motion, so that we will skip it in what follows (it is unphysical). Notice that we denote with the same symbol the whole multiplet and its scalar component. A supersymmetric transformation "rotates" the "components" ϕ and ψ into each other. The infinitesimal "angle" of such a "rotation" has to be peculiar. It has to be a "fermionic" parameter since it transforms fermions into bosons and vice versa

$$\phi \rightarrow \phi + \theta\psi \qquad (38a)$$

$$\psi \rightarrow \psi + i \partial\!\!\!/\bar\theta\phi \qquad (38b)$$

where θ_α ($\alpha = 1,2$) is the fermionic (anticommuting) infinitesimal para-meter of the SUSY transformation. From Eq. (38a) one observes that the θ_α parameters have mass dimension $= -1/2$ to match the different dimensions of ϕ (=1) and ψ(=3/2). In (38b) the "rotation" of ψ into ϕ requires a derivative ∂ in order to get the correct dimensions. The chiral multiplet is the building block for the supersymmetric matter, i.e., quarks and leptons will be contained in chiral representations. Thus, e.g., for the left-handed electron, we will have the chiral multiplet

$$\phi_{e_L} = (\tilde{e}_L, e_L) \qquad (39)$$

where $\tilde{e}_L$ denotes the scalar partner of the left-handed electron (the "slectron"). Notice again that a chiral multiplet contains a fermion bispinor (Weyl spinor) with definite chirality and a complex scalar super-partner. The natural object to describe a SUSY multiplet is a "super-field". These superfields[28] unify in a single mathematical object the different bosonic and fermionic components of a supersymmetric multiplet. This simplifies very much the writing of SUSY actions as well as perturba-tive calculations. However, due to the elementary nature of these lectures, I will not describe here the "superspace" approach to N = 1 SUSY. Furthermore, once SUSY is broken (as it eventually should), the superfield approach is not particularly useful for phenomenological computations.

How are the interactions between supersymmetric multiplets? There are two different types of interactions which can be written for a renor-malizable N = 1 globally supersymmetric theory: a) "chiral interactions" which are interactions between chiral multiplets including Yukawa couplings, $\lambda\phi^4$ interactions and mass terms; b) "gauge interactions" which we will see also include new gauge Yukawa couplings and gauge ϕ^4 couplings.

Let us describe first the chiral interactions. Take a set of chiral multiplets $\phi_i = (\phi_i, \psi_i, F_i)$ where i denotes any index numerating the different chiral multiplets or a gauge group label. The chiral interactions amongst those multiplets are completely determined by a function of the scalars ϕ_i which is called superpotential $W(\phi_i)$. It is important that W does not depend on ϕ_i^* (it is an analytic function). The most general form for $W(\phi_i)$ compatible with renormalizability is

$$W(\phi_i) = \lambda_{ijk}\phi^i\phi^j\phi^k + m_{ij}\phi^i\phi^j + f_i\phi^i \tag{40}$$

This superpotential gives rise to a fermionic Lagrangian and a scalar potential. The Lagrangian relevant to fermions may be obtained in the following manner:

a) Fermionic chiral Lagrangian:

$$L_F = \Sigma_{i,j} \frac{\partial W}{\partial\phi_i\partial\phi_j} \psi_i\psi_j + \text{h.c.} \tag{41}$$

This gives rise to Yukawa couplings amongst the ϕ's and the ψ's and also to fermion masses. This is obvious from Eq. (40). The λ's turn out to be the Yukawa coupling constants and the m's the fermion masses. As an example, let us consider the superpotential $W_e(\tilde{e}_L, \tilde{e}_R, H)$ which gives rise to the usual Yukawa coupling of the Weinberg-Salam model for the electron. This is

$$W_e = h_e\tilde{e}_L\tilde{e}_R H \tag{42}$$

where $\tilde{e}_L(\tilde{e}_R)$ is the scalar partner of the left-handed (right-handed) electron and H is the Higgs field. The corresponding term in the Lagrangian from Eq. (41) will be

$$L_F = h_e(e_Le_R)H + h_e(e_R\tilde{H})\tilde{e}_L + h_e(e_L\tilde{H})\tilde{e}_R + \text{h.c.} \tag{43}$$

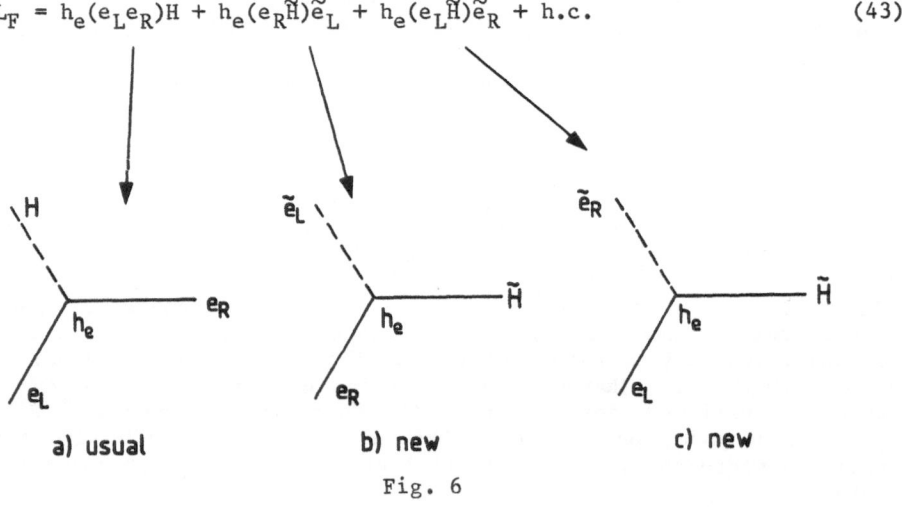

Fig. 6

where $\tilde{H}$ is the fermionic partner of the Higgs scalar. Thus we observe that the usual Yukawa coupling comes along with two new Yukawa couplings with (this is important) the same coupling constant h_e. These new couplings are required by supersymmetry. In fact, the three vertices are obtained one from another by changing two of the three particles into their SUSY partners.

The scalar chiral interactions are obtained from the superpotential W as follows.

b) Scalar potential (chiral part)

$$V(\phi_i) = \Sigma_i \; | \; \frac{\partial W}{\partial \phi_i} \; |^2 \qquad (44)$$

where the sum runs over all the scalars in the theory. In the example of the electron Yukawa superpotential one gets

$$V_e = |h_e \tilde{e}_R H|^2 + |h_e \tilde{e}_L H|^2 + |h_e \tilde{e}_L \tilde{e}_R|^2 \qquad (45)$$

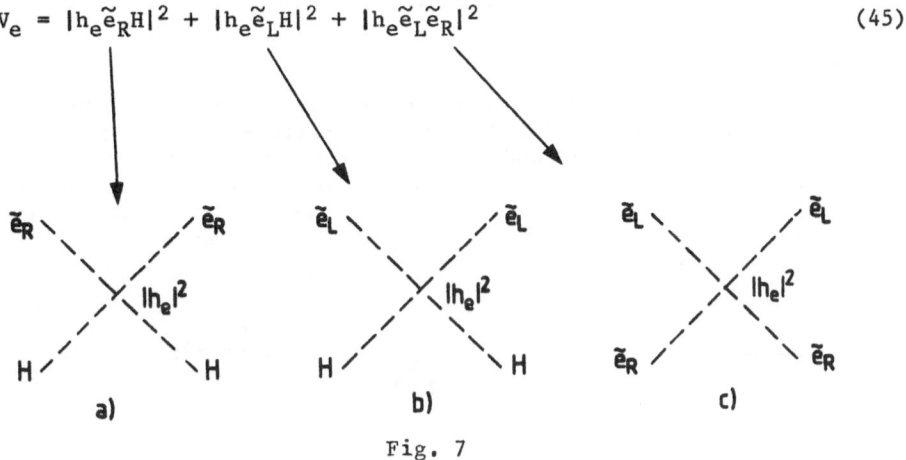

Fig. 7

From formulae (43) and (45) one observes that the supersymmetrization of the usual Yukawa for the electron requires the introduction of five new couplings (Figs. 3b-c and 4a-c) but there is only one independent coupling constant h_e, as in the non-SUSY version.

Formulae (41) and (44) are completely general and the specific dynamics is given by the particular W that we are given. We will see later on what is the superpotential of the SUSY standard model. Notice that in a SUSY theory Yukawa couplings and ϕ^4 couplings come together and one cannot consider them separately.

The second type of renormalizable interactions present in N = 1 supersymmetric theories are SUSY gauge interactions. This involves usual gauge interactions plus new SUSY interactions as we will show below. The gauge bosons in a SUSY theory are contained in "vector multiplets", the next to simplest multiplets after the chiral ones. Vector multiplets have the following field content:

$$V(\lambda_\alpha, V^\mu, D; c, \chi_\alpha, N) \qquad \begin{array}{l} \lambda_\alpha = \text{bispinor ("gaugino")} \\ V^\mu = \text{gauge boson} \\ D = \text{real scalar (auxiliary)} \end{array} \qquad (46)$$

The fields c, χ_α and N are unphysical fields which can be made to disappear in a convenient "gauge", the so-called "Wess-Zumino gauge". One can introduce a generalization of the usual concept of gauge transformations of fields and gauge bosons to chiral multiplets and vector multiplets. These SUSY gauge transformations allow for the possibility of "gauging away" the unphysical fields c, χ_α and N. Concerning the auxiliary field D, it has again dimension =2, as it happens with the chiral auxiliary field F. Thus D is also a non-propagating unphysical field and only the gauge boson V^μ and its partner the gauginos λ_α remain as physical fields. All the rest are unphysical fields needed for technical reasons.

What are the SUSY gauge interactions? As an example, let us consider $SU(n)$ gauge interactions of a vector multiplet V with chiral multiplets ϕ_i:

$$V^a = (\lambda_\alpha^a, V_\mu^a; D^a) \qquad a = 1, \ldots, n^2-1$$

$$\phi_i = (\phi_i, \psi_i^\alpha; F_i) \tag{47}$$

The vector multiplet transforms as an adjoint of $SU(n)$ and the ϕ_i as any $SU(n)$ multiplet. Recall that N = 1 SUSY partners always transform the same under the gauge group (SUSY commutes with gauge symmetry). The SUSY gauge interactions are the following. First, there are the usual gauge couplings of gauge bosons amongst themselves and also coupled to fermions and scalars. In diagrammatic form, the usual couplings are

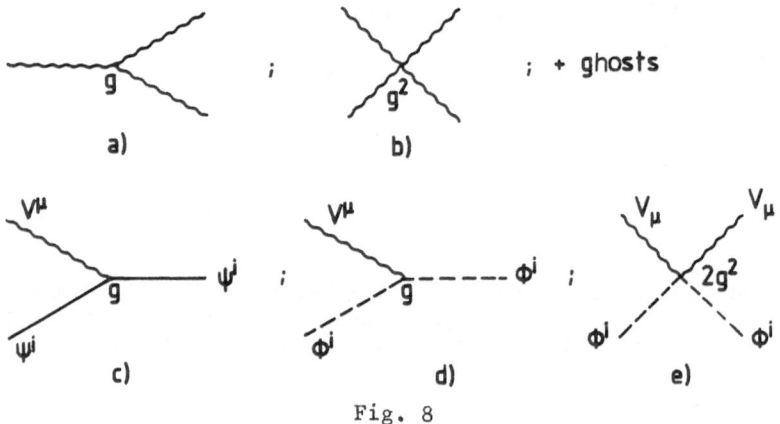

Fig. 8

In order that the gauge interactions be invariant under supersymmetry, one has to add the following new SUSY gauge couplings:

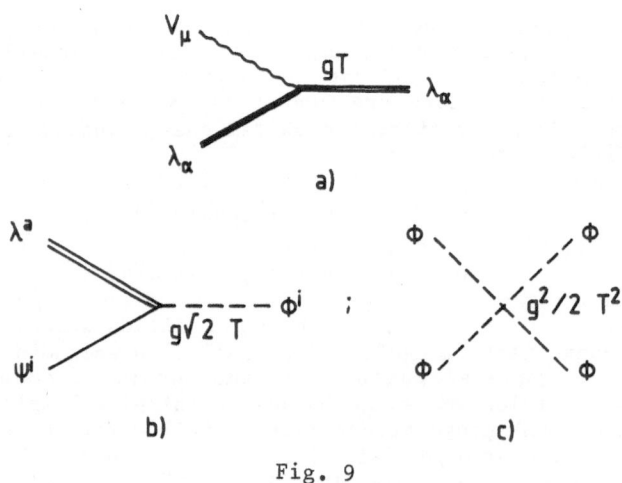

Fig. 9

The coupling of Fig. 9a is just the obvious gauge coupling of gauge bosons to gauginos, where the SU(N) factors are the same as in Fig. 8a, [T is the SU(N) generator matrix]. The coupling of Fig. 9b is a sort of "gauge Yukawa" where the SU(N) factors are as those in Figs. 8c,d. Finally, Fig. 9c represents a "gauge contribution to the scalar potential"

$$V_g(\phi_i) = g^2/2 \; |\phi_i^* \, T_{ij}^a \, \phi_j|^2 \tag{48}$$

where T_{ij}^a are the SU(N) generators written in the representation under which the ϕ_i transform.

These complete the list of interactions present in a supersymmetric gauge theory. The reader will get more acquainted with them later on when we apply them to the case of the supersymmetric standard model. Now we can write down the complete scalar potential for a supersymmetric theory with both chiral and gauge multiplets. From Eqs. (44) and (48), one gets

$$V(\phi_i) \;\; = \Sigma_i \; | \frac{\partial W}{\partial \phi_i} |^2 + \frac{g^2}{2} \; |\phi_i^* \, T_{ij}^a \, \phi_j|^2 \tag{49}$$

This is sometimes written in the literature as

$$V(\phi_i) = \Sigma_i \; |F_i|^2 + \frac{1}{2} \; |D^a|^2 \tag{50}$$

This is done because one can show that the equations of motion of the complete Lagrangian give

$$F_i^* = - \frac{\partial W}{\partial \phi_i} \tag{51a}$$

$$D^a = -g\phi_i^* \, T_{ij}^a \, \phi_j \tag{51b}$$

for the auxiliary fields F_i and D^a. Equations (51a,b) are precisely the ones which show that those fields are unphysical since they can be written in terms of other (physical) fields, i.e., they are redundant degrees of freedom. Notice the important fact that the scalar potential in (50) is positive definite, which is not the case in a generic non-supersymmetric theory.

4ii. Supersymmetry breaking

Now that we have learnt how to build interacting theories which are invariant under SUSY transformations, we have to face the fact that SUSY is not an exactly conserved symmetry of nature. Exact SUSY would require that particles and SUSY partners should be degenerate in mass and this has not been observed experimentally. Thus, we have to consider the breaking of supersymmetry[29]. There are two interesting ways for breaking SUSY: spontaneously or explicitly. From what happens with other symmetries in physics, the spontaneous breaking of supersymmetry seems quite attractive. It consists in having a Lagrangian explicitly SUSY invariant but a vacuum state which is not

$$Q_\alpha |0> \neq 0 \tag{52}$$

The "order parameter" for SUSY breaking is not the v.e.v. of a scalar but the energy of the vacuum state. To see this, recall that making use of the anticommutation relations in Eq. (36) for Q and $\bar{Q}$ one can write

$$P_0 \equiv H = \frac{1}{4} \; (\bar{Q}_1 Q_1 + Q_1 \bar{Q}_1 + \bar{Q}_2 Q_2 + Q_2 \bar{Q}_2) \tag{53}$$

and hence $<0|H|0> \geqslant 0$. Then Eq. (53) implies

$$\langle 0|H|0\rangle > 0 \rightarrow Q_\alpha|0\rangle \neq 0 \text{ (broken SUSY)} \tag{54a}$$

$$\langle 0|H|0\rangle = 0 \rightarrow Q_\alpha|0\rangle = 0 \text{ (conserved SUSY)} \tag{54b}$$

and the vacuum energy is the order parameter of supersymmetry. Notice that all this is consistent with the fact we discussed above that the scalar potential is positive definite

$$\langle 0|H|0\rangle = \langle 0|V|0\rangle = (\Sigma_i|F_i|^2 + \tfrac{1}{2}|D^a|^2)|_{\text{at the minimum}} > 0 \tag{55}$$

We thus observe that in order to break SUSY spontaneously, one must have $\langle 0|F_i|0\rangle \neq 0$ or $\langle 0|D^a|0\rangle \neq 0$ (or both). In other words, one breaks SUSY by arranging that some of the auxiliary fields F or/and D get a v.e.v.

One can show the existence of a SUSY version of the Goldstone theorem which implies the existence of a massless particle with the transformation properties of the broken generators. Since the broken generator Q_α is a fermionic spin = 1/2 object, the associated Goldstone particle is a spin = 1/2 fermion, the "goldstino". In general, the latter is the SUSY partner of the non-vanishing auxiliary field F or D which has broken SUSY. Thus, the goldstino is a fermion of a chiral multiplet if we have broken SUSY through a $\langle F\rangle \neq 0$ or a "gaugino" if we broke SUSY through a $\langle D\rangle \neq 0$. In any case, if SUSY is broken, a massless goldstino must exist. We anticipate however that when one makes a "local" version of N = 1 SUSY ("N=1 supergravity"), there is a "SUSY Higgs" mechanism in which the goldstino is swallowed by the spin = 3/2 gravitino (SUSY partner of the graviton) which then acquires a mass. No massless particle is thus left.

As we stated above, there is a second interesting way of breaking SUSY[30]. One may break SUSY "explicitly but softly". This consists in adding to the SUSY Lagrangian terms which are explicitly non-supersymmetric but are such that they do not induce quadratic divergences in loop calculations. As we mentioned before, this is required if we want to use SUSY to solve the hierarchy problem. This method for breaking SUSY looks quite ad hoc and unappealing at first sight. However, as we will see later on, it turns out to be most natural if we couple the SUSY standard model to N = 1 supergravity. I just briefly discuss this type of SUSY breaking here since we will use it later on. One can see that the SUSY-breaking terms one can add to a SUSY Lagrangian without inducing quadratic divergences in loop calculations are the following[30]:

i) Scalar mass terms of the two different types

a) $m^2 \phi^*\phi$ \hfill (56a)

b) $m^2(\phi\phi+\text{h.c.})$ \hfill (56b)

where ϕ is the scalar in a chiral multiplet.

ii) Gaugino mass terms:

c) $M(\lambda\lambda) + \text{h.c.}$ \hfill (56c)

iii) Trilinear scalar couplings

d) $Am\phi^3 + \text{h.c.}$ \hfill (56d)

where A is a dimensionless coupling constant.

80

Any of these terms may be added to the SUSY Lagrangian without generating quadratic divergences in loop calculations. However, e.g., other dim=3 terms like explicit chiral multiplet fermion masses induce those divergences, i.e., they are not soft.

4iii. <u>No renormalization theorems</u>

The improved degree of divergence of SUSY theories is what makes them interesting in trying to solve the hierarchy problem. For the case of $N = 1$ supersymmetry these properties may be stated very easily. Let us have theory with chiral interactions determined by a superpotential W of a general form like Eq. (40). Then one has the following theorem: "the parameters in W (λ's,m's,f's) do not suffer any renormalization from loop corrections". This is true both for finite and infinite loop corrections. This theorem implies that, e.g., if m = 0 at the tree level, m will be still vanishing in any order of perturbation theorem. This is particularly useful for the Higgs doublet of the WS model: if it is massless to start with, it will still be massless even after radiative corrections. This is a great improvement on the situation concerning the gauge hierarchy problem. We only (!!) have to explain now why the WS Higgs doublet was massless to start with. Let us remark that the fact that the parameters λ and m do not get loop corrections does not mean that they do not "run". They depend on the energy scale at which they are measured because there is still wave function renormalization of the fields since the kinetic energy term in the Lagrangian does receive (logarithmically) divergent loop corrections.

4iv. <u>The supersymmetric standard model</u>

With the tools described in the previous sections, we can now build a SUSY extension of the standard model. It will include, apart from the usual particles (gauge bosons, quarks, leptons and Higgs doublets), their supersymmetric partners. Eventually, we will have to consider two symmetry breakings: i) SUSY breaking and ii) SU(2)×U(1) symmetry breaking. It will turn out that in the simplest scenario, these two symmetry breakings are related. But before studying this double process of symmetry breaking, let us specify the (minimal) particle content of the SUSY-SU(3)× SU(2)×U(1) theory[31]. As we anticipated, the gauge bosons and their fermionic partners ("gauginos") will be contained in vector multiplets whereas the quarks, leptons and Higgses and their supersymmetric partners ("s-quarks", "s-leptons", and "Higgsinos") will be contained in chiral multiplets. The particle content is shown in Table 3.

Table 3

VECTOR MULTIPLETS

$J = 1$ $J = 1/2$

g $\rightarrow$ $\tilde{g}$	
(gluon) (gluino)	
$W^{\pm}W^0$ $\rightarrow$ $\tilde{W}^{\pm}\tilde{W}^0$	
(winos)	
B $\rightarrow$ $\tilde{B}$	
(bino)	

CHIRAL MULTIPLETS

$J = 1/2$ $J = 0$

q_L,q_R $\rightarrow$ $\tilde{q}_L,\tilde{q}_R$	
(quarks) (squarks)	
ℓ_L,ℓ_R $\rightarrow$ $\tilde{\ell}_L,\tilde{\ell}_R$	
(leptons) (sleptons)	
$\tilde{H}_1,\tilde{H}_2$ $\leftarrow$ H_1,H_2	
(Higgsinos) (Higgses)	

In the table, each particle and its SUSY partner transform in the same way under $SU(3) \times SU(2) \times U(1)$. Notice that one introduces two different WS doublets H_1 and H_2. This is easily understood if we recall how one gives masses to quarks in the usual WS model:

$$d\text{-quarks:} \quad h_d(\bar{d}_R d_L)H + h.c. \tag{57a}$$

$$u\text{-quarks:} \quad h_u(\bar{u}_R u_L)H^* + h.c. \tag{57b}$$

in Dirac spinor notation. Once H acquires a non-vanishing v.e.v. $\langle H \rangle \neq 0$, the quarks get a mass. A SUSY version of Eqs. (57a,b) would require [see Eqs. (40)-(42)] superpotentials

$$W_d = h_d \, \tilde{d}_L^c \, \tilde{d}_L \, H \tag{58a}$$

$$W_u = h_u \, \tilde{u}_L^c \tilde{u}_L H^* \tag{58b}$$

However, W_u cannot be an admissible superpotential because it depends on H^* (superpotentials cannot depend on complex conjugate fields). To solve this problem, one introduces two doublets H_1 and H_2 with opposite hypercharge so that one uses H_2 in Eq. (58b) (instead of H^*) and H_1 in (58a). Notice that this reasoning also implies that one cannot use, e.g., the scalar partners of (ν_L, e_L) as Higgs particles since they would only give masses to d-quarks and leptons but not to u-quarks. [Adding just one Higgs doublet H_2 will not do because, although one could then give masses to u-quarks, the fermionic Higgsino $\tilde{H}_2$ would induce $U(1)$ anomalies.]

The supersymmetric $SU(3) \times SU(2) \times U(1)$ gauge interactions amongst the particles in the SUSY standard model are easy to obtain from the rules we gave in previous sections. The explicit Lagrangian may be found, e.g., in Ref. 32. Concerning the chiral interactions they are completely determined by the superpotential W. The superpotential of the (minimal content) SUSY standard model is

$$W = \sum_{\text{generations}} (h_u \tilde{q}_L \tilde{u}_L^c H_2 + h_D \tilde{q}_L \tilde{d}_L^c H_1 + h_L \mathcal{X}_L \mathcal{X}_{LH_1}^c) + \mu H_1 H_2 \tag{59}$$

where $\tilde{q}_L$ and $\mathcal{X}_L$ denote the $SU(2)_L$ doublets $(\tilde{u}_L, \tilde{d}_L)$ and $(\tilde{\nu}_L, \tilde{e}_L)$ and $\tilde{u}_L^c$, $\tilde{d}_L^c$, $\mathcal{X}_L^c$ are the scalar partners of the right-handed quarks and leptons. $SU(3)_C$ and $SU(2)_L$ indice contractions are understood. The Yukawa coupling constants h_U, h_D and h_L are, in general, non-diagonal 3×3 matrices in flavour space leading to the usual masses and Cabibbo angles between quarks. The superpotential (59) is the most general compatible with baryon and lepton number conservation. Notice that it just includes the SUSY version of the usual Yukawas plus a possible mass term for Higgses.

4v. SUSY and grand unification

The SUSY standard model may be easily unified[33], e.g., inside SU(5). In this case, the multiplet content would be as in Table 4 where each family of quarks and leptons is contained in a $F_{\bar{5}} + T_{10}$. One introduces Higgses ϕ_{24} (and partners $\tilde{\phi}_{24}$) in order to break SU(5). The usual WS doublets are contained in $H_5 + H_{\bar{5}}$.

Table 4

VECTOR MULTIPLET			CHIRAL MULTIPLETS	
$J = 1$	$J = 1/2$		$J = 1/2$	$J = 0$
A^μ_{24}	$\rightarrow \quad \lambda_{24}$		$3(F_{\bar 5} + T_{10}) \quad \rightarrow \quad 3(\tilde F_{\bar 5} + \tilde T_{10})$	
			$\tilde\phi_{24} \quad \leftarrow \quad \phi_{24}$	
			$\tilde H_5 + \tilde H_{\bar 5} \quad \leftarrow \quad H_5 + H_{\bar 5}$	

Most of the features of usual GUTS remain true in the supersymmetric case. The existence of the SUSY partners of the usual particles cause the $SU(3) \times SU(2) \times U(1)$ gauge coupling constants to join at a larger mass scale[34]. This is due to the fact that the coefficients of the β-functions b_1, b_2, b_3 in Eq. (24a) change in the supersymmetric case ($b_1 = 11$, $b_2 = 1$, $b_3 = -3$). One thus finds $M_X^{SS} \simeq 3 \times 10^{16}$ GeV. The value of the coupling constant at the GUT scale also increases, $\alpha^{SS} \sim 1/24$ compared to $\alpha \sim 1/40$ in the non-SUSY case. Interestingly enough, the prediction for the Weinberg angle only changes slightly[34]:

$$\sin^2\theta_W = \frac{3}{8} - \frac{5}{4} \frac{\alpha_e(M_W)}{(4\pi)} \; 6 \; \text{Log} \left(\frac{M_X^{SS}}{M_W}\right) \simeq 0.23 \qquad (60)$$

compared to the ~0.20 result in the non-SUSY case (leading log). This result is, however, very sensitive to the particle content and changes sharply[35] if one adds some new particle to the standard model (like extra Higgs doublets). The prediction for the m_b/m_τ ratio is practically unchanged also[34]

$$m_b/m_\tau \simeq \left(\frac{\alpha_3(m_b)}{\alpha_3(M_W)}\right)^{4/7} \left(\frac{\alpha_3(M_W)}{\alpha_{SS}}\right)^{8/9} \simeq 3.0 \qquad (61)$$

In all these computations we have assumed that SUSY is broken in an effective way around the weak scale so that the scheme of mass scales is as in Fig. 10. This has to be so in order to preserve the masslessness of the Higgs doublets down to the M_W scale where they are used to break the $SU(2) \times U(1)$ symmetry.

The situation concerning baryon number violation in the SUSY case is more uncertain. There are three types of mechanisms potentially contributing to proton decay, as exemplified in Fig. 11.

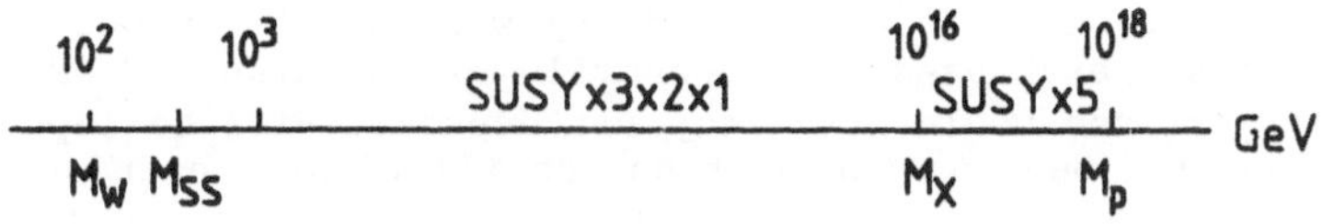

Fig. 10

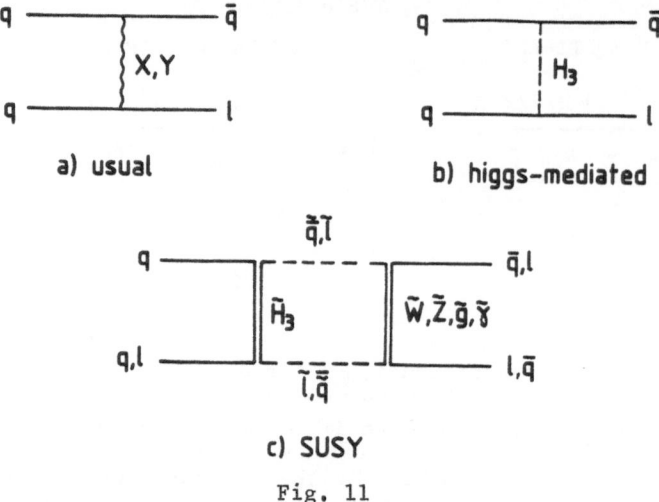

a) usual   b) higgs-mediated

c) SUSY

Fig. 11

The first mechanism is the usual one already discussed for the non-SUSY case. This one is very much suppressed in general since the unification mass (and hence the masses of X and Y) is very large and the lifetime is proportional to the fourth power of M_X^{SS}:

$$\tau_P^{SS} \sim (\frac{M_X^{SS}}{M_X^{NS}})^4 \ \tau_P^{NS} \sim 10^{38\pm2} \ \text{years} \tag{62}$$

Mechanism b) is also present in the non-SUSY case and the rate depends on the unknown mass of the Higgs H_3. The latter is the colour triplet of Higgs fields which completes the five-plet of Higgses along with the WS doublets H_2 (i.e, $H_5 = H_2 + H_3$). The decay is undetectable unless $m_{H_3} \sim 10^{11}$ GeV. Modes including strange particles would be dominant ($P \rightarrow K^0 \mu^+$, $K^+ \nu^\tau$, etc.). Finally, the mechanism in Fig. 3 is genuinely supersymmetric[36]. Unfortunately, unknown parameters (gaugino masses, squark-slepton masses, $\tilde{H}_3$ masses) are required to compute the graph. Reasonable choices of those parameters show that graphs like the one in Fig. 3 would be dominant in the SUSY case. However, due to the uncertainties mentioned, no precise prediction for the proton lifetime can be given. Again the decay modes involve predominantly strange particles.

Let us remark now a very important point concerning baryon stability. In the supersymmetric version of the SM, one can add to the Lagrangian terms compatible with SU(3)×SU(2)×U(1) and Poincaré symmetries, but explicitly violating baryon and/or lepton number. The following terms could in principle be added[37] to the superpotential in Eq. (59):

a) $(\tilde{u}_L^c \ \tilde{d}_L^c \ \tilde{d}_L^c)$ b) $(\tilde{q}_L \ \tilde{\ell}_L \ \tilde{d}_L^c)$

c) $m(\tilde{\ell}_L \ H_2)$ d) $(\tilde{\ell}_L \ \tilde{\ell}_L \ \tilde{\ell}_R)$ (63)

The first violates baryon number and the second, third and fourth lepton number. If both terms a) and b) are allowed, graphs like the one in Fig. 12 will give rise to fast proton decay within a few minutes since the mass of the squark exchanged is necessarily of the order of $\sim M_W$. If only term a) exists, proton would not decay (it does not violate lepton number) but there can be neutron-antineutron oscillations, although one can compute that the rate of these oscillations would not be ruled out by experimental limits for reasonable choices of the parameters[38].

84

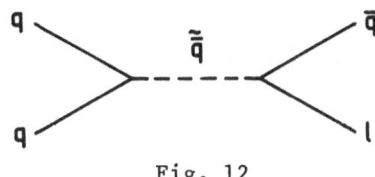

Fig. 12

The existence of possible renormalizable terms [like those in Eq. (63)] explicitly violating baryon and/or lepton number is a new feature of supersymmetric models. No such terms can be written in the minimal standard model. Of course, one can just forbid these terms in the superpotential and forget about it. This is perfectly reasonable since there is no principle within the context of SUSY-GUTs which tells us what terms should be in the superpotential and which terms should not. However, eventually, if we have a theory which is able to give us information about superpotential terms, it will have to explain why the superpotential of Eq. (59) does not contain the terms in Eq. (63). We will comment again on this problem when we talk about string models. For the moment, we will assume that none of the terms in Eq. (63) are allowed in the superpotential.

4vi. SUSY-breaking in the SUSY standard model

As is well known, no supersymmetric particles have been detected experimentally up to now (see below). $p\bar{p}$ and e^+e^- collider results imply $m_{\tilde{q},\tilde{g}} \gtrsim 80$ GeV, $m_{\tilde{\chi}} \gtrsim 25$ GeV and similar limits apply for the rest of the charged SUSY particles. How can we obtain this in a natural manner? One can try to break SUSY spontaneously in the manner we discussed before. However, it turns out to be quite complicated (if not impossible) to break SUSY spontaneously without adding a large number of new ad hoc particles. There is, however, a very elegant mechanism which gives us what we want (i.e., masses for SUSY partners) in a very economical way. Furthermore, it is much more satisfactory theoretically since it is based on the extension of the scheme to include gravity (N = 1 supergravity). We first describe the effects of this type of SUSY breaking on the SM Lagrangian and then we will briefly discuss their origin.

Once this type of SUSY-breaking has taken place the standard supersymmetric Lagrangian gets new terms as follows:

i) equal positive (mass)^2 for all scalars (squarks, slepton, Higgses) in the theory, $m^2 |\phi_i|^2$;

ii) new purely scalar interactions proportional to the superpotential, $\sim m W(\phi_i)$;

iii) equal gaugino masses M at the GUT scale.

The mass parameter m is chosen to be $m \sim M_W$ and the gaugino mass M is an independent parameter also of order M_W. Applying this to the SUSY SM one obtains the following Lagrangian[39-42]

$$L = L(SUSY) + \delta L \ (SUSY\text{-breaking})$$

$$\delta L = m^2 \sum_{i=\tilde{q},\tilde{\chi},H_{1,2}} |\phi_i|^2 + Am W_3(q,\ell,H_{1,2}) + B\mu m H_1 H_2 + h.c. \qquad (64)$$

$$+ M\lambda_i \lambda_i + h.c.$$

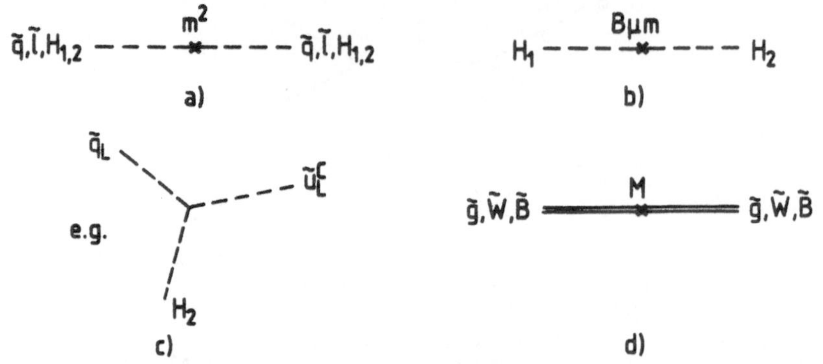

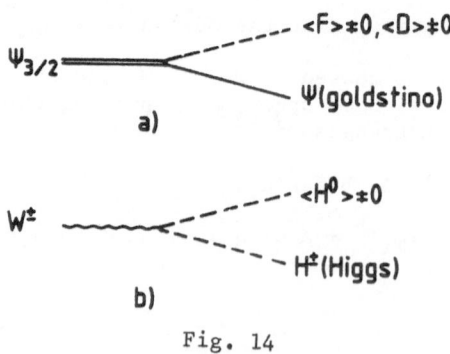

Fig. 13

where W_3 is the trilinear part of the superpotential in Eq. (59), A and B are dimensionless constants $\sim$ unity and λ_i are the gaugino fields. Thus, Eq. (64) gives us additional terms to the Lagrangian which can diagrammatically be represented as in Fig. 13. Interestingly enough, these terms correspond to the list of soft breaking terms which do not generate quadratic divergences given in Eqs. (56).

What is the magnitude of m? It has to be heavier than say $\sim$20 GeV, in order to explain why we do not see $\tilde{q}$ or $\tilde{\chi}$'s. However, it cannot be much bigger than $M_W \sim 10^2$ GeV since then the Higgs doublet would have a very large mass and it would not be possible to give a non-vanishing v.e.v. to H_1 and H_2 to break the SU(2)×U(1) symmetry. Notice that the Higgs (mass)^2 is positive and at this level it is not easy to see how one can break the SU(2)×U(1) symmetry. We will discuss that below. Before doing that, let us say a few words about the origin of these SUSY-breaking couplings.

The general idea is that it makes no sense to have supersymmetric SU(3)×SU(2)×U(1) interactions but to have non-supersymmetric gravitation. In fact, since the N = 1 SUSY algebra [Eq. (36)] involves the generator of translations P^μ, a local version of SUSY would contain general co-ordinate transformations, i.e., the invariance group of Einstein gravitation. Thus, a local version of SUSY is equivalent to a supersymmetric version of gravity, i.e., N = 1 supergravity[43]. We thus have to consider the coupling of the SM to N = 1 supergravity. The spin = 2 graviton $g^{\mu\nu}$ has a supersymmetric spin = 3/2 SUSY partner ψ^μ, the gravitino. The gravitino is the "gauge particle" corresponding to local supersymmetry transformations. When supersymmetry is broken (i.e., any of the auxiliary fields <F> or <D> $\neq$ 0) the gravitino field "swallows" the massless "goldstino" and becomes massive much in the same way the $W^\pm$'s become massive by swallowing the usual Higgs field (Fig. 14).

Fig. 14

The second important ingredient in this mechanism of supersymmetry breaking is to assume it takes place in a "hidden sector" of the theory[39-41,44]. This means to assume the existence of further singlet chiral multiplets (or maybe new gauge interactions) not coupling directly with the "observable sector" of (s)quarks, (s)leptons, Higgses, etc. The details of this "hidden world" are irrelevant and the only important point is the assumption that in this sector supersymmetry breaking takes place[45] at a scale $M_{SB} \sim \sqrt{M_W M_P} \sim 10^{10}$ GeV. Then one can show that the gravitino gains a mass

$$m_{3/2} \sim \frac{|M_{SB}|^2}{M_P} \sim M_W. \tag{65}$$

Since the "observable world" is isolated from the hidden world, it does not feel that SUSY has been broken directly (i.e., there are no mass splittings between SUSY partners $\sim M_{SS}$). However, gravitational (and SUSY-gravitational) interactions transmit the SUSY-breaking from the "hidden" to the "observable" sector. Since gravitational interactions are suppressed by powers of $(1/M_P)$, the effective SUSY breaking scale in the observable sector is only $m \sim M_{SB}^2/M_P \sim m_{3/2}$. Thus the parameter m in Eq. (64) is essentially given by the gravitino mass. That is why in Eq. (65) we chose $m_{3/2} \sim M_W$, in order not to give too large a mass to the scalars in the standard SUSY model. Let us remark that, since the "hidden matter" and the gravitino couple only gravitationally to the quark-lepton world, they are experimentally unobservable. They may, however, have cosmological importance and may also be candidates for the "dark matter" in the Universe. Summarizing this brief discussion on the mechanism of supersymmetry breaking, the important point is that the SUSY partners of the standard particles get the couplings in Eq. (64) with $m \sim m_{3/2} \sim M_W$. For the purposes of low-energy physics the details of the N = 1 super-gravity theory can be skipped.

4vii. SU(2)×U(1) symmetry breaking

Let us now discuss how the electroweak symmetry may be broken in the SUSY SM. We have to study the scalar potential for the Higgs fields H_1 and H_2 and check if there is a minimum for which $\langle H_1 \rangle$, $\langle H_2 \rangle \neq 0$. Let us start by writing down the scalar potential for H_1 and H_2[46-49]

$$V(H_1 H_2) = \frac{1}{2} \left((H_1^* \frac{\tau^a}{2} H_1 + H_2^* \frac{\tau^a}{2} H_2)^2 g^2 + (|H_2|^2 - |H_1|^2)^2 g'^2 \right) + \\ + \mu_1^2 |H_1|^2 + \mu_2^2 |H_2|^2 - \mu_3^2 (H_1 H_2 + hc.) \tag{66}$$

where g and g' are the WSG gauge couplings and the τ^a's (a=1,2,3) are $SU(2)_L$ Pauli matrices. This is obtained from formulae (44), (48) and (64) after defining

$$\mu_1^2 \equiv m^2 + \mu^2 \;\; ; \;\; \mu_2^2 \equiv m^2 + \mu^2 \;\; ; \;\; \mu_3^2 = Bm\mu \tag{67}$$

Equation (66) is the SUSY version of the celebrated "bottom of a bottle" Higgs potential which induces SU(2)×U(1) breaking in the usual SM. However, the potential in Eq. (66) is problematic because $\mu_1^2 = \mu_2^2 > 0$. A minimum with non-vanishing $\langle H_{1,2} \rangle$ may only be obtained if there is a negative (mass)^2 eigenvalue in the Higgs mass matrix, i.e., if $\mu_1^2 \mu_2^2 - \mu_3^4 < 0$. However, if $\mu_1^4 = \mu_2^4 < \mu_3^4$, the scalar potential is unbounded below in the direction $\langle H_1 \rangle = \langle H_2 \rangle \to \infty$. There is something wrong though in the above analysis. μ_1^2 and μ_2^2 are equal only at the Planck scale. Below that point they will get renormalized differently if the scalars H_1 and H_2 couple with different strengths to quarks and leptons. This is true in general for all the soft breaking appearing in Eq. (64). Thus at the Planck scale we would have

$$m_{\tilde{q}}^2 = m_{\tilde{\chi}}^2 = \mu_1^2 - \mu^2 = \mu_2^2 - \mu^2 = m^2 \tag{68a}$$

$$M_{\tilde{g}} = M_{\tilde{W}} = M_{\tilde{B}} = M \tag{68b}$$

$$A_U = A_D = A_L = A \tag{68c}$$

but below that mass scale, all these parameters may evolve differently according to their corresponding renormalization group equations. We will not write down those equations which may be found in Refs. 50,47 and 48, but we will give a general idea of how these parameters evolve with energy. Equations (68a-c) also assume that a grand unification occurs not very far from the Planck scale [notice, e.g., Eq. (68b) which implies that all the gauginos are unified in mass]. As an example, it turns out that gaugino masses M evolve with energy just like the 3×2×1 coupling constants

$$M_i = \frac{\alpha_i(Q^2)}{\alpha_i(M_X^2)} \; M. \tag{69}$$

This allows us to predict that $M_{\tilde{g}} \sim (\alpha_3/\alpha_e)M_{\tilde{\gamma}}$ where $\tilde{g}$ and $\tilde{\gamma}$ are the photino and gluino respectively. For the (masses)^2 of squarks and leptons, one gets an evolution with energy of the type shown in Fig. 15 (in that figure m = M = 100 GeV and A = 1 is taken). We thus find that $\tilde{q}$'s are usually heavier than $\tilde{\chi}$'s (one can see that they have similar masses in the limit M → 0).

What happens with the Higgs scalar H_1 and H_2? We know that a negative (mass)^2 term would be welcome in order to break SU(2)×U(1). The miracle occurs and the Higgs masses may run in a natural way as shown in Fig. 16 and μ_2^2 gets negative whereas μ_1^2 remains positive. Notice that this is what we want since if μ_1^2 were also negative, the potential Eq. (66) would be unbounded below.

The mechanism in Fig. 16 works only if the top quark is heavy enough[51]. What has the t-quark to do with all this? It is simply that there is a negative contribution to μ_2^2 in its renormalization group equation from the graphs in Fig. 17.

The negative sign comes from the fact that supergravity breaking gives mass to squarks but not to quarks and then the diagrams 17a and 17b fail to cancel giving a net negative contribution[51]

$$\delta\mu_2^2 \simeq -\frac{3}{8\pi^2} h_U^2 \, m^2 \, \text{Log}(M_X/Q) \tag{70}$$

In order for this negative contribution to overwhelm the original positive μ_2^2, h_U has to be large enough. Detailed numerical computations show that in order to get the appropriate negative μ_2^2 to break SU(2)×U(1) one needs $h_t(M_X) \gtrsim 0.1$. It is obvious that the U-quark of a possible fourth quark-lepton generation would also do the job. We will specify below how heavy the top quark has to be.

Let us come back now to the $H_{1,2}$ scalar potential in Eq. (66). One can see that the potential is minimized for[46,47]

$$v^2 \equiv v_1^2 + v_2^2 = \frac{2(\mu_1^2 - \mu_2^2 - (\mu_1^2 + \mu_2^2)\cos 2\theta)}{(g^2 + g'^2)\cos 2\theta} \tag{71}$$

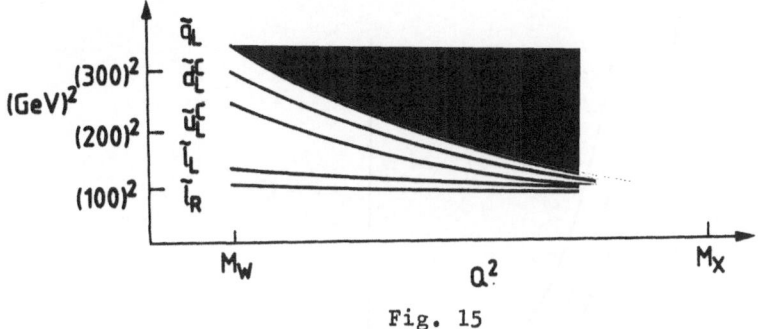

Fig. 15

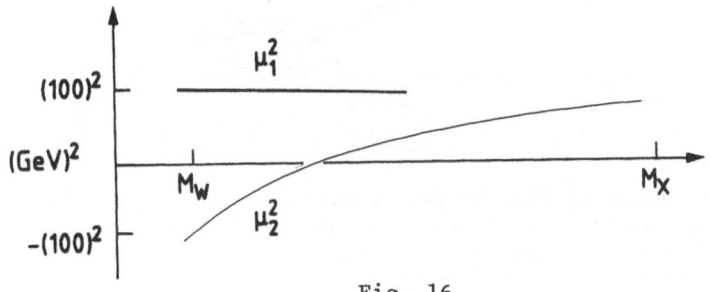

Fig. 16

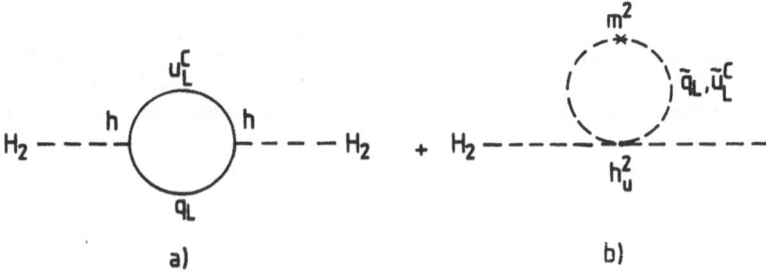

a)   b)

Fig. 17

where $\nu_{1,2} \equiv \langle H^0_{1,2}\rangle$ and $\sin 2\theta \equiv 2\mu^2_3/(\mu^2_1+\mu^2_2)$. Now we have to impose that the SU(2)×U(1) symmetry is broken at the appropriate scale, i.e., it must be $\nu^2 = 2M^2_W/g^2$, where M_W is the experimentally measured $W^\pm$-mass. This condition and Eq. (71) lead to the following constraint[47]

$$\frac{\mu^2_1-\omega^2\mu^2_2}{\omega^2-1} = \frac{1}{2} \ M^2_Z \tag{72}$$

where $\omega = \text{tg}\theta = \nu_2/\nu_1$ and M_Z is the Z^0 boson mass. The quantities μ^2_1, μ^2_2 and ω^2 in Eq. (72) should be evaluated at the M_W scale. An interesting limit is obtained in the case in which $\mu^2_3 \ll (\mu^2_1+\mu^2_2)$. In that case $\cos 2\theta \to 1$ and $\omega \to \infty$ and the condition for SU(2)×U(1) breaking at the right scale reads

$$\mu^2_2(Q^2 = M^2_W) = -\frac{1}{2} M^2_Z \tag{73}$$

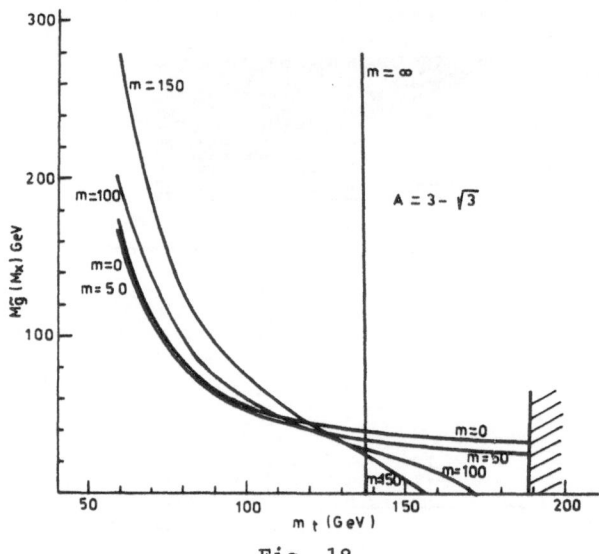

Fig. 18

which implies that μ_2^2 has to get a negative value, as indicated in Fig. 16. This implies that the top quark has to be heavy enough. However, Eq. (72) is more general than Eq. (73) and one can see that it is not strictly necessary to drive μ_2^2 negative in order to get SU(2)×U(1) breaking, although, as we will comment below, the former is the most natural possibility.

The SU(2)×U(1) breaking implies certain relationships between the parameters in the model in such a way that $\mu_1^2(M_w)$, $\mu_2^2(M_w)$ and $\omega(M_w)$ can fulfil Eq. (72). These quantities depend on the free parameters of the model. Recalling Eqs. (59), (64) and (67), we see that those free parameters are

$$m, \ M, \ A, \ \mu_3^2, \ \mu \qquad\qquad (74)$$

plus the Yukawa couplings, in particular h_t which is the only one that can be large enough to affect the renormalization group equations for μ_1^2, μ_2^2 and ω^2. The condition of SU(2)×U(1) breaking implies the existence of constraints amongst the parameters in (74) and h_t (or, equivalently, the top quark mass $m_t \simeq h_t(M_w)v_2$). One can study those constraints numerically. For example, in the limit $\mu_3^2, \mu \to 0$ (relevant in an interesting class of models) one finds numerical results as in Fig. 18.

It shows the relationship between the parameters M (gaugino masses at M_X), m (scalar masses at M_X) and m_t for a certain value of the A parameter. One observes that one has always $m_t \gtrsim 60$ GeV and that the larger M is, the lighter is the value of m_t required. Similar results are obtained for other values of A. At this point, it is interesting to recall that in the supersymmetric SM the top quark (or any possible new quark for that matter) has an upper bound[50,47,48] for his mass coming from the renormalization group equation for h_t. One can find the following numerical formula for the top quark mass in terms of its Yukawa coupling at the unification scale $h_t(0)$[47]:

$$m_t \simeq \left(\frac{3.6 h_t(0)}{\sqrt{1+11 h_t^2(0)}} \right) v_2 \qquad\qquad (75)$$

As $h_t(0) \to \infty$ (as a formal limit), one has $m_t \sim v_2 \sim 200$ GeV. Thus one can say that in the minimal SUSY SM, one has the limits for the top quark mass for the class of models in which $\mu_3^2 \ll \mu_1^2$, μ_2^2. For the case with $\mu^2_3 \sim \mu^2_{1,2}$, the lower bound disappears. However, in many SUSY GUTs or string models, one expects $\mu_3^2 \ll \mu_1^2, \mu_2^2$. This is because $\mu_3^2 = B m \mu$ and a natural value for μ [see Eq. (59)] is zero. Notice that μ and $m \cong m_{3/2}$ are completely independent parameters which, on the other hand, have to be (for phenomenological reasons) of the same order of magnitude $m, \mu \sim M_W$. It is rather unnatural that two completely independent parameters turn out to have similar values without any obvious reason. It is most reasonable to assume that $\mu = 0$ to start with and a value for $\mu_3^2 \neq 0$ is generated radiatively once supersymmetry is broken. One can easily build models in which this is the case. In this large class of models, the lowest bound in Eq. (76) will then apply. Let us finally remark that one cannot have $\mu_3^2 = 0$ because then one can see, examining the scalar potential Eq. (66), that a massless pseudoscalar ("axion") would appear of the type introduced by Peccei and Quinn. That type of axions is ruled out experimentally.

4viii. The supersymmetric particle spectrum

a) Masses of squarks and sleptons

There are four different contributions (a,b,c,d) to the masses of squarks and sleptons. The "helicity" structure of these four terms is as follows ($\phi = \tilde{q}, \tilde{\chi}$)

$$
\begin{array}{c}
\phi_L^* \\
\phi_R^*
\end{array}
\begin{bmatrix}
m_a^{L2} + m_b^{L2} + m_c^{L2} & m_d^2 \\
m_d^2 & m_a^{R2} + m_b^{R2} + m_c^{R2}
\end{bmatrix}
\qquad \begin{array}{cc} & \\ \phi_L & \phi_R \end{array}
\tag{77}
$$

The leading contributions are a) and b). Contribution a) is the one coming from supergravity (Fig. 13a) after renormalization group equations for M_X to the M_W scale are taken into account. Contribution b) comes from the $1/2 \, \Sigma_a \, |D^a|^2$ term in the SUSY scalar potential (Fig. 9c) once the doublets $H_{1,2}$ acquire a v.e.v. Solving the relevant renormalization group equations for the masses of squarks and sleptons, one can express their masses in terms of the parameters of the model m, M and $\cos 2\theta$. We give here for completeness the sum $m_a^2 + m_b^2$ for all the squarks and sleptons of a generation[47,52]

$$
m_{\tilde{U}_L}^2 = m^2 + 2M^2 \left(\tfrac{4}{3} \tilde{\alpha}_3 f_3 + \tfrac{3}{4} \tilde{\alpha}_2 f_2 + \tfrac{1}{36} \tilde{\alpha}_1 f_1\right) + \cos 2\theta M_Z^2 \left(-\tfrac{1}{2} + \tfrac{2}{3} \sin^2\theta_w\right) \tag{78a}
$$

$$
m_{\tilde{D}_L}^2 = m^2 + 2M^2 \left(\tfrac{4}{3} \tilde{\alpha}_3 f_3 + \tfrac{3}{4} \tilde{\alpha}_2 f_2 + \tfrac{1}{36} \tilde{\alpha}_1 f_1\right) + \cos 2\theta M_Z^2 \left(\tfrac{1}{2} - \tfrac{1}{3} \sin^2\theta_w\right) \tag{78b}
$$

$$
m_{\tilde{U}_R}^2 = m^2 + 2M^2 \left(\tfrac{4}{3} f_3 \tilde{\alpha}_3 + \tfrac{4}{9} \tilde{\alpha}_1 f_1\right) - \cos 2\theta M_Z^2 \left(\tfrac{2}{3} \sin^2\theta_w\right) \tag{78c}
$$

$$
m_{\tilde{D}_R}^2 = m^2 + 2M^2 \left(\tfrac{4}{3} \tilde{\alpha}_3 f_3 + \tfrac{1}{9} \tilde{\alpha}_1 f_1\right) + \cos 2\theta M_Z^2 \left(\tfrac{1}{3} \sin^2\theta_w\right) \tag{78d}
$$

$$
m_{\tilde{\chi}_L}^2 = m^2 + 2M^2 \left(\tfrac{3}{4} \tilde{\alpha}_2 f_2 + \tfrac{1}{4} \tilde{\alpha}_1 f_1\right) + \cos 2\theta M_Z^2 \left(\tfrac{1}{2} - \sin^2\theta_w\right) \tag{78e}
$$

$$
m_{\tilde{\nu}_L}^2 = m^2 + 2M^2 \left(\tfrac{3}{4} \tilde{\alpha}_2 f_2 + \tfrac{1}{4} \tilde{\alpha}_1 f_1\right) - \cos 2\theta \, \frac{M_Z^2}{2} \tag{78f}
$$

$$
m_{\tilde{E}_R}^2 = m^2 + 2M^2 \left(\tilde{\alpha}_1 f_1\right) \tfrac{1}{4} + \cos 2\theta \, M_Z^2 \sin^2\theta_w \tag{78g}
$$

Here θ_W is the Weinberg angle, M_Z is the Z^0 mass and m, M, θ are the free parameters in Eq. (74). One also has

$$\tilde{\alpha}_i \equiv \frac{\alpha_i(M_X)}{(4\pi)} \quad ; \quad f_i \equiv \frac{1}{\beta_i}\left(1 - \frac{1}{(1+\beta_i t)^2}\right)$$

$$\beta_i \equiv \tilde{\alpha}_i \, b_i \quad ; \quad t = \log(M_X^2/M_W^2) \simeq 67$$

(79)

and $b_i = (-3,1,11)$ for $i = 3,2,1$ are the coefficients of the Callan-Symanzik $SU(3) \times SU(2) \times U(1)$ β-functions. From the above equations we see that, unless M is negligible, the squarks will be heavier than the sleptons since $\tilde{\alpha}_3 \gg \tilde{\alpha}_{2,1}$. We also see that (for non-vanishing $\cos 2\theta$) the scalar neutrino will be the lightest s-fermion[53]. Contribution c) in Eq. (77) is just equal to the mass of the corresponding quark or lepton partner and thus is essentially negligible except for the quarks of the third generation. Finally, contribution d) mixes left and right squarks and sleptons and is proportional to $m_i mA$, where m_i is the corresponding quark (lepton) mass. Again, it is negligible for the first two generations.

Notice that the $\tilde{q}$ and $\tilde{\chi}$ spectrum depends essentially only on three parameters m, M and $\cos 2\theta$. Since there are $7 \times 3 = 21$ different $\tilde{q}$'s and $\tilde{\chi}$'s, it is obvious that this scheme has strong predictive power. Furthermore, finding a scalar spectrum obeying Eqs. (78) would be a verification not only of supersymmetry but also of the unification concept since a common scalar mass m is assumed at the GUT scale. Notice also that m, M and $\cos 2\theta$ are not arbitrary since they have to be such that the $SU(2) \times U(1)$ symmetry is broken at the appropriate scale. This further restricts the pattern of $\tilde{q}$ and $\tilde{\chi}$ masses. We show below examples of consistent supersymmetric spectra (Tables 5 and 6).

TABLE 5

Some allowed spectra for $m_t \sim 50$ GeV. Masses are in GeV

	a)	b)	c)
m	30	20	20
M	− 25	1	−100
A	− 1	− 2	− 4
μ_0	10	5	103
$\sin 2\theta(M_X)$	0.62	0.67	0.94
$\sin 2\theta(M_W)$	0.99	0.99	0.98
m_t	52	50	55
$\tilde{q}$	73	20	269
$\tilde{t}_\ell$	43	39	230
$\tilde{t}_h$	116	64	303
$\tilde{\chi}$	38	21	79
$\tilde{g}$	75	3	301
$\tilde{\gamma}$	12	0.5	46
N°	88	92	172
	92	85	143
	14	6	106
$\tilde{W}_\ell$	75	74	105
$\tilde{W}_h$	84	81	166
$H^\pm$	89	82	216
H_a	99	92	219
H_b	4	0.4	17
H_c	45	27	201

92

TABLE 6

Some allowed spectra for $m_t \gtrsim 60$ GeV. Masses are in GeV

	a)	b)	c)	d)	e)	f)	g)	h)	i)
m	20	20	40	40	20	30	60	30	120
M	−60	100	100	−120	−120	−60	−60	−25	−25
A	−3	2	2	−1	0	−1	−2	−1	3
μ_0	2	40	6	5	12	2	2	3	17
$\sin 2\theta(M_X)$	0.41	−0.40	−0.15	0.23	0.45	0.10	0.12	0.17	−0.28
$\sin 2\theta(M_W)$	0.88	−0.83	−0.38	0.30	0.48	0.79	0.52	0.96	−0.59
$m_{\tilde{t}}$	60	62	69	69	68	77	83	90	107
$\tilde{q}$	161	268	270	323	321	162	169	71	134
$\tilde{t}_\ell$	102	219	204	264	269	102	64	56	58
$\tilde{t}_h$	212	312	324	374	369	222	241	146	193
$\tilde{e}_L$	91	83	94	106	100	64	86	42	124
$\tilde{e}_R$	42	54	69	74	65	51	75	38	123
$\tilde{\nu}_L$	22	60	58	75	68	21	47	16	117
$\tilde{g}$	180	301	301	361	361	180	180	75	107
$\tilde{\gamma}$	30	52	48	57	57	30	30	12	12
N^0	113	125	133	145	144	112	113	97	105
	71	98	63	58	64	71	71	82	76
	3	43	4	2	10	2	2	3	13
$\tilde{W}_\ell$	50	74	22	16	32	44	28	65	28
$\tilde{W}_h$	109	127	135	146	145	112	117	90	110
H^-	82	120	98	108	106	81	92	80	139
H_a	91	122	93	95	99	90	93	91	131
H_b	12	37	52	68	57	14	40	5	63
H_c	26	92	59	76	73	23	49	20	116

b) Masses of Higgs particles

The supersymmetric SM contains two doublets of Higgses $H_1 = (H_1^+, H_1^0)$, $H_2 = (H_2^-, H_2^0)$ with altogether eight degrees of freedom. Three of them are swallowed by the $W^\pm$ and Z^0 to get a mass. There remain as physical scalar fields one charged Higgs $H^\pm$ and three neutral Higgses that we will call $H_{a,b,c}$. Using the scalar potential Eq. (66), one arrives at the following expressions for their masses[46]

$$m_{H^\pm}^2 = M_W^2 + \mu_1^2 + \mu_2^2 \tag{80a}$$

$$m_{H_c}^2 = \mu_1^2 + \mu_2^2 \tag{80b}$$

$$m_{H_{a,b}}^2 = \tfrac{1}{2}\,[m_{H_c}^2 + M_Z^2 \pm ((m_{H_a}^2 + M_Z^2)^2 - 4 m_{H_a}^2 M_Z^2 \cos 2\theta)^{1/2}] \tag{80c}$$

Examining these formulae, one observes the following facts. First, the charged Higgs is very massive, $m_{H^\pm} \gg M_W$ [see Eq. (80a)]. Concerning the neutral Higgses, two of them are relatively massive (H_a and H_c) whereas the other (H_b) is bounded above by M_Z. In fact, one can derive the following inequalities[46,55]:

$$(m_{H_a}^2 + M_Z^2)^{1/2} > m_{H_a} > M_Z \tag{81a}$$

$$\cos 2\theta \, M_Z > m_{H_b} \gtrsim 0 \tag{81b}$$

$$m_{H_a} > m_{H_c} > m_{H_b} \tag{81c}$$

Typically $\cos 2\theta \simeq 0.1-1.0$ so that the scalar H_b may be rather light. This is to be compared with the Higgs field in the usual SM for which there is no bound. Here the bound comes about because the self-couplings of the Higgses are fixed (by gauge invariance and supersymmetry) to be proportional to the gauge coupling constant. Notice how Eqs. (81a-c) are fulfilled by the examples of consistent spectra shown in Tables 5 and 6.

c) Masses of SUSY fermions

Here we call SUSY fermions the gauginos and Higgsinos. For the case of the gluino, the result is very simple

$$M_{\tilde{g}} = (\alpha_3/\alpha_{GUT})M \tag{82}$$

but the mass spectrum of the $\tilde{W}$, $\tilde{B}$ and Higgsinos is more complicated because once $SU(2)\times U(1)$ is broken they mix amongst themselves. This is due to the existence of the SUSY gauge couplings of Fig. 8b which gives rise to mass terms which mix the charged winos with the charged Higgsinos and $\tilde{W}^0$ and $\tilde{B}$ with the neutral Higgsinos. Apart from these, there are the usual Majorana mass for the gauginos (Fig. 13d) and the direct Higgsino mass μ coming from the superpotential Eq. (59). In this way, we have a mass matrix for the charged winos-Higgsinos[56]

$$
\begin{array}{c}
\tilde{W}^- \\
\tilde{H}^-
\end{array}
\begin{array}{cc}
\tilde{W}^+ \quad\quad \tilde{H}^+ \\
\begin{pmatrix}
M_2 & g_2 v_2 \\
g_2 v_1 & \mu
\end{pmatrix}
\end{array}
\tag{83}
$$

where M_2 is the direct wino Majorana mass and $v_{1,2} = \langle H_{1,2}\rangle$. If both M_2 and μ are large (larger than M_w) the eigenvalues of this matrix will both be large and will correspond to the wino ($\tilde{W}^+,\tilde{W}^-$) and Higgsino ($\tilde{H}^+,\tilde{H}^-$) states. In the limit $M_2 \to \infty$ and $\mu \to 0$, there will be two eigenstates with masses $(M_2, g_2^2 v_1 v_2/M_2)$, the latter being lighter than the $W^\pm$ boson. In the limit $M_2 \to 0$, $\mu \to 0$ there are two eigenstates with masses $(g_2 v_1, g_2 v_2)$. The latter would correspond to the Dirac fermions $\psi 1 = (\tilde{W}^+, \tilde{H}_1^-)$, $\psi_2 (\tilde{W}^-, \tilde{H}_2^+)$ respectively. Notice that in this case one has $M_w^2 = 1/2(M_{\psi_1}^2 + M_{\psi_2}^2)$ and ψ_1 has to be lighter than the W-boson. Thus, in the interesting class of models in which μ is negligible, there is a charged fermion $\lambda^\pm$ which will be lighter than the W-boson. This fermion has good chances to be the lightest charged supersymmetric particle in the SUSY standard model. Values of this mass for several consistent choices of the parameters are shown in Tables 5 and 6. Notice how in many cases the lightest chargino $\tilde{W}_\ell$ will be available in forthcoming accelerators.

The mass matrix for the neutral "fermioninos" is more complicated[56]

$$
\begin{array}{c}
\tilde{W}^0 \\[6pt]
\tilde{B}^0 \\[6pt]
\tilde{H}_1^0 \\[6pt]
\tilde{H}_2^0
\end{array}
\begin{array}{cccc}
\tilde{W}^0 \quad\quad & \tilde{B}^0 \quad\quad & \tilde{H}_1^0 \quad\quad & \tilde{H}_2^0 \\
\begin{bmatrix}
M_2 & 0 & \dfrac{-g_2 v_1}{\sqrt{2}} & \dfrac{g_2 v_1}{\sqrt{2}} \\[8pt]
0 & M_1 & \dfrac{g_1 v_1}{\sqrt{2}} & \dfrac{-g_1 v_2}{\sqrt{2}} \\[8pt]
\dfrac{-g_2 v_1}{\sqrt{2}} & \dfrac{g_1 v_1}{\sqrt{2}} & 0 & \mu \\[8pt]
\dfrac{g_2 v_2}{\sqrt{2}} & \dfrac{-g_1 v_1}{\sqrt{2}} & \mu & 0
\end{bmatrix}
\end{array}
\tag{84}
$$

where M_2 and M_1 are related through Eq. (69) implying $M_1 = (3\alpha_1/5\alpha_2)M_2$. There is a linear combination of $\tilde{W}^0$ and $\tilde{B}^0$ (the photino $\tilde{\gamma}$) which does not mix directly with the Higgsinos and is given by with mass

$$\tilde{\gamma} \simeq g_1 \tilde{W}^0 + g_2 \tilde{B}^0 \tag{85}$$

$$M_{\tilde{\gamma}} \simeq \frac{8}{3} \left(\frac{g_1^2}{g_1^2 + g_2^2} \right) M_2. \tag{86}$$

Notice that $M_{\tilde{g}}/M_{\tilde{\gamma}} \simeq \alpha_3/\alpha_e \simeq 6$. The photino is the lightest SUSY particle in many models [when I say "model", I refer to possible ranges of the free parameters in Eq. (74)], particularly when M is not too large.   On the other hand, a neutral Higgsino is also very often the lightest SUSY particle when μ is very small.   In general, the mass matrix (84) has to be diagonalized numerically case by case.   Again Tables 5 and 6 show examples of consistent spectra.   Notice how often a neutralino N^0 is lighter than the photino.   In some cases also, the sneutrino may be the lightest super-symmetric particle[53].   We should remark that some of the details given above rest on the assumption of a "minimal content" of the SUSY SM.   For example, some of the features of particle spectra (particularly regarding neutralinos and neutral Higgses as well as the lower bound on the top quark) change if an extra singlet multiplet S^0 coupling to Higgses with a Yukawa as $(S^0 H_1 H_2)$ exists.   This possibility seems to appear in some string-inspired models in which direct Higgsino mass terms $\mu H_1 H_2$ seem difficult to get.   However, most of the general features survive also in this case and a v.e.v. for S^0 acts as an "effective" μ parameter.

4ix. Supersymmetry and experiment

One of the most attractive points of the supersymmetric version of the SM is that this theory can be directly checked experimentally (it is not clear if this is the case for grand unification of superstrings).   If no supersymmetric particles are found below ~1 Tev, low-energy supersymmetry would be ruled out.   In fact, this upper limit is clearly pessimistic.   Using naturality arguments, one may estimate "reasonable" upper bounds on the masses of squarks, sleptons, gluinos and "fermioninos" as in Fig. 19 (Barbieri et al. in Ref. 58).   Again, one does not expect those limits to be saturated and one would expect slepton masses ~100 GeV and heavier squark and gluinos ~200-300 GeV although, of course, these are matters open to bet.   Unfortunately, the range of masses explored by present accelerators is rather small compared to the expected range of SUSY particles masses.

The effect of SUSY particles may also appear indirectly through radiative corrections to SM processes or through SUSY contributions to rare decays.   In this connection, it is important to remark that in the SUSY SM discussed above, the masses of squarks and sleptons are practically (up to Yukawa couplings) independent of the generation.   This degeneracy between $\tilde{q}$'s and $\tilde{\mathcal{X}}$'s of different families is very much welcome, since otherwise there would be sizeable flavour-changing neutral currents (FCNC).   There is a SUSY version[59] of the Glashow-Iliopoulos-Maiani (GIM) mechanism at work which suppresses FCNC to the appropriate limits if the $\tilde{q}$'s and $\tilde{\mathcal{X}}$'s are almost degenerate.   Let us remark however that if some rare decay rate or process incompatible with the SM were to be detected experimentally, most probably it could be "explained" by many other possible models apart from supersymmetry.   A reliable check of supersymmetry will require the production of SUSY particles at accelerators.

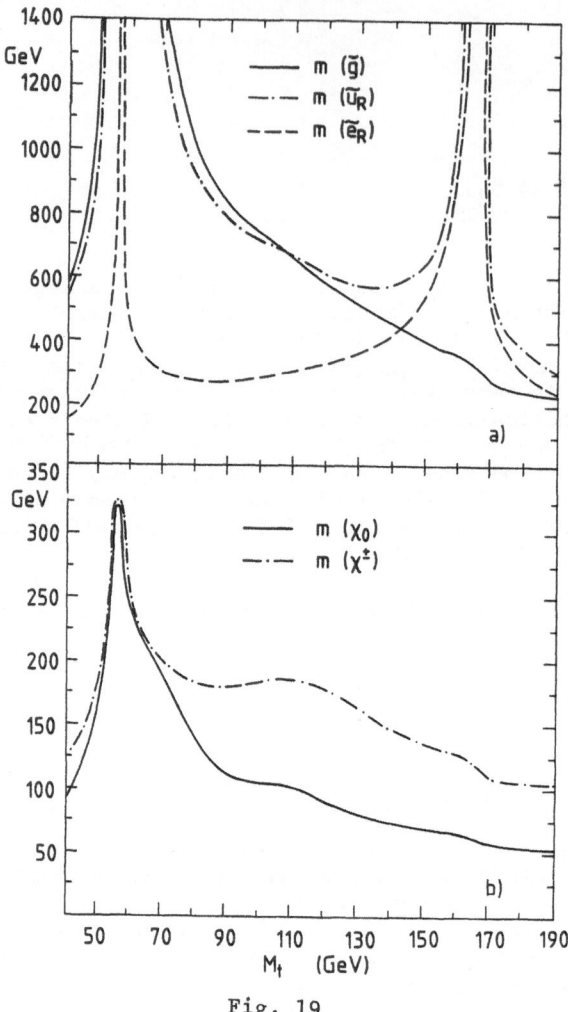

Fig. 19

In order to look for sparticles at accelerators, one needs to know something about the possible signatures. An important point to note is that, with a superpotential like the one in Eq. (59), there is a discrete symmetry ("R-parity") in the theory under which sparticles are "odd" and usual particles are "even"

$$
\begin{array}{lll}
& \text{R-parity} & \\
q,\ell,H,g,\ldots & \rightarrow & q,\ell,H,g\ldots \\
& & \\
\tilde{q}, \tilde{\chi}, \tilde{H}, \tilde{g}, \ldots & \rightarrow & -\tilde{q}, -\tilde{\chi}, -\tilde{H}, -\tilde{g} \ldots .
\end{array}
\tag{87}
$$

This means that sparticles can only be produced in pairs. Using these facts and the SUSY interactions described in Figs 6, 8 and 9, one can find the expected decay modes of each sparticle. Typical decays are depicted in Fig. 20 under the assumption that the gauginos are lighter than the squarks and sleptons (this is not always the case but the intelligent reader should be able to find the decay modes in the opposite situation). Of course, the decays in Fig. 20 are not the only possibilities. Notice also that, in general, the fermioninos ($\tilde{W}^{\pm}, \tilde{H}^{\pm}, z^0, \tilde{\gamma}, \tilde{H}^0$) are not mass eigen-

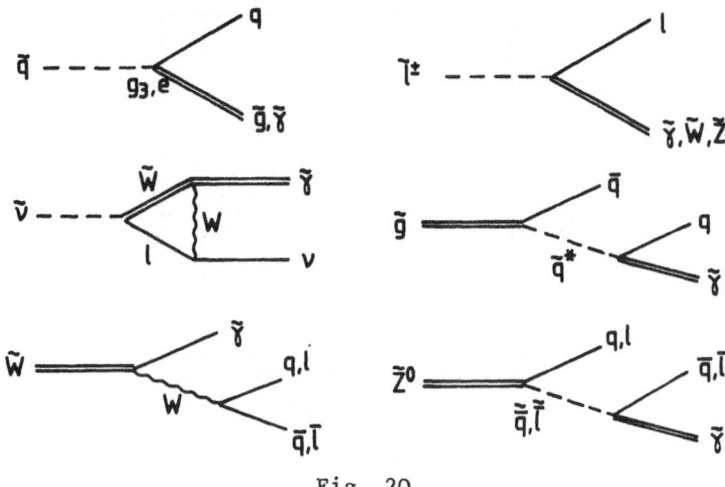

Fig. 20

states and one has to consider the eigenstates of matrices (83) and (84)
as the physical fields. Notice that in all the decays in Fig. 20 there is
a photino amongst the final state particles. For some choices of the
supersymmetry breaking parameters (particularly when M is not too large),
the photino is the lightest supersymmetric particle. Then R-parity
conservation would imply that it is absolutely stable and will appear at
the end of any decay of a supersymmetric particle. On the other hand, the
photino interacts very weakly with matter with cross-sections typical of
weak interactions (this is because they require the exchange of squarks
and sleptons). Then the photino will not be detected by a conventional
detector and will give rise to large missing momentum when produced in the
final state. This is a typical signature which also appears in the more
general case in which the lightest supersymmetric particle (LSP) is a
Higgsino or a scalar neutrino.

In the last few years, there has been an important experimental
effort looking for SUSY particles and a number of useful limits have been
put. The e^+e^- machines[60] give lower bounds for the masses of charged
sleptons $m_{\tilde{\chi}\pm} \gtrsim 25$ GeV. For the selectron the limit $m_{\tilde{e}} \gtrsim 32$ GeV may be put
from the non-observation of the decay $W \to \tilde{e}\tilde{\nu}$ (UA1)[61]. A nice indirect
limit on the mass of the $\tilde{e}$ comes from the search for events produced by
the (hypothetical) process depicted in Fig. 21. The signature is a photon
plus missing energy and there is a well-understood conventional background
from the $e^+e^- \to \nu\bar{\nu}\gamma$ process. The cross-section from Fig. 21 is proportional to $(m_{\tilde{e}})^{-4}$ and thus one obtains a limit on $m_{\tilde{e}}$ depending on $m_{\tilde{\gamma}}$
(provided $m_{\tilde{\gamma}} \lesssim \sqrt{S}/2$). The results from ASP[62] (there are also similar
limits from MAC[63]) are shown in Fig. 22.

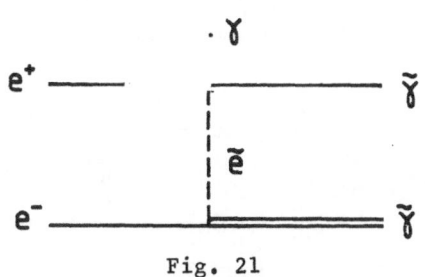

Fig. 21

97

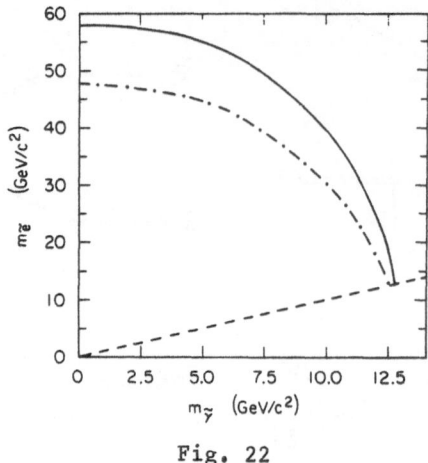

Fig. 22

One finds, e.g., $m_{\tilde{e}} \gtrsim 50$ GeV for $m_{\tilde{\gamma}} \lesssim 5$ GeV. The best limits on the masses of squarks and gluino come from the $p\bar{p}$ colliders through[64] diagrams like those in Fig. 23. The decay of squarks and/or gluinos involve photinos giving rise to missing E_T events. There is a soft spectrum conventional background from the $W \to \tau\nu$ and $q\bar{q} \to Z^0(\nu\bar{\nu})g$ processes amongst others. The present UA1 data[65] exclude an area in the $m_{\tilde{\gamma}}$–$m_{\tilde{g}}$ plane as shown in Fig. 24. The tevatron will soon improve these limits, at least up to the ~80 GeV range. Concerning the "charginos" $\tilde{W}^{\pm}$, $\tilde{H}^{\pm}$, e^+e^- machines put a limit[60] $m^{\pm} \gtrsim 22$ GeV and a slightly better (though more model-dependent) limit may be obtained from the non-observation in UA2[66] of decays like $W^{\pm} \to C^{\pm}\tilde{\gamma}$ ($\tilde{C}^{\pm} = \tilde{W}^{\pm}$–$\tilde{H}^{\pm}$ physical linear combination). The limits on neutral particles like the three neutral physical Higgses, neutralinos (physical linear combinations of Higgsinos, photino and zeeno) and sneutrinos are much weaker and the best (although very model-dependent) come from cosmological considerations. Due to the presence of too much dark matter, one gets from the non-closure of the Universe, a lower bound $m_{\tilde{\gamma}} \gtrsim 5$ GeV for a stable photino and $m_{\tilde{N}0} \gtrsim 5$ GeV for the lightest supersymmetric particle (LSP) being a Higgsino[67]. I should remark at this point that many experimental limits on the masses of SUSY particles (like squarks and gluinos) assume that the LSP is the photino. A complete reanalysis of all experimental data for the hypothesis of a Higgsino being the LSP is still to be done but a general feature is that present experimental limits are in general worse since then photinos decay into Higgsinos + hadrons (leptons) and the missing energy signature is less efficient (unless the photino is sufficiently stable to escape the detector).

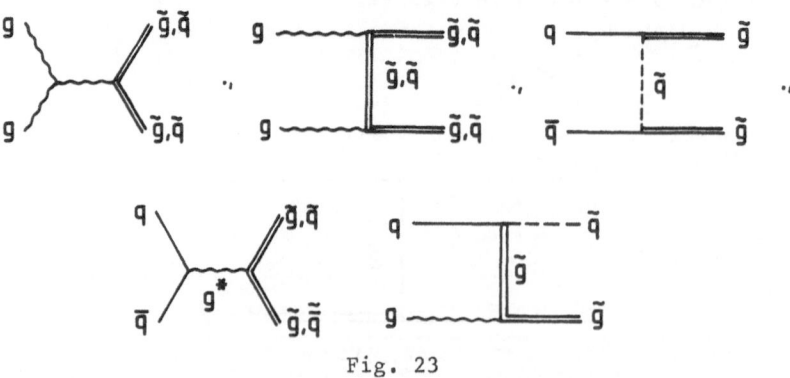

Fig. 23

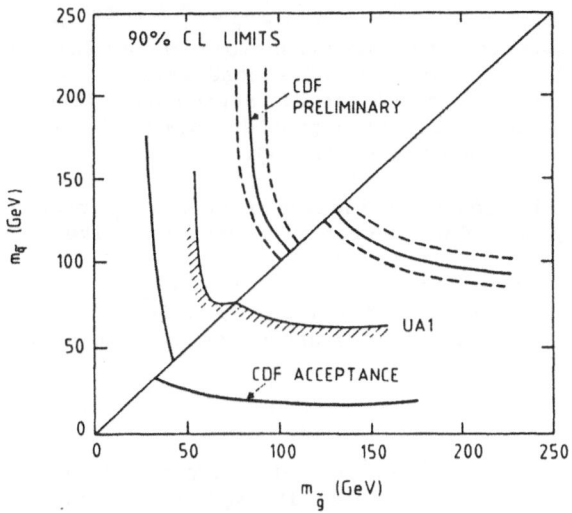

Fig. 24

A word of caution should be given concerning the conservation of "R-parity". If terms like those of Eq. (63) are present in the superpotential of the SUSY-standard model, R-parity is explicitly broken and SUSY-particles need no longer be produced in pairs[68,69]. Thus, e.g., if term (63a) exists, then the photino would decay mostly into strange particles, $\tilde{\gamma} \rightarrow u^c d^c s^c$. If a term like (63d) exists, one could produce sneutrinos in $e^+ e^-$ collisions, $e^+ e^- \rightarrow \tilde{\nu}$, etc. In the presence of these R-parity violating terms in the Lagrangian much of the expected SUSY-signatures would change. Notice, however, that, as we remarked above, not all the terms in Eq. (36) can be present simultaneously since they violate baryon or lepton number and when both things are simultaneously violated the proton would decay too fast[68]. Thus, although R-parity violation may exist, to make it compatible with the observed conservation of baryon and lepton numbers would require that some of the terms in Eq. (63) were present but others do not. This makes this possibility rather unattractive but still certainly possible.

Let us end this section on supersymmetry with a few comments. The main goal of the introduction of a SUSY version of the Standard Model is to avoid the naturality problem. Scalars may be relatively light (compared with the Planck mass) in supersymmetric theories. Furthermore, since the self-coupling of Higgs particles (at least in the minimal model) is governed by the gauge coupling constant, the Higgs particles of the SUSY-Standard Model are much more constrained than in the non-supersymmetric case. Moreover the breaking of the electroweak symmetry $SU(2) \times U(1)$ occurs in a much more elegant way as a consequence of supersymmetry breaking. Radiative corrections rend a negative $(\text{mass})^2$ contribution to one of the Higgses which then triggers $SU(2) \times U(1)$ breaking. This is to be compared with the standard model procedure in which one puts by hand a negative $(\text{mass})^2$ term in the scalar potential.

The natural extension of $N = 1$ supersymmetry to a local symmetry ($N = 1$ supergravity) provides also for the possibility of an elegant supersymmetry breaking mechanism. This is assumed to take place in a "hidden sector" of the theory which couples to the "observable" world only through gravitational interactions . The details of the structure of the hidden world are not relevant as long as supersymmetry breaking takes place in that sector at a scale $\sim\sqrt{M_W M_p} \sim 10^{10}$ GeV. This type of SUSY-breaking allows for the construction of a SUSY version of the Standard Model obeying all experimental constraints. We remark that this is a non-trivial result since other approaches like technicolour or composite models do in general have problems with experimental constraints (e.g., flavour-changing neutral currents).

In spite of all these attractive features, supersymmetry does not provide us with any new clues concerning questions like fermion masses and mixings or why there is family replication. Furthermore, one of the ingredients is N = 1 supergravity, which is known to be non-renormalizable as a field theory. Thus again, N = 1 supergravity coupled to the Standard Model (or to a GUT for that matter) cannot be the whole story.

5. STRINGS

The introduction of strings[3] is certainly a more daring step than the ones we have been discussing before. It implies a change in many concepts usually taken for granted. Thus the fundamental objects are no longer particles but extended one-dimensional objects ("strings") and the observed particles correspond to some "vibration modes" of the string. Space and time (the number of dimensions) are now dynamical variables instead of the "arena" in which physics takes place. Many points are still poorly understood in string theory (e.g., there is no consistent field theory of closed strings) although in the last few years there has been much progress. Of course, in these lectures I will not pretend to give an introduction to string theory but will just try to give a flavour of the topic for the non-experts. I will try to emphasize how the topics we have been discussing in previous sections could fit within the string context. The interested reader is referred to the abundant reviews on the subject[3].

We are all familiar with the dynamics of a relativistic free particle. Its co-ordinates X^μ are a function of the proper time τ (Fig. 25) and the action is given by

$$S = m \int_{\tau_i}^{\tau_f} d\tau \sqrt{-\dot{x}^2(\tau)} = m \int_{\tau_i}^{\tau_f} ds \qquad (88)$$

where $\tau_{i,f}$ are the initial and final times respectively and ds is the infinitesimal interval. We see from Eq. (88) that the action of the free particle is given by the length of the world-line which is described when it propagates. An open string is a one-dimensional extended object parametrized by a real parameter σ (Fig. 25b). When it propagates in space it sweeps a "world-sheet" parametrized by σ and a time-like variable τ. Then each point in the string is described by a co-ordinate $X^\mu(\sigma,\tau)$. In analogy with Eq. (88) one can show that the action of a free relativistic string is given by the area swept by the world sheet. The simplest expression for the string action is obtained as[70]

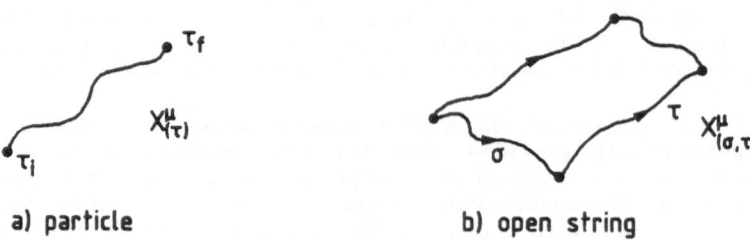

a) particle   b) open string

Fig. 25

$$S = \int d\sigma \, d\tau \, \sqrt{-g} \, g^{ab} \, \partial_a X^\mu \, \partial_b X_\mu \qquad (89)$$

where g^{ab} is a two-dimensional metric on the world-sheet ($a,b = \sigma,\tau$) and $\partial_a = \partial/\partial\sigma$, $\partial_b = \partial/\partial\tau$ ($g = \det|g^{ab}|$). This metric behaves as an auxiliary field and may be eliminated from Eq. (89) by making use of the equations of motion. In fact the action is invariant under reparametrizations of the world-sheet $\sigma,\tau \to \sigma',\tau'$ ($S \sim$ area, independently of how we choose to parametrize the surface) and one can use this invariance to choose a flat metric in Eq. (89):

$$S = \int d\sigma \, d\tau \, \eta^{ab} \, \partial_a X^\mu \, \partial_b X_\mu \qquad (90)$$

with $\eta = \text{diag}(1,-1)$. An important point to realize is that Eq. (90) looks like the action of a massless bosonic field in a space-time with two dimensions (σ,τ). Thus the movement of a string along with time may be represented by a two-dimensional massless field theory. These theories are rather remarkable due to the fact that they possess a huge amount of symmetry ("conformal invariance"). Conformal invariance[71] is the symmetry of conformal mappings of the two-dimensional plane. It is convenient to work with a Euclidean version ($\tau \to -i\tau$) of the world-sheet and to use the complex variables z, $\bar{z}$: ($z \equiv \tau + i\sigma$) instead of σ and τ. Then conformal invariance implies invariance of the action under transformations

$$z \to f(z) \qquad\qquad \bar{z} \to g(\bar{z}) \qquad (91)$$

where $f(g)$ are holomorphic (antiholomorphic) arbitrary functions. Co-ordinate transformations of this type generate the "conformal group". For more than two dimensions the conformal group is finite-dimensional. Two dimensions (like in our case σ,τ) are special and in this case the group is infinite-dimensional. This enormous symmetry makes it possible to get rid of unphysical negative norm states.

The energy momentum tensor plays a special rôle since it is the generator of local conformal transformations. In fact one can see that of the four possible components T^{ab} ($a,b = \sigma,\tau$) of this tensor only two $T_{zz} \equiv T$ (and the complex conjugate $T_{\bar{z}\bar{z}} = \bar{T}$) are non-vanishing. For the action (90) the corresponding energy-momentum component is:

$$T = \,:-\tfrac{1}{2} \, (\partial_z X)^2: \qquad (92)$$

where the dots indicate normal ordering product of the fields. Thus T is a "composite" operator. The behaviour of fields $\phi(z)$ under a conformal transformation are given by the operator product expansions[71] (OPE) (we omit normal ordering dots from now on)

$$T(w) \, \phi(z)_{w-z\to 0} \sim \frac{d_\phi}{(w-z)^2} \, \phi(T) + \frac{1}{(w-z)} \, \partial_z \, \phi(z) + (\text{non-singular}) \qquad (93)$$

d_ϕ is the "conformal dimension" of the field $\phi(z)$. One finds that the conformal dimension of the field ($\partial_z X$) is $d = 1$. Recalling the form of T in Eq. (92) one would expect T to have conformal dimension $d = 2$. However, when computing the relevant operator product expansion [using Wick's theorem and the form of the propagator of the field, $\langle X^\mu(z) X^\nu(w) \rangle = -\eta^{\mu\nu} \log(z-w)$] one finds[3]

$$T(z)T(w) \underset{w-z\to 0}{\sim} \frac{c/2}{(z-w)^4} + \frac{2}{(z-w)^2} \, T(w) + \frac{1}{(z-w)} \, \partial_w T + \dots \qquad (94)$$

where in the present case $c = 1$. The last two terms are the ones one would expect for a field for conformal dimension $d = 2$. One can see that the first most singular term, which is not expected at the classical level, appears as a quantum mechanical effect. The coefficient c in Eq. (94) is called "central charge". The behaviour in Eq. (94) occurs in general for any conformal field theory and not only for the simple case of Eqs. (90) and (92). The appearance of such a term leads to a violation of conformal invariance at the quantum level (an "anomaly"). In fact one can consider the Fourier components of T(z) on the circle centered in the origin[3]

$$L_n = \int \frac{dz}{2\pi i} \ z^{n+1} \ T(z) \tag{95}$$

They generate the infinitesimal conformal transformations corresponding to $f(z) = z^{n+1}$. The infinite operators L_n ($n = 0, \pm\infty$) constitute the Virasoro operators associated to the conformal invariance. Using Eqs. (94) and (95) one can obtain the commutators

$$[L_n, L_m] = (n-m) \ L_{n+m} + \frac{c}{12} \ n(n+1) \ (n-1) \ \delta_{m+n} \tag{96}$$

leading to the Virasoro algebra. The last term corresponds to the first in Eq. (94) and one can see that it has the typical aspect of an "anomalous" quantum term violating a classical (conformal) invariance. To have a consistent conformal invariant theory (which we need in order not to have ghosts), the overall central charge c (the sum over all the central charges of each independent conformal field) has to vanish.

In the case of an action like that in Eq. (90) the central charge would be c = D, D = number of dimensions, since c = 1 for a bosonic field X(z), and thus the system would not be conformally invariant at the quantum level. In fact this is not completely correct. In analogy with what happens when quantizing a usual non-Abelian Yang-Mills theory in which in order to avoid "double counting" due to the "gauge volume" one introduces the Faddeev-Popov ghosts, one also has to add "ghosts" to the system of a bosonic string. These "ghosts" are associated to the "gauge volume" of reparametrizations of the world sheet (ghosts "b" and "c"). They are scalars which anticommute, as usually happens with ghosts. They have conformal dimensions d(c) = -1, d(b) = 2. The associated piece of the energy-momentum tensor is $T(z) = -2b\partial_z C - \partial_z bC$. Again, using Wick's theorem and the propagator $<b(z)C(w)> = (z-w)^{-1}$ one gets the operator product expansion

$$T_{ghosts}(z) \ T_{ghosts}(w) \sim \frac{(-13)}{(z-w)^4} + \ldots \tag{97}$$

so that the ghost system has central charge c = -26. Thus the central charge of a bosonic string in D dimensions is

$$c = (D - 26) \tag{98}$$

and the theory is consistent only in 26 space-time dimensions. This is the well-known result that the purely bosonic string[73] is only consistent in 26 dimensions. Let us remark, however, that what c = 0 implies is the necessity of having 26 $X^\mu(\sigma,\tau)$ ($\mu = 0,\ldots,25$) fields but not necessarily all of them need to have an interpretation as physical space-time dimensions. As we will see later on, by appropriately choosing boundary conditions for the fields $X^\mu(\sigma,\tau)$ one can make that only a subset of them (e.g., 4 = 3+1) admits an interpretation as observable space-time dimensions.

What is the mass spectrum of the bosonic string? The equations of motion for an action as that in Eq. (90) are the ones of a two-dimensional Klein-Gordon field

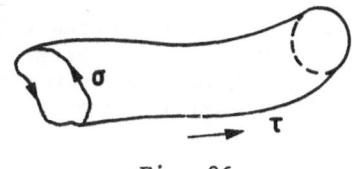

Fig. 26

$$\left(\frac{\partial^2}{\partial\sigma^2} - \frac{\partial^3}{\partial\tau^2}\right) X^\mu(\sigma,\tau) = 0 \tag{99}$$

The freedom to make conformal transformations allows one to work in the "light-cone gauge" in which only the D-2 transverse $X^i(\sigma,\tau)$ ($i = 1 \rightarrow D-2$) fields appear as independent physical fields (this is reminiscent of what happens with the physical transverse degrees of freedom of a usual gauge boson like, e.g., the photon). Let us consider the case of a closed string as in Fig. 26. The world-sheet is now a cylinder. The general solution of Eq. (99), subject to the boundary condition $X^i(\sigma,\tau) = X(\sigma+\pi,\tau)^i$ (so that the string is really closed), is

$$X^i(\sigma,\tau) = X_0^i + p^i\tau + i/2 \sum_{n\neq 0} \left\{\frac{\alpha_n}{n} e^{-2in(\tau-\sigma)} + \frac{\tilde{\alpha}_n}{n} e^{-i2n(\tau+\sigma)}\right\} \tag{100}$$

[here σ and τ have again a Minkowskian metric as in Eq. (90)]. Here X_0^i and p^i are the centre-of-mass position and momentum and the two terms in brackets correspond to string vibrations going in one direction ("right-moving") and in the opposite ("left-moving"). In complex (Euclidean) notation they would correspond to holomorphic ($X^i(z)$) and antiholomorphic ($X^i(\bar{z})$) fields. Upon quantization one has

$$[\alpha_n,\alpha_m] = n\delta_{n+m,0} \quad ; \quad [\tilde{\alpha}_n,\tilde{\alpha}_m] = n\delta_{n+m,0} \tag{101}$$

One can explicitly compute the mass operator $m^2 = p^\mu p_\mu$ and one obtains

$$\frac{m^2}{4} = \frac{m_L^2}{4} + \frac{m_R^2}{4} = (\tilde{N}-1) + (N-1) . \tag{102}$$

where N, $\tilde{N}$ are the "number operators"

$$N = \sum_{n=1}^\infty \alpha_{-n}^i \alpha_n^i \quad ; \quad \tilde{N} = \sum_{n=1}^\infty \tilde{\alpha}_{-n}^i \tilde{\alpha}_n^i, \quad (i = 1\text{-}24) \tag{103}$$

for "left" and "right" vibrations.

A general state in the Hilbert space of the string is of the usual Fock space form:

$$\Pi \; \alpha_{-n}^i \cdots \alpha_{-m}^j \; |0\rangle \tag{104}$$

The eigenvalues of N and $\tilde{N}$ are just positive integers counting the number of oscillators acting on the vacuum. One can see that the fact that there is no privileged point (origin for measuring σ) in the closed string imposes the constraint $(N-\tilde{N})|\psi\rangle = 0$ for any string state $|\psi\rangle$. Now from Eq. (102) one sees that the lightest states are

$$N = \tilde{N} = 0 \rightarrow m^2 = -8, \quad |0\rangle$$
$$N = \tilde{N} = 1 \rightarrow m^2 = 0, \quad \alpha_{-1}^i \tilde{\alpha}_{-1}^j |0\rangle \tag{105}$$

The vacuum state $|0>$ is tachyonic. This is the well-known problem of the purely bosonic string. Then for $N = \tilde{N} = 1$ there are massless states with two space indices (i,j). A symmetric combination of the indices leads to a state which can be identified with the graviton g^{ij}. Thus the bosonic string is a theory of gravity! This is a remarkable fact since we have put gravitation nowhere to start with. There are more massless fields, the trace $\alpha^i_{-1}\alpha^i_{-1}|0>$ is a scalar field ("dilaton") and the antisymmetric combination of the two oscillators correspond to a two-index antisymmetric field.

The bosonic string we have been discussing does not seem very promising: it contains a tachyon and has no massless fermions (quark-lepton candidates) nor gauge bosons. It turns out that in order to extend the string concept to fermions one has to extend the simple Lagrangian in Eq. (90) in a supersymmetric way. Thus one introduces fermionic two-dimensional fields[74] $\psi^\mu(z)$ and $\overline{\psi}^\mu(\overline{z})$ $(\mu = 0,\ldots,(D-1))$. The relevant action is

$$S = \int dz d\overline{z} \ \{\partial_{\overline{z}}X^\mu \partial_z X_\mu - \psi^\mu \partial_{\overline{z}}\psi^\mu - \overline{\psi}^\mu \partial_z \overline{\psi}^\mu\} \tag{106}$$

This action, as we discussed when we talked about supersymmetry, possesses an $N = 1$ SUSY in the two-dimensional world-sheet (not necessarily implying space-time supersymmetry in the D-dimensional space!). (X^μ, ψ^μ) form SUSY multiplets and the bosonic energy momentum tensor T_B has now a SUSY-partner T_F:

$$T_B = -\tfrac{1}{2} \ (\partial_z X^\mu)^2 + \tfrac{1}{2} \ \psi^\mu \partial_z \psi^\mu \ ; \quad T_F = -\tfrac{1}{2} \ \psi^\mu \partial_z X^\mu \tag{107}$$

and similarly for $\overline{T}_B$, $\overline{T}_F$. As it happened with the purely bosonic case, one has to introduce "ghosts" upon quantization of fermions. One introduces "superconformal ghosts" γ (conformal dimension $d = -\tfrac{1}{2}$) and β ($d = 3/2$). One can see that the contribution of the (β,γ) system to the central charge is $c = 11$. Since the contribution to each $\psi(z)$ fermion to c is $\tfrac{1}{2}$, one gets an overall central charge for the "superstring" from Eq. (106):

$$c = D + \tfrac{1}{2}D -26 + 11 \ = \ 3/2 \ (D-10) \tag{108}$$
$$\hspace{-4em} \downarrow \quad \downarrow \ \ \downarrow \quad \downarrow$$
$$\hspace{-4em} X \quad \psi \ \ b,c \ \ \gamma,\beta$$

Thus the superstring is truly consistent only in $D = 10$ space-time dimensions. In (107), T_F is the generator of local supersymmetry on the world-sheet. $\overline{T}_F$ generates also a local supersymmetry for antiholomorphic fields and one says that the superstring has (1,1) local world-sheet supersymmetry, referring to the $N = 1$ SUSY generated by T_F and $\overline{T}_F$ on left- and right-movers. The superstring has two types of "fermionic" states depending[74] on the boundary conditions of the ψ^μ fields. If $\psi^\mu(\sigma) = (-)\psi^\mu(\sigma+\pi)$, one has the "Neveu-Schwarz" sector of the string and for $\psi^\mu(\sigma) = \psi^\mu(\sigma+\pi)$ one gets the "Ramond" sector. One can see that the string states in the first sector are bosons and the ones in the second are fermions. A remarkable point is that the theory can be shown to be inconsistent if it contains different numbers of fermions and bosons. This means that one must have a complete matching of bosonic and fermionic states, i.e., one must have a space-time supersymmetry[75,76] (and not only world-sheet SUSY). Thus the spectrum of the ten-dimensional superstring must be supersymmetric. Due to this fact this theory has no tachyons and the massless fields include a supersymmetric version of gravity, ten-dimensional "N = 2 supergravity". We will not discuss here why we are forced to have a fermion-boson matching in the superstring. Let me only remark that when one considers a one-loop amplitude for the superstring, a property called "modular invariance" must be requested for the string amplitudes[3]. Otherwise "global anomalies" to some extent similar to the ones we discussed for gauge theories appear. When one imposes modular invariance of one-loop amplitudes one can see that necessarily the spectrum of the fermionic string is supersymmetric (a projection killing states which are not supersymmetric appears)[75]. Let me also remark that this property of "modular invariance", which plays a prominent rôle in string theory, also makes that all one-loop amplitudes are ultra-violet finite both in the bosonic string and in the superstring.

Although the superstring described above has no tachyons and contains massless fermions, it still is not suitable to describe observed physics (apart from the fact that D = 10). There are no massless gauge bosons one could identify with gluons, W's or the photon. However, one can construct new more promising types of supersymmetric strings, the heterotic strings. To do this one makes use of another remarkable property of closed strings: left-moving and right-moving states are completely independent and follow separate dynamics. This we have already observed above and it allows one to have different strings for left- and right-movers. The heterotic strings[77] are right-moving superstrings and left-moving bosonic strings:

$$
\left[
\begin{array}{ll}
\text{R-M} & \text{L-M} \\
\psi^\mu(z) & \\
& X^\mu(\bar{z}), \quad \mu = 0-25 \\
X^\mu(z), \quad \mu = 0-9 &
\end{array}
\right] \tag{109}
$$

This sounds a bit puzzling since the superstring is ten-dimensional and the bosonic string is 26-dimensional. However, one can see that for such a theory to be consistent, 16 of the 26 dimensions of the bosonic string have to be curled up ("compactified") on a 16-dimensional torus of fixed radi, so that the actual number of dimensions of a heterotic string is 10. "Compactified" on a torus means that one identifies the co-ordinates X^I, I = 10-25 with their translations by some "lattice vectors" $\{e_i^I\}$:

$$
X^I = X^I + R \sum_{i=1}^{16} n_i \, e_i^I \tag{110}
$$

where R is the "radius of compactification". To have an idea consider the case of a two-dimensional torus as in Fig. 27a. It may be obtained from Fig. 27b by gluing the sides A (identification $X = X + \vec{e}_1$) and the sides B (identification $X = X + \vec{e}_2$). When we force the co-ordinates X^I (I = 10-25) to move only on a 16-dimensional torus, the momenta associated to these co-ordinates p^I are quantized. One can further show that the "modular invariance" property we mentioned above requires the p^I to be very specific types of vectors[77] belonging to the "root lattices" of the Lie algebras of $E_8 \times E_8$ and $SO(32)$[78]. I will not explain how this works out since it is rather technical. Let me just comment on the spectrum of these heterotic strings. It turns out that the modulus squared of p^I must be an even number (in string units, $R^2 = \alpha'$, where α' is the string tension which we have taken $\alpha' = \frac{1}{2}$ all over). The mass formula for the heterotic strings has the form (e.g., Ramond R-M sector)

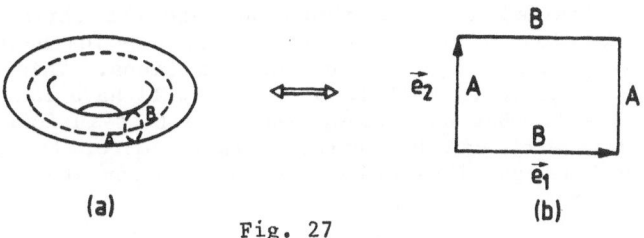

(a) Fig. 27 (b)

$$m^2 = m_R^2 + m_L^2 \; ; \quad \begin{cases} \dfrac{m_R^2}{4} = N \\[2mm] \\ \dfrac{m_L^2}{4} = \tilde{N} + \dfrac{p^{I^2}}{2} - 1 \end{cases} \tag{111}$$

where N, $\tilde{N}$ are the number operators. Notice there is no tachyon in the right-handed (superstring) contribution. The condition $(m_R^2 - m_L^2)|\psi\rangle = 0$ (analogous to the $N = \tilde{N}$ condition of the bosonic string) makes then that no tachyons may appear in the heterotic string. There are massless states, e.g., for $N = \tilde{N} = 0$ and $(p^I)^2 = 2$. One can see that the vectors of the $E_8 \times E_8$ root-lattices with $(p^I)^2 = 2$ are just the roots. As is well known, each charged generator of a group has associated a "root vector". This is just telling us that the set of generators of a group $E_8 \times E_8$ (gauge bosons) will be massless. Thus in the heterotic strings there are massless gauge bosons. Furthermore, as happened with the "superstring", there is space-time supersymmetry. Altogether one can see that the massless spectrum is $N = 1$ supergravity coupled to an $E_8 \times E_8$ [or $SO(32)$] super Yang-Mills theory.

The heterotic strings seem more promising than unification theories since they have massless gauge bosons and also charged fermions (the SUSY partners of the gauge bosons). However, the theories are formulated in ten dimensions instead of four. This is easy to mend since not necessarily all the co-ordinates in Eq. (109) must admit a space-time interpretation. Thus it could well happen that the six extra unwanted dimensions were also "curled up" (compactified) on some torus (or other space). As an example, one could compactify the six extra right and left co-ordinates in Eq. (109) $X^i(Z)$, $X^i(\bar{Z})$ ($i = 4$-9) in three copies of the 2-torus of Fig. 27

$$X^i = X^i + \sum_{i=5}^{9} R_a n_a e_a^i \tag{112}$$

where $e_{4,5}^i$, $e_{6,7}^i$, $e_{8,9}^i$ could be orthogonal unit vectors as in Fig. 27b. R_a, the radi of compactification, are arbitrary since there are both left and right co-ordinates. However, it turns out that in order to get the correct size for the gravitational ($\sim 1/M_p^2$) and gauge ($g \sim 1$) interactions all the radi R_a of compactification have to be $R_a \sim 1/M_p$ and hence the size of the extra six dimensions is tremendously small. This means that to "probe" those extra dimensions we would need energies of the order of the Planck mass, $M_p \sim 10^{18}$ GeV. Thus at ordinary energies the physical number of dimensions will be just four.

What we have just been discussing is called toroidal string compactification and this method gives us the simplest four-dimensional strings. However, it is not yet what we want. The reason is that the theories so obtained in four dimensions are not chiral. This happens because the fermions (spinors) of an $N = 1$, $D = 10$ theory from which we started split into four spinors in four dimensions. This means that if we had one supersymmetry ($N = 1$) in $D = 10$ we will have four supersymmetries ($N = 4$) in $D = 4$. But we already mentioned that in four dimensions only $N = 1$ supersymmetry can be chiral. Furthermore, the groups $E_8 \times E_8$ or $SO(32)$ are not amongst the interesting unification groups we discussed in the first lecture.

One way to solve this problem is to compactify the six extra dimensions on an "orbifold"[79]. Orbifolds of the type we are discussing are relatives of tori[80]. You start from a torus and not only identify points which differ by some lattice vector but you identify points which are related by some discrete symmetry P of the torus. Consider as an example the two-torus in Fig. 27b. Now we are going to identify all the points on the torus which can be obtained from each other as a reflection with respect to the origin in Fig. 27b. Thus, e.g., the points E and E' in Fig. 28 are identified since they can be obtained from each other after a reflection with respect to the origin (plus a shift by $\vec{e}_1 + \vec{e}_2$). In this way all the points in the original torus are identified with points in the shadowed area. One expresses this identification mathematically by saying that the above two-dimensional orbifold is the result of "dividing" the torus T^2 by an isometry P, $O_2 = T_2/P$. Again one can obtain a four-dimensional string by compactifying the six unwanted dimensions, e.g., on a six-dimensional orbifold with three copies of O_2, i.e., $O_6 \equiv O_2 \times O_2 \times O_2$. Of course there are many more tori and orbifolds one can construct. The important point is that only the fields of the compactified string which are invariant under the symmetries P will remain as physical fields in the string spectrum. Thus we may get rid of three of the four supersymmetries of toroidal compactifications by appropriately choosing a symmetry P which kills three of the four existing massless gravitini. There are plenty of possibilities to do so leading each one to different four-dimensional strings.

We thus have four-dimensional strings with N = 1 supersymmetry and massless gauge bosons and fermions. Again this is not enough since the gauge groups we have [$E_8 \times E_8$ or SO(32)] are not adequate (recall our discussion in the first lecture). Thus we want to break the gauge group down to some more promising subgroup. A natural process is to do something similar to the "dividing by a symmetry P" of the orbifold. One can "mode" the gauge degrees of freedom of the heterotic string by some subgroup G of $E_8 \times E_8$ [or SO(32)]. Then only the string states invariant under the combined action on space and gauge degrees of freedom P × G remain as physical states. It turns out that "modular invariance" constraints strongly correlate the possible forms of P and G. The simplest solution is to have similar forms for the symmetries P acting on the six compactified co-ordinates and the symmetry G acting on the gauge degrees of freedom. This type of compactifications is called (2,2) or "standard embedding" compactifications. The case of the $E_8 \times E_8$ string is particularly interesting. By acting with an action G inside one of the E_8's, a (2,2) compactification leads to a gauge group

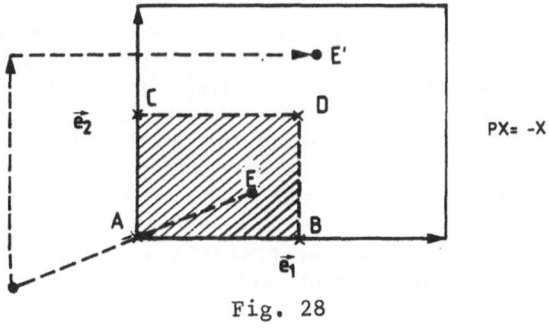

Fig. 28

and E_6 is one of the interesting groups for unification discussed in the first lecture. We are thus left with a D = 4, N = 1 supersymmetric theory coupled to a gauge group $E_6 \times E'_8$. In fact one can see that the E_6 and E_8 sectors of the theory are decoupled from each other (only interact gravitationally amongst themselves) so that the E'_8 world is a "hidden world". It is remarkable how similar this situation is to the one we discussed for the supersymmetric standard model. Thus the existence of a "hidden sector" of the theory is a generic fact in four-dimensional strings. One can also see that the chiral matter obtained in this type of strings is just a number of E_6 generations (27's) and antigenerations (27). The number of quark-lepton families will be the net number $(N_{27} - N_{\overline{27}})$ since pairs $27 \times \overline{27}$ can in general gain large masses and disappear from the physical low energy spectrum (although this may be problematic in practice).

There are two other known methods to construct four-dimensional (2,2) strings. One of them is the compactification of the six extra dimensions in a particular class of six-dimensional manifolds called "complete-intersection Calabi-Yau manifolds" (CICY)[81]. The problem with this method is that the conformal field theory associated to these four-dimensional strings is not known. This was, however, historically the first method attempted and much of the string "phenomenology" done up to now has a "Calabi-Yau inspiration". There is a third method[82] which is also very interesting. It just constructs a conformal invariant theory with the usual constraints of modular invariance and overall vanishing central charge, c = 0, forgetting about the possible interpretations of the unwanted degrees of freedom as extra dimensions. Consider first the right-moving (superstring) sector of the heterotic string. The degrees of freedom we want to get rid of (or at least do not want to have as physical dimensions) are $X^i(Z)$, i = 4-9 and $\psi^i(Z)$ (i = 4-9). The central charge of this superconformal system is c = 9 = 6×1 + 6×½. Thus the idea is that any c = 9 superconformal system leading to a modular invariant string is in principle acceptable (the overall central charge will vanish as required). The simplest superconformal models in the market are the so-called "minimal models". These models are labelled by an integer K = 1...∞ and have central charges

$$c = \frac{3K}{K+2}, \qquad K = 1...\infty \qquad (114)$$

so that they have 1 < c < 3. The idea[82] now is to combine a set of models of this type so that the overall central charge is c = 9. For example, nine copies of the K = 1 model or six copies of the K = 2 model, etc. All this refers to the right-moving sector. There is then a standard way[82] to build the left-moving sector of the heterotic string so that the result is a (2,2) modular invariant for dimensional string.

These three methods to construct (2,2) four-dimensional strings are not unrelated. There are string models which may be constructed in more than one way[83,84]. The situation concerning the relationship among these constructions is summarized in Fig. 29. How do these models look compared to observed physics? A first selection criterium is the number of generations. A complete catalogue of all these (2,2) models is still to be made but a lot of classification work has already been done. The situation at present is as follows. Concerning orbifolds, models with an Abelian P symmetry ("Abelian orbifolds") have been studied in some detail[85,83]. The number of net generations found is always a multiple of six. Concerning non-Abelian orbifolds, although a complete classification

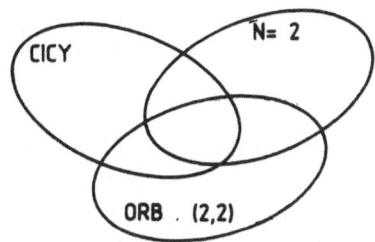

Fig. 29

has not been done, it has been shown that no three-generation model[79] can be found [in the case of (2,2) compactifications!]. There is a unique complete-intersection Calabi-Yau model with net number of generations = 3 (nine generations and six antigenerations)[86]. Its phenomenology has been studied in some detail[87] although the lack of a known conformal field theory did not allow for a string formulation of the model. Finally, the analysis of the N = 2 superconformal constructions mentioned above is still incomplete[88]. Some of the models seem to be equivalently obtained from orbifolds[83]. However, and most interestingly, some of the models seem to correspond to some known CICY[82,84]. In particular, the three-generation model mentioned above has a formulation as an N = 2 superconformal construction[82]. This is interesting since one can in principle compute all the details of the model, including Yukawa couplings. Still, the possibility that new three-generation (2,2) models exist is open.

We have seen that amongst the relatively large number of (2,2) models known to exist up to now, only one has a net number of generations equal to three; it is not so easy to get three-generation models. However, this model has also six pairs of $(27+\overline{27})$ matter fields and it is still not clear whether they get a large mass and decouple from low-energy physics. This should be the case since otherwise the gauge coupling constants of the standard model would get too large before unification. Moreover, preliminary field-theoretical studies seem to show that the breaking of E_6 down to the standard model while maintaining sufficient proton stability and other experimental constraints may perhaps be possible but in a rather contrived way. Are there any other types of four-dimensional strings which may lead to realistic physics? The answer is yes. The (2,2) four-dimensional strings are the simplest solution to the conditions of modular invariance but certainly not the only ones[79]. In orbifold language, there are many other possible actions G on the gauge degrees of freedom which may be consistent. Their common feature is that the gauge group is no longer $E_6 \times E'_8$ but many other possibilities [including $SO(10)$, $SU(5)$ or even the standard model group $SU(3) \times SU(2) \times U(1)$] are now available. These other constructions are called (2,0) models. Very little is known about (2,0) models obtained from CICY or N = 2 superconformal models. On the other hand, many such orbifold models have been constructed[85,89,90], some of them with promising phenomenology. Other techniques in order to get (2,0) models also exist[92-94].

The simplest quasi-realistic four-dimensional strings constructed are based on a (2,0) version of the "Z_3" orbifold[91]. It is called so because the orbifold P symmetry is an Abelian Z_3 group. I will not give the details here and refer the reader to the original literature[89,91,85].

This class of models has three generations of quarks and leptons plus additional Higgs doublets and singlets. One of the most promising models has gauge group[95,96]

$$SU(3)_C \times SU(2)_L \times U(1)_Y \times [U(1)_L \times U(1)' \times SO(10)'] \qquad (115)$$

where the $SO(10)'\times U(1)'$ is "hidden" from the standard model physics (it only interacts with usual particles through gravitational interactions). The extra $U(1)_L$ gauge boson may or may not be there depending on the version of the model. The particle content is three generations of quarks and leptons plus a number of $SU(2)_L$ doublets and singlets. Preliminary studies of the Yukawa couplings have also been done. A different formalism ("four-dimensional fermionic strings")[93] has been used to obtain other (2,0) strings leading to a grand unified "flipped" $SU(5)\times U(1)^3$ model[97].

The examples of quasi-realistic four-dimensional strings found are encouraging but we are still far away from a situation in which (a) a given model is consistent with absolutely all the known phenomenological constraints and (b) any new prediction can be given. A point which is encouraging is that it is not so easy to find three-generation four-dimensional strings. If one insists on mimicking more and more details of the observed physics of quarks and leptons, more and more string candidates are ruled out. At the moment I do not think that anybody can seriously claim that a completely realistic string exists. Points like sufficient stability of the proton are hard to fulfil. Once the gauge symmetry is broken down to the standard model there is in general no symmetry forbidding the B- and L-violating operators we discussed in Eq. (63). One can "cook" the models so that these kinds of couplings do not appear but there does not seem to exist any symmetry principle based on string physics which may prefer a conservation of baryon or lepton number. In fact, strings do not like continuous global symmetries, they prefer gauge symmetries. A possible solution to this baryon stability problem of four-dimensional strings is the conservation of an unbroken extra gauge symmetry forbidding the dangerous couplings of Eq. (63). That is precisely what happens if an unbroken $SU(2)_L\times SU(2)_R\times U(1)_{B-L}$ symmetry remains close to the weak scale. This may be obtained in some four-dimensional string models[98].

In spite of the difficulties mentioned, four-dimensional strings are a big improvement over previous schemes. First of all, superstrings are believed to be finite theories. They are really unified theories, including for the first time a consistent theory of quantum gravity. Furthermore they are highly constrained theories. Thus, given a four-dimensional string, the particle content and gauge interactions are completely fixed. As an example, no exotic massless representations (e.g., colour sextet quarks or similar objects) are in general allowed. Also no adjoint Higgs fields ever appear in the massless sector of a string. Yukawa couplings are also determined by the corresponding conformal field theory associated to the model. This makes string models very easy to rule out. Let us remark, however, that all string models have some neutral scalar fields ("moduli" and "marginal operators") whose (unknown) vacuum expectation values govern the precise numerical values of quantities like Yukawa couplings. Then one may expect from a given model a prediction on what couplings are allowed and which ones are forbidden or qualitative features of the quark-lepton spectrum but no precise prediction on quark and lepton masses is to be expected as long as we do not find a dynamical reason which could explain why some four-dimensional strings are dynamically selected.

How do the topics of the previous lectures, grand unification and low-energy supersymmetry, fit in the string context? As we mentioned already, some of the grand unification ideas are also present in four-dimensional strings. Thus there are bigger gauge groups [E_8 in $D = 10$ or maybe even some GUT like $SU(5)$, E_6 or $SO(10)$ in some models] although the way in which they appear is more sophisticated. Some models (but not all) maintain the standard predictions for $\sin^2\theta_W$. In general quark-lepton mass relations disappear and the Yukawa couplings are given by correlation functions between the conformal fields ("vertices") associated to each particle. Concerning low-energy SUSY, it fits quite naturally inside the four-dimensional string context. $N = 1$ supersymmetry in $D = 4$ guarantees the absence of tachyons in the string. Thus this provides us with another argument (independent of the use for solving the "naturality" problem) in favour of the existence of a space-time supersymmetry in four dimensions. Furthermore, the existence of a "hidden sector" in which SUSY-breaking may occur appears automatically in string models. A natural possibility for the SUSY-breaking mechanism is a dynamical SUSY-breaking triggered by condensation of the gauginos[99] of the hidden gauge sector (i.e., $\langle\lambda\lambda\rangle \neq 0$). The size of this breaking can be hierarchically small compared to the Planck mass and may explain why the weak scale is so small compared to the Planck mass. Thus one would expect as the low-energy limit of a four-dimensional string something similar to the supersymmetric standard model described in the previous lecture. Of course the possibility exists of having some extra particles like new gauge interactions or heavy leptons, etc. The experimental detection of these hypothetical particles would be very important. Meanwhile, unfortunately, there is no serious string argument which predicts the existence of these new states.

6.   FINAL COMMENTS

Four-dimensional strings seem to provide us with the first really unified theories of all interactions including gravity. Unfortunately strings are theories which reveal themselves at energies of the order of the Planck mass, $\sim10^{18}$ GeV and we can only guess at their existence (if they are there) indirectly. Since direct experimental tests of strings seem out of question the risk exists of getting very little progress (if any) in the complete construction of the theory. Past experience tells us that experimental input has played a key rôle in finding the correct field theory of interactions amongst elementary particles. However, string theories are so far superior to the previously considered unifying schemes that one just cannot throw them away in the dustbin. An effort must be made both in understanding the theory better and trying to make contact with observed physics.

Fortunately enough, the prospects for low-energy SUSY to be checked experimentally are much brighter since new particles within the reach of present or future accelerators are predicted. Certainly, if SUSY particles are not found at supercollider energies, low-energy supersymmetry would be ruled out. If SUSY particles are found the detailed study of their properties would give us precious information about the SUSY breaking process and, maybe, about Planck mass physics.

ACKNOWLEDGEMENTS

I warmly thank Tom Ferbel for the nice atmosphere created at the School. Special thanks go to Captain Guido Martinelli whose skills as navigator were clearly demonstrated in our attempted trip to Buck island. His providential command saved us from disaster. I also thank C. Albajar and F. Pauss for providing me with some experimental information.

REFERENCES

1. For a review, see:
   P. Langacker, Physics Reports 72C:185 (1981);
   R. Slanski, Physics Reports 79:1 (1981);
   G.G. Ross, "Grand Unified Theories", Benjamin Inc. (1984).
2. J. Wess and J. Bagger, "Supersymmetry and Supergravity", Princeton University Press, Princeton (1983);
   P. West, "Introduction to Supersymmetry and Supergravity", World Scientific (1986).
   For the phenomenology of supersymmetry, see:
   H.P. Nilles, Physics Reports C10:1 (1984);
   H. Haber and G. Kane, Physics Reports C117:75 (1985);
   R. Barbieri and S. Ferrara, Surveys in High Energy Physics 4:33 (1983).
3. M.B. Green, J.H. Schwarz and E. Witten, Superstring Theory, Vols. I and II, Cambridge University Press (1986);
   D. Gross, in Proceedings of the 1986 ASI School (Virgin Islands). Plenum Press (1987).
   J. Schwarz, "Introduction to Superstrings", CALT-68-1290 (1985);
   M. Peskin, "Introduction to String and Superstring Theory", SLAC-PUB-4251 (1987).
4. S. Adler, Phys.Rev. Vol. 177:2426 (1969);
   J.S. Bell and R. Jackiw, Nuovo Cimento 60A:47 (1969).
5. L. Alvarez-Gaumé and E. Witten, Nucl.Phys. B234:269 (1983).
6. See, e.g.:
   M. Luscher and P. Weisz, Nucl.Phys. B295:65 (1988);
   A. Hasenfratz et al., Phys.Lett. 199:531 (1987);
   M. Grzadowski and M. Lindner, Phys.Lett. 178B:81 (1986).
7. E. Gildener, Phys.Rev. D14:1667 (1976).
8. G. 't Hooft, in "Recent Developments in Gauge Theories", ed. by G. 't Hooft et al., Plenum Press, New York (1980), p. 135.
9. S. Weinberg, Phys.Rev. D13:974 (1976) and D19:1277 (1979);
   L. Susskind, Phys.Rev. D20:2619 (1979);
   E. Farhi and L. Susskind, Physics Reports 74C:277 (1981).
10. S. Dimopoulos and J. Ellis, Nucl.Phys. B182:505 (1981);
    See, however:
    S. Dimopoulos, H. Georgi and S. Raby, Phys.Lett. 127B:101 (1983);
    B. Holdom, Phys.Rev. D24:1441 (1981);
    T. Appelquist and L.C.R. Wigewardhana, Phys.Rev. D35:774 (1987).
11. L. Abbot and E. Farhi, Nucl.Phys. B189:547 (1981);
    M. Claudson, E. Farhi and R. Jaffe, Phys.Rev. D34:873 (1986).
12. M. Chanowitz and M.K. Gaillard, Nucl.Phys. B261:379 (1985);
    M. Chanowitz, M. Golden and H. Georgi, Phys.Rev. D36:149 (1987);
    Phys.Rev.Lett. 57:2344 (1986).
13. H. Georgi and S.L. Glashow, Phys.Rev.Lett. 32:32 (1974).
14. J. Pati and A. Salam, Phys.Rev. D10:275 (1974).
15. H. Georgi, H. Quinn and S. Weinberg, Phys.Rev.Lett. 33:451 (1974).
16. H. Georgi, unpublished;
    H. Fritzsch and P. Minkowski, Ann.Phys. 93:193 (1975);
    H. Georgi and D.V. Nanopoulos, Nucl.Phys. B155:59 (1979).
17. H. Georgi and D.V. Nanopoulos, Nucl.Phys. B159:16 (1979);
    F. Del Aguila and L.E. Ibañez, Nucl.Phys. B177:60 (1981).
18. F. Gursey and M. Serdaroglu, Nuovo Cimento 65A:337 (1981) and references therein;
    Y. Achiman and B. Stech, Phys.Lett. 77B:389 (1978);
    P. Ramond, in Proceedings of Sanibel Symposium (1979);
    R. Barbieri and D.V. Nanopoulos, Phys.Lett. 91B:369 (1980).
19. A.J. Buras, J. Ellis, M.K. Gaillard and D.V. Nanopoulos, Nucl.Phys. B195:66 (1978).
20. See, e.g.: A. Savoy-Navarro, in Proceedings of ASI Summer School (1986), Plenum Press (1987).

21. M. Gell-Mann, P. Ramond and R. Slansky, unpublished (1979);
    T. Yanagida, Proceedings Workshop on the Unified Theory and the Baryon Number in the Universe, KEK, Japan (1979).

22. See, e.g.: G.G. Ross, Proceedings of the 1987 International Symposium on Lepton and Photon Interactions at Higher Energies, DESY, Hamburg (1987), CERN Preprint TH. 4881 (1987);
    See also: A. Savoy-Navarro in Ref. 20.

23. G. Blewilt et al., Phys.Rev.Lett. 54:22 (1985).

24. L.E. Ibañez, Nucl.Phys. B181:105 (1981).

25. W. Marciano and A. Sirlin, Phys.Rev. D22:2095 (1980) and Nucl.Phys. B189:442 (1981);
    C.H. Llewellyn Smith, G.G. Ross and J.F. Wheater, Nucl.Phys. B177:263 (1981);
    G. Costa et al., CERN Preprint TH. 4675 (1987), LBL-23271 (1987);
    U. Amaldi et al., Phys.Rev. D36:1385 (1987).

26. Y.A. Gol'fand and E.P. Likhtman, JETP Lett. 13:   (1971);
    D.V. Volkov and V.P. Akulov, JETP Lett. 16:438 (1972);
    J. Wess and B. Zumino, Nucl.Phys. B70:39 (1974).

27. M. Veltman, Acta Phys.Polon. B12:437 (1981);
    L. Maiani, Proceedings Summer School of Gif-Sur-Yvette (Paris 1980);
    E. Witten, Nucl.Phys. B188:513 (1981).

28. A. Salam and J. Strathdee, Nucl.Phys. B76:477 (1974);
    S. Ferara, J. Wess and B. Zumino, Phys.Lett. 51B:239 (1974);
    S. Gates, M. Grisaru, M. Rocek and W. Siegel, "Superspace", Frontiers in Physics, Benjamin Inc. (1983).

29. P. Fayet and J. Iliopoulos, Phys.Lett. 51B:461 (1974);
    L. O'Raifeartaigh, Nucl.Phys. B96:331 (1975);
    P. Fayet, Phys.Lett. 58B:67 (1975).

30. L. Girardello and M. Grisaru, Nucl.Phys. B194:65 (1982);
    K. Harada and N. Sakai, Progr.Theor.Phys. 67 (1982) 1877.

31. P. Fayet, Phys.Lett. 69B:489 (1977); 84B:416 (1979); 78B:417 (1978);
    G. Farrar and P. Fayet, Phys.Lett. 79B:442 (1978); 89B:191 (1980).

32. H. Haber and G. Kane, Physics Reports C117:75 (1985).

33. S. Dimopoulos and H. Georgi, Nucl.Phys. B93:150 (1981);
    N. Sakai, Z.Phys. C11:153 (1982).

34. S. Dimopoulos, S. Raby and F. Wilczek, Phys.Rev. D24:1681 (1981);
    L.E. Ibañez and G.G. Ross, Phys.Lett. 105B:439 (1982);
    M.B. Einhorn and D.R.T. Jones, Nucl.Phys. B196:475 (1982).

35. L.E. Ibañez, Phys.Lett. 126B:196 (1983).

36. S. Weinberg, Phys.Rev. D26:287 (1982);
    N. Sakai and T. Yanagida, Nucl.Phys. B197:533 (1982);
    S. Dimopoulos, S. Raby and F. Wilczek, Phys.Lett. 112B:133 (1982);
    J. Ellis, D.V. Nanopoulos and S. Rudaz, Nucl.Phys. B202:43 (1982).

37. L. Hall and M. Suzuki, Nucl.Phys. B231:419 (1984);
    I.H. Lee, Nucl.Phys. B246:120 (1984).

38. F. Zwirner, Phys.Lett. B132:103 (1983);
    R. Barbieri and A. Masiero, Nucl.Phys. B267:679 (1986);
    S. Dimopoulos and L. Hall, Phys.Lett. B196:135 (1987).

39. L.E. Ibañez, Phys.Lett. 118B:73 (1982); Nucl.Phys. B218:514 (1983).

40. R. Barbieri, S. Ferrara and C. Savoy, Phys.Lett. 119B:343 (1982).

41. P. Nath, R. Arnowitt and A. Chamseddine, Phys.Rev.Lett. 49:970 (1982).

42. H.P. Nilles, M. Srednicki and D. Wyler, Phys.Lett. 120:346 (1982).

43. R. Barbieri and S. Ferrara, Surveys in High Energy Physics 4:33 (1983);
    E. Cremmer, S. Ferrara, L. Girardello and A. Van Proeyen, Nucl.Phys. B212:413 (1983).

44. H.P. Nilles, Phys.Lett. 115B:193 (1982).

45. S. Dimopoulos and S. Raby, Nucl.Phys. B219:479 (1983);
    M. Dine and W. Fischler, Nucl.Phys. B204:346 (1982);
    J. Polchiski and L. Susskind, Phys.Rev. D26:3661 (1982).

46. K. Inoue et al., Progr.Theor.Phys. 67:1859 (1982).
47. L.E. Ibañez and C. Lopez, Phys.Lett. 126B:54 (1983); Nucl.Phys.
     B233:511 (1984).
48. L. Alvarez-Gaumé, J. Polchinski and M. Wise, Nucl.Phys. B221:495
     (1983).
49. J. Ellis, J. Hagelin, D.V. Nanopoulos and K. Tamvakis, Phys.Lett.
     125B:275 (1983).
50. K. Inoue et al., Progr.Theor.Phys. 68:927 (1982).
51. L.E. Ibañez and G.G. Ross, Phys.Lett. 110B:227 (1982).
52. L.E. Ibañez, C. Lopez and C. Muñoz, Nucl.Phys. B256:218 (1985).
53. L.E. Ibañez, Phys.Lett. 137B:160 (1984);
     J. Hagelin, G. Kane and S. Raby, Nucl.Phys. B241:638 (1984).
54. C. Muñoz, Ph.D. Thesis, Universidad Autonoma de Madrid (1986),
     unpublished;
     L.E. Ibañez and C. Muñoz, Paper in preparation (1989).
55. R. Flores and M. Sher, Ann.Phys. 148:95 (1983);
     H.P. Nilles and M. Nusbaumer, Phys.Lett. 145B:73 (1984);
     P. Majumdar and P. Roy, Phys.Rev. D30:2432 (1984).
56. J. Ellis and G.G. Ross, Phys.Lett. 117B:397 (1982);
     A. Chamseddine, P. Nath and R. Arnowitt, Phys.Lett. 129B:445 (1983);
     P. Dias, S. Nandi and X. Tata, Phys.Lett. 129B:451 (1983).
57. S. Weinberg, Phys.Rev.Lett. 50:387 (1983);
     B. Grinstein, J. Polchinski and M. Wise, Phys.Lett. 130B:285 (1983).
58. R. Barbieri and G.F. Giudice, Nucl.Phys. B296:75 (1988).
59. J. Ellis and D.V. Nanopoulos, Phys.Lett. 110B:44 (1982);
     R. Barbieri and R. Gatto, Phys.Lett. 110B:211 (1982);
     T. Inami and C.S. Lim, Nucl.Phys. B207:593 (1982);
     M. Suzuki, Phys.Lett. 115B:46 (1982);
     M. Duncan, Nucl.Phys. B221:285 (1983);
     J. Donoghue, H.P. Nilles and D. Wyler, Phys.Lett. 128B:55 (1983).
60. M. Davier, Proceedings 23rd International Conference on High-Energy
     Physics, Berkeley (1986), ed. S.C. Loken, World Scientific (1987),
     p. 25;
     S. Whitaker, ibid., p. 602.
     F. Pauss, CERN EP/88-158 (1988).
61. G. Arnison et al. (UA1 Collaboration), Europhys.Lett. 1:327 (1986);
     C. Albajar et al. (UA1 Collaboration), CERN preprint CERN-EP/88-168
     (1988).
62. C. Hearty et al. (ASP Collaboration), Phys.Rev.Lett. 58:1711 (1987).
63. E. Fernandez et al. (MAC Collaboration), Phys.Rev.Lett. 54:1118
     (1983).
64. G. Kane and J. Leveillé, Phys.Lett. 112B:227 (1982);
     P. Harrison and C.H. Llewellyn Smith, Nucl.Phys. B213:223 (1983);
     M.J. Herrero, L.E. Ibañez, C. Lopez and F.J. Yndurain, Phys.Lett.
     132B:199 (1983); 145B:430 (1984);
     E. Reya and P. Roy, Phys.Lett. 141B:442 (1984); Phys.Rev. D32:645
     (1985);
     J. Ellis and H. Kowalski, Phys.Lett. 142B:441 (1984); Nucl.Phys.
     B259:109 (1985).
65. C. Albajar et al. (UA1 Collaboration), Phys.Lett. B198:261 (1987);
     See also: F. Pauss, CERN-EP/88-158 (1988) and references therein.
66. R. Ansari et al. (UA2 Collaboration), Phys.Lett. B195:613 (1987).
67. H. Goldberg, Phys.Rev.Lett. 50:1419 (1983);
     J. Ellis, J. Hagelin, D.V. Nanopoulos and M. Srednicki, Phys.Lett.
     127B:233 (1983);
     J. Ellis et al., Nucl.Phys. B238:453 (1984).
68. S. Dimopoulos and L. Hall, Phys.Lett. B207:216 (1988).
69. L. Hall, LBL Preprint LBL-26059 (1988).
70. A. Polyakov, Phys.Lett. 103B:207 (1981).
71. A. Belavin, A. Polyakov and A. Zamolodchikov, Nucl.Phys. B241:333
     (1984).

72. D. Friedan, E. Martinec and S. Shenker, Nucl.Phys. B271:93 (1986).

73. V. Alessandrini, D. Amati, M. Le Bellac and D. Olive, Physics Reports 1C:170 (1971);
S. Mandelstam, Physics Reports 13C:259 (1974);
J. Scherk, Rev.Mod.Phys. 47:123 (1975).

74. P. Ramond, Phys.Rev. D3:2415 (1971);
A. Neveu and J. Schwarz, Nucl.Phys. B31:86 (1971); Phys.Rev. D4:1109 (1971).

75. F. Gliozzi, D. Olive and J. Scherk, Nucl.Phys. B122:253 (1977).

76. M. Green and J. Schwarz, Phys.Lett. 136B:367 (1984).

77. D. Gross, J. Harvey, E. Martinec and R. Rohm, Nucl.Phys. B256:468 (1985) and B267:75 (1986).

78. M.B. Green and J.H. Schwarz, Nucl.Phys. B255:93 (1985); Phys.Lett. 149B:117 (1984).

79. L. Dixon, J. Harvey, C. Vafa and E. Witten, Nucl.Phys. B261:651 (1985); Nucl.Phys. B274:285 (1986).

80. L.E. Ibañez, CERN Preprint TH. 4769 (1987), to be published in the Proceedings of the XVIIIth International GIFT Seminar on Theoretical Physics, World Scientific, Singapore (1988).

81. P. Candelas, G. Horowitz, A. Strominger and E. Witten, Nucl.Phys. B258:46 (1985);
P. Green, T. Hübsch and C.A. Lütken, CERN Preprint TH. 4933 (1987), to appear in Class.Quant.Gravity.

82. D. Gepner, Nucl.Phys. B296:757 (1988); Phys.Lett. 199B:380 (1987).

83. A. Font, L.E. Ibañez and F. Quevedo, CERN Preprint TH. 5217 (1988), to appear in Phys.Lett. B.

84. T. Eguchi, H. Ooguri, A. Taormina and S. Yang, Preprint UT-536 (1988);
K. Li and N. Warner, CERN Preprint TH. 5047 (1988);
B. Greene, C. Vafa and N. Warner, HUTP-88/A047 (1988).

85. L.E. Ibañez, J. Mas, H.P. Nilles and F. Quevedo, Nucl.Phys. B301:157 (1988).

86. S.T. Yau, in Proceedings of the Argonne Symposium on Anomalies, Geometry and Topology, World Scientific, Singapore (1985).

87. B. Greene, K. Kirklin, P. Miron and G. Ross, Nucl.Phys. B278:667 (1986); B279:574 (1986).

38. C.A. Lütken and G.G. Ross, CERN Preprint TH. 5058 (1988);
M. Lynker and R. Schimmirgk, Texas Preprint UTTG-17-88 (1988).

89. L.E. Ibañez, H.P. Nilles and F. Quevedo, Phys.Lett. B187:25 (1987).

90. L. Dixon, Ph.D. Thesis, Princeton (1985), unpublished.

91. L.E. Ibañez, J.E. Kim, H.P. Nilles and F. Quevedo, Phys.Lett. B191:282 (1987).

92. W. Lerche, A.N. Schellekens and N. Warner, CERN Preprint TH. 5155 (1988), to appear in Physics Reports.

93. I. Antoniadis, CERN Preprint TH. 5199 (1988), and references therein.

94. K. Narain, M. Sarmadi and C. Vafa, Nucl.Phys. B288:951 (1987).

95. A. Font, L.E. Ibañez, H.P. Nilles and F. Quevedo, Phys.Lett. B210:101 (1988).

96. J.A. Casas and C. Muñoz, Phys.Lett. 209B:214 (1988); Oxford-TP-32/88 (1988); CERN Preprint TH. 5052 (1988);
M. Quiros, IEM-Madrid Preprint (1988), and references therein.

97. I. Antoniadis, J. Ellis, J. Hagelin and D.V. Nanopoulos, CERN Preprint TH. 4931 (1987), 5005 (1988).

98. A. Font, L.E. Ibañez, F. Quevedo and A. Sierra, in preparation.

99. J.P. Derendinger, L.E. Ibañez and H.P. Nilles, Phys.Lett. 155B:65 (1985);
M. Dine, R. Rohm, N. Seiberg and E. Witten, Phys.Lett. 156B:55 (1985).

Z° PHYSICS AT THE SLC*

JONATHAN M. DORFAN

Stanford Linear Accelerator Center

Stanford University, Stanford, California 94309

1. ABSTRACT

This report summarizes the material covered in two lectures given during the summer of 1988. The same two lectures were given at both the Banff Summer Institute and the Advanced Study Institute on Techniques and Concepts of High Energy Physics at St. Croix. In both cases the audience was a mixture of recent and soon-to-be Ph.D.'s. The original intent of these lectures was to discuss the early physics results from the MARKII program at the SLC. Given the delayed start of this program, the focus of the lectures was changed to discuss the detailed studies which have been carried out on how to extract the physics. To put the discussion of the measurements in context, a brief summary of the Standard Model is presented along with a discussion of the MARKII detector and the energy measurement spectrometers. In areas where there has been little or no change from previous write-ups which I have done,[1] I have lifted the text from these write-ups. However, where the techniques have improved, this report reflects those changes.

2. INTRODUCTION

The discovery at CERN[2] of the $Z°$ in $\bar{p}p$ collisions was a spectacular achievement. Background-free signals in both the $Z° \rightarrow e^+e^-$ and $Z° \rightarrow \mu^+\mu^-$ channels were seen which provided the first direct observation of the neutral weak force carrier. However, because of the difficulties inherent in the $\bar{p}p$ environment, detailed studies of the decays of the $Z°$, and hence, detailed studies of the weak interaction, were not possible. The main difficulties are two-fold (refer to Fig. 1):

---

* Work supported by the Department of Energy, contract DE–AC03–7600515.

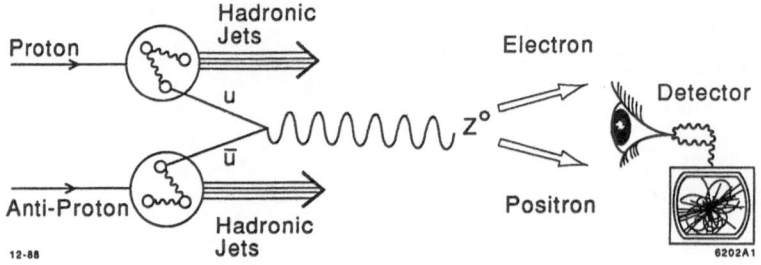

Fig. 1. A pictorial view of p̄p collision in which a u quark from the proton and a ū quark from the anti-proton combine to form a Z°. The remaining partons produce debris in the form of hadronic jets. The Z° is envisaged to decay to an electron-positron pair which will be distinctive enough to unravel them from the hadrons, thereby forming a tag for the Z°.

1. $Z°$ production is a small part of the total $\bar{p}p$ collision cross section. There is no way to "tune" the hard collision, constituent subenergy $\hat{s}$, to the $Z°$ mass. Rather one is at the mercy of the overlap of the distribution functions of the two partons to conspire to provide $\hat{s} = M_Z^2$.

2. Once a $Z°$ is produced it must be detected in the presence of the large hadronic debris which results from the partons which did not participate in the hard collision. The only practical method for beating down these large backgrounds is to tag the $Z°$ using its leptonic decay modes. This has the disadvantage of a small yield $(BF(Z° \rightarrow e^+e^-, \mu^+\mu^- = 3\%)$ and does not permit an unbiased and systematic study of all the decay modes of the $Z°$.

The collider results provide us with a relatively crude measurement of the $Z°$ mass:

$$M_{Z°} = 91.5 \pm 1.2 \pm 1.7 \ \ \text{GeV/c}^2 \ \ \text{(UA2)} \ \ ,$$

$$M_{Z°} = 93.1 \pm 1.0 \pm 3.1 \ \ \text{GeV/c}^2 \ \ \text{(UA1)} \ \ .$$

This measurement is systematics limited, the major problem being the lack of precise knowledge of the calorimeter energy scale (the measurement comes from the $Z° \rightarrow e^+e^-$ mode).

The next major step in improving our understanding of the $Z°$ will come from studies at the SLC and LEP which will provide a background-free data set of $Z°$ produced via e^+e^- collisions. This environment overcomes the problems of the $\bar{p}p$

colliders; one is able to tune the collision energy precisely to the $Z^\circ(\hat{s} \equiv s)$, Z° particle production totally dominates all other processes and Z°'s can be studied in an unbiased, debris-free environment.

3. THE STANDARD MODEL AND ITS APPLICATION TO $e^+e^- \to Z^\circ \to f\bar{f}$

For most of these lectures we will assume the Standard Model. When we look beyond the Standard Model, we will develop whatever formalism we need. The goal of this Section is not to be complete or detailed—but merely to build a foundation from which we can extract useful experimental tests at the Z°.

The Standard Model is characterized by the gauge group

$$\text{SU}(3)_{color} \wedge \text{SU}(2) \wedge \text{U}(1).$$

Leptons are pointlike particles which couple to the gauge bosons of SU(2) through their weak charge and to the photon of U(1) through their electric charge. There are six leptons e, μ, τ, and their zero mass partners ν_e, ν_μ, and ν_τ. There are six quarks u, d, s, c, b and t which carry color and there are three color states for each quark. Leptons have no color charge and are therefore "blind to the strong interaction.

The left-handed fermions are arranged in weak iso-doublets

$$\begin{pmatrix}\nu_e \\ e\end{pmatrix}_L \quad \begin{pmatrix}\nu_\mu \\ \mu\end{pmatrix}_L \quad \begin{pmatrix}\nu_\tau \\ \tau\end{pmatrix}_L \qquad \begin{matrix}T_3 = 1/2 \\ 1/2\end{matrix} \quad,$$

$$\begin{pmatrix}u \\ d'\end{pmatrix}_L \quad \begin{pmatrix}c \\ s'\end{pmatrix}_L \quad \begin{pmatrix}t \\ b'\end{pmatrix}_L \qquad \begin{matrix}T_3 = 1/2 \\ -1/2\end{matrix} \quad,$$

where T_3 is the 3rd component of the weak charge. The primes on the quarks indicate that flavor conservation in the quark sector is not perfect. This generation mixing can be summarized by the elements of the Kobayashi–Maskawa matrix—the most familiar component being the Cabibbo angle which tells us that the d quark has $\sim 5\%$ strange quark admixture. More succinctly, in the quark sector the weak eigenstates are related by a rotation matrix to the mass eigenstates. We notice in passing, the peculiarity of the three generations; the ν_e, e, u and d being the members of the lightest generation. The Standard Model does not explain why nature chooses to replicate itself in this peculiar manner.

Right-handed fermions appear in singlets, u_R, d_R ...t_R, e_R, μ_R, τ_R and, since the ν's are massless, there are no right-handed ν's. $T_3 = 0$ for all right-handed fermions.

There are nine massless bosons in the Standard Model—eight gluons and the photon. There are three massive vector bosons W^+, W^- and Z^0 and, in the minimal model with one Higgs doublet, there is one neutral scalar, H^0. Gluons carry color (unlike photons which don't carry charge) and hence $SU(3)_{\text{color}}$ is non–Abelian. Since gluons carry color, they can couple to other gluons. The polarization of the QCD vacuum by virtual quark and gluon pairs results in an *anti-screening* of color charge. This can be contrasted with the *screening* of electric charge by virtual e^+e^- pairs in QED. This anti-screening leads to the notion of confinement of quarks and the decrease of the strong coupling constant, α_s, with increasing q^2. Free quarks should not be seen, and this notion will be tested at the Z^0 although not discussed further in these lectures.

The Standard Model does not predict masses for the fundamental particles. The $W^\pm$, Z^0 masses, ignoring electroweak radiative effects, are given in terms of the parameter $\sin^2 \theta_w$:

$$M_W^2 = \frac{\pi \alpha}{\sqrt{2} G_F} \left(\frac{1}{\sin^2 \theta_w} \right)$$

$$M_{Z^0}^2 = \frac{M_W^2}{\cos^2 \theta_w} = \frac{\pi \alpha}{\sqrt{2} G_F} \left(\frac{1}{\sin^2 \theta_w \cos^2 \theta_w} \right) \quad,$$

where α is the fine structure constant and G_F is the Fermi coupling constant. The H^0 mass is expected to fall in the range $7.5 \lesssim M_{H^0} \lesssim 10^3$ GeV. This, however, is of no consolation to the experimentalist searching for the H^0. The presence of the neutral Higgs is crucial to the success of the Standard Model.

There is nothing fundamental about the minimal Higgs scheme. The model is constrained by the measured value of $\rho = 1$ to contain explicitly doublets (as opposed to triplets). A non-minimal model with two doublets (eight fields) is perfectly acceptable. In this case, three of the fields are needed to provide mass for the $W^\pm$ and Z^0 leaving five physical Higgs particles. These are two neutral scalars, H_1^o and H_2^o; one pseudoscalar h^o (often called an axion); and two charged pseudoscalars H^+ and H^-. In such a model, the lower mass limit of 7.5 GeV/c^2 no longer pertains.

The electroweak interactions of all the gauge fields are specified by the model and are determined by e, the electric charge, and one free parameter θ_w. Spinors couple to the photon field with strength e and to the Z^0 with strength

$$-\frac{e}{\sin \theta_w \cos \theta_w} \left(T_3^{R/L} - Q \sin^2 \theta_w \right) = 2\sqrt{2} \left(\frac{M_Z^2 G_F}{\sqrt{2}} \right)^{1/2} \left(T_3^{R/L} - Q \sin^2 \theta_w \right)$$

where R/L indicates left and right couplings and Q is the charge of the fermion.

120

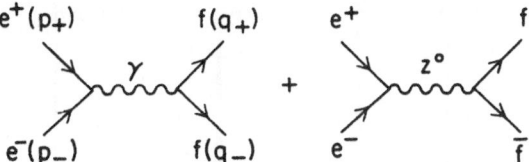

Fig. 2. The basic $e^+e^- \rightarrow \gamma$, Z° process.

Aside from Higgs and fermion masses, the Electroweak theory is totally specified if we know α, G_F and M_Z; α and G_F are extremely accurately known (better than one part in 10^5), whereas M_Z is known only to about 2%. A precise measurement of M_Z will constrain considerably the Standard Model.

For almost all the physics discussed in these lectures, we are interested in the basic process $e^+e^- \rightarrow f\bar{f}$ where the symbol f signifies a fundamental fermion, either a quark or a lepton. There are two processes which contribute to the cross section as shown in Fig. 2, namely $e^+e^- \rightarrow \gamma \rightarrow f\bar{f}$ and $e^+e^- \rightarrow Z^0 \rightarrow f\bar{f}$. The Standard Model specifies all the couplings and hence the cross section for these processes can be calculated. If θ is the fermion polar angle, the differential cross section has the form[3]

$$\frac{d\sigma_{f\bar{f}}}{d\cos\theta} = \frac{\pi\alpha^2 Q_f^2 D}{2s} (1 + \cos^2\theta) - \frac{\alpha Q_f D G_F M_z^2(s - M_z^2)}{8\sqrt{2}[(s - M_z^2)^2 + M_z^2\Gamma_z^2]}$$

$$[(R_e + L_e)(R_f + L_f)(1 + \cos^2\theta) + 2(R_e - L_e)(R_f - L_f)\cos\theta]$$

$$+ \frac{DG_F^2 M_z^4 s}{64\pi[(s - M_z^2)^2 + M_z^2\Gamma_z^2]}$$

$$[(R_e^2 + L_e^2)(R_f^2 + L_f^2)(1 + \cos^2\theta) + 2(R_e^2 - L_e^2)(R_f^2 - L_f^2)\cos\theta] \quad ,$$

(1)

where Q_f is the fermion charge, $s = E_{c.m.}^2$, M_Z the mass of the Z^0 and D takes into account the number of color degrees of freedom. For $f \equiv$ quark, $D = 3$; otherwise $D = 1$. The left- and right-handed weak coupling constants are given by

$$L_f = T_3^f - Q_f \sin^2\theta_w$$

$$R_f = -Q_f \sin^2\theta_w \quad .$$

The three terms in the cross section are the purely electromagnetic contribution, the interference between the weak and electromagnetic diagrams and the purely weak con-

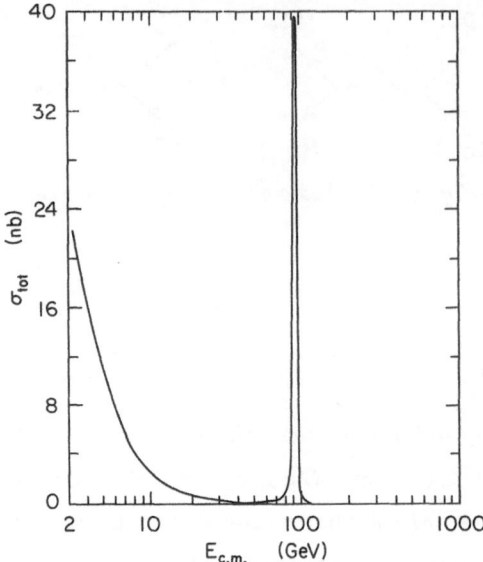

Fig. 3. The cross section for $e^+e^- \to \gamma,\ Z^\circ \to f\bar{f}$ as calculated in the Standard Model. The $c\bar{c}$ and $b\bar{b}$ threshold behavior are omitted from the plot.

tribution. Notice that (a) the interference term disappears at $\sqrt{s} = M_Z$ as it should (b) the first term is just the point QED differential cross section and (c) at $\sqrt{s} = M_Z$, the purely weak term dominates.

It is illustrative to integrate over $\cos\theta$ and plot the cross section as a function of $E_{c.m.} = \sqrt{s}$. This is shown in Fig. 3 for hadronic final states. One sees that below the region of the Z^0 mass, the purely electromagnetic cross section dominates as is reflected by the $E_{c.m.}^{-2}$ behavior. On the Z° pole however the weak cross section dominates, providing 10^3 times more particle production than the electromagnetic process. This is part of the magic of running at the Z^0—the Z° provides an enormous enhancement in event rate over running in the continuum (*i.e.*, off resonance). Studying e^+e^- interactions at ~ 92 GeV in the absence of the Z° with the SLC or LEP would be ex- tremely painful if not in many cases impossible. The presence of the Z°, however, renders these relatively low luminosity machines capable of very high event rates.

Let us now return to $d\sigma/d\cos\theta$ and consider running at $s = M_Z^2$, namely on the Z° pole. Changing notation to axial and vector coupling constants

$$a = \frac{1}{2}\,(L - R) \quad \text{and} \quad v = \frac{1}{2}\,(L + R) \quad ,$$

one finds $L = a + v$, $R = v - a$, $L^2 + R^2 = 2(a^2 + v^2)$, $L^2 - R^2 = 4av$ and

$$\frac{d\sigma_{f\bar{f}}}{d\cos\theta} = \frac{DG_F^2 M_Z^4}{16\pi\Gamma_Z^2} \left[(a_e^2 + v_e^2)(a_f^2 + v_f^2)(1 + \cos^2\theta) + 8a_e v_e a_f v_f \cos\theta\right] \quad . \qquad (2)$$

It is useful to tabulate the couplings and the sum of their squares. Assuming $\sin^2\theta_W = 0.23$ (which we will do throughout for convenience) we find the values in Table I.

Table I

	Q	T_3	a	v	$a^2 + v^2$
e, μ, τ	-1	$-1/2$	$-1/2$	$-.04$	$.2516$
ν_e, ν_μ, ν_τ	0	$1/2$	$1/2$	$1/2$	$1/2$
d, s, b	$-1/3$	$-1/2$	$-1/2$	$-.347$	$.370$
u, c, t	$+2/3$	$+1/2$	$1/2$	$.193$	$.287$

We turn our attention back to Eq. (2). The term linear in $\cos\theta$ contributes a front–back asymmetry, A_{F-B}. $A_{F-B} \propto v_e v_f$ which, for charged leptons, is a very small number. However, a measurement of A_{F-B} for charged leptons has great sensitivity to $\sin^2\theta_w$ as we will see later in this section. Since $\int_0^\pi \cos\theta d\theta = 0$, the term linear in $\cos\theta$ does not contribute to the total cross section.

Integrating the term in $(1 + \cos^2\theta)$ yields the total cross section for producing a final state $f\bar{f}$ at the Z^0 :

$$\sigma_{f\bar{f}} = \frac{DG_F^2 M_Z^4}{6\pi\Gamma_Z^2} (v_e^2 + a_e^2)(v_f^2 + a_f^2) \quad .$$

We omit here the derivation of Γ_Z, but note that

$$\Gamma_Z = \frac{G_F M_Z^3}{24\sqrt{2}\pi} \sum_i (v_i^2 + a_i^2)D_i \quad , \qquad (3)$$

where i ranges over all fundamental fermions and D_i is the color factor (three for quarks, one for leptons). We can obtain σ_{point}, which is the lepton point QED cross section, from the first term in Eq. (1):

$$\sigma_{\text{point}} = \frac{\pi\alpha^2}{2s} \int (1 + \cos^2\theta) \, d\cos\theta$$

$$= \frac{4\pi\alpha^2}{3s} \simeq \frac{87 \text{ nb}}{s} \quad .$$

Hence, we can write

$$R_{f\bar{f}} = \frac{\sigma_{f\bar{f}}}{\sigma_{\text{point}}} = \frac{D(a_f^2 + v_f^2)(a_e^2 + v_e^2)}{16\alpha^2(1 - 2x_w + 8x_w^2/3)^2} \quad . \tag{4}$$

where $x_w = \sin^2\theta_w$.

Assuming five quarks and $x_w = 0.23$, we find the total cross section

$$\sigma_{tot} = \sum_f \sigma_{f\bar{f}} = \frac{DG_F^2 M_Z^4}{6\pi\Gamma_Z^2}(v_e^2 + a_e^2)\sum_f(v_f^2 + a_f^2)$$

$$= 47 \ nb \ \text{(no radiative corrections)}$$

$$= 37 \ nb \ \text{(with radiative corrections)}.$$

Table II shows the R values (not radiatively corrected) and the branching fraction for each process. We notice that the $Z°$ decays predominantly to hadrons ($BF(Z° \rightarrow hadrons) = 72\%$), and that the weak interaction (modulo the color factor) is rather democratic. One therefore produces roughly equal amounts of $Z° \rightarrow b\bar{b}$ and $Z° \rightarrow c\bar{c}$, in contrast to the electromagnetic interaction where quark production rates go like the electric charges squared and, hence, one gets four times as much $c\bar{c}$ production as $b\bar{b}$ production.

Table II

CHANNEL $(f\bar{f})$	$R_{f\bar{f}}$	$\Gamma_{f\bar{f}}/\Gamma_{Z°}$ (%)
each $\nu\bar{\nu}$	319	6.1
$\mu^+\mu^-$, $\tau^+\tau^-$, e^+e^{-*}	160	3.1
$u\bar{u}$, $c\bar{c}$, $t\bar{t}$	550	10.6
$d\bar{d}$, $s\bar{s}$, $b\bar{b}$	709	13.6

* We have ignored t channel diagrams which are only important at small values of θ.

The physics at the $Z°$ cannot be extracted without paying careful attention to the effects of radiative corrections. In lowest order, the $Z°$ line shape is a Breit–Wigner which is characterized by three parameters: mass, width and peak cross section. Radiative corrections, and most notably initial state $e^{\pm}$ Bremsstrahlung, will alter these quantities significantly. A complete treatment of this subject goes way beyond these lectures, but is covered very clearly and thoroughly in Ref. 4. Here we will simply

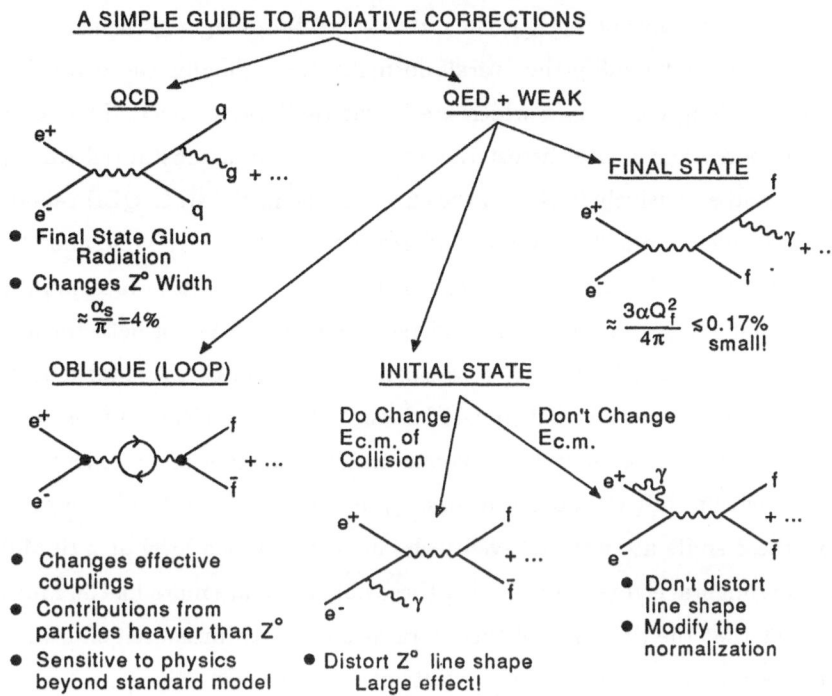

Fig. 4. A diagramatic guide to radiative corrections to $e^+e^- \to Z^\circ \to f\bar{f}$.

indicate the results as they effect the measurements and comment that, due to the considerable effort of the past few years (as summarized in Ref. 4), no measurements will be limited in precision by our present understanding of the radiative corrections. Other measurement errors will always dominate.

In the absence of a resonance, initial state Bremsstrahlung in e^+e^- collisions lowers the effective collision energy, which raises the cross section ($\sigma_{e^+e^-} \sim E_{cm}^{-2}$). Given a giant resonance like the Z°, one must convolute the Bremsstrahlung spectrum with the resonance line shape, altering significantly the resonance parameters. As discussed in Ref. 4 there are two classes of corrections to consider — electroweak and QCD. These are shown schematically in Fig. 4. QCD corrections occur, of course, only in final states involving hadronic production.

They are proportional to $\alpha_s/\pi \approx 4\%$ and directly effect the Z° width. Final state electromagnetic corrections are at the level of $3\alpha Q_f^2/4\pi < 0.17$ and can safely be ignored. Oblique corrections (see Fig. 4) contain internal loops of leptons and bosons and have the effect of changing the effective couplings. They can involve particles in the loops which are heavier than the Z°, and hence have sensitivity, in principle, to physics beyond the Standard Model. Initial state electromagnetic radiation can occur without a change in collision energy via vertex correction diagrams which do not distort the line shape but modify the overall normalization. Finally, there are the initial state Bremsstrahlung corrections which are by far the largest effect. They do change the collision energy and hence distort the $Z°$ line shape in a substantial way. As discussed above and extensively in Ref. 4, the effects of the initial state QED radiation are now very well understood. The two second-order calculations available and the exponentiated first-order calculation agree remarkably well. Generators incorporating this theoretical input are also available for understanding the effects of detector inadequacies. Figure 5 shows the $Z° \to \mu^+\mu^-$ cross section ($M_z = 93$ GeV/c^2) for four different calculations. One sees clearly that first-order would be insufficient, but the agreement between the two second-order and the exponentiated first-order are excellent. So, while the radiative effects shift the observed mass up by about 100 MeV/c^2 and raise Γ_z by about 3%, these shifts are now believed to be understood at a level of <10 MeV. The intrinsic experimental errors in measuring these quantities are more like 30 MeV; hence the statement that the precision of the $Z°$ mass and width measurements will not be subject to the present accuracy of the effects of radiative corrections.

Returning then to the predictions of the Standard Model, Γ_z is given by

$$\Gamma_z = \frac{G_F M_Z^3}{24\sqrt{2}\pi} \sum_f D_f(v_f^2 + a_f^2)(1 + \delta_f) \quad ,$$

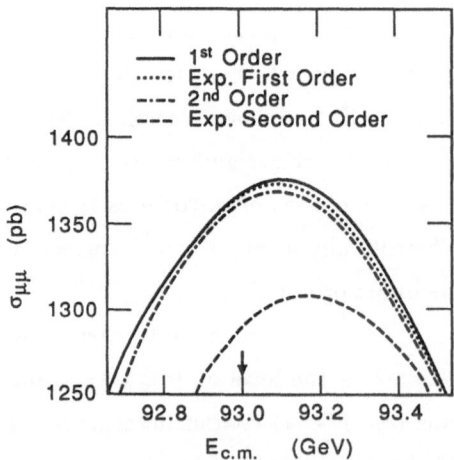

Fig. 5. The effect of electroweak radiative corrections on the $Z° \to \mu^+\mu^-$ cross section. Four different calculations are shown as indicated by the figure key.

where δ_f = radiative corrections; $\delta_f = 3\alpha Q_f^2 \ 4\pi$ for leptons, $= \alpha/\pi$ for quarks. Hence for five quarks, $\sin^2 \theta_w = 0.23$ and including the radiative corrections, the Standard Model predicts

$$\Gamma_z(5 \text{ quarks}) = 2.54 \ \text{GeV} \quad .$$

If Γ_z is significantly larger than this value, there must exist physics beyond the Standard Model. The bench mark for measuring increases in the width is the contribution of one massless ν, namely

$$\Gamma(Z^\circ \rightarrow \nu\bar{\nu}) = 160 \ \text{MeV} \quad .$$

Now, leptons or quarks produced at the Z° will necessarily be heavy given the limits from TRISTAN, PETRA and the $\bar{p}p$ collider at CERN. In this case—$m_f/M_z \approx 1$—we must take into account threshold effects. In a general way we can write

$$\frac{d\sigma_{f\bar{f}}}{d\theta} = f(\beta_f, \theta) \ \sigma(m_f = 0)$$

and

$$\sigma_{f\bar{f}} = f(\beta_f)\sigma(m_f = 0)$$

where β_f is the fermion velocity and m_f is the fermion mass. For *vector* couplings

$$f(\beta_f, \theta) = \frac{3}{16\pi} \ \beta_f[(1 + \cos^2 \theta) + (1 - \beta_f^2)\sin^2 \theta]$$

and

$$f(\beta_f) = \frac{1}{2} \ \beta_f(3 - \beta_f^2) \quad .$$

For *axial-vector* couplings

$$f(\beta_f, \theta) = \frac{3}{16\pi} \ \beta_f^3(1 + \cos^2 \theta)$$

and

$$f(\beta_f) = \beta_f^3 \quad .$$

Therefore, for the t quark with velocity β_t, the correct form of the contribution to the Z^0 width is [see Eq. (3)]

$$\Gamma(Z^0 \rightarrow t\bar{t}) = \frac{G_F M_Z^3}{8\sqrt{2}\pi} \left(v_t^2 \ \frac{1}{2} \ \beta_t(3 - \beta_t^2) + a_t^2\beta_t^3 \right) \quad .$$

Figure 6 shows the suppression of $t\bar{t}$ relative to a full strength (light) charge two-thirds quark as a function of the t quark mass. Since we know from TRISTAN that

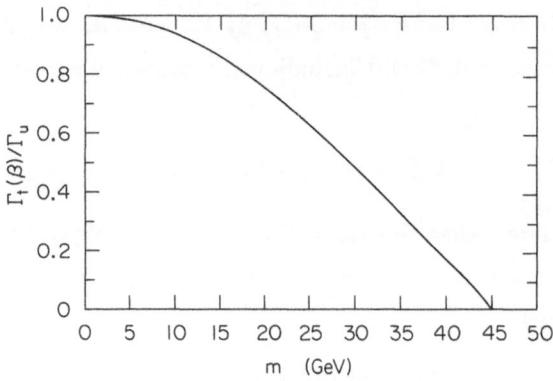

Fig. 6. The suppression factor of $t\bar{t}$ decays of the Z^0 as a function of M_t.

$M_t \gtrsim 27.5$ GeV/c^2, the $t\bar{t}$ final state at the Z^0 is suppressed at least to 0.6 of the $u\bar{u}$ rate.

Because the weak interaction is parity violating, the reaction $Z^\circ \to f\bar{f}$ has a forward–backward charge asymmetry. The number of f's produced at θ do not equal the number of f's produced at $\pi - \theta$. From the master Eq. (2) we see this manifested in the terms linear in $\cos\theta$, which lead to the asymmetry:

$$A^f_{F-B} = \frac{\int_0^1 \frac{d\sigma}{dz}dz - \int_{-1}^0 \frac{d\sigma}{dz}dz}{\int_0^1 \frac{d\sigma}{dz}dz + \int_{-1}^0 \frac{d\sigma}{dz}dz} = \frac{N^F_f - N^B_f}{N^F_f + N^B_f} \quad ,$$

where $N^F_f (N^B_f)$ is the number of fermions in the forward (backward) hemisphere relative to the incoming e^- direction. On the Z^0 pole

$$A^f_{F-B} = \frac{3a_e v_e a_f v_f}{(a_f^2 + v_f^2)(a_e^2 + v_e^2)} \quad ,$$

which is quadratic in the vector coupling. Since the vector coupling for the leptons is small, A_{F-B} is a very small number for the charged leptons. For μ-pairs, which is experimentally a very attractive channel,

$$A^\mu_{F-B} = \frac{3(1 - 4\sin^2\theta_w)^2}{4(1 - 4\sin^2\theta_w + 8\sin^4\theta_w)^2}$$

$$= 1.9\% \ \ for \ \ \sin^2\theta_w = 0.23, \ \ 4.2\% \ \ for \ \ \sin^2\theta_w = 0.22$$

However, despite the smallness of A^μ_{F-B}, it has considerable sensitivity to $\sin^2\theta_w$, namely

$$\frac{d\sin^2\theta_w}{\sin^2\theta_w} = \frac{1}{23}\frac{dA^{\mu}_{F-B}}{A^{\mu}_{F-B}}.$$

Hence, large errors in A^{μ}_{F-B} are significantly beaten down by the large factor of 23. Quark asymmetries are much larger because $v_q >> v_{\mu}$: in particular for $Z^{\circ} \rightarrow b\bar{b}$,

$$A^b_{F-B} = 11\% \quad \text{and} \quad \frac{d\sin^2\theta_w}{\sin^2\theta_w} = \frac{1}{12}\frac{dA^b_{F-B}}{A^b_{F-B}} \quad .$$

An additional problem in measuring charge asymmetries is the rapid dependence of the asymmetry on the collision energy E_{cm}, as shown in Fig. 7 for $Z^{\circ} \rightarrow \mu^+\mu^-$ and $Z^{\circ} \rightarrow b\bar{b}$. Hence, intrinsic errors in E_{cm} (expected to be ~ 30 MeV) contribute significantly to the error in A_{F-B}. For $Z^{\circ} \rightarrow \mu^+\mu^-$, $dA^{\mu}/dE_{cm} \approx 1\%/100$ MeV and for $Z^{\circ} \rightarrow b\bar{b}$, $dA^b/dE_{cm} \sim 0.2\%/100$ MeV. Again, the μ-pair channel is much more problematical than the $b\bar{b}$ final state.

This, then, summarizes the theoretical predictions of the Standard Model. Figure 8 is a cartoon depiction of what Standard Model Z° decays will look like in the detector. This will be the menu of *expected* events.

Despite its tremendous success, the Standard Model is sorely lacking as our ultimate theory. Presumably it is an excellent low energy approximation for the ultimate theory; few people believe that it will not be eclipsed.

There are many problems with the Standard Model. There are too many parameters (18), numbers which must be inserted by hand. It does not unify the forces, nor does it explain the pattern of masses or the presence of the generations. The Higgs mechanism is ad hoc and very unnatural, requiring exquisitely fine tuning to achieve its aim. These are but a few of the objections. It is most likely, that without further experimental clues, we will not make rapid progress in selecting the appropriate direction in which to depart from this model. Therefore, one of the major thrusts of the Z° studies will be to look for physics beyond the Standard Model. A large part of this write-up, then, is directed towards measurements which test the validity of the 3-generation Standard Model or which search for physics which is not directly contained in this model. The more pedestrian, but extremely rich, measurements which extend our knowledge of the presently known quarks and leptons are largely ignored.

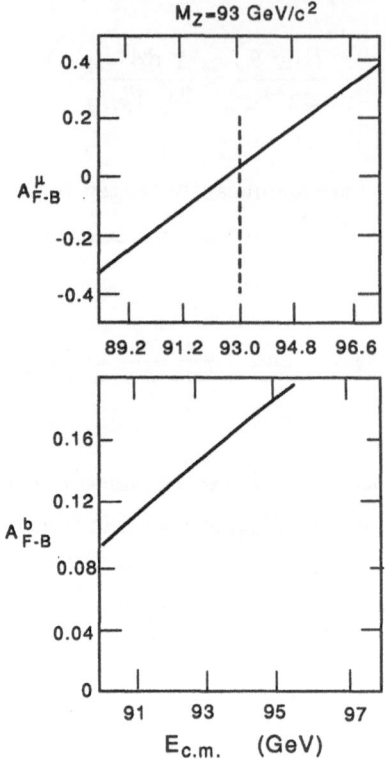

Fig. 7. The dependence of the foward backward asymmetry on E_{cm} and for $Z^\circ \to \mu^+\mu^-$ and $b\bar{b}$.

It should not escape the reader's attention that unprecedently large samples of quarks and leptons will be produced in Z° decays which will greatly enhance our knowledge about their properties.

As is well known, there do exist extensions and alternatives to the Standard Model. Typically, these models are aimed at a much more natural solution to the mass hierarchy problem. Supersymmetry (SUSY) and Technicolor are leading examples of such models, and it is entirely possible that these ideas are important pieces in the ultimate solution of the puzzle. These two models have a richer Higgs spectrum than the minimal Standard Model; both charged and neutral Higgs scalars must exist, if these ideas are correct. In addition, new constituents are predicted: sleptons and squarks for SUSY, technipions for Technicolor. With an increased particle spectrum then, these models are amenable to experimental verification.

What will be the role of the Z° in this quest for a better theoretical understanding?

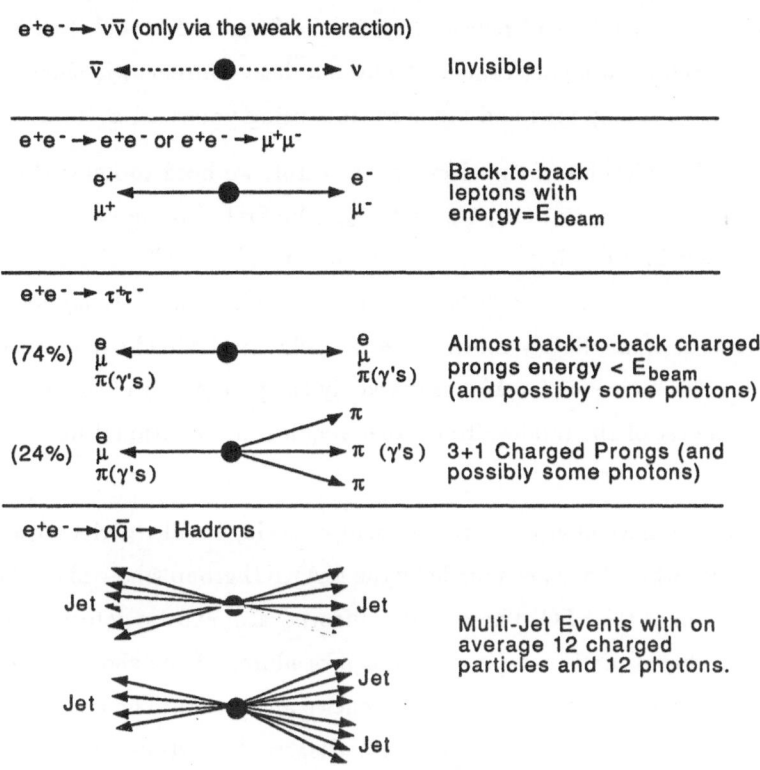

Fig. 8. A schematic representation of the primary expectations for Z° decays in the Standard Model.

1. We will be able to make much more precise tests of the electroweak sector of the Standard Model than are known today. Perhaps under such close scrutiny, chinks will begin to appear in the now formidable armor.

2. The top quark can be sought. While such a discovery would not violate anything in the Standard Model, a t quark mass as low as $M_z/2$ is getting very hard to accommodate in the parameter space of the 3-generation Standard Model. Recent measurements, particularly $B\bar{B}$ mixing, tend to indicate a top mass >50 GeV/c^2 in a 3-generation model.

3. The generation problem can be confronted, either by direct searches for 4th generation charged fermions or by neutrino counting. Neutral heavy leptons are also easily found if they have masses below $M_Z/2$.

4. The Higgs sector can be studied both in the neutral and charged domain.

5. Searches for indications of physics beyond the Standard Model (SUSY or Technicolor objects) can be performed. Additional heavy bosons will show up as loop corrections to tree graphs and are accessible using polarized e^- beams.

To pin down the predictions of the electroweak sector, we need to know three parameters which can be chosen to be α, G_F and M_z. The first two are known to exquisite accuracy (<1 part in 10^7), but M_z is only known to about 2%. Fortunately this is measured with great precision (<35 MeV/c^2 at the SLC) with relatively few Z°'s and hence the model is immediately much more severely constrained. It is worth noting that while such a measurement sharpens markedly the predictions of the model, it does not constitute a test of the model. To do that requires an additional measurement as discussed later.

To illustrate how physics at the Z° will precede, I will restrict myself almost entirely to studies which have been made by the MARKII group who will be the first detector group to run at the SLC. These studies incorporate realistic detector simulations (raw data is produced) and real data analysis procedures. They should therefore represent realistic measures of the expected efficiencies and errors. The analyses discussed here are more thoroughly covered in Ref. 5(d). Clearly these studies have applicability to the LEP experiments and SLD (the MARKII replacement detector), although in most cases these detectors will bring more powerful tools to bear on the problem.

4.   DETECTOR REQUIREMENTS; THE UPGRADED MARKII DETECTOR

The detector requirements must be well matched to the rigors of the environment. The Z° environment has been studied at great length and the interested reader can find summaries of these studies in Ref. 5. We mention here the essential elements as they pertain to the design of a Z° detector. Figure 9 summarizes the main spectrum of Z° decays which we are interested in studying, where the final state naming convention is given in the figure caption. We need to detect these final states efficiently and preserve the essential properties of the physics. We also need to monitor the integrated luminosity so that we can normalize the measurements. Precise measurement of the Z° mass and width require precise knowledge of the collision energy. For the SLC this is not provided by the machine and, hence, dedicated spectrometers had to be built. Characteristic decays of the unstable particles in Fig. 9 are given in Fig. 10. We notice that the environment is characterized by the presence of high energy jets, leptons, and missing energy. General detector requirements which follow then are:

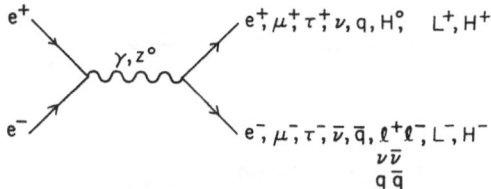

Fig. 9. The basic e^+e^- process where final states are produced via an intermediate photon or Z°. The notation is obvious except that q stands for a quark, H° the neutral Higgs scalar, $\ell^\pm$ a charged lepton, $H^\pm$ a charged Higgs scalar and $L^\pm$ a (new) heavy charged lepton.

1. The ability to measure the detailed properties of high multiplicity, high energy jets. These jets are typically comprised of 10 charged particles and 10 photons all contained within a cone of half angle 5°.

2. The ability to measure and identify electrons and muons over a wide range of momenta (1–50 GeV/c). These leptons must be tagged both in isolation and in the center of the dense jets.

3. The ability to measure as much of the collision energy as possible. Missing energy is a powerful tool for discovering new physics.

4. The ability to tag charm and bottom jets which is likewise important for the discovery and exploitation of new physics.

In addition, there are many event topologies involving multiple jet and multiple leptons which demands that one instrument the detector uniformly over a large solid angle. To satisfy these demands requires large solid angle tracking with multihit capability, large solid angle calorimetry, good hermeticity down to small ($\approx$50 mrad) angles relative to the beam axes, muon and electron coverage over a large solid angle with hadron rejection $> 10^3$ for momenta above 1 GeV/c and an excellent vertex detector placed at as small a radius as possible.

The MARKII detector was upgraded from its PEP configuration with an eye to satisfying as many of the above criteria as possible. The thrust of the MARKII program at the SLC was to begin with a well understood detector. The upgrades were made with sufficient haste so as to permit full checkout at PEP prior to installation at the SLC. Figure 11 shows an isometric view of the detector, Fig. 12 shows a cut through the transverse plane. The detector incorporates excellent tracking, hermetic calorimetry down to 15 mrads, excellent hadron/lepton separation and high resolution vertex detectors. It lacks hadron calorimetry and has no useful $\pi/K/p$ separation above 2 GeV/c.

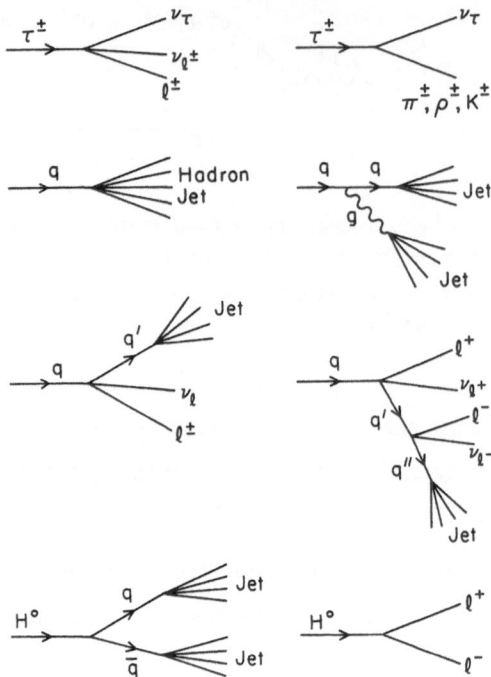

Fig. 10. Typical decays which result from the process in Fig. 9. The symbol g stands for a gluon.

Tracking is achieved with a 72-layer drift chamber in combination with a 5 kg magnetic field (non-cryogenic). The drift chamber has multihit electronics and a measured spatial resolution of 160 μm. The tracking system is fully efficient out to $cos\theta \simeq 0.85$ and drops to 50% around $cos\theta \simeq 0.9$. The intrinsic momentum resolution, incorporating the SLC vertex constraint, is $\sigma_p/p^2 = 0.2$, ignoring the effects of multiple scattering. Pulse height information is also available from the drift chamber, providing dE/dx information for electron identification with a resolution of 7.5%. Combining time-of-flight information with dE/dx provides excellent hadron rejection for momenta in the range 500 MeV/c to 5 GeV/c. Above 5 GeV/c, the calorimetry is used with comparable rejection. All three systems provide hadron rejection $> 10^3$ for momenta above 500 MeV/c.

The barrel plus endcap shower counters cover 95% of 4π and have an energy resolution of $13\%/\sqrt{E}$ and $20\%/\sqrt{E}$ (E in GeV), respectively. For high energy (>1GeV) isolated electrons they provide hadron rejection of $> 10^2$. The calorimeters have poorer rejection when the electrons are within a jet—in this case the power of the dE/dx systems is utilized. Figure 13 shows the calorimetry in the forward direction. The small

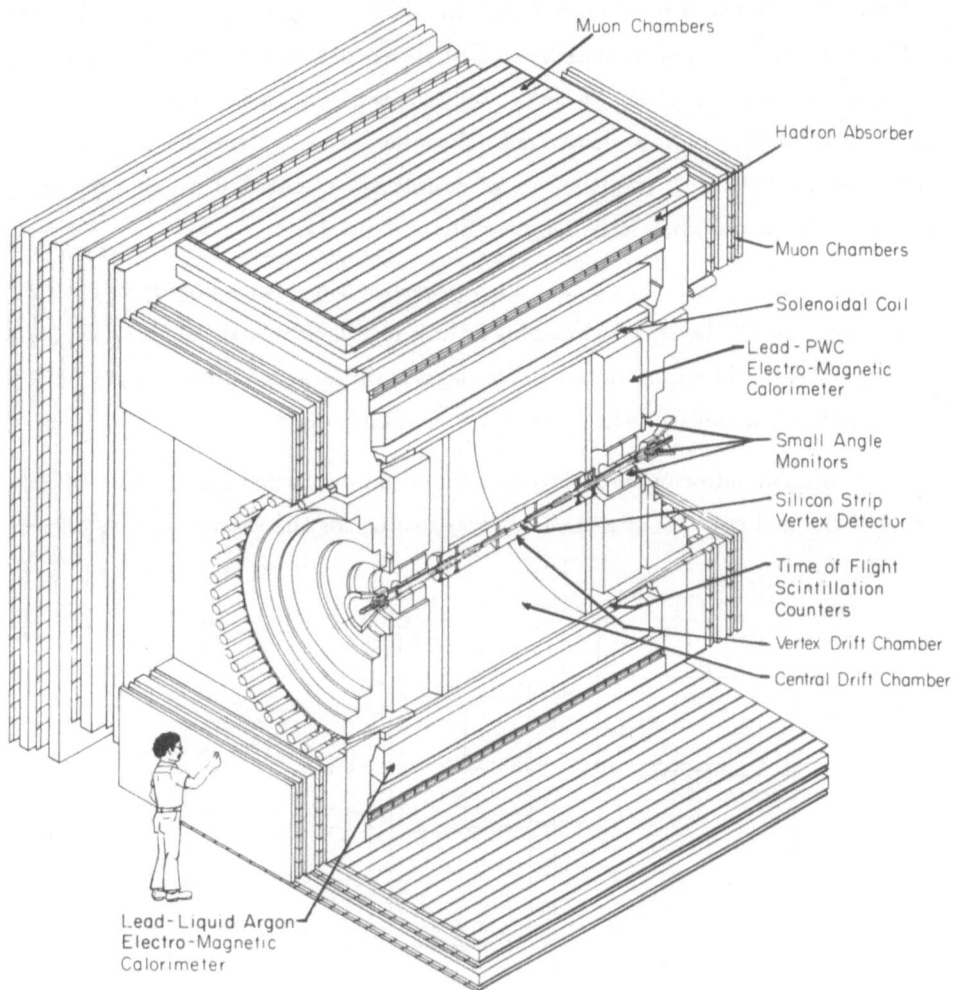

Fig. 11. An isometric view of the MARKII.

angle monitor (SAM) provides hermeticity in the 50–200 mrad region, the mini-SAM in the 15–25 mrad region and an instrumented mask in the region between these monitors. Additional cracks, like those between the endcap and SAM (see Fig. 13) and between adjacent liquid argon barrel modules are covered by relatively crude shower counters, whose main design goal was efficient detection of electromagnetic energy, albeit with poor energy and spatial resolution. These devices therefore act as effective veto counters so that events with electromagnetic energy leaking through the cracks don't masquerade as missing energy events.

Muons are detected in a 4–layer system which covers 75% of the solid angle. Above 2 GeV/c, hadron rejection is at the 10^3 level in this system. The system is useful at momenta between 1 and 2 GeV/c only for isolated tracks. The time-of-flight system has a resolution of 250 picoseconds. Its role is to augment the dE/dx in the "overlap" regions, help with the isolation of cosmic ray events and provide discovery potential for slow (heavy) particles like possibly free quarks.

There are two vertex detectors, a high pressure drift chamber with 30 μm resolution (per wire) and 38 layers of wires plus a three layer silicon strip device with 5 μm resolution per layer. The inner-most silicon layer will be at a radius 2.5 cm. These devices are not yet installed in the MARKII—installation is expected in mid-1989.

Luminosity monitoring is achieved in the usual way using small angle Bhabha scattering (t channel $e^+e^- \rightarrow e^+e^-$). The cross section for the scattering falls off very

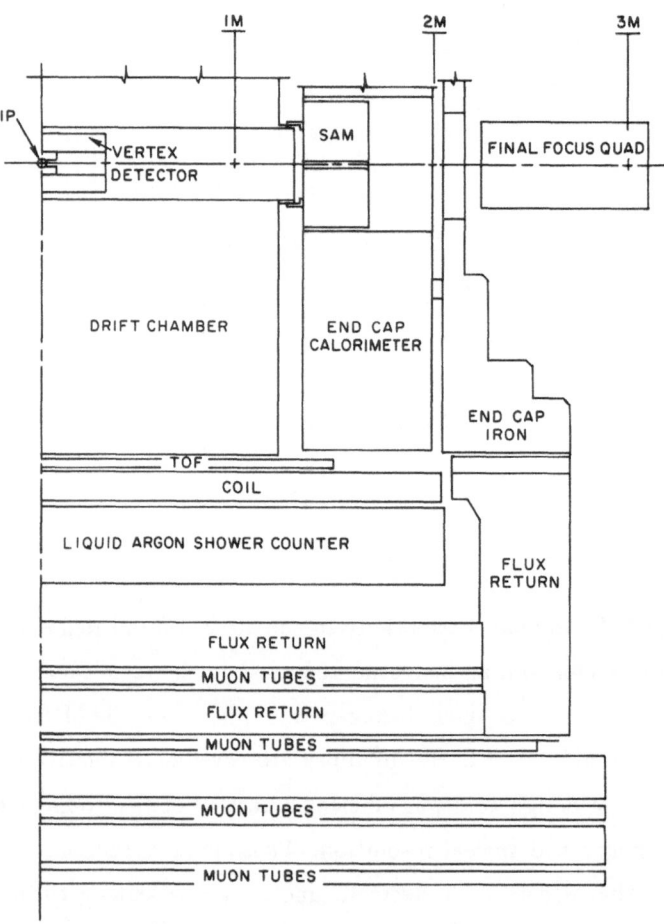

Fig. 12. A conventional view of one quadrant of the MARKII.

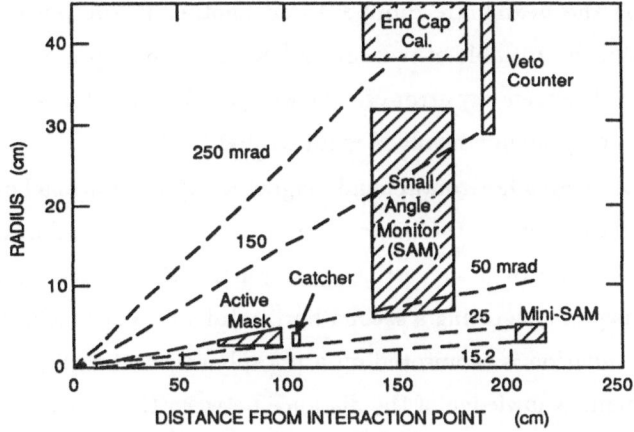

Fig. 13. Placement of the MARKII small angle detectors.

rapidly with the angle relative to the incoming beam, $d\sigma/d\theta \sim \theta^{-3}$, and to lowest order in QED

$$\sigma_{LO} = \frac{16\pi\alpha^2}{5}(\theta_{min}^{-2} + \theta_{max}^{-2}) \quad .$$

In order to achieve a useful counting rate one must therefore place luminosity monitors at small angles. This placement has a further advantage of avoiding contamination from weak production, namely $Z^\circ \to e^+e^-$.

The MARKII detector has two luminosity monitors. The SAM is a precise monitor which has nine layers of proportional wire tracking preceding the shower counters. The tracking chambers permit one to define a well understood solid angle and thereby control systematics. The SAM covers the angular range of $50 < \theta < 200$ mrads and has a counting rate equal to the Z° decay rate. Systematic effects are expected to limit the measurement at the <2% level. A second, less precise monitor, the mini-SAM, covers the solid angle $15 < \theta < 25$ mrads and has a counting rate of approximately seven times the Z° decay rate. It provides a cross-check of the SAM and a higher rate which will make it useful as an on-line monitor.

The limiting error for the Z° mass and width measurements comes from the knowledge of the collision energy, E_{cm}. Typical systematic errors coming from other sources are at the <10 MeV level. The SLC machine itself does not have provision for a collision energy measurement with anything approaching this precision—the best absolute accuracy is < 300 MeV. To overcome this, the MARKII group has constructed and commissioned a pair of energy measuring spectrometers, one in the electron dump line (after collision the beams are dumped) and another in the positron dump line. These spectrometers are capable of measuring both beam energies on a pulse-to-pulse basis providing a collision energy error of <35 MeV/c^2 absolute (for M_z) and <30 MeV relative (for Γ_z). The principle involved is illustrated in Fig. 14.

The beam passes through a horizontal bend magnet which generates a horizontal sweep of synchrotron radiation. Next, the beam encounters a precisely calibrated spectrometer dipole magnet which provides a vertical kick. Finally, the beam encounters a second horizontal bend magnet generating a second horizontal sweep of synchrotron radiation. The synchrotron radiation is monitored on a phosphorescent screen viewed by a digitizing camera system. Knowledge of the distance between the synchrotron stripes, the distance from the magnetic center of the spectrometer magnet to the digitizing screens and the spectrometer magnet $\int B d\ell$ provides the beam energy. Figure 15 shows the phosphorescent screen digitizing system. The distance between the stripes is 27 cm and the accuracy of location of each strip is 80 μm. A wire array directly in front of the screens provides fiducial markers for monitoring the system. The spectrometer magnet was very carefully made and a precision of $< 10^{-4}$ in $\int B d\ell$ was achieved in the laboratory. The field is monitored in situ with two independent systems (a flip coil and an NMR).

Table III summarizes the contributions to the absolute and relative errors arising from the spectrometers. All the entries are self-explanatory except possibly the last one. Motion of the beam coupled with the energy/position correlation of particles in the beam provide a systematic error correlated with the luminosity. Where the beams overlay fully, the total energy spectrum (as seen by the spectrometers) is contributing to the luminosity. However, if the beams move apart somewhat, lowering the luminosity, the energy spectrum of the collision is not the same as measured by the spectrometer. This is the origin of the last contribution given in the table.

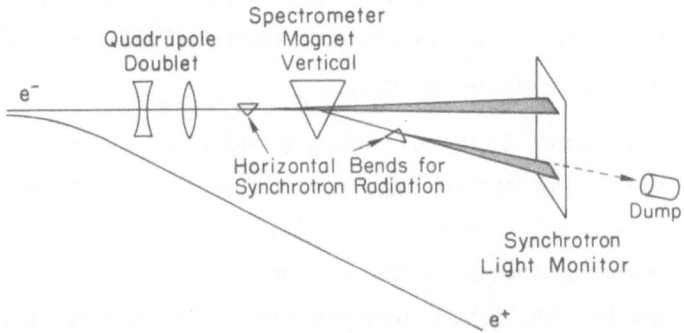

Fig. 14. Schematic drawing of the extraction line spectrometer for measuring the beam energy.

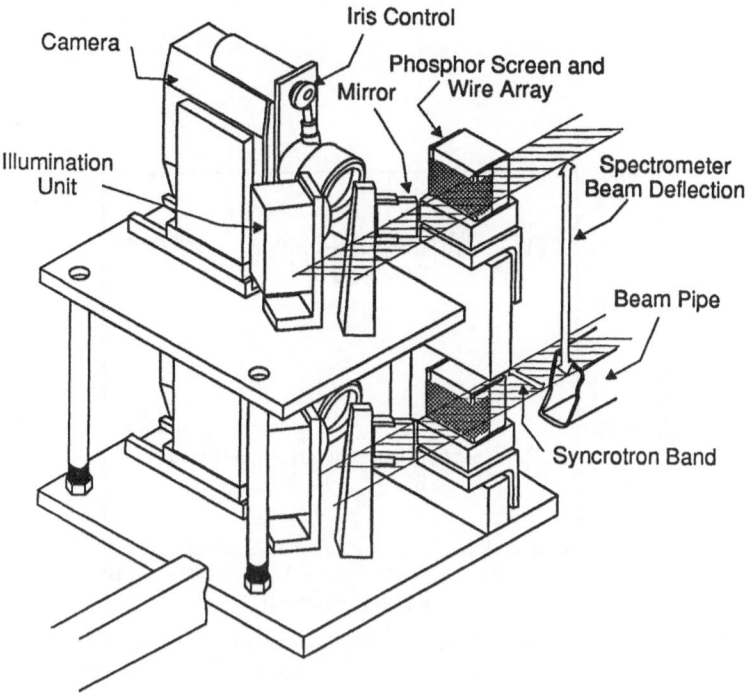

Fig. 15. Phosporescent screen synchroton light monitor system for the energy spectrometer.

The spectrometers are fully operational devices and are read into the MARKII data acquisition system each event. On-line displays of the energy of each beam are available. A typical example of such a plot is given in Fig. 16. These spectrometers will play a central role in the initial SLC measurements.

Another SLC tool, not yet operational, is the availability of a polarized e^- beam. An electron beam with 45% polarization will be available at the collision point for running in 1990. The physics advantages of such a tool are discussed later.

5. THE PHYSICS MEASUREMENTS

We have developed the theoretical background and discussed the measurables for testing the Standard Model. We have also given a brief description of the hardware available for the initial SLC running. It is now time to consider how the measurements will proceed and establish the precision with which information about Nature can be gleaned. As stated before, the simulations of the measurements have been done with the MARKII detector incorporating realistic analysis techniques and the generation of raw data which should therefore provide estimates of efficiencies and measurement errors which are close to what will be realized when data is available. Detailed discussions of these measurements can be found in Ref. 5(d).

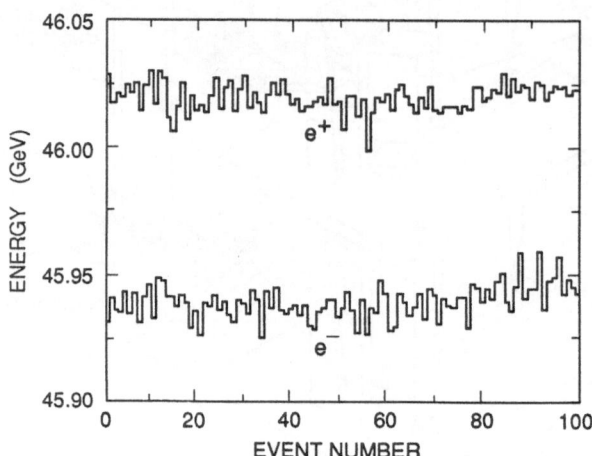

Fig. 16. Energy of the electron and positron beams as measured by the extraction line energy spectrometer over a period of a couple of minutes.

Table III. Systematic errors in beam energy measurements.

Source of Error	Error in E_{cm}	
	Relative MeV	Absolute MeV
Laboratory field map	–	5
Monitoring field	5	5
X-ray detector localization of image (80 μ/27 cm)	15	15
Magnet alignment	–	5
Error in the single beam measurement	15 MeV	20 MeV
Error in the center-of-mass measurement	$\sqrt{2} \times 15$	$\sqrt{2} \times 20$
Beams at IP have an energy/position correlation	15	15
Total Error	30 MeV	35 MeV

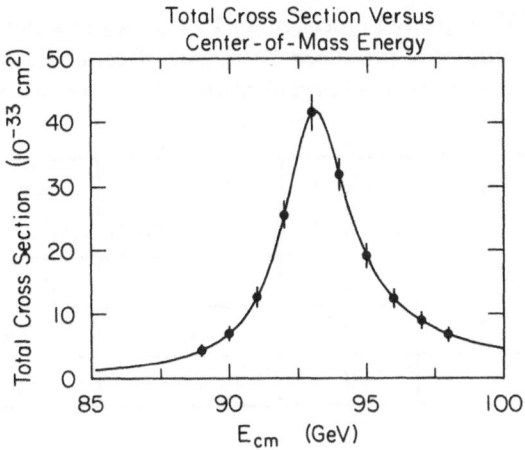

Fig. 17. A simulation of the data which would constitute a scan to map out the Z° line shape. The fit exhibits the clear asymmetry which arises from the initial state Bremsstrahlung.

5.1   The Z° Mass and Width

Clearly the first order of business is to measure the resonance line shape and extract the resonance parameters. This is relatively straightforward involving measuring the normalized Z° decay yield at a series of different collision energies, namely a scan in E_{cm} is performed. Presumably the scan energy will begin with our best knowledge of the Z° mass which at present is close to 92 GeV. This is based on the recent analysis of all the world's data, both from deep inelastic lepton scattering and the $\bar{p}p$ collider, which has been done by Amaldi et al.[6] They find a value of M_z (world average) = 91.8±0.9 GeV/c^2 which is very close to the UA2 measurement of 91.5± 1.2±1.7 GeV/c^2.

Because the Z° is rather wide ($\approx$2.5 GeV), large scan steps (1–2 GeV) are appropriate. Figure 17 shows simulated experimental data for a scan. As discussed earlier, these data must be fit taking into account the effects of radiative corrections. Fortunately (see Ref. 4) Cahn[7] has come up with an analytic fitting function which extracts the resonance parameters, accounting for the radiative effects. Application of this fit introduces errors smaller than the measurement errors due to E_{cm}. Aside from statistics, the dominant error comes from the measurement of E_{cm}. With 100,000 Z° one achieves this systematic limit. The precision achieved for smaller data-sets is summarized in Table IV. One sees that even a modest sized data-set of 10,000 Z°'s provides an impressive measurement of both M_z and Γ_z; Γ_z will be known to almost a third of a neutrino generation.

Table IV. Expected precision for Γ_z and M_z as a function of the number of produced Z°'s. Statistical and systematic errors are added in quadrature. The error in $\sin^2 \theta_w$ coming from δM_z is shown assuming $\Delta r = 0.07$.

#Z° Produced	Width $\delta\Gamma_z(\text{MeV})$	Mass $\delta M_z(\text{MeV}/c^2)$	$\delta \sin^2 \theta_w$ $(\Delta r = 0.07)$
5–10 K	60	65	0.0004
10–20 K	45	50	0.0003
100 K	30	35	0.0002

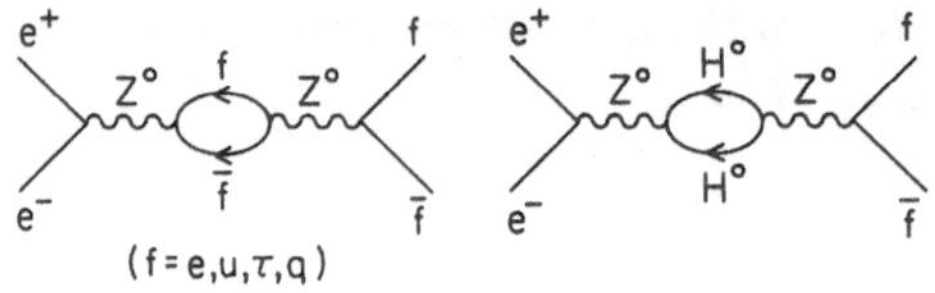

$(f = e, u, \tau, q)$

Fig. 18. One-loop weak radiative corrections to the process $e^+e^- \to Z^\circ \to f\bar{f}$.

With this level of precision for M_z, what do we learn about $\sin^2 \theta_w$? Taking into account electroweak radiative corrections, as typified by the loop diagrams in Fig. 18, one finds

$$\sin^2 \theta_w = \frac{1}{2} \left\{ 1 - \left[1 - \frac{1}{(1-\Delta r)} \left(\frac{74.56}{M_z(\text{GeV})} \right)^2 \right]^{1/2} \right\}$$

where $\Delta r = \Delta r \, (M_t, M_{H^\circ}, \dots$ loop masses) is the contribution from the radiative corrections. The largest uncertainty arises from the t quark as discussed in Ref. 8 and summarized in Table V. Hence, lack of knowledge of Δr impedes extracting $\sin^2 \theta_w$ from a measurement of M_z. However, if the t quark mass is known, the next largest effect is due to the Higgs mass, but this contributes a much smaller uncertainty to Δr. As an indication of the effective precision is $\sin^2 \theta_w$ resulting from the M_z measurement, $\delta \sin^2 \theta_w$ is given in Table V assuming $\Delta r = 0.07$ ($M_t = 50$ GeV/c^2). So, an additional measurement of $\sin^2 \theta_w$ will have to be made at the Z° in order to untangle the effects of radiative corrections.

It is interesting to surmise what we can learn about M_t from an accurate measurement of M_z. Figure 19 is taken from Ref. 8 and plots contours of M_t versus M_z assuming the results from low energy data combined with a measurement of M_z good to ± 100 MeV/c^2. If one uses the UA1 top quark mass limit, and if $M_z > 93.5$ GeV/c^2, this scenario implies there must be physics beyond the Standard Model. If $M_z < 90.5$ GeV/c^2, M_t would have to be > 100 GeV/c^2.

Table V. Effect of the top quark mass on the size of the electroweak radiative corrections and thereby on $sin^2\theta_w$ (Taken from Ref. 8).

M_t GeV	Δr	$sin^2\theta_w$
45	0.0713	0.230
90	0.0606	0.226
150	0.0412	0.219
200	0.0180	0.213

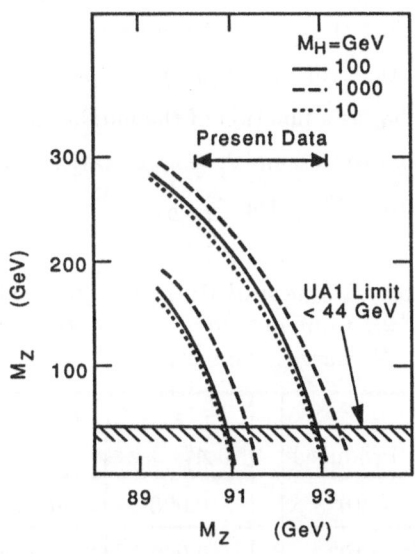

Fig. 19. *Shown is a 90% C.L. range allowed for M_t by combining existing data with a measurement $M_Z = M_Z^{expt} \pm 100 \ MeV/c^2$, shown as a function of M_Z^{expt} for three values of the Higgs-boson mass. Also shown are the UA1 limit $M_t > 44 \ GeV$ and the 90% C.L. range $M_Z = 91.8 \pm 1.5 \ GeV$ allowed by existing data. (Taken from Ref. 8.)*

5.2   Measuring $\sin^2 \theta_w$

We have seen that M_z does not provide a direct measurement of $\sin^2 \theta_w$ because of the uncertainty in the size of the radiative corrections. Additional measurements of $\sin^2 \theta_w$ are needed. Any measurement of a vector coupling measures $\sin^2 \theta_w$:

$$v_f = T_3^f - 2Q_f \sin^2 \theta_w \quad .$$

We recall the forward-backward asymmetry provides such a measurement.

Consider first using the reaction $Z^\circ \to \mu^+\mu^-$. This has the advantage of a very simple, background-free topology. There is no need to identify the muons in the muon system because the only competing topological background is from $Z^\circ \to e^+e^-$. But this has a very distinctive signal in the electromagnetic calorimeter (a 50 GeV electron deposits about 42 GeV in the MARKII calorimeters while a 50 GeV muon deposits only 300 MeV/c) and hence the solid angle available for this measurement is set by the calorimeters (95% in the case of MARKII). The disadvantages of this channel are the small branching fraction (3%) and the small asymmetry (2% for $\sin^2 \theta_w = 0.23$, 4% for $\sin^2 \theta_w = 0.22$). In addition, the effects of the uncertainty in the collision energy are a large source of systematic error (see Fig. 7). Propagating the statistical errors one finds the precision in $\sin^2 \theta_w$ as a function of the number of produced Z°'s in Table VI, where $\sin^2 \theta_w = 0.22$ has been assumed. Large statistics are needed for a precision measurement of $\sin^2 \theta_w$ from A^μ_{F-B} (or A^e_{F-B}).

Table VI. The statistical error obtained for $\sin^2 \theta_w$ from measurements of the $\mu^+\mu^-$ and $b\bar{b}$ charge asymmetries.

# Z°'s Produced	$\delta \sin^2 \theta_w$ from	
	A^μ_{F-B}	A^b_{F-B}
10^5	0.006	0.008
10^6	0.002	0.003

How about using quark pair asymmetries? The reaction $Z^\circ \to b\bar{b}$ is the most practical. The asymmetry is fairly large—11%—and the branching fraction for $Z^\circ \to b\bar{b}$ is also respectable at 13.6%. In contrast to the $\mu^+\mu^-$ channel, it is about a factor of 2-3 times more difficult to get a cleanly tagged $b\bar{b}$ sample. Also, as discussed in Chap. 3, this channel is only half as sensitive to $\sin^2 \theta_w$ as $\mu^+\mu^-$, but the systematic error from δE_{cm} (see Fig. 7) is not large ($< 1\%$).

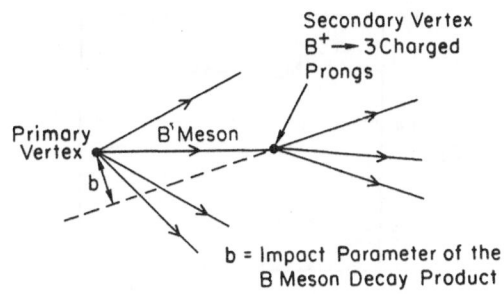

Fig. 20. The production and subsequent decay of a B meson indicating the primary vertex, secondary vertex and the impact parameter b of one of the B decay tracks.

So the question arises as to how one tags $Z° \rightarrow b\bar{b}$. One utilizes the fact that B mesons, which are produced when the $b, \bar{b}$ quarks fragment are long lived ($\tau \approx 1$ picosecond). The long B meson lifetimes generate secondary vertices in the detector. The presence of these secondary vertices can be used as an event tag by observing that the decay products at the secondary vertex do not extrapolate back to the primary vertex (see Fig. 20), but have a finite impact parameter: b. Simulations of the MARKII vertex detector system indicate that the efficiency for tagging a $b\bar{b}$ event using the large impact parameter secondaries is 40%. The fraction of (charm) background in this sample is less than 10% of the signal. The algorithm used requires the event to have $\geq$ 3 charged tracks with a measured impact parameter larger than 3 σ_b (where σ_b is the impact parameter measurement error) and an invariant mass larger than 1.9 GeV/c^2 (to eliminate charm decays from $c\bar{c}$ events).

In order to measure A^b_{F-B}, one must be able to distinguish the b from the $\bar{b}$. This is done by requiring a lepton in the event, the sign of the electric charge of the lepton tags that hemisphere as b or $\bar{b}$. From simulations which incorporate all of these requirements, one obtains errors for $\sin^2 \theta_w$ given in Table VI. Again, large statistical samples are needed to get the required precision.

The path to a high precision measurement of $\sin^2 \theta_w$ with relatively low statistics is via the use of a polarized e^- beam where one gains a statistical advantage of a factor of $\sim$50. The measurement of interest is the left-right polarization asymmetry which is the difference between the total $Z°$ cross section for left-handed electrons colliding with unpolarized positrons and for right-handed electrons colliding with unpolarized positrons:

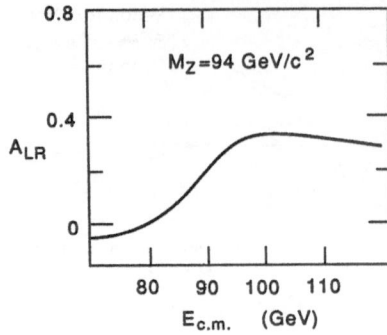

Fig. 21. The dependence of A_{L-R} on E_{cm}.

$$A_{L-R} = \frac{\sigma_L - \sigma_R}{\sigma_L + \sigma_R}$$
$$= \frac{2P_{e^-}a_e v_e}{(a_e^2 + v_e^2)}$$

where P_{e^-} is the e^- polarization (45%). $A_{L-R} = 7\%$ for $\sin^2\theta_w = 0.23$, 12% $\sin^2\theta_w = 0.22$. A_{L-R} is independent of the couplings of the final state fermions and hence it has the maximum statistical power since one can use all the visible Z° decays for this measurement. The measurement is relatively simple, namely measuring a total cross section, and the limiting systematic error turns out to be the error in the polarization. Two polarimeters are being constructed, one uses Möller scattering another Compton scattering. A 5% error in the knowledge of P_{e^-} is considered easy, 1% is possible but will be hard to achieve. As shown in Fig. 21, errors arising from the uncertainty in E_{cm} are negligible, $\Delta A_{L-R}/\Delta E_{cm} \approx .15\%/100$ MeV. The precision with which one measures $\sin^2\theta_w$ from A_{L-R} is given in Fig. 22 (along with the expectation from A_{F-B}^μ) as a function of the number of produced Z°'s. For δP_{e^-} of 3%, an impressive systematic limit is reached at $\delta\sin^2\theta_w \approx 0.0007$ with a sample of 10^5 Z°'s.

5.3   Additional Methods for Indirectly Sensing New Physics

The Z° width is a crucial indicator of physics beyond the 3-generation Standard Model—an anomolous width with respect to the Standard Model prediction immediately signals new physics. So far we have discussed one way to measure Γ_z, namely by mapping out the Z° line shape. It is interesting to ask whether we might have any cross checks of comparable accuracy as a way of verifying the direct measurement. It turns out that there are two measurements which can play this role; a)the so-called invisible width, Γ_{invis} and b) Γ_{tot} extracted from the cross section for $Z^\circ \rightarrow \ell^+\ell^-$ where $\ell^\pm$ are charged leptons.

146

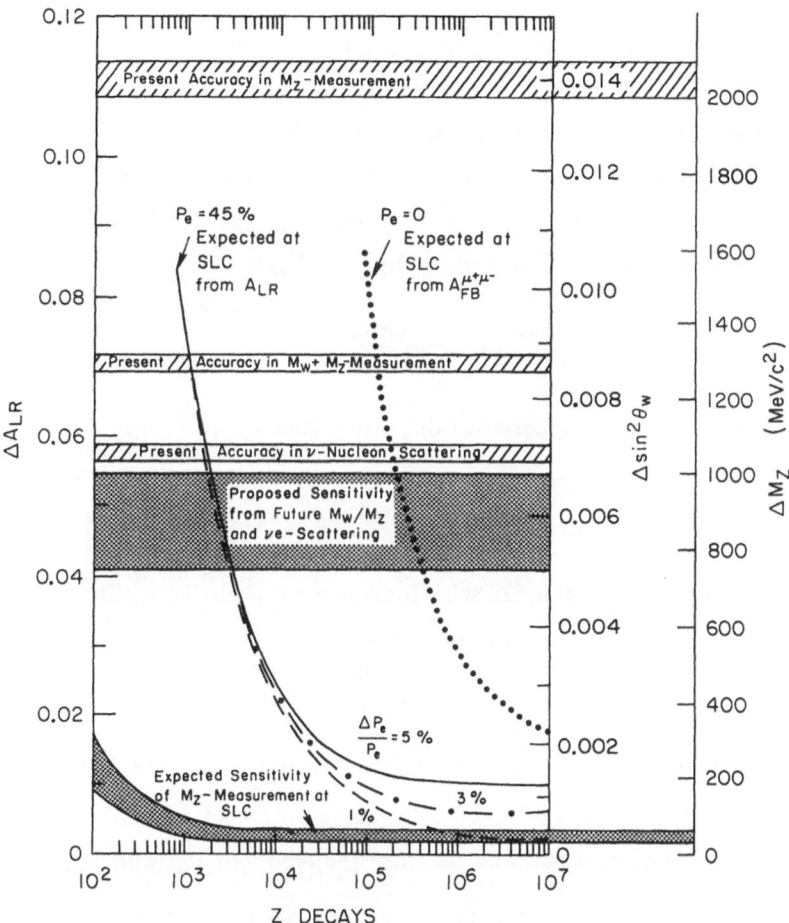

Fig. 22. The uncertainties in the determination of M_Z, $\sin^2\theta_w$, or A_{L-R} for various measurements which can be made at the SLC as a function of the number of observed Z° decays.

Γ_{invis} is defined as the difference between the total width and the visible width:

$$\Gamma_{invis} = \Gamma_{tot} - \Gamma_{vis} \quad .$$

(Note that if Γ_{invis} is entirely due to neutrinos $N_\nu = 12\sqrt{2}\pi/G_F M_Z^3 \Gamma_{invis}$, where N_ν is the number of massless neutrino species in the world.) Γ_{vis} will get contributions from the charged leptons and the hadronic events, namely

$$\Gamma_{vis} = \Gamma_{e^+e^-} + \Gamma_{\mu^+\mu^-} + \Gamma_{\tau^+\tau^-} + \Gamma_{had} \quad ,$$

where Γ_{had} may well contain "new physics." We will assume (quite legitimately) that we can use the Standard Model to calculate $\Gamma_{\ell^+\ell^-} = G_F M_Z^3/24\sqrt{2}\pi(v_e^2 + a_e^2)$, leaving Γ_{had} to be measured. If we run on the Z° peak, Γ_{had} can be obtained by measuring the yield of hadronic events (N_{had}) and mu-pairs ($N_{\mu\mu}$):

$$\Gamma_{had} = \frac{N_{had}\epsilon_{\mu\mu}}{N_{\mu\mu}\epsilon_{had}}\Gamma_{\mu\mu} \quad .$$

In addition, recall from our theoretical discussion that

$$\Gamma_{tot} = \frac{\sqrt{12\pi}\Gamma_{\mu\mu}}{M_{\hat{Z}}}\sigma_{\mu\mu}^{-1/2} \quad ,$$

where we can measure $\sigma_{\mu\mu}$ in a run with luminosity $\mathcal{L}$ as

$$\sigma_{\mu\mu} = \frac{N_{\mu\mu}}{\mathcal{L}\epsilon_{\mu\mu}} \quad .$$

Hence, if one measures the five quantities $N_{\mu\mu}, N_{had}, \epsilon_{\mu\mu}, \epsilon_{had}$ and $\mathcal{L}$, one can extract both Γ_{tot} and Γ_{invis}.

With some manipulation and patience the errors in Γ_{tot} and Γ_{invis} can be obtained:

$$\Delta\Gamma_{tot} = 0.5 \; \Gamma_{tot} \left\{ \left(\frac{\Delta N_{\mu\mu}}{N_{\mu\mu}} \oplus \frac{\Delta\epsilon_{\mu\mu}}{\epsilon_{\mu\mu}} \right) \oplus \frac{\Delta\mathcal{L}}{\mathcal{L}} \right\}$$

$$\Delta\Gamma_{invis} = \left(-\frac{1}{2}\Gamma_{tot} + \Gamma_{had} \right) \left(\frac{\Delta N_{\mu\mu}}{N_{\mu\mu}} \right) \oplus \frac{\Delta\epsilon_{\mu\mu}}{\epsilon_{\mu\mu}} \oplus$$

$$\frac{1}{2}\Gamma_{tot}\frac{\Delta\mathcal{L}}{\mathcal{L}} \oplus \Gamma_{had} \left(\frac{\Delta N_{had}}{N_{had}} + \frac{\Delta\epsilon_{had}}{\epsilon_{had}} \right)$$

where the $\oplus$ symbol means add in quadrature. Because of the small branching fraction for $Z^\circ \to \mu^+\mu^-$, $\Delta N_{\mu\mu}/N_{\mu\mu}$ is the dominant error. Fortunately, it does not end up contributing significantly to $\Delta\Gamma_{invis}$ because of the small weighing factor $(-1/2\Gamma_{tot} + \Gamma_{had}) < 1/4\Gamma_{tot}$. Simulations have been performed to establish the precision with which the efficiencies will be known and yield $\Delta\epsilon_{Had} = 1\%$, $\Delta\epsilon_{\mu\mu} = 2\%$. We expect $\Delta\mathcal{L} = 2\%$; 3% has been used for the estimates which are summarized in Table VII as a function of the number of produced Z°'s. For these estimates it is assumed that all three lepton states, $\mu^+\mu^-$, e^+e^- and $\tau^+\tau^-$ can be used for the measurement. One sees that for a 10,000 Z° sample size one has two powerful adjuncts to the 60 MeV width measurement obtained from fitting the line shape.

Table VII. The measurement errors for Γ_{invis} and Γ_{tot} as a function of the number of produced Z°'s.

# Z°'s Produced	$\Delta\Gamma_{invis}$ MeV	$\Delta\Gamma_{tot}$ MeV
1,000	105	156
5,000	62	82
10,000	54	67

If there is an indication of an anomolous, width, we will know that there must be new physics. But of course we don't know specifically what the new physics channel(s) is. One has to look directly for the visible topologies in the detector. In addition it should be evident that heavy particles which are produced in pairs in the usual way, whose masses are close to (but smaller than) $M_z/2$, will make relatively small contributions to Γ_z. These contributions can easily be less than the experimental resolution. As examples, a 45 GeV/c^2 fourth generation b quark would only contribute 37 MeV to Γ_{tot}, a 45 GeV/c^2 t quark, 13 MeV. To discover such physics likewise requires topological searches. If the measured width is anomolous, and no evidence is found for non-standard events, one will then know that the additional width must arise from weakly coupled, stable neutral objects (i.e., neutrinos or sneutrinos) and confirmation would come from the Γ_{invis} measurement.

6.   SEARCHING FOR THE TOP QUARK

There is not much phase space left for $Z^0 \rightarrow t\bar{t}$ searches since we have an unambiguous limit from TRISTAN of $M_t > 27.5\ GeV/c^2$ and a somewhat less direct limit from UA1 of $M_t > 41\ GeV/c^2$. If indeed there are only three generations, the Standard Model is pushing us in the direction of higher top quark masses; certainly $M_t < M_z/2$ is hard to accommodate in the three-generation Standard Model. Nonetheless, unambiguous experimental measurements are the final arbiter and hence we will certainly search for the top quark. Indeed, if nature is kind and $M_t < M_z/2$, a relatively small data set of several thousand Z°'s will provide a clean signal.

There are several topological search procedures which have been studied. These procedures rely on the fact that the t quark is necessarily much heavier than the five known quarks. Clearly the background for the $Z^\circ \rightarrow t\bar{t}$ searches comes from the hadronic decays of the Z° into the five light quarks and more specifically from events which contain gluon radiation. These events can simulate the "fatter" $Z^\circ \rightarrow t\bar{t}$ kinematics. Two search methods are presented here: the use of event-shape parameters as an example of a poor technique and isolated leptons as an example of the search method of choice.

The reason that event-shape parameters are "dangerous" is that they are subject to our lack of understanding of the fragmentation process. Different Monte Carlo models, all tuned to adequately fit the PEP/PETRA data, do not provide reliable or consistent background predictions in the kinematic region (multi-jet events) of interest to these searches. The variable most useful for isolating $Z^\circ \rightarrow t\bar{t}$ is the aplanarity as defined in the sphericity tensor scheme. Aplanarity is a measure of the amount of momentum "out" of the event plane. Two-jet events from light quarks have very small aplanarity; light quark events can have large aplanarities due to gluon radiation. For the $Z^\circ \rightarrow t\bar{t}$ one naturally expects large aplanarities because of the heavy t-quark mass.

Table VIII summarizes the results of an analysis based on the use of aplanarity. For this simulation 10^4 events of the type $Z^\circ \rightarrow hadrons$ via the five known quarks were produced using three different QCD/fragmentation models and the events were reconstructed in the detector. A sphericity analysis was performed and events were excluded if the aplanarity was less than 0.12. The number of (background) events passing this cut are given in the first column of Table VIII. The problem alluded to above is now rather clear, namely the background estimates of the different Monte Carlo models vary by a large amount, especially when compared with the expected signal yields also given in Table VIII for different t quark mass assumptions. The number of $Z^\circ \rightarrow t\bar{t}$ events is normalized to the 10^4 Z° hadronic events using the branching fractions in Table II (with M_z assumed to be 93 GeV) augmented by the QCD radiative correction outlined in Ref. 9 and shown graphically in Fig. 23. One might argue that with enough hadronic Z decays, the Monte Carlos could be optimized to give a proper description of the Z° hadronic environment for the five known quarks. However, this cannot be done until one has a complete understanding of all the possible sources of hadrons, i.e., searches for new hadron sources must necessarily precede this optimization.

Table VIII. The search for top using the shape parameter aplanarity. The first column summarizes the contributions from the background for three different Monte Carlo models. Note the large variation in the predictions of the different models. The rest of the columns are the signal assuming different top masses.

M_t	# of Events Produced	# of Events with Aplanarity > 0.12
Background:		
Lund $0(\alpha_s^2)$	10^4 udscb	10 ± 3.5
Lund Leading Log	$(1.4 \times 10^4 Z^\circ)$	37 ± 5.6
Webber Shower		76 ± 9.0
40 GeV/c^2	512	112
42.5 GeV/c^2	372	82
45 GeV/c^2	240	40

In summary then, the method of shape parameters is a poor way to proceed. This statement is not just true for the aplanarity—all the potentially useful shape parameters suffer the same fate.

The use of large transverse momentum leptons, from quark leptonic decays, does not suffer from the uncertainties of fragmentation and is a clean, high efficiency method of finding top at the Z°. The signal topology involves tagging isolated electrons and/or muons coming from the decay sequence $Z^\circ \rightarrow t\bar{t}$; $t \rightarrow b + e(\text{or } \mu) + \nu, \bar{t} \rightarrow$ hadron jets. Because of the large t quark mass, the resulting high momentum e and μ are often well isolated from the hadronic jets.

The background comes potentially from $Z^\circ \rightarrow b\bar{b}$; $b \rightarrow c + e(\text{or } \mu) + \nu, \bar{b} \rightarrow$ hadron jets. However, in this case, even when the e or μ have high momentum, they are not isolated from the hadronic jets.

A clean separation of the signal is obtained with relatively simple cuts which have good efficiency. Multiparticle events are selected which have a lepton (electron or muon) with transverse momentum (P_t) relative to the thrust axis larger than 3 GeV/c. The hadrons are then partitioned into jets using a cluster algorithm. The lepton

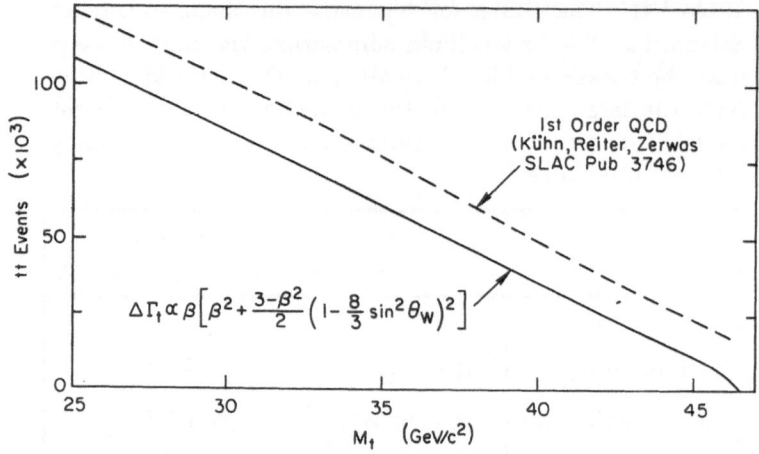

Fig. 23. *The effects of QCD radiative corrections to the yield of $Z^\circ \to t\bar{t}$ events following the calculations of Ref. 9.*

isolation parameter is then calculated for each jet (j) as follows:

$$\rho_j = \{E_\ell(1 - \cos\theta_{\ell j})\}^{\frac{1}{2}} \quad ,$$

where E_ℓ is the lepton momentum and $\theta_{\ell j}$ is the angle between the lepton and the axis of the jth jet: ρ_j is effectively the invariant mass of the lepton-jet system assuming the jet mass to be 1 GeV/c^2. The lepton isolation parameter for the event is chosen as

$$\rho = min\{\rho_j\} \quad .$$

Figure 24 shows dn/dρ for a background sample of 10^4 Z°'s decaying to the five light quarks and for a sample of appropriately normalized $Z \to t\bar{t}$ events with $M_t = 40$ GeV/c^2. Both electrons and muons are used in this analysis and both the signal and background are subject to the selection criteria given above. A cut at $\rho > 1.8$ GeV$^{1/2}$ provides an efficient and clean $Z^\circ \to t\bar{t}$ signal. Predictions of the background spectrum have been verified to be independent of the QCD/fragmentation model as indicated in Table IX which summarizes the sensitivity of the selection technique for different t-quark masses. Again, $M_z = 93$ GeV/c^2 was assumed for these simulations.

The isolated lepton search procedure for $Z^\circ \to t\bar{t}$ is robust and particularly free of background. The efficiency is high enough that with 1,000 Z° events one would have a significant excess of events for $M_t \leq 43$ GeV/c^2, with 5,000 events one could explore the region very close to the kinematic threshold of $M_z/2$.

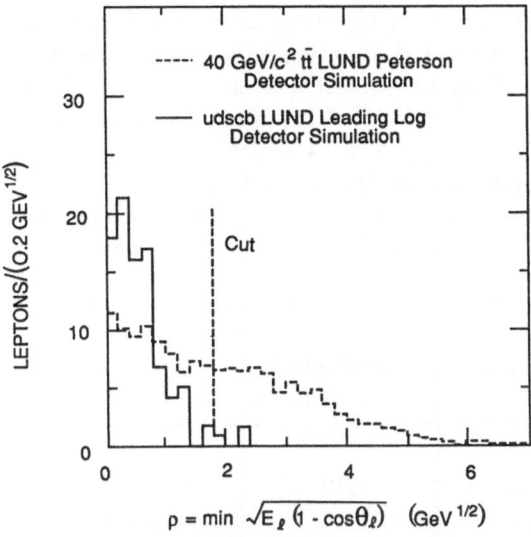

Fig. 24. Isolation criterion for leptons with $P_t > 3$ GeV/c. Distribution of ρ (defined in text) for leptons with $P_t > 3$ GeV/c for 10,000 udscb events from the Lund leading log model with full detector simulation and for 512 $t\bar{t}$ events with $M_t = 40$ GeV/c² from the Lund model with Peterson fragmentation.

Having found the signal described above, how does one know that the source is $t\bar{t}$ as opposed to say $b'\bar{b}'$? The rate is not a useful means of separating these two possibilities, unless the mass is well known, QCD corrections are understood, and one has large statistics. It turns out that it is possible to distinguish these two scenarios as discussed in the next section.

There are several possible ways to measure M_t once one has a signal. These include counting the yield of high P_t leptons, fitting the shape of the P_t distribution, reconstructing the hadronic jet mass in the isolated lepton events.... All these methods suffer from one deficiency or another and yield typical mass uncertainties of about 2 GeV/c² for the mass range and event sample sizes ($\sim 10^4 Z°$'s) discussed here. Presumably if a precise measurement of M_t was needed, one could lower the beam energy and scan for toponium using the crude mass measurement as an indicator of where to scan. A relatively large luminosity ($\approx 10^{30}$cm^{-2}sec^{-1}) will be needed to find toponium in a reasonable time. In addition, if the toponium mass is very close to $M_z/2$, interference effects greatly distort the toponium shape and make it impossible to find.[5d]

Table IX. The search for top using the isolation criteria described in the text. The first column summarizes the contributions from the background for two different Monto Carlo models. The rest of the columns are the signal assuming different top masses. As described in the text, $\rho > 1.8$ GeV/c^2 and $P_t > 3$ GeV/c are the primary analysis cuts for the isolated lepton.

	# of Events Produced	# of Isolated Lepton Events	Signal: Background
Background: Lund Leading Log Webber	10,000 udscb ($1.4 \times 10^4 Z^\circ$'s)	2.6 ± 1.5 3.3 ± 1.9	
$M_t = 40$ GeV/c^2: Lund Symmetric Webber	512	76 ± 2.2 74 ± 4.3	25:1
$M_t = 42.5$ GeV/c^2 Lund Symmetric Lund Petersen	372	61 ± 4.7 62 ± 1.5	20:1
$M_t = 45$ GeV/c^2	240	38 ± 3.1	13:1
$M_t = 46$ GeV/c^2	195	30 ± 2.4	10:1

In conclusion, the Z° resonance is an excellent place to search for top as long as it is sufficiently low in mass to be produced. With 10,000 Z° events one would have sensitivity to masses up to $M_z/2$. The possible confusion between a top quark and a b' quark is easily resolved. Mass estimates in the range of ± 2 GeV/c^2 are possible.

7.   IS THERE A FOURTH GENERATION?

There are four obvious ways to search for a fourth generation of quarks and leptons:

1. Find a $Q = -1/3$ quark and demonstrate that indeed it is a $Q = -1/3$ quark (i.e., its not the top quark).

2. Find a 4th charged lepton.

3. Measure the number of massless neutrino species to be > 3.

4. Find a massive, neutral lepton.

It is entirely possible that a 4th generation exists and that all of its charged members are too heavy to produce at the Z°. In this case one would have to rely on the ν counting experiments discussed in the previous section. This is also not infallible if the neutrino for the 4th generation is massive. Indeed the only way for a 4th generation to escape detection at the SLC would be if all its members had masses $> M_z/2$. We now consider the four possibilities suggested above.

7.1   Searching for a Fourth Generation $Q = -1/3, b'$ Quark

Clearly the isolated lepton technique discussed for the top quark search in the previous section works equally well for the b' quark. These studies have been done and the same level of efficiency and cleanliness is achieved for b' as for t. The issue then becomes whether these two possibilities are distinguishable?

In the absence of a good measurement of the quark mass and reliable QCD radiative corrections to the production cross section, using the absolute rate will not be useful. However, if we assume that $M_{b'} < M_t$ (which is a safe assumption for this scenario), b' decays are distinguishable from t decays because they result in a lot of leading charm (D*'s) which is not true for t decays:

$$b' \to c + W \qquad\qquad t \to b + W \quad .$$

The b from the t will decay to charm, but these charm jets will not produce leading D*'s. So the trick for distinguishing b' jets from t jets is to tag $D^{*\pm}$'s which carry a large fraction of the beam energy. D*'s can be tagged using the famous ΔM technique, but this method has a very low efficiency since specific low branching fraction modes of the D^o enter. As discussed in Ref. 10, an inclusive D* tag is possible if one recalls that the bachelor pion in the decay $D^{*\pm} \to \pi^\pm D^o$ has very little momentum transverse to the D* flight direction ($< P_t^\pi > \sim 30$ MeV/c). This can be contrasted with the typical fragmentation pion which has $< P_t >$ of 300 MeV/c. We use this low P_t as an inclusive tag for charm.

In order to make this tag useful for separating t and b', one must remove contamination coming from $Z \to b\bar{b}, c\bar{c}$. This can be done by making a series of cuts which favor the heavy quark events and discriminate against slow D*'s; the full details can be found in Ref. 10. Multihadronic events are partitioned into jets using a cluster algorithm. Events with the event thrust > 0.9 are rejected. This cut favors the heavy quark events and discriminates strongly against $Z^o \to c\bar{c}$. Each charged track's P_t is measured relative to the axis of the jet to which it belongs. A further cut is made for candidate bachelor pions requiring them to have $Z = E^\pi/E_{cluster}$ between 0.04 and 0.08 where E^π and $E_{cluster}$ are the charged particle and cluster energy, respectively. This requirement discriminates against D* produced in b quark decays in which the bachelor pions are softer than these cuts permit. Figure 25(a) shows the P_t^2 spectrum for charged particles in a sample of 500 events of the type $Z \to b'\bar{b}'$ ($M_{b'} =$

45 GeV/c^2). The events and charged particle candidates satisfy the criteria discussed above. One sees the clear excess of low P_t tracks coming from the D*'s superimposed on the typical fragmentation spectrum with slope ~ 300 MeV/c. Figure 25(b) shows the same distribution for a sample of $Z° \rightarrow t\bar{t}$ events (M_t = 45 GeV/c^2) for which no hard D* component can be seen. Finally, Fig. 25(c) shows the P_t^2 distribution for a sample of 10^4 decays of $Z°$ to the five light quarks plus 500 $Z° \rightarrow b'\bar{b}'$ decays. One sees that, even in the presence of the "standard physics", the tagging technique has sufficient signal-to-noise to distinguish between a $Z \rightarrow b'\bar{b}'$ and $Z \rightarrow t\bar{t}$ scenario. Thus, these two scenarios are distinguishable. It is worth noting that this inclusive method is about 10 times more efficient than the more standard ΔM method.[10] In the same data set of 500 $b'\bar{b}'$ events, 125 tagged $D^{*\pm}$ are found [Fig. 25(a)]. Applying the ΔM method and using as many D^o decay modes as possible one finds 12 exclusively tagged events.

If $M_{b'} < M_z/2$, the b' quark is easily discovered at the SLC and it is distinguishable from the t quark.

7.2. Searching for a Conventional Fourth Generation Charged Lepton

We consider, for the moment, a conventional charged heavy lepton with a massless neutrino partner. The production rate is relatively small, $BF(Z° \rightarrow L^+L^-) = 3\% f(\beta)$ where $f(\beta)$ is given in Sec. 3 and β is the $L^{\pm}$ velocity. $f(\beta) \rightarrow \beta^3$ as $M_L \rightarrow M_z/2$, which greatly limits the production for $L^{\pm}$ masses close to the kinematic limit. We will assume that the $L^{\pm}$ decays via the standard weak interaction which means (ignoring QCD corrections which are small) that $BF(L^{\pm} \rightarrow \ell^{\pm}\nu\nu) = 11\%$ and $BF(L^{\pm} \rightarrow hadrons) = 67\%$.

Events containing a pair of acoplanar leptons ($e^{\pm}\mu^{\mp}$ for example) would constitute an unambiguous signal. For small data sets this is impractical because of the low statistical yield; only a handful of events are produced per 10^5 $Z°$'s. Hence, one must utilize the larger branching fraction modes involving hadrons.

Two topologies are envisaged as shown in Fig. 26. We refer to them as the 1+N and N+N topologies as described in the figure caption. Clearly these topologies (more particularly the N+N topology) will potentially have large background from the standard $Z° \rightarrow$ hadrons events. In order to limit this background the analysis makes use of the substantial energy carried away by the neutrinos in the $Z° \rightarrow L^+L^-$ events by requiring large missing energy. This is effective but the level of background is now model dependent. Applying an analysis of this type to the MARKII detector, one finds

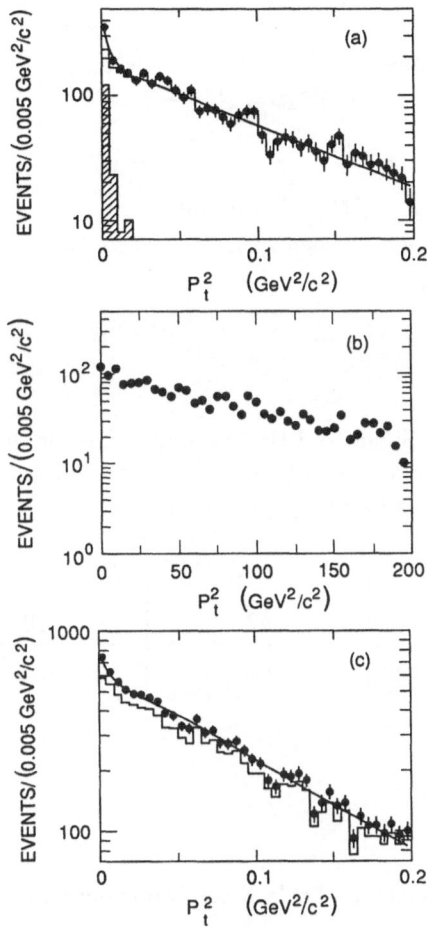

Fig. 25. (a). P_t^2 distribution for bachelor pion candidates (data points) obtained in the decay of a 45 GeV b' quark. The solid histogram corresponds to candidates not coming from a D* and the hatched histogram to candidates coming from a D*. The solid line is a fit to the data with two Gaussians, one of slope ∼ 30 MeV/c, the other ∼ 300 MeV/c. (b). The same as 25(a) except for a sample of $Z° \rightarrow t\bar{t}$ with $M_t = 45$ GeV/c^2. (c). P_t^2 distribution for bachelor pion candidates (data points) obtained in 10,000 Z° decays containing 500 b'b̄' events. The histogram corresponds to only normal Z° decays (udscb). The solid line corresponds to the fit of two Gaussians to the total sample.

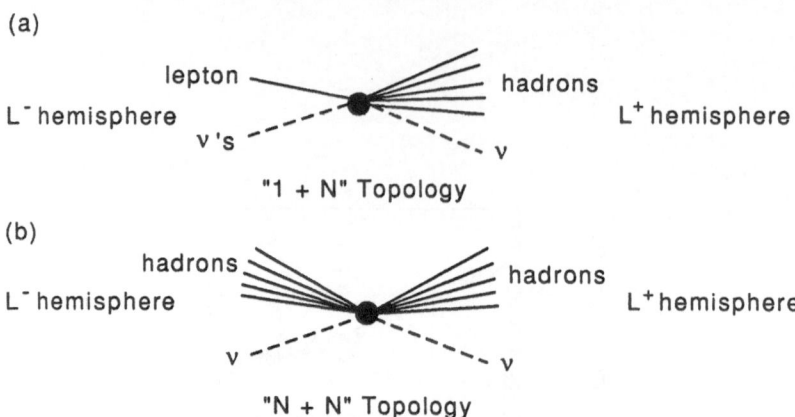

(a)

lepton ——— hadrons

L⁻ hemisphere   L⁺ hemisphere

ν 's ---   -- ν

"1 + N" Topology

(b)

hadrons   hadrons

L⁻ hemisphere   L⁺ hemisphere

ν ---   --- ν

"N + N" Topology

Fig. 26. Decay topologies used to search for $Z^\circ \to L^+L^-$.

Table X. Signal and Background yields for the $Z^\circ \to$ L^+L^- search described in the text. These numbers correspond to 20,000 produced Z°'s.

	$M_L = 30$ GeV/c^2	$M_L = 40$ GeV/c^2
Signal	42	10
Background	1.5	< 1

the signal to noise ratios given in Table X for two choices of $L^\pm$ mass. Both the N+N and 1+N topologies were used for this analysis and the data sample size was 20,000 produced Z°'s.

Another analysis was done using only the 1+N topology, thus alleviating the concern about the model dependence of the background. Figure 27 gives the 90% confidence upper limit obtained for the $L^\pm$ mass as a function of the number of produced Z°'s.

If a 4th generation $L^\pm$ exists with a mass below $M_z/2$, it can be discovered at the SLC. A data set of $\sim$ 50,000 Z°'s is needed to cover the full mass range.

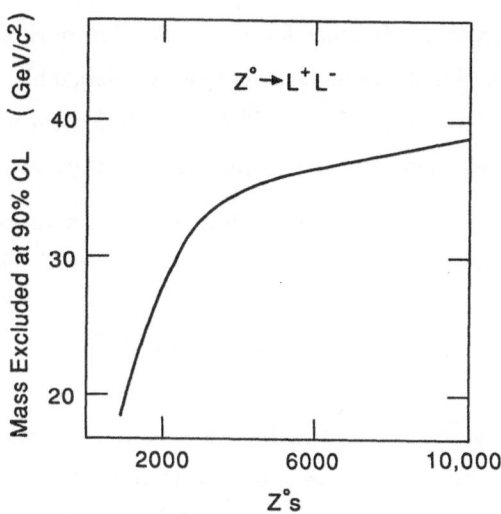

Fig. 27. *90% confidence limits on the* $L^{\pm}$ *mass as a function of the number of produced* Z°'s. *The analyses strategy is explained in the text.*

7.3. Searching for a Heavy Neutral Lepton

We consider now the example of a 4th generation doublet

$$\begin{pmatrix} L^{\pm} \\ L^{\circ} \end{pmatrix} \quad ,$$

in which the L° is massive. The production of the neutral partner is in pairs, namely $Z^{\circ} \to L^{\circ}\bar{L}^{\circ}$, for which the branching fraction is $BF(Z^{\circ} \to L^{\circ}\bar{L}^{\circ}) = 6\%\ [0.25\beta(3+\beta^2)]$, where β is the L° velocity.

There are several scenarios to consider depending on the L° and L^{-} relative masses.

1. If $M_{L^{\circ}} < M_{L^{\pm}}$, the L° is stable and will contribute to Γ_{invis} as discussed earlier. The amount it contributes depends on its mass, $\Delta\Gamma_{invis} = 1600.25\beta(3 + \beta^2)]$ MeV.

2. If $M_{L^{\circ}} > M_{L^{-}}$ then $L^{\circ} \to L^{-}W^{+}$ is possible and L^{-} will be stable. If this were indeed the scenario the $Z^{\circ} \to L^{\circ}\bar{L}^{\circ}$ events would have a striking signature in which each hemisphere would contain the hadronic fragments of the W decay plus as penetrating heavy "muon-like" particle. This "heavy muon" would easily

be distinguished from the familiar low mass muon in both the TOF and dE/dx detectors. Such a signal topology would be unmistakable. This scenario also produces events of the type $Z^\circ \rightarrow L^+L^-$ which for the same reasons of two back-to-back "heavy muons", has an unmistakable signature.

3. The L° could also decay by mixing with a lighter generation(s) as indicated in Fig. 28(a). The mixing strength ($U_{\ell 4}$) determines the L° lifetime:

$$\tau_{L^\circ} = \tau_\mu \left(\frac{M_\mu}{M_{L^\circ}}\right)^5 \frac{BF(L^\circ \rightarrow L^+ \nu \ell^-)}{|U_{\ell 4}|^2} \quad ,$$

where ℓ^- stands for τ^-, μ^- or e^-. It would be most probable that the mixing would occur with the τ, but there is no reason to exclude e or μ. The measured properties of the τ (or e and μ) constrain the amount of mixing allowed which in turn puts lower limits on τ_{L°. It is not unreasonable to expect the L° to have a decay length of at least several millimeters. Of course if the mixing is very weak (and/or the mass is relatively small) the decay length could exceed the size of a detector. This decay scenario leads to striking signatures with no conventional backgrounds. The distinctive features are low multiplicity multiple and mixed type leptons with displaced vertices [see Fig. 27(b) and (c)]. The two interesting topologies are (a) four-charged particle events in which the W's both decay leptonically which occurs about 10% of the time and (b) events with two charged particles in one hemisphere and a jet of hadrons in the other hemisphere which occurs about 40% of the time. The unusual mix of lepton types on the low-multiplicity side, coupled with the displaced vertices, will make these events unmistakenly anomolous. If indeed these heavy neutral leptons exist with masses below $M_z/2$, a few thousand produced Z°'s is all that one needs to discover them.

We can summarize the search for a 4th generation as follows:

1. b': is easily discovered in $< 10^4$ Z° if it can be produced.

2. $L^\pm$: if stable easily discovered; if unstable clear search topologies exist and $\lesssim$ 50,000 Z°'s are needed to cover the full mass range.

3. L°: if stable it will contribute to Γ_{invis}; if unstable it is easily discovered with $\sim$ few 1000 Z°'s.

4. $\nu_{L\pm}$: a conventional, massless 4th generation neutrino will show up in the neutrino counting measurements of Γ_{tot} and Γ_{invis}.

The only way for the 4th generation to elude detection at the SLC is if *all four* of the members have masses $> M_z/2$.

160

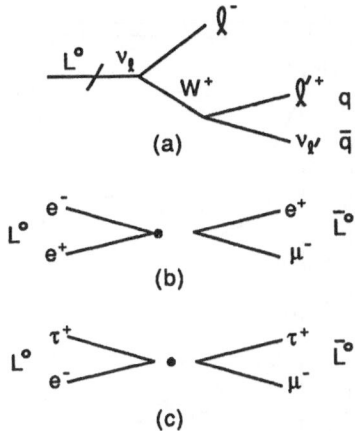

Fig. 28 (a) Schematic for L° decay by mixing with a lighter generation, (b) and (c) Event topologies used in the search for $L^\circ \bar{L}^\circ$.

8. SEARCHING FOR HIGGS SCALARS

At the Bonn Conference in 1981, Okun said that in his mind the outstanding experimental challenge was the search for scalars. He urged experimentalists to "drop everything" and devise cunning searches for the elusive scalars. To date no search has proven successful and it is interesting to speculate how one could search for the Higgs particles running at the Z^0.

The H^0 will couple to the heaviest fermions available and this feature will be used in any search for the H^0. The decay rate for $H^0 \to f\bar{f}$ is given by:

$$\frac{d\Gamma}{d\Omega} = \frac{G_F M_{H^\circ} m_f^2}{16\pi^2 \sqrt{2}} \quad .$$

The decay rate depends on m_f^2 (m_f is the fermion mass) and is isotropic. So if $M_{H^\circ} < 2M_b$, the H^0 will decay mostly to $c\bar{c}$ and $\tau^+\tau^-$. If $2m_t < M_{H^\circ} < 2M_b$ then the H^0 will decay mostly to $b\bar{b}$. These conclusions are summarized in Fig. 29.

How can we search for the H^0? The process $e^+e^- \to Z^0 \to H^0 H^0$ is forbidden by spin-statistics. The process $Z^0 \to H^0\gamma$ vanishes in first order because the Z° and γ are "orthogonal"—in second order the rate is too small to be of any practical use. The most promising search channel seems to be $Z^0 \to H^0 Z^{0*} \to H^0 \ell^+\ell^-$ (see Fig. 30) which was first discussed[11] by Bjorken and is also discussed in Ref. 12. The rate for this process is given by:

$$\frac{1}{\Gamma(Z^0 \to \mu^+\mu^-)} \frac{d\Gamma(Z^0 \to H^0 \ell^+\ell^-)}{dM_{\ell^+\ell^-}} = \frac{\alpha F}{4\pi \sin^2\theta_w \cos^2\theta_w}$$

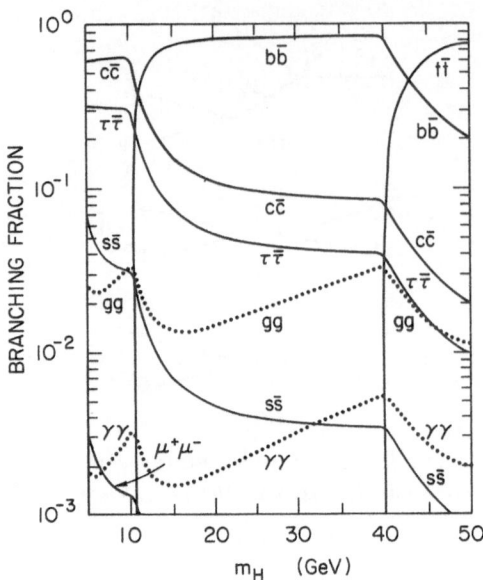

Fig. 29. Decay modes of the neutral Higgs boson as a function of its mass.

where

$$F = \frac{10k^2 + 10\lambda^2 + 1 + (k^2 - \lambda^2)[(1 - k^2 - \lambda^2) - 4k^2\lambda^2]^{1/2}}{(1 - k^2)^2}$$

$$M_{\ell^+\ell^-} = \text{lepton pair mass}$$

$$k = M_L/M_{Z^0}$$

and $\quad\quad \lambda = M_{H^0}/M_{Z^0}\quad.$

This relative rate, integrated over $M_{\ell^+\ell^-}$, is plotted as a function of M_{H^0} in Fig. 31. Also shown for comparison is the rate for $Z^0 \rightarrow H^0\gamma$. $BF(Z^0 \rightarrow \mu^+\mu^-) = 3\%$, so one sees that for $M_{H^0} \approx 20$ GeV/c^2, $BF(Z^0 \rightarrow H^0\ell^+\ell^-) \approx 3 \times 10^{-5}$, a yield of 30 events for 10^6 Z^0 events. Unfortunately, the rate drops off very rapidly with increasing H^0 mass and for masses above ~ 40 GeV/c^2 the measurement becomes severely rate limited.

The $H^0\ell^+\ell^-$ signal must be sought in the presence of an enormous background from $Z^0 \rightarrow$ hadrons. For $M_{H^0} \approx 20$ GeV/c^2 there are $\approx 10^4$ $Z^0 \rightarrow$ hadron events per $Z^0 \rightarrow H^0\ell^+\ell^-$ event! Luckily the event topology is very favorable and a measurement indeed seems possible. Many of the detector groups[5] at SLC and LEP have studied

162

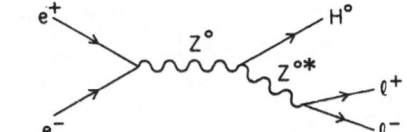

Fig. 30. The process $e^+ e^- \to Z^0 \to H^0 \ell^+ \ell^-$.

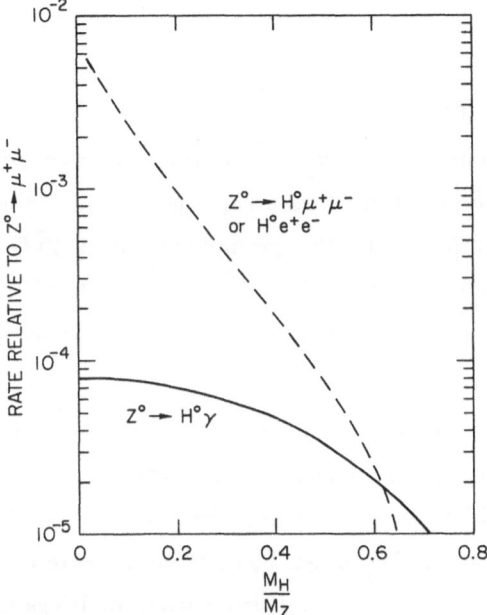

Fig. 31. The decay rate for $Z^0 \to H^0 e^+ e^-$ or $Z^0 \to H^0 \mu^+ \mu^-$ relative to $Z^0 \to \mu^+ \mu^-$ which has a branching fraction of 3%.

the experimental problems and their conclusions are pretty uniform. We chose here in the MARK II study.

The favorable topology arises from the fact that most of the energy in the process $Z^0 \to H^0 \ell^+ \ell^-$ goes to the virtual Z^0 and hence the two leptons which result from the decay of the virtual Z^0 have very high invariant mass and momenta. The H^0 is produced with a fairly small fraction of the available energy and will decay mostly into two quark jets. In addition there is very little correlation between the H^0 direction and the e^+ or e^- direction and in most events the $e^\pm$ will be well separated from the H^0 decay products. The topology is schematically shown in Fig. 32.

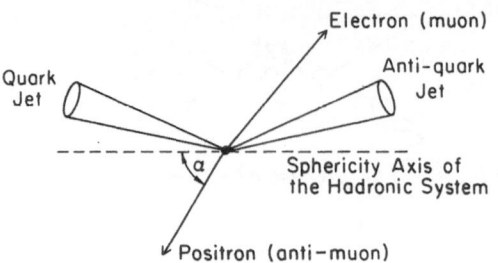

Fig. 32. A schematic representation of the topology of the $Z^0 \to H^0 \ell^+ \ell^-$ events.

The main source of background comes from the process $Z^0 \to t\bar{t}$ where both the t and $\bar{t}$ decay semi-leptonically. However, requiring the angle between the sphericity axis of the hadronic system (all particles except the ℓ^+ and ℓ^-) and the leptons to be $\gtrsim 200$ mrad virtually eliminates this background for $M_{H^0} \lesssim 40$ GeV/c^2. This cut loses very little signal ($\approx 6\%$) because there is virtually no correlation between the direction of the leptons and the hadronic sphericity axis. A small residual background from two photon production exists as discussed below.

The mass of the hadronic system (the H^0) is obtained from the missing mass recoiling against the lepton pair. The experiment can be done with either a e^+e^- or $\mu^+\mu^-$ lepton pair providing that the energy resolution of the leptons is sufficiently good to see a peak in the missing mass. The missing mass recoiling against the e^+e^- pair is shown in Fig. 33 for the MARK II simulation for Higgs masses of 10, 25, and 35 GeV/c^2. Clear signals are seen. Also shown is the background from the two photon process. A similar result is obtained from the $\mu^+\mu^-$ channel, although the missing mass resolution is somewhat worse. Ten events in the combined e^+e^- and $\mu^+\mu^-$ channels would constitute a discovery. Table XI summarizes the number of Z°'s needed for ten detected events as a function of the Higgs mass.

Assuming the search was successful and we found a peak in the recoil mass spectrum, how do we know that we have discovered the Higgs scalar? We would have to verify that the signal decayed isotropically and that the couplings favored the heaviest fermion pair available.

We can measure the decay angular distribution as follows. First, we would reconstruct the two-jet directions from the particles associated with the jets. From the ℓ^+ and ℓ^- momenta we can reconstruct $\vec{P}_{H^0}$. Knowing M_{H^0} and $\vec{P}_{H^0}$, we can transform the jet directions into the H^0 center of mass and plot the decay angular distribution. (This method will work as long as we can make the assumption that the

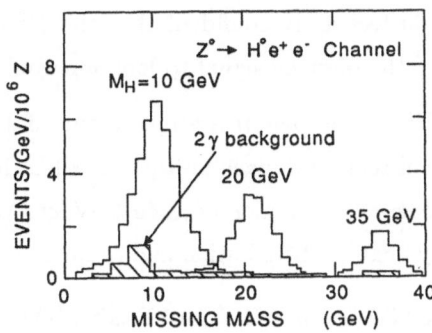

Fig. 33. The Higgs signal from $Z^0 \rightarrow H^0 e^+ e^-$. The expected backgrounds are also shown.

Table XI. The number of Z^o's required to produce 10 detected events as a function of the Higgs mass. The missing mass resolution for the $Z^o \rightarrow e^+ e^- H^o$ channel is given.

M_{H^o}(GeV)	# Z/10 Events	$\sigma_m(e^+ e^-)$ GeV/c^2
4	2×10^4	1.2
10	2×10^5	1.1
25	6×10^5	0.9
35	2×10^6	0.7

decay angular distribution is symmetric about $\theta^* = 90°$. This is because we don't know how to distinguish the jet from the anti-jet (θ^* from $\pi - \theta^*$) and hence by plotting both we are assuming a symmetric decay distribution. Realistically the major problem with this procedure will be the limited statistics. Optimistically one might have ≈ 50 events to play with.

Now, how about measuring if the coupling is proportional to m_f^2? Here the procedure would depend on M_{H^o}. Suppose, as is likely, that $M_{H^o} > 10$ GeV/c^2 in which case $H^0 \rightarrow b\bar{b}$ almost exclusively (see Fig. 29). As discussed in Sec. 5.2, using a vertex detector one can expect to tag events containing two b jets with an efficiency $\gtrsim 40\%$, and this with very little contamination from c jets. This can be done because the b quark has a long measured (~ 1 psec) lifetime. So one would subject the $H^0 \ell^+ \ell^-$ candidate events to this test and if indeed half (= tag efficiency) the events were tagged as having a b jet, one would feel fairly confident that the H^0 decayed predominantly to $b\bar{b}$. If $M_{H^0} < 10$ GeV/c^2 the obvious signal to look for would be $H^0 \rightarrow \tau^+\tau^-$.

To summarize the H^0 search then, it is probable that if $M_{H^0} \lesssim 40$ GeV/c^2 it can be found at the Z^0. We will require a machine with excellent luminosity $-$ $\langle \mathcal{L} \rangle > 10^{30}$ cm^{-2} sec^{-1} to achieve a mass search region of $\lesssim 40$ GeV/c^2. With sufficient statistics ($\gtrsim 50$ events) the H^0 decay angular distribution and coupling can probably be inferred.

9. PHYSICS BEYOND THE MINIMAL STANDARD MODEL

9.1 The Higgs Sector

There is no compelling reason to assume the minimal Higgs scheme with one doublet (four fields); a two doublet scheme as discussed in Sec. 3 is quite permissable. In such a symmetry breaking scheme one has five physical scalars—two neutral scalars H_1^o, H_2^o, one pseudoscalar h^o, and two charged Higgs particles, $H^{\pm}$. For decay purposes the usual rule applies—couplings are largest for the heaviest fermion decay products permissable.

The search for H_1^o, H_2^o proceeds exactly as descirbed before except that in the non-minimal case the 7.5 GeV/c^2 lower limit on the H^o mass no longer applies. Searching for H^o below 7.5 GeV/c^2 has the advantage of increasing production rate (see Fig. 31). However, for masses below a few GeV/c^2, two-photon backgrounds begin to present a significant problem.

$H^{\pm}$ are produced in pairs via the reaction $Z^o \rightarrow H^+H^-$ with a branching fraction $BF(Z^o \rightarrow H^+H^-) = 1.5\%$ β^3, where β is the velocity to the $H^{\pm}$. From PETRA experiments, it is known that $M_{H^{\pm}} > 15$ GeV/c^2 and therefore the dominant decay mode, will be $H^{\pm} \rightarrow b\bar{c}$ ($\approx 75\%$) with the next most favored decay mode being $H^{\pm} \rightarrow \tau^{\pm}\nu_{\tau}$. Therefore, most of the events ($\approx 55\%$) coming from $Z^o \rightarrow H^+H^-$ will be four-jet events. The topologically more attractive modes involving τ's are considerably less probable. Given the relatively small production rate (especially at larger masses) one will most likely focus attention on the four-jet topology. Higher order QCD will produce four-jet events in standard hadronic decays at a rate of $\approx BF(Z^o \rightarrow hadrons)\alpha_s^2 \approx 1.5\%$, comparable to the rate of signal events. However, a series of cuts have been developed which provide sufficient suppression of this background, that statistically significant signals can be extracted with $\gtrsim 20,000$ Z^o's.

The analysis[5d] requires that the events be partitioned into four jets. Using a χ^2 optimization procedure one can chose the best combination of two pairs of jets which fits the hypothesis of pair production of two equal mass objects. The jets are

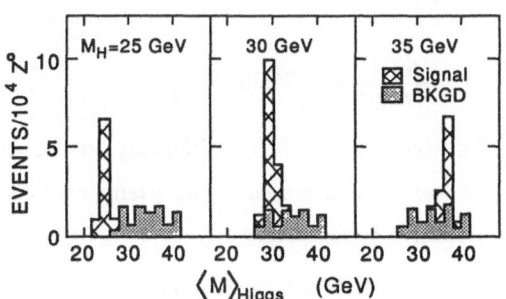

Fig. 34. Dijet invariant mass for the $Z° \rightarrow H^+ H^-$ search. The analysis procedure is described in the text.

required to be relatively well separated (not true for the majority of QCD four-jet events) and the calculated $H^+(H^-)$ momentum vector is required to point into the well instrumented region of the MARKII. Given the assignment of the jets to the H^+ and H^- hemispheres, an average $H^\pm$ mass can be calculated for each event. The distribution of this mass is shown in Fig. 34 for the background and three different $H^\pm$ mass hypotheses. Ten thousand $Z°$'s were assumed in the simulation. At the 10,000 $Z°$ event level, indications of the $H^\pm$ signals appear above the background. With 50,000 $Z°$'s, $H^\pm$ in the mass range from $15 \rightarrow M_z/2$ GeV/c^2 would not be missed.

9.2. Supersymmetry

Supersymmetric theories (SUSY) provide a natural solution to the hierachy problem of the Standard Model by introducing a bosonic (fermionic) partner for every fermionic (bosonic) particle in the Standard Model spectrum. This enlarged fermion and scalar constituent spectrum provides an opportunity to test Nature for the validity of SUSY. SUSY also contains, at a minimum, two Higgs doublets and therefore five physical Higgs particles, as discussed earlier.

Production cross sections for the partners of the normal fermions are characteristic of scalars, namely:

$$R_{\tilde{s}\tilde{s}} = \frac{1}{4}R_{f\bar{f}} \quad .$$

Here, $\tilde{s}$ indicates a SUSY scalar whose normal partner is denoted by f. However, there are two SUSY partners for each normal fermion; so in reality

$$R_{\tilde{s}\tilde{s}} = \frac{1}{2}R_{f\bar{f}}$$

and

$$\frac{d\sigma_{\tilde{s}\tilde{s}}}{d\cos\theta} = \frac{1}{2}\beta^3 \sin^2\theta \sigma_{f\bar{f}} \quad .$$

So, if $M_{\tilde{s}} < M_Z/2$, SUSY scalars could add considerably to the width of the Z^0. As we said previously, if Γ_Z is too wide there could be many reasons for it. One would have to search for each possibility separately.

Scalar leptons with $M_{\tilde{l}} < M_z/2$ will be produced at a rate $BF(Z^0 \to \tilde{l}^+\tilde{l}^-) = 1\frac{1}{2}\% \ \beta^3$ where β is the scalar lepton velocity. Presumably $\tilde{l}^{\pm} \to l^{\pm} \ \tilde{\gamma}$ and, assuming the $\tilde{\gamma}$ is stable, one gets a very distinctive signature—namely events at the Z^0 which have two high energy leptons (e^+e^-, $\mu^+\mu^-$ or $\tau^+\tau^-$) with large missing P_t and energy. The presence of a stable, light particle ($\tilde{\gamma}$) in the decay chains of all the SUSY particles implies that SUSY events are characterized by missing P_t and energy. This is a key element in the search for SUSY signatures.

Backgrounds arise from normal dilepton ($e^+e^-, \mu^+\mu^-, \tau^+\tau^-$) production, but these are relatively easily eliminated by requiring large missing P_t and energy. The main problem is statistics, the β^3 factor provides an increasing barrier as one probles closer to $M_z/2$. Since the present limit on $M_{\tilde{e}}$ is larger than $M_z/2$, the most sensible channel to search for is $Z^\circ \to \tilde{\mu}^+\tilde{\mu}^-$. For $M_{\tilde{\mu}} = 35$, 40 and 43 Gev/c^2 one produces 41, 18 and 7 $\tilde{\mu}\tilde{\mu}$ events respectively per 10^4 Z°; detected events will be half these numbers due to the analysis cuts. There is sensitivity up to the kinematic limit of $M_z/2$ given a data set of $\lesssim 50{,}000$ Z°'s; far fewer Z°'s are needed to discover a $\tilde{\mu}$ with a mass of 40 GeV/c^2. Note that $Z^\circ \to \tilde{\tau}^+\tilde{\tau}^-$ can be similarly pursued—the main difference being that the detector efficiencies are somewhat lower.

The story is similar for the scalar quarks where instead of acoplanar two-particle events comprising the signal, acoplanar two-jets are sought. The background comes from standard $Z^\circ \to hadrons$ which can be removed with cuts in missing P_t and energy. Background estimates are relatively small, but do depend to some extent on the details of the QCD simulation models. Production rates are relatively large; $BF(Z^\circ \to \tilde{u}\tilde{u}) = 6.6\% \ \beta^3$, $BF(Z^\circ \to \tilde{d}\tilde{d}) = 5.3\% \ \beta^3$ where β is the quark velocity. The scalar quark will decay to a quark and a gluino or $\tilde{\gamma}$ and hence one has events with two jets which are not back-to-back but have substantial missing P_t and energy. A sample of 20,000 Z°'s is sufficient to cover searches over the full kinematic range.

For scalar neutrinos $BF(Z^0 \to \tilde{\nu}\tilde{\nu}) = 3\% \ \beta^3$ where β is the $\tilde{\nu}$ velocity. In order to discuss this channel further requires a decay scheme for the $\tilde{\nu}$. The schemes are complicated by the fact that one has no idea of the scalar electron, scalar $\nu, \ldots$ masses. Certainly a prominent decay mode will be $\tilde{\nu} \to \nu\tilde{\gamma}$, an invisible mode which could have

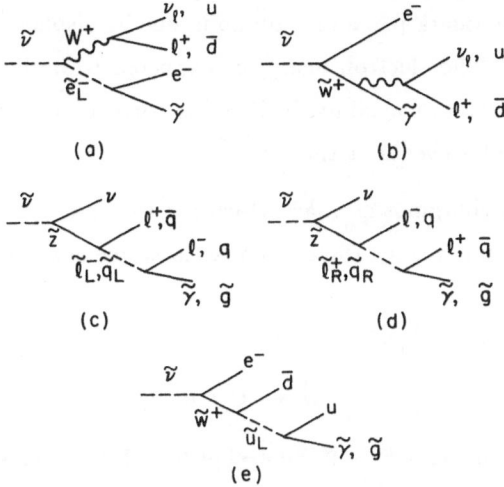

Fig. 35. Possible decay modes for $\tilde{\nu} \to$ multiple charged particles taken from the model of Barnett et al., Ref. 13.

a branching fraction $\simeq 0.6$. There are also multiple charged particle modes possible as shown in Fig. 35 taken from Barnett et al.[13] How much will $Z^0 \to \tilde{\nu}\tilde{\bar{\nu}}$ contribute to the ν counting experiment? The contribution per SUSY species relative to a ν species will be

$$N_{\tilde{\nu}}/N_{\nu} = BF^2(\tilde{\nu} \to \nu\tilde{\gamma}) \frac{\Gamma(Z^0 \to \tilde{\nu}\tilde{\bar{\nu}})}{\Gamma(Z^0 \to \nu\bar{\nu})} \approx 0.2 \quad,$$

where I have used $BF(\tilde{\nu} \to \nu\tilde{\gamma}) \simeq 0.4$. In all likelihood then, it will be hard to see a scalar ν species in the neutrino counting experiment. However, the possibility that scalar neutrinos exist could place systematic limits on how well one would measure N_{ν}.

The multicharge decays shown in Fig. 35 could generate some spectacular events at Z^0. The topology

$$Z^0 \to \tilde{\nu}\tilde{\bar{\nu}}$$
$$\quad \quad \hookrightarrow \nu\tilde{\gamma}$$
$$\quad \quad \hookrightarrow \nu_e e^- e^+ \tilde{\gamma}$$

would yield an electron and positron in one hemisphere of the detector and nothing else! Even if $BF(\tilde{\nu} \to \nu_e e^+ e^- \tilde{\gamma}) \approx 10^{-3}$, 10^5 Z^0's would yield ~ 4 such events! Another interesting topology would be

$$Z^0 \to \tilde{\nu}\tilde{\bar{\nu}}$$
$$\quad \quad \hookrightarrow \nu\tilde{\gamma}$$
$$\quad \quad \hookrightarrow e^- q\bar{q}\tilde{g} \quad,$$

yielding an electron, two quark jets and a gluino in one hemisphere and nothing visible recoiling against them. The electron energy is expected to be large ($\langle P_e \rangle \approx 8$ GeV) making them easy to detect. Certainly, if SUSY is correct, there is a chance that we could see some spectacular events at the Z^0 .

What about the charginos $\omega^{\pm}$, $h^{\pm}$ which are the spin 1/2 partners of the $W^{\pm}$ and $H^{\pm}$. Since they couple weakly, these particles look like heavy leptons $L^{\pm}$ discussed earlier. They decay via

$$h^{\pm}, w^{\pm} \rightarrow W^{\pm} \tilde{\gamma}$$
$$\quad\quad \hookrightarrow \ell^{\pm}\nu \text{ or } q\bar{q} \quad .$$

The decay will be the same as $L^{\pm} \rightarrow W^{\pm}\nu$ except for effects arising from large $\tilde{\gamma}$ mass. How are they distinguished from $L^{\pm}$? Consider for the moment the unmixed case for which the weak couplings are

$$v = (T_{3L} + T_{3R} - 2Q\sin^2\theta_w)$$
$$a = T_{3L} - T_{3R}$$

with

$$T_{3L} = T_{3R} = \pm 1 \quad\quad \text{for } w^{\pm}$$
$$T_{3L} = T_{3R} = \pm 1/2 \quad \text{for } h^{\pm} \quad .$$

Hence $v_{w^{\pm}} = 1.56$, $v_{h^{\pm}} = 0.56$, and $a_{w^{\pm}} = a_{h^{\pm}} = 0$. So, for this unmixed case

$$\frac{\Gamma(Z^0 \rightarrow w^+w^-)}{\Gamma(Z^0 \rightarrow \tau^+\tau^-)} = 9.6$$

and

$$\frac{\Gamma(Z^0 \rightarrow h^+h^-)}{\Gamma(Z^0 \rightarrow \tau^+\tau^-)} = 1.2 \quad .$$

Hence, charginos could add as much as $\sim 33\%$ to Γ_Z. The search for $w^{\pm}$, $h^{\pm}$ proceeds exactly as the search for $L^{\pm}$ discussed earlier.

How does one distinguish the $L^{\pm}$ from the $w^{\pm}$, $h^{\pm}$? Their weak interactions are very different! The charge asymmetry is

$$A_{F-B} = \frac{3v_e a_e v_f a_f}{(v_e^2 + a_e^2)(v_f^2 + a_f^2)}$$
$$= 4.3\% \text{ for } L^{\pm} \, ,$$
$$= 0 \text{ for } w^{\pm},\ h^{\pm} \text{ in the unmixed case.}$$

We have considered the simplest unmixed case. Suppose the $w^{\pm}$ and $h^{\pm}$ are maximally mixed in states $\tilde{w}_1$ and $\tilde{w}_2$. Then

$$\frac{\Gamma(Z^0 \to \tilde{w}_1^+ \tilde{w}_1^-)}{\Gamma(Z^0 \to \tau^+ \tau^-)} = \frac{\Gamma(Z^0 \to \tilde{w}_2^+ \tilde{w}_2^-)}{\Gamma(Z^0 \to \tau^+ \tau^-)} = 5.5$$

and

$$A_{F-B}^{\tilde{w}_1} = 14\%$$

$$A_{F-B}^{\tilde{w}_2} = -14\% \quad .$$

Of course, we have no guidance from the theory as to what level of mixing, if any, there is.

10. CONCLUSIONS

I have chosen in these lectures to highlight the more speculative or discovery-oriented measurements because they exemplify the potential of the SLC physics program to make a significant impact on our understanding of Nature. In particular what emerges is the diversity of central issues which can be confronted with a data set as small as 10^4 Z°'s. Many leading questions are probed at this level, the least impressive being the neutral Higgs sector where sensitivity to masses in the 40 GeV/c^2 range require a data set on the order of 10^6 Z°'s. As one moves towards data sets in the 10^5 Z° range, there is the potential to study any new physics which has emerged at the 10^4 level, as well as benefit from what will be very substantial samples of Z° decays to the conventional, known leptons and quarks. Considerable "bread and butter" physics will be possible which will greatly add to our present knowledge of the details of the Standard Model.

In the absence of e^- polarization, very precise tests of the electroweak structure require large ($10^6 Z^\circ$'s) statistics. However, having an electron beam with 45% polarization enhances one's sensitivity by a factor of about 50.

Finally, the Z° is a new frontier and surprises could well be lurking—let's hope so.

11. ACKNOWLEDGMENTS

I wish to thank my hosts at both the St. Croix and Banff schools for their kind invitation and their wonderful hospitality. I also wish to thank the students and fellow lecturers who enriched the lecture experience with their perceptive questions and interesting discussions. I also wish to thank Sharron Lankford and the staff of the SLAC publications department for preparing this document.

REFERENCES

1. (a) *Future Frontier for e^+e^- Collisions: Physics of SLC and LEP*, SLAC–PUB–3928 and *Proceedings of New Frontiers on Particle Physics*, Lake Louise, Canada, February 1986 and (b) *Z^0 Decay Modes – Experimental Measrement*, SLAC–PUB–3407 and *Proceedings of the 1984 Theoretical Advanced Study Institute on Elementary Particle Physics.*

2. G. Arnison *et al.*, Phys. Lett. **126B**, 398 (1983). (UA1) and P. Bagnaia *et al.*, Phys. Lett. **129B**, 130 (1983) (UA2).

3. For a derivation of this cross section and most of the formulae quoted in this chapter see C. Quigg, *Gauge Theories of the Strong, Weak and Electromagnetic Interactions.*

4. J. Alexander *et al.*, Phys. Rev. D **37**, 56 (1988).

5. (a) CERN Yellow Reports 76–18 (1976) and 79–01 (1979). (b) Proceedings of the SLC Workshop on Experimental use of the SLC; SLAC–247 (1982). (c) Proceedings of the Cornell Z^0 Theory Workshop; CLNS 81–485. (d) Proceedings of the 2nd and 3rd MARKII Workshop on SLC Physics, SLAC–REPORT–306, and SLAC–REPORT–315.

6. U. Amaldi *et al.*, Phys. Rev. D **36**, 1385 (1987).

7. R. N. Cahn, Phys. Rev. D **36**, 2666 (1987).

8. W. Marciano *et al.*, Phys. Rev. D **29**, 1810 (1984).

9. J. Kuhn *et al.*, Nucl. Phys. B **272**, 560 (1986).

10. G. Wormser, SLAC–PUB–3509.

11. J. D. Bjorken, SLAC–198 (1976).

12. J. Finjord, Physica Scripta, Vol. **21**, 143 (1980).

13. R. M. Barnett et al., Phys. Rev. D **29**, 1990 (1984).

TOPICS IN KAON PHYSICS

Bruce Winstein

Enrico Fermi Institute and the Department of Physics
The University of Chicago, 5640 South Ellis Avenue
Chicago, Illinois   60637

In these lectures, we will treat some selected topics in the field of
kaon physics. Today there is much activity in the field with about 250
experimentalists and many theorists engaged in many precise, incisive, and
search oriented studies of Kaon decays. We will begin with a brief review
of the properties of the kaon and follow with a discussion of "sources"
(i.e., beams) of kaons that are currently used in experiments. Dynamical
effects in the decays of the kaons which are important in current
experiments will then be treated. This will be followed by a treatment of
the $K\bar{K}$ mixing process and the phenomenon of CP non-conservation. The
successful Kobayashi-Maskawa model for the weak charged current is
presented and the predictions for this model in the area of CP violation
are discussed.

We will then discuss the three current experiments which are attempting
to measure ϵ'/ϵ in the neutral kaon system. The substantial progress in
the active program of rare kaon decay experiments is reviewed. Finally, we
close with two related topics: CPT tests in the Kaon system and CP
violation in heavy quark systems.

The attempt has been made to be pedagogical where appropriate; as a
result, there will not be a complete historical discussion of the treated
topics nor will there be many references to the original literature. I ask
my colleagues to accept my apology for these omissions.

I. KAON PROPERTIES

For the charged kaons, we have

mass: 493.667(14) MeV

lifetime: $1.2371(26) \times 10^{-8}$ sec

spin: 0

while for the neutral ones,

mass: 497.72(7) MeV

lifetime(s): $0.8923(22) \times 10^{-10}$ sec

$5.183(40) \times 10^{-8}$ sec

We are indicating the experimental error on the least significant digits by the number in parentheses. We note that CPT symmetry guarantees the equality of particle and antiparticle masses and lifetimes; this is, of course, subject to experimental tests.

The quark structure of the kaon is given by:

K^0 $\qquad\qquad\qquad\qquad\qquad$ K^+

$(d\bar{s})$ $\qquad\qquad\qquad\qquad\qquad$ $(u\bar{s})$

K^- $\qquad\qquad\qquad\qquad\qquad$ $\bar{K}^0$

$(\bar{u}s)$ $\qquad\qquad\qquad\qquad\qquad$ $(\bar{d}s)$

For completeness, we mention that the K^0 has a charge radius, since it is composed of two oppositely charged quarks of unequal mass:

$$\langle R^2 \rangle_{K^0} = -0.054(26) \text{ fm}^2 .$$

This is substantially smaller than that for the K^-:

$$\langle R^2 \rangle_{K^-} = -0.28(5) \text{ fm}^2 .$$

Probably the most important feature of the kaons that permits the observation of a host of phenomenon is their very long lifetime. This allows relatively clean beams to be formed far from the production point and rare processes to be studied.

Consider the dominant decay mode of the K^+:

$$K^+ \rightarrow \mu\nu$$

How do we understand this decay process?

The diagram is shown in Figure 1.

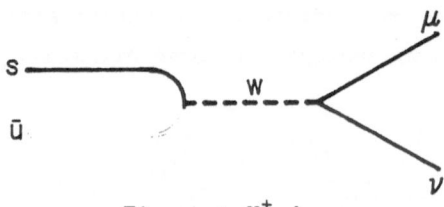

Figure 1 K^+ decay.

The rate W is given by

$$W = \frac{G_F^2}{8\pi} f_K^2 |V_{us}|^2 m_\mu^2 \ m_K \left(1 - \frac{m_\mu^2}{m_K^2}\right)^2$$

where G_F, which is proportional to $1/M_W^2$ and is taken from muon decay, is

$$G_F = 1.16632(4) \times 10^{-5} \quad GeV^{-2} \ c^4$$

and V_{us}, essentially the Cabibbo angle, is given by

$$V_{us} \overset{\sim}{=} 0.22.$$

Given that

$$W(K \rightarrow \mu\nu) = 5.12 \times 10^7 /sec$$

$$= 3.37 \times 10^{-17} \ GeV$$

We find

$$f_K = 0.159 \ GeV.$$

f_K is called the kaon decay constant and is related to the "wave function at the origin" of the initial state.

Note that a process with a branching ratio of 10^{-11} can probe a mass scale of

$$80 \ GeV \times (10^{11})^{1/4} = 45 \ TeV!$$

II. KAON BEAMS

We will now discuss the "sources" of the studies of kaon decays. By this we mean the various beams that have been configured. We will consider four sources: a neutral K_L beam made from a high energy proton beam striking a stationary target, a K^+ "beam" at rest, having been stopped, a K^0 ($\bar{K}^0$) "beam" arising from $\bar{p}p$ collisions at rest, and a "beam" of K_L and K_S from ϕ meson production in e^+e^- collisions.

II-1. High Energy Neutral Beam

Kaons are produced in the beam fragmentation region in high energy collisions. The kaon yield can be expressed as a yield per incident proton per GeV/c kaon momentum per str in the beam. A convenient parameterization of the yield at angle θ by Malensek is given by

$$\frac{d^2\sigma}{dpd\Omega} = \frac{B}{400.} \ \frac{P \ (1 - x)^A \ (1 + 5e^{-Dx})}{(1 + P_T^2/M^2)^4} = pf(x, P_T)$$

where $x = P_{kaon}/P_{beam}$ and $P_T = \theta P_K$.

We see that the cross-section for particle production is "invariant" i.e.,

$$\cdot \frac{d^2N}{dpd\Omega} \equiv \text{Yield} \stackrel{\sim}{-} \frac{p\sigma_{inv}}{\sigma_{inel}}$$

The relevant values of the constants are here shown for $K^\pm$:

	K^+	K^-
A	2.90	6.10
B	14.20	12.30
M^2	1.16	1.10
D	19.90	17.80

Notice that the most significant difference in the production is in the x dependence:

$$K^+ \propto (1 - x)^3$$
$$K^- \propto (1 - x)^6$$

This difference is typical for particle/antiparticle production.

Clearly, the maximum in the production is in the forward direction as a result of the P_T dependence.

For neutral beams, it is a good assumption that K^0's are produced like K^+'s and $\bar{K}^0$'s like K^-'s. Thus one in general has produced a mixture which depends upon energy and angle. At high energy (large x), the K^0 dominates and experiments situated close to the production target can study the time evolution of the K^0 state, as we will see later. Experiments situated far from the production target by and large have a pure K_L beam and the initial composition becomes irrelevant.

It is interesting to point out that the weak interaction (mixing) produces a beam with, from the point of view of the strong interactions, an equal component of particle and antiparticle. This permits studies of the difference in the interactions of the two which are otherwise not possible because of the steep fall off in antiparticle production. We will study the mixing process in detail later.

For a neutral beam, it is difficult to work in the forward direction because of the other stable particles produced: photons and neutrons.

Photons are the result of π^0 decay for the most part and they are easily removed with a lead converter near the target. Generally about 10 rl (~6 cm) is sufficient, the probability of a photon surviving being $e^{-7/9 \times 10} \stackrel{\sim}{-} 4 \times 10^{-4}$. Subsequent sweeping magnets remove the (charged) products of the resulting electromagnetic shower.

The neutrons are a more serious problem and there are two techniques for relatively enhancing the kaon flux. First, the neutrons are produced (from interactions with a Be target) with an invariant cross-section as follows:

$$\sigma_{inv} \simeq \frac{25 \text{ mb}}{(\text{GeV}/c)^2} \, e^{-4.6 \, P_T}$$

Note the absence of any x dependence: the neutrons are therefore a much harder spectrum. The neutron flux is many times the kaon flux at 0^0.

The first technique is differential absorption in an absorber of low atomic number. The neutron and kaon total cross-sections have the following atomic number dependence

$$
\left.
\begin{aligned}
\sigma_N &\simeq 49 \text{ mb } A^{0.77} \\[2em]
\sigma_K &\simeq 24 \text{ mb } A^{0.84}
\end{aligned}
\right\} \quad A \gtrsim 7
$$

Thus the cross-section ratio is greatest for small A. Also elastic scattering in such an absorber is better spread in angle for small A:

$$\frac{d\sigma}{dt} \propto e^{-bt} \qquad \text{where } b \propto A^{2/3}$$

As an example, a Be absorber of 50 cm provides

$$\sigma_N \simeq 270 \text{ mb} \qquad\qquad \sigma_K \simeq 150 \text{ mb}$$

$$x_N = \text{\# of neutron interaction lengths} \qquad x_K = 0.93$$

$$= \sigma_N \, L \, N_o \, d/A = 1.67$$

where N_o = Avogadro's Number
$\quad\quad d$ = density = 1.85 g/cm^3
$\quad\quad A$ = 9 (Atomic Number)
$\quad\quad$ neutron transmission = 0.19
$\quad\quad$ kaon tranmission = 0.40 $\left.\right\}$ → x2 increase in the K/n ratio

The second, and most important, method of enhancing the kaon to neutron ratio is to target away from the 0^0 direction.

The mean neutron momentum in a (0^0) beam is given by

$$\frac{\int p \, \frac{dN}{dp d\Omega} \, dp}{\int \frac{dN}{dp d\Omega} \, dp} \;=\; \frac{\int_0^{E_0} p^2 \, dp \, \sigma_{inv}}{\int_0^{E_0} p dp \, \sigma_{inv}} \;=\; \frac{2}{3} \, E_0$$

since σ_{inv} has little p dependence.

The corresponding mean for kaon production is given approximately by

$$\frac{\int_0^{E_0} p^2 \, dp \, (1-x)^3}{\int_0^{E_0} p \, dp \, (1-x)^3} \;=\; \frac{\int_0^1 E_0^2 \, x^2 \, E_0 \, dx \, (1-x)^3}{\int_0^1 E_0 \, x \, E_0 \, dx \, (1-x)^3}$$

$$=\; E_0 \, \frac{\int_0^1 x^2 \, (1-x)^3 \, dx}{\int_0^1 x \, (1-x)^3 \, dx} \;=\; \frac{1}{3} \, E_0$$

Now the <u>total</u> neutron flux into a solid angle $d\Omega$ is given by $[P_T = x E_0 \theta]$.

$$\int \frac{dN}{dx \, E_0^2 \, d\Omega} \, dx \;\alpha\; \int x \, e^{-5 \, E_0 \, x \theta} \, dx$$

$$= 1 \text{ at } 0^0$$

$$=\; \frac{2}{(5 \, E_0 \, \theta)^2} \left[1 - (1 + 5 \, E_0 \, \theta) \, e^{-5 \, E_0 \, \theta} \right] \;\sim\; \frac{2}{(5 \, E_0 \, \theta)^2}$$

where the last approximation is good for $\theta \gtrsim 1/E_0$ (i.e. beyond 1 mr at FNAL). For example, if one wants to measure at $x \sim 0.1$ at Fermilab, the fluxes at several angles are shown in the table.

$$x = 0.1$$

θ [mr]	n flux	K flux	Enhancement
0	$\equiv 1.0$	$\equiv 1.0$	---
3mr	0.014	0.82	59
5mr	0.005	0.58	116
7mr	0.0026	0.37	145

Thus a factor of over 100 enhancement can be obtained with less than a factor of 2 loss in K flux.

There are two other factors which are important in the design of neutral beams. The first involves beam halo, or a soft component of neutrals in and around the beam. This component can be reduced significantly by making sure that the collimator walls subtend little solid angle compared to the beam solid angle. In practice, this means that the first collimator should not be too close to the production target.

The second factor has to do with how the non-interacting beam, as well as the forward "jet" are disposed of. The dump should be thick enough to completely absorb the hadron shower and should be close enough to the target so that muons from π and K decay are reduced.

In spite of all the effort to the contrary, it is usually the case that neutral kaon experiments are not limited by the available beam intensity but rather by instantaneous rates in the detector. Then the duty cycle becomes very crucial. For example, Fermilab used to deliver beam for about 1 sec every 12 sec while now, with the Tevatron, beam is delivered for 23 sec every 58 sec. Thus the duty cycle has increased from 8% to about 40% so that an experiment can acquire five times more data per hour with the same instantaneous load on the apparatus (if the protons are available).

The characteristics of the MC beam line at Fermilab are shown in the table. This is the beam in which CP nonconservation studies are being performed. We will see that in this beam, one can study the CP violating neutral kaon decays to two pions at the level of 10^6 events.

Fermilab MC Beam Characteristics

Intensity:	3×10^{12} 800 GeV protons
Spill:	20 sec every 60 sec
Targeting angle:	5 mr (nominal) + 1 mr, -3 mr vertical
$d\Omega$/beam:	4.5×10^{-7} str (4" X 4" at 500')
Absorber:	18" Be
neutrons:	6×10^6/beam
K_L:	1.4×10^7/beam (50 to 150 GeV/c)
K_L/hour:	8.4×10^8 (50 to 150 GeV/c)
muons:	$\approx 1.5 \times 10^6$/sec

II-2. Stopped K^+ Beam

We next consider experiments with charged kaons. Later we will discuss experiments on charged kaon decay "in flight." Here we will mention another technique, that of a stopped K^+ beam.

To be useful, the beam is to be stopped in an instrumented target and thus only a very low energy beam with a small momentum range is used. Typically $p \approx 800$ meV/c and $\Delta p/p \approx 15\%$.

At such energies, an electrostatic separator can be used so that, by means of collimation and timing, a beam of high purity "stops" can be created. At such low energies, it is desirable to stop the kaons as close to their production target as possible because losses from decays are important. In practice, a beam line is about 14 m long (see Figure 2) and this is about 3 kaon lifetimes.

The attainable rates are about 200×10^5 stops/5×10^{12} protons on target. In contrast to decay in flight experiments, where only a very

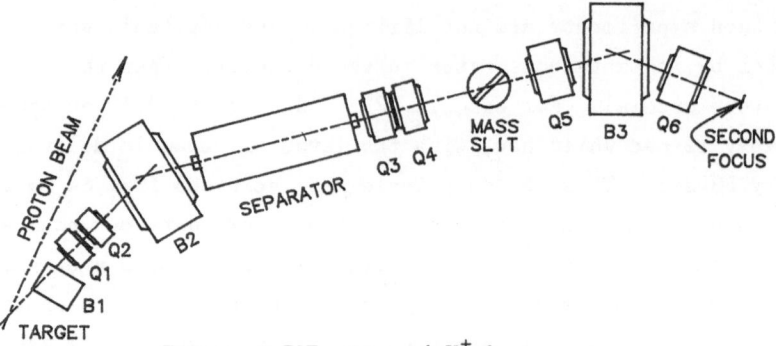

Figure 2 BNL stopped K$^+$ beam.

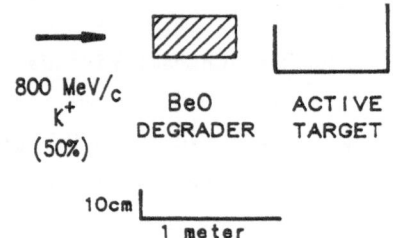

Figure 3 Schematic of target region.

small fraction of the beam decays, <u>all</u> of the stopped kaons will decay and the decay products are studied then in "the center-of-mass" in as nearly a 4π detector as possible.

A schematic of the region near the target is shown in Figure 3. In the optimization of this beam, one considers that the particle yield will increase at higher energies as a result of less decay inflight. However, the beam must be degraded so that it can be stopped in a manageable length of the instrumented target.

The degrader "slows" the beam to 300 meV/c but it also appreciably broadens it so that the kaons effectively fill up the target which is ~10 cm in cross-section. Note that only the kaon component of the beam (~ 50%) will be appropriately degraded to stop at the middle of the target.

II-3. $\bar{p}p$ Annihilation

The source of kaons that we will next consider is that created at the LEAR facility at CERN.

LEAR provides an intense antiproton source. The experiment (CPLEAR) will use a beam of about 2 × 10^6 $\bar{p}$'s per second at a momentum of about 175 meV/c which will stop and annihilate in a hydrogen target at 15 atmospheres. This requires about 20% of the available beam. While the bulk of the annihilation cross-section will produce pions, there is still copious production of kaons. The relevant reactions are the following:

<u>BR</u>

$$\bar{p}p \rightarrow K^+ \pi^- \bar{K}^0 \qquad\qquad 1.5 \times 10^{-3}$$

$$\rightarrow K^- \pi^+ K^0 \qquad\qquad 1.5 \times 10^{-3}$$

$$\rightarrow \bar{K}^0 K^0 * \qquad\qquad 0.5 \times 10^{-3}$$

$$\rightarrow K^+ \pi^-$$

$$K^0 \bar{K}^0 * \qquad\qquad \underline{0.5 \times 10^{-3}}$$

$$\rightarrow K^- \pi^+$$

$$4.0 \times 10^{-3}$$

Two factors are important here. First, the annihilation conserves strangeness so that by measuring the strangeness of "one side" one uniquely determines that of the other. Second, the initial state is symmetric with respect to particle/antiparticle and this feature is particularly convenient for the study of any symmetry violation (C, CP, CPT) in the subsequent weak decays of the produced kaons.

In particular, let's concentrate on the study of any difference in K^0 vs $\bar{K}^0$ decays. For each particle, there will be approximately

$$\underset{\text{annih's/sec}}{2 \times 10^6} \cdot \underset{\text{BR}}{2 \times 10^{-3}} \times \frac{3600 \text{ sec}}{\text{hour}} \times \frac{10 \text{ hrs}}{\text{day}} \times \frac{100 \text{ days}}{\text{year}} = \frac{1.44 \times 10^{10} \text{ decays}}{\text{year}}$$

II-4. ϕ Production

Finally, we consider the production of K^0's through $e^+ e^-$ annihilation. At Novosbirsk, such a facility exists (VEPP-2m) for the study of the following reaction

$$e^+ e^- \rightarrow \phi \rightarrow K_L K_S$$

Symmetry arguments based upon Bose statistics guarantee only the above combination of kaons so that one can in effect have a tagged pure K_S beam which is very difficult to obtain in any other way. This could be useful for the study of rare K_S (as opposed to K_L) decays.

The cross-section for the process is about 1.4 ub so that, with $L \overset{\sim}{-} 10^{31} \text{ cm}^{-2} \text{ sec}^{-1}$, about 10^8 $K_L K_S$ pairs are produced in a year of running. The $K_{L(S)}$ decay length is 3.4m (6mm). With substantially more luminosity, CP violation could be studied at such a facility.

III. DECAY DYNAMICS

Consider a K_0 produced at $t = 0$. The principal decay rates of the K^0 are given below:

$$K^0 \to \pi^+\pi^- \qquad\qquad 7.75 \times 10^9 /\text{sec}$$

$$K^0 \to \pi^0\pi^0 \qquad\qquad 3.53 \times 10^9 /\text{sec}$$

$$K^0 \to \pi^+\pi^-\pi^0 \qquad\qquad 2.39 \times 10^6 /\text{sec}$$

$$K^0 \to \pi^0\pi^0\pi^0 \qquad\qquad 4.15 \times 10^6 /\text{sec}$$

nonleptonic

$$K^0 \to \pi^- e^+ \bar{\nu}_e \qquad\qquad 7.47 \times 10^6 /\text{sec}$$

$$K^0 \to \pi^- \mu^+ \bar{\nu}_\mu \qquad\qquad 5.23 \times 10^6 /\text{sec}$$

semileptonic

$$K^0 \to \mu^+\mu^- \qquad\qquad 0.18/\text{sec} \qquad \} \qquad \text{pure leptonic}$$

Let us sketch out the quark diagrams (representative) for these processes.

III-1. Nonleptonic Decay

Note that two pions can be in a state of isospin $I = 0$, $I = 1$, or $I = 2$. However, with spin 0, the (total) symmetry of the wave function eliminates $I = 1$. Thus we have $I = 0$ and $I = 2$. Now, the $\pi^+\pi^-$ state, decomposed into states of $I = 0$ and 2, is

$$|\pi^+\pi^-\rangle = \sqrt{1/3}|2\rangle + \sqrt{2/3}|0\rangle \text{ and}$$

$$|\pi^0\pi^0\rangle = \sqrt{2/3}|2\rangle - \sqrt{1/3}|0\rangle$$

Thus, if the final state were pure $I = 0$, we would expect a 2/1 ratio for the rate ratio $K^0 \to \pi^+\pi^- / K^0 \to \pi^0\pi^0$ which is close to the experimental value. We note also that the rate for $K^+ \to \pi^+\pi^0 \stackrel{\sim}{-} 17 \times 10^6 /\text{sec}$. This is much smaller than the K^0 rate but this transition must be to a pure $I = 2$ final state. From the ratio

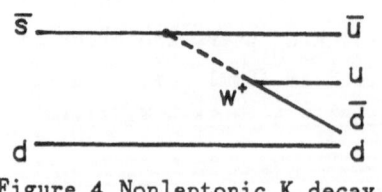

Figure 4 Nonleptonic K decay.

$$\sqrt{\frac{K^+ \to \pi^+ \pi^0}{K^0 \to \pi^+ \pi^-}} \stackrel{\sim}{=} 0.05,$$

we note that the I = 2 final state is suppressed by about a factor of 20 (in amplitude).

What about the 3π decay? Our 2π results indicate that $\Delta I = 1/2$ transitions dominate over $\Delta I = 3/2$. If this were the case for the decays

$$\left.\begin{array}{ll} K^+ \to \pi^+ \pi^+ \pi^- & 4.52 \times 10^6 \\[2em] K^+ \to \pi^+ \pi^0 \pi^0 & 1.40 \times 10^6 \end{array}\right\} \quad \text{ratio} = 0.310$$

then we would expect that the final state three pions were pure I = 1. (I = 0 is not allowed by charge.) This is consistent with the observed branching ratio when phase space corrections are applied. The ratio is expected to be

$$0.311 = 1/4 * 1.243 \text{ for } \frac{\Gamma(K^+ \to \pi^+ \pi^0 \pi^0)}{\Gamma(K^+ \to \pi^+ \pi^+ \pi^-)} .$$

For the K^0 decay into three pions, we would again expect that the 3π state would be pure I = 1 if the $\Delta I = 1/2$ rule is correct. Why is this so, i.e., why is there not an I = 0 component to the final state? The energy release is so little that we may assume that all pions are in a relative S-state. Then, by Bose-Einstein symmetry, the $\pi^+ \pi^-$ and $\pi^0 \pi^0$ must be either I = 2 or I = 0 with $I_Z = 0$ as well. The additional π^0 will then yield, for the total isotopic spin,

$$(2,0)(1,0) = \sqrt{3/5}(3,0) + \sqrt{2/5}(1,0)$$

$$(0,0)(1,0) = (1,0)$$

so that there is no I = 0 state with the appropriate symmetry.

The expected ratio is given by

$$\frac{\Gamma(K^0 \to \pi^+ \pi^- \pi^0)}{\Gamma(K^0 \to \pi^0 \pi^0 \pi^0)} = \frac{2}{3}$$

where the data gives the value of 0.58. The agreement is better after phase space corrections are made.

What is the origin of this $\Delta I = 1/2$ rule? There is another (penguin) diagram shown in Figure 5 responsible for the nonleptonic decays. The spectator diagram (Figure 4) contributes to both $\Delta I = 1/2$ and $\Delta I = 3/2$ transitions: the s quark (I = 0) turns into three I = 1/2 quarks which can have I = 1/2 or I = 3/2. However, the penguin only creates one I = 1/2 quark. At present, there is a good prospect that the penguin diagram is sufficiently enhanced to explain the $\Delta I = 3/2$ "suppression."

The enhancement is due to a QCD effect. The question, however, is still open and is an active area of theoretical research. We will later see why this is important for CP violating effects.

III-2. Semileptonic Decays

Let us now look at the semileptonic decays. Diagrammatically, they proceed as in Figure 6:

The $\Delta S = \Delta Q$ rule in K^0 semileptonic decay can be seen to be a result of the fact that it is the S quark which decays: K^0's will make e^+'s and μ^+'s while $\bar{K}^0$'s will produce e^-'s and μ^-'s. Note that the semileptonic rates are comparable to the 3π rates even though there is significantly more phase space available. This again points to a $\Delta I = 1/2$ enhancement (rather than a $\Delta I = 3/2$ suppression).

III-3. Fully Leptonic Decays

The (fully) leptonic decays are diagrammed in Figure 7.

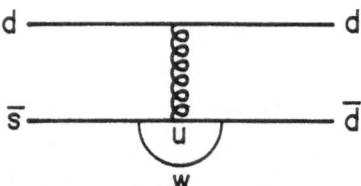

Figure 5 Penguin amplitude.

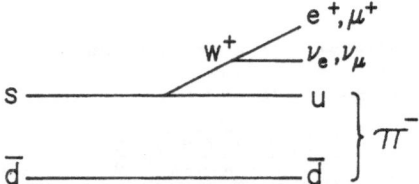

Figure 6 Semileptonic K decay.

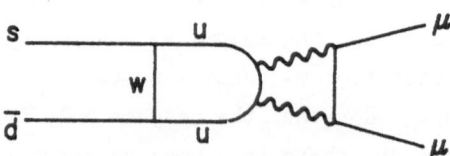

Figure 7 Fully leptonic K decay.

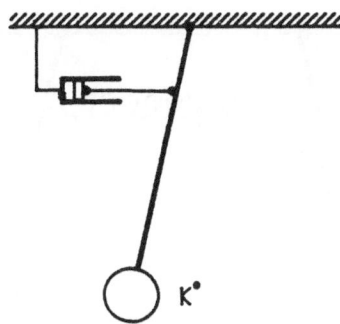

Figure 8 K^0 analogue.

The rate is suppressed by the additional electromagnetic interactions (and by the cancellation with other "top" quarks). The e^+e^- mode is further suppressed by helicity conservation. We will come back to the leptonic decays later on in these lectures.

IV. MIXING IN THE K^0-$\bar{K}^0$ SYSTEM

We will now discuss the effects associated with the fact that K^0 and $\bar{K}^0$ have common decay channels, the most common of which is 2π. The "mixing" effect was predicted by Gell-Mann and Pais in 1956.

Since each has access to many of the same channels, the presence of one will result in the appearance of the other. To gain insight into this phenomenon, we will develop our intuition using a mechanical analogy (see Figure 8).

The pendulum has a natural frequency which will correspond to the K^0 rest "frequency" of ~10^{24} Hz. Imagine that there is a velocity dependent dissipation at the support which causes the amplitude of the motion to decay at a rate of ~10^{10} Hz or one part in 10^{14} per cycle.

This damping is provided, in the mechanical analogy, by an air dashpot. Now we imagine that the $\bar{K}^0$ looks identically. This follows from the CPT theorem and is the subject of experimental verification to which we will return.

Now we must introduce some coupling between the two. (See Figure 9.)

The introduction of the spring between the two balls couples their motions in a familiar way. Now there are two normal modes of oscillation, an anti-symmetric and a symmetric mode as shown in Figure 10.

The modes are split in frequency (the higher being the antisymmetric mode). Note that the modes will have the same lifetime as before.

Next, we introduce some dissipative coupling as in Figure 11.

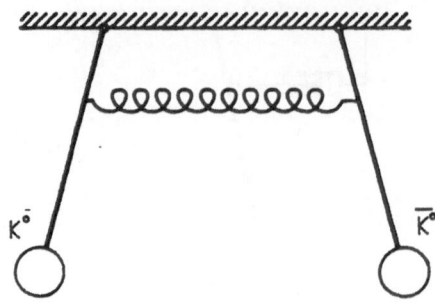

Figure 9 $K^0 \bar{K}^0$ system with conservative coupling.

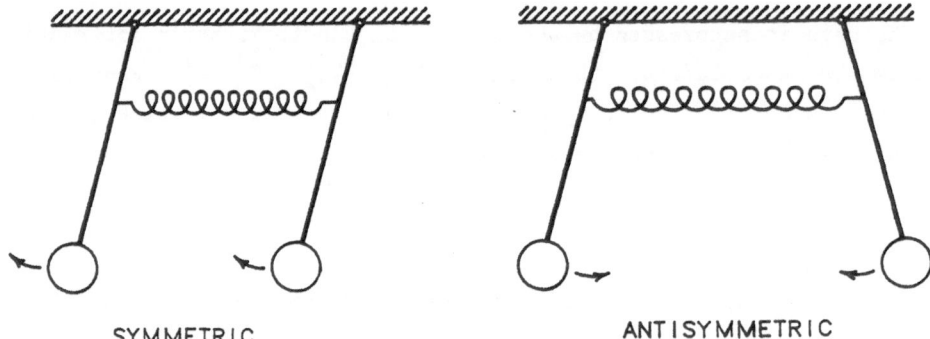

SYMMETRIC   ANTISYMMETRIC

Figure 10 Normal modes.

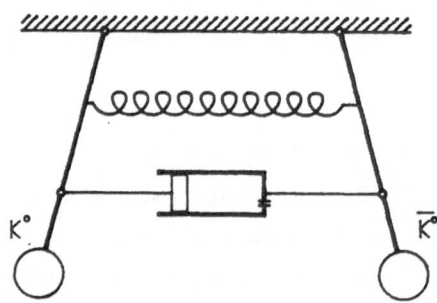

Figure 11 $K^0 \bar{K}^0$ system with conservative and dissipative coupling.

Note that this coupling will not mix the modes but it will split the lifetimes of the two. The anti-symmetric mode will decay away most quickly.

Now, how does this apply to the K^0, $\bar{K}^0$ system? Let us perform an experiment beginning with a K^0 and measuring semileptonic decays vs time. Our semileptonic decay is a perfect measure for the mixing process. Lets calculate what we would expect, based upon our model. We will let $\omega_{s(A)}$

and $t_{s(A)}$ be the frequency and lifetime, respectively, for the symmetric (anti-symmetric) mode.

$$K^0 = S + A$$

$$S(t) = \frac{(K^0 + \bar{K}^0)}{2} \, e^{-t/t_s} \, e^{i\omega_s t}$$

$$A(t) = \frac{(K^0 - \bar{K}^0)}{2} \, e^{-t/t_A} \, e^{i\omega_A t}$$

$$K^0(t) = \frac{K^0}{2} \left[e^{-t/t_S} \, e^{i\omega_S t} + e^{-t/t_A} \, e^{i\omega_A t} \right]$$

$$+ \frac{\bar{K}^0}{2} \left[e^{-t/t_S} \, e^{i\omega_S t} - e^{-t/t_A} \, e^{i\omega_A t} \right]$$

Now, let us form the "charge asymmetry"

$$\delta(t) \equiv \frac{K^0(t) - \bar{K}^0(t)}{K^0(t) + \bar{K}^0(t)} = \frac{e^{-t/t_A} \, e^{i\omega_A t}}{e^{-t/t_S} \, e^{i\omega_S t}}$$

$$\delta = e^{-t/t_A + t/t_S} \, e^{i(\omega_A - \omega_S)t}$$

Thus we expect an oscillating term (at $\omega = \omega_A - \omega_S$) with an envelope which decays with a decay rate of $1/t_A - 1/t_S$.

If we now look at the experiment, we note the expected general features. Indeed such experiments give very good determinations of the decay rate and mass difference for the appropriate neutral kaon "eigenstates."

The world average values are

$$\Delta\omega = \Delta M = 3.521(14) \times 10^{-12} \text{ MeV}$$

$$= 0.5349(22) \times 10^{10}/\text{sec}$$

$$t_S = \tau_L = 5.183(40) \times 10^{-8} \text{ sec}$$

$$t_A = \tau_S = 0.8923(22) \times 10^{-10} \text{ sec}$$

However, there is one striking feature that is not reproduced in our mechanical analogue: the oscillations do not die down. The charge asymmetry reaches an asymptotic value of $\delta(\infty) = 3.30(12) \times 10^{-3}$. Note that this error corresponds to a statistical sample of about 7×10^7 events.

V. CP VIOLATION

Thus the long lived state is not exactly an equal mixture of K^0 and $\bar{K}^0$. With this phenomenon, one can clearly distinguish a world made of antimatter from that of matter - (an initial $\bar{K}^0$ evolves to the same final state, with an excess of K^0) independent of whether the coordinate system in the worlds are left or right handed. This is CP violation.

We can see that time reversal invariance is also violated in that we must have

$$\text{amp } K^0 \rightarrow \bar{K}^0 \neq \text{amp } \bar{K}^0 \rightarrow K^0$$

so that an unequal mixture will decay purely exponentially. This statement assumes CPT invariance as we will see later.

So far, there are only two manifestations of CP violation. The other is in the 2π decay mode. Measurements of the 2π rate in an initial K^0 beam show two exponentials with an interference term between them. Note that the long lived kaon state decays much less to 2π than does the short lived one indicating that the short lived state has positive CP (predominantly). The decay rate difference is approximately a factor of 200.

We will define the CP eigenstates to be

$$K_2 = \frac{1}{\sqrt{2}} \left[\; |K^0\rangle + |\bar{K}^0\rangle \; \right] \qquad CP = -1$$

$$K_1 = \frac{1}{\sqrt{2}} \left[\; |K^0\rangle - |\bar{K}^0\rangle \; \right] \qquad CP = +1$$

Evidently the short lived component is predominately the K_1 state.

We can find the form of the eigenstates by solving the Schroedinger equation for the coupled system.

$$H\psi = id\psi/dt$$

where ψ is a column vector $\binom{a}{b}$ with a = the K^0 amplitude and b = the $\bar{K}^0$ amplitude.

In the absence of any coupling, H would have the form

$$H = \begin{bmatrix} M - i\Gamma/2 & 0 \\ 0 & M - i\Gamma/2 \end{bmatrix}$$

where M is the particle (anti-particle) mass and Γ is its decay width.

With coupling, we have

$$H = \begin{bmatrix} M - i\Gamma/2 & M_{12} - i\Gamma_{12}/2 \\ M_{12}^* - i\Gamma_{12}^*/2 & M - i\Gamma/2 \end{bmatrix} \equiv M - \frac{i\Gamma}{2}$$

Here M_{12} corresponds to our spring and Γ_{12} to our dash pot.

The M and Γ matrices are separately Hermitian, as they correspond to observables.

If we diagonalize this matrix, we find that the eigenstates are given by

$$K_{L,S} = \frac{K^0 \pm \alpha \, \bar{K}^0}{\sqrt{1 + |\alpha|^2}}$$

where

$$\alpha = \sqrt{\frac{M^*_{12} - i\Gamma^*_{12}/2}{M_{12} - i\Gamma_{12}/2}}$$

and

$$\Delta M = 2|M_{12}|$$

$$\Delta\Gamma = -2|\Gamma_{12}|.$$

Note the measurable quantities ΔM and $\Delta\Gamma$ are expressed independently of phase convention. We are defining

$$CP|K^0\rangle = -|\bar{K}^0\rangle.$$

If we introduce a relative phase between $|K^0\rangle$ and $|\bar{K}^0\rangle$:

$$|K^0\rangle \rightarrow e^{i\delta}|K^0\rangle$$

$$|\bar{K}^0\rangle \rightarrow e^{-i\delta}|\bar{K}^0\rangle,$$

then

$$M_{12} = \langle K^0|M|\bar{K}^0\rangle$$

$$\rightarrow e^{-2i\delta} \langle K^0|M|\bar{K}^0\rangle$$

$$\Gamma_{12} \rightarrow e^{-2i\delta} \langle K^0|\Gamma|\bar{K}^0\rangle.$$

However, the mixing parameter α is not independent of phase convention:

$$\alpha \rightarrow \sqrt{\frac{e^{2i\delta}}{e^{-2i\delta}}} \, \alpha = e^{2i\delta}\alpha$$

The phase of α is then arbitrary since the relative phase between the $|K^0\rangle$ and $|\bar{K}^0\rangle$ is not measurable. Hence only the modulus of α is physically meaningful. The modulus is given by

$$|\alpha| = 1 + \frac{|\Gamma_{12}|^2/2}{|M_{12}|^2 + \frac{1}{4}|\Gamma_{12}|^2} \quad \text{Im} \frac{M_{12}}{\Gamma_{12}}$$

$$= 1 + \frac{x}{2(x^2 + \frac{1}{4})} \quad \text{arg} \left(\frac{M_{12}}{\Gamma_{12}}\right) \qquad \text{where } x = \frac{\Delta M}{\Delta \Gamma}$$

Now, using the experimental value for x:

$$x = 0.477$$

and defining

$$\phi_{\Delta S=2} \equiv \text{arg} \left(\frac{M_{12}}{\Gamma_{12}}\right)$$

we have

$$|\alpha| \overset{\sim}{=} 1 + \frac{\phi_{\Delta S=2}}{2} \; .$$

We thus see that the semileptonic charge asymmetry results from an angle between the two off-diagonal "couplings": Γ_{12} and M_{12}.

How big is this angle? The K_L state is given by

$$K_L = K^0 + \alpha \, \bar{K}^0 \; .$$

$$\text{Rate } (K_L \to \pi \ell \bar{\nu}) \sim 1$$

$$\text{Rate } (K_L \to \pi \bar{\ell} \nu) \sim |\alpha|^2$$

$$\to \delta(\infty) = \frac{1 - |\alpha|^2}{1 + |\alpha|^2} = 3.30(12) \times 10^{-3}$$

$$\to |\alpha| = 0.99671(13)$$

and therefore

$$\phi_{\Delta S=2} = -6.58(26) \times 10^{-3}$$

To understand the origin of this phase angle, which is a purely quantum mechanical effect, we will consider the dynamics.

First, note that our ability to see such a small phase difference manifested is a result of the relative sizes of the mass and lifetime difference between the two kaon eigenstates. For heavier systems (e.g. $B^0 \bar{B}^0$), the lifetime difference is much smaller which we can see significantly dilutes the effect.

Consider the elements of the mass matrix, expanded in H_W as a perturbation:

$$M_{11} = M_{K^0} + \langle K^0|H_W|K^0\rangle + \sum_n \frac{\langle K^0|H_W|n\rangle \langle n|H_W|K^0\rangle}{M_{K^0} - E_n}$$

$$M_{22} = M_{\bar{K}^0} + \langle \bar{K}^0|H_W|\bar{K}^0\rangle + \sum_n \frac{\langle \bar{K}^0|H_W|n\rangle \langle n|H_W|\bar{K}^0\rangle}{M_{\bar{K}^0} - E_n}$$

Here M_{K^0} and $M_{\bar{K}^0}$ are the "strong masses." Now it is easy to see that CPT invariance guarantees that $M_{11} = M_{22}$.

Assume $(CPT)^{-1} H_W CPT = H_W$.

Then

$$\langle K^0|H_W|n\rangle = \langle K^0|(CPT)^{-1} H_W CPT|n\rangle$$

Now $\langle K^0|(CPT)^{-1} = -|\bar{K}^0\rangle$

and $CPT|n\rangle = \langle \bar{n}'|$ (spins reversed)

so that we have

$$\langle K^0|H_W|n\rangle = -\langle \bar{n}'|H_W|\bar{K}^0\rangle = -\langle \bar{K}^0|H_W|\bar{n}'\rangle$$

Thus, in the sum over all states $|n\rangle$, we see that $M_{11} = M_{22}$.

The off diagonal elements of the mass matrix are

$$M_{12} = M_{21}^* = \sum_n \frac{\langle \bar{K}^0|H_W|n\rangle \langle n|H_W|K^0\rangle}{M_{K^0} - E_n}$$

$$+ \langle \bar{K}^0|H_W|K^0\rangle .$$

Consider the decay matrix. From the conservation of probability, we can write

$$\frac{d}{dt} \langle \psi|\psi\rangle + 2\pi \sum_F \rho(F)|\langle F|H_W|\psi(t)\rangle|^2 = 0$$

Here $\psi(t)$ is the wave function (linear combination of K^0, $\bar{K}^0$) at time t

F = a real final state

$\rho(F)$ = density of states for state F

$\langle F|H_W|\psi(t)\rangle$ = decay amplitude into state F.

Now

$$\frac{d}{dt} \langle \psi|\psi\rangle = \langle \psi|\dot{\psi}\rangle + \langle \dot{\psi}|\psi\rangle$$

and since

$$H\psi = i\frac{d}{dt}\psi, \quad \dot{\psi} = \frac{H}{i}\psi = \frac{M - \frac{i\Gamma}{2}}{i} = -iM - \frac{\Gamma}{2} , \quad \text{then}$$

191

$$\frac{d}{dt} \langle \psi | \psi \rangle = -\langle \psi | iM + \frac{\Gamma}{2} | \psi \rangle + \langle \psi | iM - \frac{\Gamma}{2} | \psi \rangle$$

$$= -\langle \psi | \Gamma | \psi \rangle$$

Now let us substitute for ψ:

$$\psi = A|K^0\rangle + B|\bar{K}^0\rangle$$

Then

$$\frac{d}{dt} \langle \psi | \psi \rangle = -\langle A^*K^0 + B^*\bar{K}^0 | \Gamma | AK^0 + B\bar{K}^0 \rangle$$

$$
\begin{aligned}
= & - A^* A \langle K^0 | \Gamma | K^0 \rangle \\
& - B^* B \langle \bar{K}^0 | \Gamma | \bar{K}^0 \rangle \\
& \left.
\begin{aligned}
& - A^* B \langle K^0 | \Gamma | \bar{K}^0 \rangle \\
& - B^* A \langle \bar{K}^0 | \Gamma | K^0 \rangle
\end{aligned}
\right\} \quad -2\text{Re}(A^*B \langle K^0 | \Gamma | \bar{K}^0 \rangle)
\end{aligned}
$$

Thus

$$A^*A\, \Gamma_{11} + B^*B\, \Gamma_{22} + 2\text{Re}A^*B\, \Gamma_{12} = 2\pi \sum \rho(F) | \langle F | H_W | \psi(t) \rangle |^2$$

$$= 2\pi \sum_F \rho(F) | \langle F | H_W | (A|K^0\rangle + B|\bar{K}^0\rangle) |^2$$

$$= 2\pi \sum_F \rho(F) | A \langle F | H_W | K^0 \rangle + B \langle F | H_W | \bar{K}^0 \rangle |^2$$

$$= 2\pi \sum_F \rho(F) [A^*A | \langle F | H_W | K^0 \rangle |^2 + B^*B | \langle F | H_W | \bar{K}^0 \rangle |^2$$

$$+ A^*B \langle F | H_W | K^0 \rangle^* \langle F | H_W | \bar{K}^0 \rangle$$

$$+ AB^* \langle F | H_W | K^0 \rangle \langle F | H_W | \bar{K}^0 \rangle^*$$

Now since H_W is Hermitian,

$$\langle F | H_W | \bar{K}^0 \rangle^* = \langle \bar{K}^0 | H_W | F \rangle$$

and, since A and B are arbitrary, we have

$$\Gamma_{11} = 2\pi \sum_F \rho(F) | \langle F | H_W | K^0 \rangle |^2$$

$$\Gamma_{22} = 2\pi \sum_F \rho(F) | \langle F | H_W | \bar{K}^0 \rangle |^2 \qquad \text{and}$$

$$\Gamma_{12} = \Gamma_{21}^* = 2\pi \sum_F \rho(F) \langle K^0 | H_W | F \rangle \langle F | H_W | \bar{K}^0 \rangle.$$

Thus, the lifetime difference is a sum over all (real) final states which K^0 and $\bar{K}^0$ have in common.

What is the source of the phase between Γ_{12} and M_{12}? Suppose all K^0 transition amplitudes (real and virtual) are relatively real. Then the only possibility is the first order term in the expansion of M_{12}:

$$\langle \bar{K}^0 | H_W | K^0 \rangle.$$

This term is $\Delta S = 2$ and there is no such interaction known. That this is the source of CP violation is known as the superweak hypothesis of L. Wolfenstein.

Let us look for a moment at the contributions to Γ_{12}. It will receive a contribution from the following states: (in order of importance)

$$2\pi \; (I = 0)$$
$$2\pi \; (I = 2)$$
$$3\pi \; (I = 1)$$

Therefore,

$$\Gamma_{12} = 2\pi \; [\rho_{2\pi} \langle 2\pi(0) | H_W | K^0 \rangle^* \; \langle 2\pi(0) | H_W | \bar{K}^0 \rangle \; +$$

$$\rho_{2\pi} \langle 2\pi(2) | H_W | K^0 \rangle^* \; \langle 2\pi(2) | H_W | \bar{K}^0 \rangle \; +$$

$$\rho_{3\pi} \langle 3\pi(1) | H_W | K^0 \rangle^* \; \langle 3\pi(1) | H_W | \bar{K}^0 \rangle].$$

Now we define

$$A_0 = \langle 2\pi(0) | H_W | K^0 \rangle$$

$$\bar{A}_0 = \langle 2\pi(0) | H_W | \bar{K}^0 \rangle$$

$$A_2 = \langle 2\pi(2) | H_W | K^0 \rangle$$

$$\bar{A}_2 = \langle 2\pi(2) | H_W | \bar{K}^0 \rangle$$

CP invariance (in the decay process) means that

$$A_0 = \langle 2\pi(0) | H_W | K^0 \rangle = \langle 2\pi(0) | (CP)^{-1} H_W \; CP | K^0 \rangle$$

$$= -\langle 2\pi(0) | H_W | \bar{K}^0 \rangle = -\bar{A}_0$$

and similarly for the other channels.

CPT invariance means that

$$A_0 = \langle 2\pi(0)(CPT)^{-1}|H_W|CPT|K^0\rangle = -\langle \bar{K}^0|H_W|2\pi(0)\rangle$$

$$= -\langle 2\pi(0)|H_W|\bar{K}^0\rangle^*$$

$$= -\bar{A}_0^*.$$

Now,

$$\Gamma_{12} = 2\pi[\rho_{2\pi}\{A_0^*\bar{A}_0 + A_2^*\bar{A}_2\} + \rho_{3\pi}A_{3\pi(1)}^*\bar{A}_{3\pi(1)}]$$

Assuming CP invariance in the decay process,

$$\Gamma_{12} = -2\pi[\rho_{2\pi}\{A_0^*A_0 + A_2^*A_2\} + \rho_{3\pi}A_{3\pi(1)}^*A_{3\pi(1)}]$$

$$\tilde{=} 1 + 2.5 \times 10^{-3} + 0.75 \times 10^{-3}$$

where the relative sizes of the three terms are indicated.

Thus we see that Γ_{12} mainly arises from the $2\pi(0)$ state. We note that there is not a big enough contribution from the other states to generate a significant imaginary part (relative to the $2\pi(0)$ contribution) in Γ_{12} to account for $\mathrm{Im}(M_{12}/\Gamma_{12}) \neq 0$. Thus we must examine the structure of the virtual states.

VI. THE KOBAYASHI-MASKAWA MODEL

In the quark model, the virtual states - at least at short distances - are combinations of the bare quarks and gluons. With three generations, we can draw a "box" diagram which is a second order weak interaction coupling K^0 to $\bar{K}^0$.

This diagram, which accounts for (or contributes to) M_{12} should have a phase with respect to the contributions to Γ_{12}; both are shown in Figure 12. Thus, if the couplings of the heavy quarks were imaginary with respect to the light ones, we could generate the desired effect.

This is the well known scheme of Kobayashi and Maskawa. They showed that three generations are needed before any non-trivial phase can enter.

Let there be an n×n matrix, V, whose elements are the charged current coupling constants from the up to the down quarks.

Then there are n^2 (complex) elements implying $2n^2$ real parameters. The unitarity relation, $V^+V = 1$, gives n^2 relations.

Now, there are 2n quarks and the relative phases of the quark fields (2n-1) are arbitrary. Thus we have

$$2n^2 - n^2 - (2n - 1) \text{ parameters}$$
$$= (n - 1)^2 \text{ parameters}$$

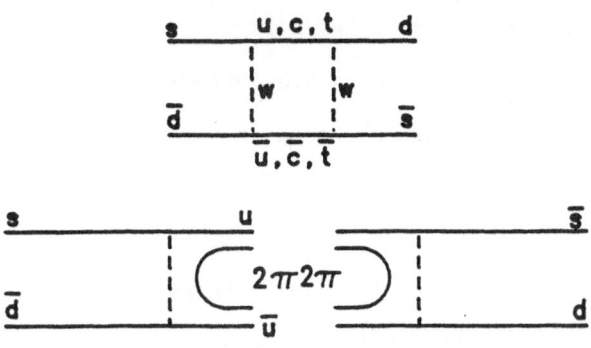

Figure 12 Contributions to M_{12} and Γ_{12}.

n	parameters
1	0
2	1 ("Cabibbo angle")
3	4 (3 "Euler" angles and 1 phase)

For the three generation case, the matrix is denoted:

$$\begin{pmatrix} V_{ud} & V_{us} & V_{ub} \\ V_{cd} & V_{cs} & V_{cb} \\ V_{td} & V_{ts} & V_{tb} \end{pmatrix}$$

We will now discuss briefly our knowledge of the various elements. (See the paper and references therein by Gilman, Kleinknecht and Renk prepared for Physics Letters B, Volume 204, for a more complete discussion of the KM matrix elements.)

First, as the t quark has not been seen, we do not have direct information on the three bottom elements. Only by invoking unitarity can we say anything about these.

V_{ud}

V_{ud} is the best known element. It is determined from Nuclear beta decay using $0^+ \to 0^+$ transitions. Radiative corrections and Coulomb corrections are important. The rates in comparison with muon decay yield:

$$|V_{ud}| = 0.9747(11)$$

Note that this value would imply the rate for pion β decay

$$\Gamma(\pi^+ \to \pi^0 e^+ \nu_e) = 0.4031(16) \text{ sec}^{-1}$$

and the recent value, measured at LAMPF is

$$= 0.3980(150) \ sec^{-1}$$

This process has a 10^{-8} branching ratio and is measured to a precision of 4%!

V_{us}

V_{us} is determined primarily from analysis of K_{e3} decays.
These yield $|V_{us}| = 0.2196(23)$.

Similar information, with larger theoretical uncertainty, comes from the Hyperon β decays:

$$\Sigma^- \rightarrow n \dot{e} \nu$$

$$\Xi^- \rightarrow \Sigma e \nu$$

$$\Lambda \rightarrow p e \nu$$

$$|V_{us}| = 0.220(1)(3)$$

The average of these values is

$$|V_{us}| = 0.2197(19)$$

V_{cd}

This element has been determined from a study of neutrino induced charm production, primarily by the CDHS collaboration. The reaction is

$$\nu d \rightarrow \mu^- c \rightarrow \mu^- \mu^+ s \nu,$$

producing opposite sign dimuons. Backgrounds to this process are from both neutrino and anti-neutrino production of charm from the sea quarks. By combining the single and di-muon cross-sections from an isoscalar target, one can determine V_{cd} and the result (which depends upon the semileptonic branching ratio for charm mesons) is

$$V_{cd} = 0.207(24).$$

There is some uncertainty due to threshold effects so that at higher energies (FNAL Tevatron) this element could perhaps be determined more accurately.

V_{cs}

Similarly, the strange-sea contribution can be isolated but only a lower bound has been determined:

$$V_{cs} \geq 0.59 \ (90\% \ confidence)$$

$\underline{V_{bc}}$

   This element is determined from the b lifetime:

$$\Gamma(b \to c\ell\bar{\nu}_e) = \frac{BF(b \to c\ell\bar{\nu}_e)}{\tau_b} = \frac{G_F^2 M_b^5}{192\pi^3} \; F\left(\frac{M_c}{M_b}\right) |V_{cb}|^2$$

where

$$\tau_b = b \text{ lifetime}$$

$$F\left(\frac{M_C}{M_B}\right) = \text{phase space factor} \overset{\sim}{-} 0.45$$

   Now, the semileptonic branching fraction has been measured in the continuum to be

$$BF(b \to x\ell\bar{\nu}_e) = 12.1(8)\%$$

and the world average b lifetime

$$\tau_b = 1.18(14) \times 10^{-12} \text{ sec}$$

The result is the following:

$$|V_{bc}| = 0.046(10).$$

Several comments are in order here:

(1) We have assumed that the semileptonic branching fraction has little
    contribution from $b \to u\ell\bar{\nu}$ compared to $b \to c\ell\bar{\nu}$. This will be
    justified when we next discuss V_{ub}.

(2) The extracted matrix element depends sensitively on the mass of the
    b quark. Letting $M_b = 5.0 \pm 0.2$ GeV results in an error of about
    10%.

(3) The rate depends upon modeling the various exclusive final states
    in semileptonic B decay for which there is considerable theoretical
    uncertainty. Also the possible difference in charged and neutral B
    decays is neglected.

$\underline{V_{ub}}$

   Here we can hope to see

$$b \to ue\nu$$

distinguished from

$$b \to ce\nu$$

by the shape of the lepton momentum spectrum. This technique is limited by
uncertainties in the overall difference in phase space. One needs
knowledge of the hadronic wave functions for such channels as $B \to D\ell\nu$,
$D^*\ell\nu$, $\pi\ell\nu$, $\rho\ell\nu$, ---.

   From the absence of a signal (CLEO and ARGUS) in the high lepton
momentum region, the following limit can be derived

$$|V_{ub}/V_{cb}| < 0.17 \qquad (90\% \text{ confidence})$$

where the theoretical uncertainty is included.

Thus

$$|V_{ub}| < 0.008$$

The question about a positive signal remains. With a purely hadronic decay:

$$b \to u + W, \ W \to \bar{u}d$$

one can see decays such as

$$B^- \to 3\pi$$

Somewhat surprisingly, the Argus group has reported seeing baryonic decay modes:

$$B \to p\bar{p}\pi \qquad BR(B^+ \to p\bar{p}\pi^+) = 5.2(1.4)(1.9) \times 10^{-4}$$

$$p\bar{p}\pi\pi \qquad BR(B^0 \to p\bar{p}\pi^+\pi^-) = 6.0(2.0)(2.2) \times 10^{-4}$$

With a good deal of uncertainty, one can set a <u>lower</u> limit of

$$|V_{ub}| \gtrsim 0.003.$$

The CLEO group has recently reported the lack of observation of the above decay channels so it is not yet clear whether this important transition has been seen.

To summarize what we have said above:

$$V = \begin{pmatrix} 0.09747(11) & 0.2197(19) & 0.0065(35) \\ 0.21(3) & > 0.66 & 0.46(10) \\ --- & --- & --- \end{pmatrix}$$

The accuracy is sufficient to check unitarity only for the top row:

$$|V_{ud}|^2 \ + \ |V_{us}|^2 \ + \ |V_{ub}|^2 \overset{?}{=} 1$$

$$0.9500(20) \ + \ 0.0483(8) \ + \ 0.00004(4) \overset{?}{=} 0.9983(22)$$

Thus, within one sigma, there is no evidence of any additional generation, although we note that the third element hardly contributes.

If we improve the constraint of unitarity, we have the following for the coupling constants:

(the errors are 90% confidence values)

$$V = \begin{pmatrix} 0.9754(6) & 0.219(2) & 0.0065(35) \\ 0.219(2) & 0.9743(10) & 0.046(16) \\ 0.012(11) & 0.046(16) & 0.9988(8) \end{pmatrix}$$

VII. CP VIOLATION IN THE KM MODEL

So far we have not included CP nonconservation. The KM matrix can be parameterized arbitrarily. We will use a convention advocated by Hararri and Leurer

$$V = \begin{pmatrix} C_{12}C_{13} & S_{12}C_{13} & S_{13}e^{-i\delta_{13}} \\ -S_{12}C_{23} - C_{12}S_{23}S_{12}e^{i\delta_{13}} & C_{12}C_{23} - S_{12}S_{23}e^{i\delta_{13}} & S_{23}C_{13} \\ S_{12}S_{23} - C_{12}C_{23}S_{13}e^{i\delta_{13}} & -C_{12}S_{23} - S_{12}C_{23}S_{13}e^{i\delta_{13}} & C_{23}C_{13} \end{pmatrix}$$

where the KM angles are θ_{12}, θ_{23}, and θ_{13} and $C_{ij} = \cos(\theta_{ij})$; $S_{ij} = \sin(\theta_{ij})$.

In the (real) case of small angles, we have $S_{ij} = \theta_i$; $C_{ij} = 1$.

$$V = \begin{pmatrix} 1 & S_{12} & S_{13}e^{-\delta_{13}} \\ -S_{12} & 1 & S_{23} \\ S_{12}S_{23} - S_{13}e^{i\delta_{13}} & -S_{23} & 1 \end{pmatrix}$$

L. Wolfenstein noticed a regularity in the sizes of the elements which at best has physical significance but at worst provides a very easy method of estimating various processes.

$$V = \begin{pmatrix} 1 - \lambda^2/2 & \lambda & \lambda^3 e^{i\delta} \\ -\lambda & 1 - \lambda^2/2 & \lambda^2 \\ \lambda^3(1 - e^{-i\delta}) & -\lambda^2 & 1 \end{pmatrix}$$

Here $\lambda = 0.22$ and $\delta \neq 0$, $\pi \rightarrow$ CP nonconservation. Certainly within current experimental error, this hierarchy for the couplings holds.

Let us see whether we can approximately account for CP violation in this model. Figure 13 shows two diagrams contributing to the matrix element M_{12}.

Thus, for these two contributions, we have

$$\frac{\text{Im } M_{12}}{|M_{12}|} \stackrel{\sim}{=} \lambda^4 \sin\delta$$

Now, since in this phase convention $\text{Im}\Gamma_{12} = 0$ and since $|M_{12}|$ and $|\Gamma_{12}|$ are of the same order, we have

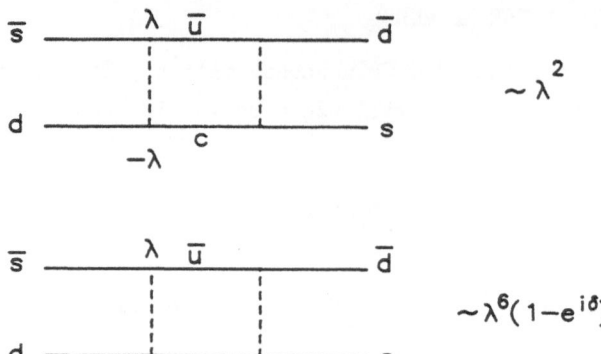

Figure 13 Contributions to M_{12}.

$$\phi_{\Delta S=2} \sim \lambda^4 \sin\delta.$$

Since $\lambda^4 = 2.3 \times 10^{-3}$, it is clear that we can accommodate CP violation with a δ value not too small. The small size of CP violation effects can be seen to be a result of the very small couplings across two generations. At this point, we have seen that the KM scheme can account for CP nonconservation. Soon we will discuss a more rigorous treatment which attempts to take into account all of the many details which we have so far not treated. However, it is clear that we have one parameter, δ, which yields one experimental number, $\phi_{\Delta S=2}$. For the KM scheme to be correct, we need another manifestation.

There is the possibility that, with particles made of the heavier quarks, CP violating effects will be significantly greater because of the lack of Cabibbo suppression. We will see that this is so. There is another manifestation in the K^0 system which is now being pursued vigorously (ϵ'/ϵ).

VIII. THE STANDARD PHASE CONVENTION

First, we must now go to the "standard" phase convention for the K^0 system. It would be desirable to express all measurable quantities in convention-independent notation, but to do this would make for difficulties in reading the literature.

We write the Kaon decay eigenstates as

$$K_{L,S} = \frac{(1 + \epsilon)|K^0> \mp (1 - \epsilon)|\bar{K}^0>}{\sqrt{2}}$$

and where the phase of ϵ is determined by the choice (Wu-Yang convention) of A_0 to be real.

Now, $\delta = \dfrac{1 - |\alpha|^2}{1 + |\alpha|^2}$ $\qquad$ $\alpha = \dfrac{1 - \epsilon}{1 + \epsilon}$

$$|\alpha|^2 = \left| \dfrac{1 - \epsilon}{1 + \epsilon} \right|^2 = \dfrac{(1 - \epsilon)(1 - \epsilon^*)}{(1 + \epsilon)(1 + \epsilon^*)} \simeq \dfrac{1 - \epsilon - \epsilon^*}{1 + \epsilon + \epsilon^*} \simeq 1 - 4\,\mathrm{Re}\,\epsilon$$

$$\delta = \dfrac{4\,\mathrm{Re}\,\epsilon}{2} = 2\,\mathrm{Re}\,\epsilon$$

We have assumed that ϵ is small. This follows from our phase convention and the dominance of $\Delta I = 1/2$ transitions.

$$K_L = \dfrac{(1 + \epsilon)\,|K^0\rangle - (1 - \epsilon)\,|\bar{K}^0\rangle}{\sqrt{2}}$$

$$\langle 2\pi(0)\,|H_W|K_L\rangle = \dfrac{(1 + \epsilon)A_0 - (1 - \epsilon)A_0}{\sqrt{2}} = \sqrt{2}\,\epsilon\,A_0$$

$$\langle 2\pi(0)\,|H_W|K_S\rangle = \dfrac{2}{\sqrt{2}}\,A_0 = \sqrt{2}\,A_0$$

Thus, with pure $\Delta I = 1/2$ transitions, we would expect

$$\eta_{+-} \equiv \dfrac{\langle \pi^+\pi^-\,|H_W|K_L\rangle}{\langle \pi^+\pi^-\,|H_W|K_S\rangle} = \epsilon = \eta_{00} \equiv \dfrac{\langle 2\pi^0\,|H_W|K_L\rangle}{\langle 2\pi^0\,|H_W|K_S\rangle}$$

The parameters η_{+-} and η_{00} are directly measurable, in both magnitude and phase. Note that the 2π decays of the K_L state come from the "wrong" CP admixture in

$$K_L \stackrel{\sim}{=} K_2 + \epsilon\,K_1$$

where $\qquad\qquad K_1 = \dfrac{K^0 + \bar{K}^0}{\sqrt{2}}$

$$K_2 = \dfrac{K^0 - \bar{K}^0}{\sqrt{2}}\ .$$

IX. MEASUREMENT OF η_{+-}

If we begin with a pure K^0, its time evolution will be

$$K^0 = (K_L + K_S)/(1 + \epsilon)$$

$$= (|K_L\rangle\, e^{-(iM_\ell + \gamma_{\ell/2})\tau} + |K_S\rangle\, e^{-(iM_S + \gamma_{S/2})\tau})/(1 + \epsilon)$$

$$\langle \pi^+\pi^-\,|H_W|K^0(t)\rangle|^2 \alpha\ (e^{-\gamma_S\tau} + |\eta_{+-}|^2\, e^{-\gamma_\ell\tau} + 2|\eta_{+-}|e^{-(\gamma_\ell + \gamma_S)\tau}\cos(\Delta M\tau + \phi_{+-}))$$

where $\eta_{+-} = |\eta_{+-}|e^{i\phi}{}_{+-}$.

The result from experiment is

$$\eta_{+-} = 2.275(21) \times 10^{-3} e^{i44.6(1.2)^0}$$

and therefore

$$2 \text{ Re } \eta_{+-} = 3.239(30) \times 10^{-3}$$

in excellent agreement with

$$\delta = 3.30(12) \times 10^{-3}$$

as is expected from the dominance of the $\Delta I = 1/2$ transition.

X. OTHER MANIFESTATIONS IN THE KAON SYSTEM: ϵ'/ϵ

Let us consider the decay process $K^0 \rightarrow 2\pi$. We have seen that if the final state is pure $\Delta I = 1/2$, then one can learn no more from a study of this decay mode. However, there is a small $\Delta I = 3/2$ transition which we will now treat.

$$\text{amp}(K^0 \rightarrow \pi^+\pi^-) = \sqrt{\frac{2}{3}} A_0 e^{i\delta_0} + \sqrt{\frac{1}{3}} A_2 e^{i\delta_2}$$

$$\text{amp}(\bar{K}^0 \rightarrow \pi^+\pi^-) = \sqrt{\frac{2}{3}} A_0^* e^{i\delta_0} + \sqrt{\frac{1}{3}} A_2^* e^{i\delta_2}$$

where $\delta_{2(0)}$ are the $\pi\pi$ phase shifts in the $I = 2(0)$ channels. Thus

$$\text{Rate}(K^0 \rightarrow \pi^+\pi^-) \propto |A_0 + \sqrt{\frac{1}{2}} A_2 e^{i(\delta_2-\delta_0)}|^2$$

$$\text{Rate}(\bar{K}^0 \rightarrow \pi^+\pi^-) \propto |A_0^* + \sqrt{\frac{1}{2}} A_2^* e^{i(\delta_2-\delta_0)}|^2$$

Thus, using $|A_2|^2 \ll |A_0|^2$, we have

$$\text{Rate }(K^0 \rightarrow \pi^+\pi^-) = |A_0|^2 + \frac{A_0}{\sqrt{2}} A_2^* e^{-i(\delta_2-\delta_0)} + \frac{A_0^*}{\sqrt{2}} A_2 e^{i(\delta_2-\delta_0)}$$

$$= |A_0|^2 + \sqrt{2} \text{ Re } (A_0 A_2^* e^{-i(\delta_2-\delta_0)})$$

$$\propto 1 + \sqrt{2} \text{ Re } \left[\frac{A_2^*}{A_0^*} e^{-i(\delta_2-\delta_0)} \right]$$

similarly

$$\text{Rate }(\bar{K}^0 \rightarrow \pi^+\pi^-) \propto 1 + \sqrt{2} \text{ Re } \left[\frac{A_2}{A_0} e^{-i(\delta_2-\delta_0)} \right]$$

Thus, if A_2 and A_0 are relatively real, the rates for K^0 and $\bar{K}^0$ are identical, independent of final state interactions.

In general,

$$\frac{\text{Rate } (K^0 \to \pi^+\pi^-)}{\text{Rate } (\bar{K}^0 \to \pi^+\pi^-)} \cong 1 + \sqrt{2}\left[\text{Re}\left(\frac{A_2^*}{A_0^*}\ e^{-i(\delta_2-\delta_0)}\right) - \text{Re}\left(\frac{A_2}{A_0}\ e^{-i(\delta_2-\delta_0)}\right)\right]$$

$$= 1 - 2\sqrt{2}\ \text{Im}\left(\frac{A_2}{A_0}\right)\ \text{Im}\ (e^{-i(\delta_2-\delta_0)})$$

$$= 1 + 2\sqrt{2}\ \text{Im}\left(\frac{A_2}{A_0}\right)\ \sin(\delta_2-\delta_0)$$

Thus we see that there would be a rate asymmetry if
(a) $\text{Im}(A_2/A_0) \neq 0$
(b) $\delta_2 \neq \delta_0 \pmod{\pi}$

This is a general result for any particle/antiparticle rate asymmetry.
The parameter ϵ' is defined as

$$\epsilon' = \frac{1}{\sqrt{2}}\ i\ \text{Im}\left(\frac{A_2}{A_0}\right)\ e^{i(\delta_2-\delta_0)}$$

so that we have

$$\frac{\text{Rate } (K^0 \to \pi^+\pi^-)}{\text{Rate } (\bar{K}^0 \to \pi^+\pi^-)} = 1 - 4\ \text{Re}\ \epsilon'.$$

Now, let's look at the KM model to see if we expect that $\epsilon' \neq 0$, i.e., that $\text{Im}(A_2/A_0) \neq 0$.

We will comment here briefly on the phase-shifts δ_2 and δ_0. They can both be determined over a broad energy range by a study of the reactions

$$\pi^-p \to \pi^-\pi^0 p$$
$$\to \pi^+\pi^- n$$

In addition, δ_0 can be determined separately up to $M_{\pi\pi} \cong 360$ MeV from a partial wave analysis of the reaction

$$K^+ \to \pi^+\pi^- e^+\nu$$

The results are consistent and a "grand" average yields

$$(\delta_2 - \delta_0)_K = -43^0 \pm 5^0$$

where the error is an estimate of the systematic uncertainty.

Thus, the phase of ϵ' is well known and is therefore

$$\arg(\epsilon') \cong 47^0 \pm 5^0$$

In the KM model, the $\Delta I = 3/2$ amplitude comes from the spectator graph, and, in the adopted phase convention, is purely real. The penguin graph which we previously introduced with a virtual top quark will contribute

$$\sim V_{ts}V_{td} = \lambda^2\lambda^3(1 - e^{-i\delta})$$

which has an imaginary part $\sim \lambda^5 \sin\delta$.

The spectator graph $\sim \lambda$ so that again

$$\frac{\text{Im } A_0}{A_0} \sim \lambda^4 \sin\delta$$

More careful estimates give (as we will see)

$$|\epsilon'| \sim 5 \times 10^{-6}$$

There is much effort to see this effect which would constitute another manifestation of CP violation and, an effect in this range would do much towards establishing the KM mechanism.

How can it be detected? We have seen that a rate asymmetry on the order of 4 Re $\epsilon' \stackrel{\sim}{-} 1.4 \times 10^{-5}$ is expected. The LEAR facility may be a good place to see such an effect. However, to establish such an effect at the 2σ level would require nearly 10^{11} detected events! In addition, there are complications due to particle mixing to which we shall return.

Let us now consider the $K_{L,S}$ decay amplitudes to two pions.

$$\text{amp}(K^0 \rightarrow \pi^+\pi^-) = \sqrt{\frac{2}{3}}\, A_0 \left[1 + \frac{1}{\sqrt{2}}\, \frac{A_2}{A_0}\, e^{i(\delta_2-\delta_0)}\right]$$

$$\text{amp}(\bar{K}^0 \rightarrow \pi^+\pi^-) = \sqrt{\frac{2}{3}}\, A_0 \left[1 + \frac{1}{\sqrt{2}}\, \frac{A_2^*}{A_0}\, e^{i(\delta_2-\delta_0)}\right]$$

$$K_L = K_2 + \epsilon K_1 \qquad\qquad K_1 = \frac{K^0 + \bar{K}^0}{\sqrt{2}}$$

$$K_2 = \frac{K^0 - \bar{K}^0}{\sqrt{2}}$$

$$\text{amp}(K_1 \rightarrow \pi^+\pi^-) \stackrel{\sim}{-} \frac{2}{\sqrt{3}}\, A_0$$

$$\text{amp}(K_2 \rightarrow \pi^+\pi^-) \stackrel{\sim}{-} \frac{1}{\sqrt{2}} \sqrt{\frac{1}{3}}\, e^{i(\delta_2-\delta_0)}\, (A_2 - A_2^*) = \sqrt{\frac{2}{3}}\, i e^{i(\delta_2-\delta_0)}\, \text{Im } A_2$$

Thus $\text{amp}(K_L \rightarrow \pi^+\pi^-)$

$$= \frac{2}{\sqrt{3}} \left[\epsilon A_0 + i\, \frac{\text{Im } A_2}{\sqrt{2}}\, e^{i(\delta_2-\delta_0)}\right]$$

$$= \frac{2}{\sqrt{3}} A_0 \left[\epsilon + i \frac{\text{Im } A_2}{\sqrt{2} A_0} e^{i(\delta_2 - \delta_0)} \right] = \frac{2}{\sqrt{3}} A_0 (\epsilon + \epsilon'),$$

Thus

$$\eta_{+-} = \frac{\text{amp}(K_L \rightarrow \pi^+\pi^-)}{\text{amp}(K_S \rightarrow \pi^+\pi^-)} = \epsilon + \epsilon' \quad \text{and}$$

$$\eta_{00} = \qquad\qquad \epsilon - 2\epsilon'$$

Thus we see that by comparing the complex parameters η_{+-} and η_{00}, we can hope to isolate a "direct" effect.

$$\frac{\eta_{+-}}{\eta_{00}} \overset{\sim}{-} 1 + 3\frac{\epsilon'}{\epsilon}$$

Since $\arg(\eta_{+-}) \overset{\sim}{-} \arg(\epsilon')$ and $\eta_{+-} \overset{\sim}{-} \epsilon$, we see that ϵ'/ϵ is essentially real. The experiments usually measure

$$\left| \frac{\eta_{+-}}{\eta_{00}} \right|^2 \overset{\sim}{-} 1 + 6\frac{\epsilon'}{\epsilon}$$

which will differ from 1 by 1-2% in the KM model. Before we discuss the experiments on ϵ'/ϵ, we want to treat two points:

1) Why is $\arg(\epsilon) \overset{\sim}{-} 45^0$?

2) What are the more reliable predictions for ϵ'/ϵ?

Consider again the conservation of probability equation.

$$\frac{d}{dt} \langle \psi | \psi \rangle + 2\pi \sum_F \rho_F |\langle F | H_W | \psi(t) \rangle|^2 = 0$$

Let $\psi = a K_L + b K_S$

$$\frac{d}{dt} \langle a^* K_L + b^* K_S | a K_L + b K_S \rangle$$

$$= a^*a \frac{d}{dt} \langle K_L | K_L \rangle + b^*b \frac{d}{dt} \langle K_S | K_S \rangle + a^*b \frac{d}{dt} \langle K_L | K_S \rangle + ab^* \frac{d}{dt} \langle K_S | K_L \rangle$$

$$= a^*a \frac{d}{dt} \langle K_L | K_L \rangle + b^*b \frac{d}{dt} \langle K_S | K_S \rangle + 2 \, \text{Re} \left[a^*b \frac{d}{dt} \langle K_L | K_S \rangle \right]$$

from the conservation of probability, this equals

$$-2\pi \sum_F \rho_F \langle a^* K_L + b^* K_S | H_W | F \rangle \langle F | H_W | a K_L + b K_S \rangle$$

$$= -2\pi \sum_F \rho_F \left[a^*a \langle K_L | H_W | F \rangle \langle F | H_W | K_L \rangle + b^*b \langle K_S | H_W | F \rangle \langle F | H_W | K_S \rangle \right.$$

$$+ a^*b \langle K_L|H_W|F\rangle \langle F|H_W|K_S\rangle + b^*a \langle K_S|H_W|F\rangle \langle F|H_W|K_L\rangle \Big]$$

Again, since this is valid for all a and b, we have

$$-\frac{d}{dt} \langle K_S|K_L\rangle = 2\pi \sum_F \rho_F \langle K_S|H_W|F\rangle \langle F|H_W|K_L\rangle$$

Now, the left side of this equation is given by

$$-\frac{d}{dt} \langle e^{iM_S - \gamma_S/2 \, t} K_S|K_L \, e^{-(iM_L + \gamma_L/2)t}\rangle = \left[i(M_L - M_S) + \frac{\gamma_S + \gamma_L}{2}\right] \langle K_S|K_L\rangle$$

$$= \left[i\Delta M + \frac{\gamma_S + \gamma_L}{2}\right] \langle K_1 + \epsilon^* K_2|K_2 + \epsilon K_1\rangle$$

$$= \left[i\Delta M + \frac{\gamma_S + \gamma_L}{2}\right] 2 \, \text{Re} \, \epsilon \; \tilde{=} \; \left[i\Delta M + \frac{\gamma_S}{2}\right] 2 \, \text{Re} \, \epsilon$$

Let us now consider the right hand side, using the dominance of the 2π modes "connecting" K_L and K_S:

$$2\pi \sum_F \rho_F \langle K_S|H_W|F\rangle \langle F|H_W|K_L\rangle$$

$$= 2\pi \left\{ |\langle K_S|H_W|\pi^+\pi^-\rangle|^2 \; \eta_{+-} \; \rho_{\pi^+\pi^-} \right.$$

$$+ |\langle K_S|H_W|2\pi^0\rangle|^2 \; \eta_{00} \; \rho_{\pi^0\pi^0} \Big\}$$

$$= \gamma_S \{\eta_{+-} \, \text{BR}(K_S \to \pi^+\pi^-) + \eta_{00} \, \text{BR}(K_S \to \pi^0\pi^0)\}$$

Thus we have

$$2 \, \text{Re} \, \epsilon \; \tilde{=} \; \frac{\gamma_S}{i\Delta M + \gamma_{S/2}} \{\eta_{+-} \, \text{BR}(K_S \to \pi^+\pi^-) + \eta_{00} \, \text{BR}(K_S \to \pi^0\pi^0)\}$$

Now, if we <u>neglect</u> ϵ' effects, the term in brackets is simply ϵ and we have

$$\frac{\text{Re} \, \epsilon}{\epsilon} = \frac{\gamma_S}{2i\Delta M + \gamma_S}$$

so that

$$\arg \epsilon = \arctan 2\Delta M/\gamma_S = 43.7^0 \, (2^0)$$

This phase is known as the superweak phase. Including ϵ' effects, we get

$$2 \, \text{Re} \, \epsilon \; \tilde{=} \; \frac{\gamma_S}{i\Delta M + \dfrac{\gamma_S}{2}} \{\epsilon + 0.06 \, \epsilon'\}$$

and the conclusion hardly changes.

XI.   THE THEORY: ONE EXAMPLE

Let's now look more closely at the theoretical situation. We are guided by the recent paper by Buras and Gerard PLB <u>203</u>, 272 (1988). The theoretical uncertainties are in

    a)   values of KM parameters M_t, θ_{13}, and M_S, the strange quark mass.

    b)   values of hadronic matrix elements of operators contributing to ϵ and ϵ'

To address b) there are 4 (approximate) techniques

    1)   lattice approach

    2)   hadronic sum rules

    3)   QCD sum rules

    4)   1/N approach

The 1/N approach used by the authors accounts for the $\Delta I = 1/2$ rule as a "standard model phenomenon." The other approaches do or have not. The result of the calculation by Buras and Gerard is expressed as

$$\frac{\epsilon'}{\epsilon} = 2.1 \times 10^{-3} \left[\frac{150 \text{ MeV}}{M_S(1 \text{ GeV})} \right]^2 \left[\frac{M_W}{M_t} \right]^{4/3} \frac{\beta^{4/5}}{1 + \frac{8}{3}\sqrt{\bar{R}}}$$

Where

   - M_S is the strange quark mass, evaluated at 1 GeV

   - M_t is the top quark mass

   - β represents the value for the b lifetime

$$\tau_B = \beta \times 10^{-12} \text{ S}$$

   - and $\bar{R}$ represents the magnitude of the b $\rightarrow$ u transition

$$\bar{R} = \frac{b \rightarrow ue\nu}{b \rightarrow ce\nu}$$

The expression includes a constraint upon $\bar{R}$, M_t, and β so that the experimental value for ϵ obtains.

Another piece of information which will further constrain the parameter space is the size of the $B^0, \bar{B}^0$ mixing.

The diagram governing this phenomenon is shown below in Figure 14.

This will go $\sim\lambda^6$ and will increase as M_t increases. Experiments measure

$$x_b \equiv \frac{\Delta M}{\Gamma} = \begin{array}{ll} 0.73(18) & \text{ARGUS} \\ 0.66 & \text{CLEO (preliminary)} \end{array}$$

The final result of Buras and Gerard is then:

$$\frac{\epsilon'}{\epsilon} = (1.6 \pm 0.6) \left[\frac{f_B}{f_K} \right]^{3/2} \left[\frac{150 \text{ MeV}}{M_S (1 \text{ GeV})} \right]^2 \times 10^{-3}$$

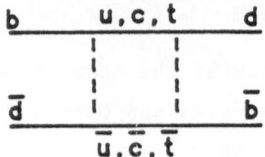

Figure 14 $B^0 \bar{B}^0$ mixing.

for any value of M_t. The error allows $\bar{R}$ to vary from 0.01 to 0.10 where the smaller $\bar{R}$ corresponds to the larger ϵ'/ϵ.

CONCLUSIONS

1. The theory is certainly getting more refined.
2. The uncertainty on the strange quark mass contributes greatly.
3. There is sensitivity to f_B, the B meson decay constant.
4. Experiments should clearly be sensitive in the 10^{-3} range.

The (large) mixing in the B system gives information on V_{td}.

$$|V_{td}| \; \overset{\sim}{=} \; \frac{1.6(4) \text{ GeV}}{M_t \text{ [GeV]}}$$

Now lets use the unitarity of the KM matrix to derive

$$\sum_{u'} V_{u'd} V^*_{u'b} = 0$$

$$= V_{ud} V^*_{ub} + V_{cd} V^*_{cb} + V_{td} V^*_{tb}$$

$$= V^*_{ub} - 0.010 + V_{td}$$

where the only "assumption" is that $V_{tb} \overset{\sim}{=} 1$. This leads to the now famous "Bj triangle" relationship (first written down by J. Bjorken) among the elements V_{td} and V_{ub} shown below.

Can we establish that the Bj triangle is indeed a triangle? In other words, is $|V_{td}| - 0.01 \overset{<}{=} |V_{ub}|$?

If the equal sign holds, then the KM model (with three generations) has no bearing on CP violation!

XII. EXPERIMENTS

We will discuss the three (modern) experiments which have sensitivity to ϵ'/ϵ. We begin with CPLEAR which, conceptually, is easiest to understand. Next we will treat CERN NA31 and finally FNAL E731.

XII-1. CPLEAR

(Basel, CERN, Saclay, Athens, Democrites, EIH, Fribourg, Liverpool, SIN, Stockholm, Thessaloniki collaboration.)

At LEAR, one is well set up to measure the particle/antiparticle differences in decay rates as both are produced symmetrically and are tagged.

We pointed out earlier that $\sim 2 \times 10^6$ $\bar{K}^0$'s are tagged per second with a mean energy $\langle E \rangle \stackrel{\sim}{-} 660$ MeV, a mean velocity $\langle \beta \rangle \stackrel{\sim}{-} 0.67$ and a mean decay length of 3 cm.

This implies that one can study decays up to about 20 or 30 lifetimes. The charged resolution is very good so that the experimenters expect

$$|\eta_{+-}| \text{ measured to } 5 \times 10^{-3}$$
$$\phi_{+-} \text{ measured to } 0.4^0 \text{ and}$$
$$\Delta M/M \text{ measured to } 1.2 \times 10^{-3}.$$

The neutral mode is much harder, requiring the detection of 4 γ's at very low energy. Both energy and position must be reconstructed.

However, there is information even without good vertex resolution.

The time integrated asymmetry, defined as

$$I_{+-} = \frac{\int_0^T (K^0 \rightarrow \pi^+ \pi^-) \, dt - \int_0^T (\bar{K}^0 \rightarrow \pi^+ \pi^-) \, dt}{\int_0^T (K^0 \rightarrow \pi^+ \pi^-) \, dt + \int_0^T (\bar{K}^0 \rightarrow \pi^+ \pi^-) \, dt}$$

can easily be shown to satisfy

$$I_{+-} = 2 \text{ Re } (\epsilon + 2\epsilon') \text{ while}$$
$$I_{00} = 2 \text{ Re } (\epsilon - 4\epsilon')$$

so that

$$\frac{I_{+-}}{I_{00}} \stackrel{\sim}{-} 1 + 6 \text{ Re } \frac{\epsilon'}{\epsilon}$$

Each asymmetry is of the order of 2 Re $\epsilon \stackrel{\sim}{-} 3 \times 10^{-3}$ and if each is measured with a precision of 1%, then ϵ'/ϵ will be determined with a precision of

$$\sigma(\epsilon'/\epsilon) = \frac{.01 \sqrt{2}}{6} \stackrel{\sim}{-} 0.0024$$

To measure an asymmetry of 3×10^{-3} to the 1% level, however, requires 10^9 events.

The required statistics can be obtained in not too long a period so that systematic effects will be important.

We will see that there are other CP parameters than can be well measured at LEAR.

The schedule is roughly as follows:

Detector Assembly	Aug → Nov 88
Test Run	Nov, Dec 88
Data Taking	Spring, and Fall 89

XII-2. CERN NA31

Next, we discuss the NA31 experiment which has reported a result for

$$\left|\frac{\eta_{00}}{\eta_{+-}}\right|^2 = 0.980 \pm 0.004 \pm 0.005 \to \text{Re } \epsilon'/\epsilon = (3.3 \pm 1.1) \times 10^{-3}$$
$$\text{statistical systematic}$$

Their technique is to measure first in a K_L beam the detected rates for both 2π modes and second in a K^0 beam, i.e. one where they target closer to the detector. In fact, data are taken at different target positions to reduce the effects of detector acceptance. A schematic view of the detector is shown in Figure 15.

The experiment determines the above double ratio in bins of momentum and reconstructed vertex.

There are 10 energy bins from 70 GeV to 170 GeV and 32 vertex bins from 10.5m to 48.9m where the vertex is measured from the final K_L collimator.

Let us briefly consider some of the reconstruction and systematic issues.

1. Acceptance and smearing corrections

The two beam configurations differ somewhat in beam divergence and as well in the K_S beam there is a 7mm lead converter which results in scattering of the incident beam. In addition, the different vertex and

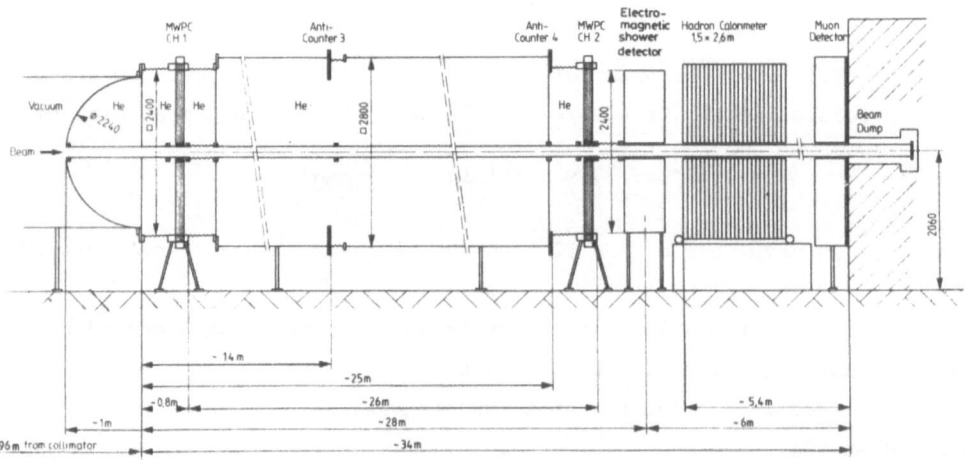

Figure 15 NA31 detector.

energy resolutions for the two modes will result in different amounts of smearing. The <u>net</u> of all of these effects is claimed to be 0.3%.

2. Energy determination, neutral

The lead/liquid Ar calorimeter used for the neutral mode provides $\sigma/E = 7.5\%/\sqrt{E}$ energy resolution so that for a 120 GeV Kaon decay, the total energy of the event should be determined with a precision of about 0.8%.

The decay position Z relative to the calorimeter can be determined by constraining the 4 detected γ rays to have the (known) K^0 mass:

$$M^2_{4\gamma} = (\Sigma P_i)^2$$

$$= 2 \sum_{i>j} P_i P_j \qquad\qquad P_i = (E_i, \vec{P}_i)$$

$$= 2 \Sigma \Sigma (E_i E_j - \vec{P}_i \cdot \vec{P}_j)$$

$$= 2 \Sigma \Sigma E_i E_i (1 - \cos\theta_{ij})$$

$$\simeq 2 \Sigma \Sigma E_i E_i \frac{\theta_{ij}^2}{2}$$

$$= \frac{2 \Sigma \Sigma E_i E_j r_{ij}^2}{Z^2}$$

where r_{ij} is the measured distance between γ's i and j in the calorimeter.

For each running period, the overall energy scale of the detector is adjusted by up to 10 or 15% so that the reconstructed Z position has its sharp edge (in the K_S running) at the correct position. The claimed accuracy is 10^{-3} or about 10cm.

3. Energy determination, charged

The charged decays are measured with wire chambers and an iron/scintillator calorimeter with energy resolution

$$\sigma/E = \frac{65\%}{\sqrt{E}}$$

The total energy resolution for a 120 GeV decay would then be about 6%. However, by using the constraint between the K^0 mass and the opening angle:

$$E^2 = \frac{(M_K^2 - M_\pi^2 R)R}{\theta^2}$$

where

$$R = 2 + \frac{E_1}{E_2} + \frac{E_2}{E_1}$$

the resolution in E, the total energy, can be improved.

It is desirable to keep R small so that its uncertainty has minimal effect on the E determination. The group cuts at

$$0.4 \leq \frac{E_1}{E_2} \leq 2.5$$

so that

$$4 \leq R \leq 4.9$$

and $\Delta E/E$ becomes 1%.

4. Backgrounds, neutral mode:

The principal background for K_L decays is the $3\pi^0$ mode. It varies across the decay region and is about 4% overall.

5. Backgrounds, charged mode:

Even with the relatively poor mass resolution, the residual (semileptonic) backgrounds can be subtracted by looking at the (perpendicular) distance between the production target and the decay plane. The combined backgrounds are $(6.5 \pm 2.0) \times 10^{-3}$.

6. Accidental effects (0.34%)

This small correction actually results from the difference of effects nearly 10 times as large.

This group is now running with some improvements to further reduce the background.

XII-3. FNAL E731

We will now discuss E731 which has a very different technique.

One major difference is the use of a regenerator to provide K_S decays. We can think of the regeneration process as altering the K^0-$\bar{K}^0$ mixture in the incident K_L.

The mixture is altered upon passage through material since K^0 and $\bar{K}^0$ have different scattering amplitudes.

Let $f = K^0 A$ forward scattering amplitude

$\bar{f} = \bar{K}^0 A$ forward scattering amplitude

Then the amplitude of K_S produced will be proportional to the difference in forward amplitudes:

$$|K_L\rangle \rightarrow |K_L\rangle + \rho \, |K_S\rangle$$

$$\rho \stackrel{\sim}{-} i\pi NL \, \frac{f - \bar{f}}{K} \qquad\qquad N = \# \text{ density} = \frac{N_0 d}{A}$$

$$L = \text{length of regenerator}$$

This expression comes from summing the amplitudes for regeneration over the entire target and neglecting the phase difference and K_S decay across the target which is valid at high energies.

The advantages of using regenerated K_S's are in the fact that the "beams" are identical and accidental and dead time effects are essentially eliminated when one employs two such beams side-by-side and decays from both modes are collected simultaneously.

We note that

$$\sigma_T NL \equiv x = \text{the number of Kaon interaction lengths,}$$

so that

$$\rho = \frac{i\pi x}{\sigma_T} \frac{f - \bar{f}}{K}$$

The number of K_S's produced is given by

$$N_S = |\rho|^2 e^{-x} \alpha x^2 e^{-x}$$

so that, for any material, the maximum rate is at $x = 2$.

Now the difference in forward amplitudes decreases with energy. The total cross-section is given by the optical theorem:

$$\sigma_T = 4\pi \text{ Im } \frac{f(0)}{K}$$

so that if we write

$$\frac{f - \bar{f}}{K} = \left| \frac{f - \bar{f}}{K} \right| e^{i\phi_{21}},$$

we find

$$\Delta\sigma = 4\pi \left| \frac{f - \bar{f}}{K} \right| \sin\phi_{21}.$$

The regeneration amplitude (magnitude and phase) can be measured as a result of the interference in $K \rightarrow 2\pi$ behind a regenerator. The 2π rate will be given by

$$\text{amp}(K \rightarrow 2\pi) = |\rho e^{-t/2 + i\Delta Mt} + \eta|^2$$

The results of many such measurements yield the following

$$\phi_{21} \cong -126^0 \qquad \text{(independent of material)}$$

$$\left| \frac{f - \bar{f}}{K} \right| \cong 2.23 \, p^{-0.614} \, A^{0.76} \text{ mb}$$

We will briefly discuss the characteristics of this amplitude.

(1) momentum dependence

For such forward scattering, a t channel analysis is appropriate. We ask what particles can be exchanged in the scattering. For K^0 scattering, we have

$$f = P + \rho + \omega + A_2 + - - -$$

and $\bar{K}^0$

$$\bar{f} = P - \rho - \omega + A_2 + - - -$$

where P represents the Pomeron, responsible for diffractive scattering. Thus

$$f - \bar{f} = \rho + \omega$$

We see that those meson 'trajectories' contributing with <u>opposite</u> sign to K and $\bar{K}$ have $J^{PC} = \text{odd}^{--}$.

Now the ω exchange will dominate the ρ, primarily because of I-spin conservation: for Carbon, which is I = 0, the ρ cannot contribute.

Regge theory provides an expression for the amplitude for such an exchange:

$$\frac{f - \bar{f}}{K} \propto P^{\alpha_\omega(0) - 1} e^{-i\pi/2(\alpha_\omega(0) + 1)}$$

where $\alpha_\omega(0)$ is known as the intercept of the Regge trajectory and is a function of the ω particles and is approximately given by

$$\alpha_\omega \simeq 0.40$$

Thus we expect

$$\frac{f - \bar{f}}{K} \propto p^{-0.60} e^{i\pi/2(1.4)}$$

$$\text{or } \phi_{21} = -126^0$$

confirming the remarkable relation between the momentum dependence and the phase of the regeneration amplitude.

Let's now consider the A dependence which we recall is proportional to $A^{0.76}$. For comparison, we note that

$$\sigma_T(K_L A) \simeq 24 A^{0.84} \text{ mb.}$$

For no nuclear shadowing, we would expect

$$\sigma_T \propto A$$

while for complete shadowing, we would expect

$$\sigma_T \propto A^{2/3}.$$

214

Now, the regeneration phenomenon is sensitive to the difference in the mean free paths and there will be a greater contribution at larger impact parameters where there is less absorption.

Thus we can understand why the power in $\Delta\sigma$ is less than that in σ. An optical model of nuclear scattering can reproduce these phenomena reasonably well. Thus we have

$$\rho = \frac{i\pi x}{\sigma_T} \; \frac{f - \bar{f}}{K}$$

$$|\rho|^2 = \frac{\pi^2 x^2 \; \left|\dfrac{f - \bar{f}}{K}\right|^2}{\sigma_T^2}$$

$$\simeq \frac{\pi^2 x^2 \; (2.23)^2}{(24)^2} \; p^{-1.23} \; \left[\frac{A^{0.76}}{A^{0.84}}\right]^2$$

$$= 0.085 \; x^2 \; p^{-1.23} \; A^{-0.16}$$

We see that the rate is maximal at low atomic number. For E731, we use a 2 interaction length regenerator of B_4C. The rate is therefore approximately

$$|\rho|^2 \; e^{-2} \simeq 0.23 \; P^{-1.23} \; e^{-2} \simeq 1.4 \times 10^{-4}$$

This is the ratio of K_S/K_L exciting the regenerator but since the decay rate to 2π is given by $|\eta|^2 \simeq 5.3 \times 10^{-6}$, we see that K_S decays still dominate.

The Fermilab-based E731 experiment - a Chicago, Elmhurst, Fermilab, Princeton, Saclay collaboration - has as its primary goal a precision measurement of the ratio of the CP non-conserving parameters ϵ'/ϵ. For this effort, a new detector and beam were constructed and a brief test run was made three years ago. A schematic of the detector is shown in Figure 16.

Both K_S and K_L decays are simultaneously recorded; usually charged and neutral 2π decays are taken at different times. A preliminary result was announced in January of 1987 which was:

$$\epsilon'/\epsilon = 0.0035(30)(20)$$

with the first error being statistical and the second systematic. This was nearly a factor of two more precise than the world average value at the time and it was based upon about 700 $K_L \rightarrow 2\pi^0$ decays where one photon conversion in an 0.1rl lead sheet was required.

Since there would be significantly greater statistics in the main data run, the group worked for another year on further understanding of the

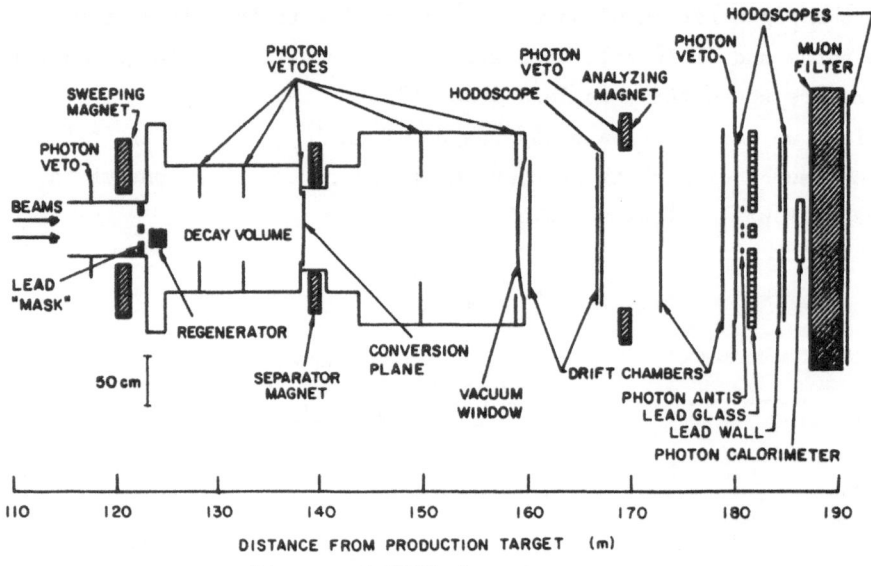

Figure 16 E731 detector.

detector and reducing the systematic error. The final result from the test run was then published as:

$$\epsilon'/\epsilon = 0.0032(28)(12)$$

At the same time as this result was being obtained, extensive upgrades were made to the detector and associated electronics to enable reaching the goal of more than 100,000 $K_L \rightarrow 2\pi^0$ decays.

The most important of these was a device to be used in the trigger for recognizing the number of photons in an event very quickly, a hardware cluster finder (HCF) processor.

E731 DATA RUN

We began taking physics quality data on August 1, 1987 and the run ended on February 15th, 1988. This was probably the most successful fixed-target run in the history of Fermilab, at least from our point of view. We wrote over 5000 data tapes and the quality is excellent with a great deal of ADC and TDC information recorded to enable careful studies of a whole variety of possible systematic effects. The HCF (which worked flawlessly throughout the run, providing a factor of 10 reduction in trigger rate) enabled us to take what we call open triggers where none of the photons is converted. Such triggers were recorded simultaneously with conversions throughout the run, and, for a period of about five weeks at the end of the run, the lead converter was removed and then all four modes were recorded simultaneously, although with considerable dead-time. It is this data set that we have been analyzing so far: it comprises about 20% of the total.

216

The possible drawback to the use of the open trigger results from the fact that scattering in the regenerator used to provide K_S decays produced "cross-over" and thereby confuses the determination of whether one has a vacuum decay or a decay of a regenerated K_S. Over the last several years, we have made good progress on reducing the amount of scattering, both by the composition of the regenerator and in improving our ability to reject events with particle production or nuclear breakup. The reconstructed profile at the lead glass of $2\pi^0$ decays for K_S and K_L are shown in Figure 17; it is seen that the two signals are very well resolved and further, the necessary small corrections can be determined and checked in several ways.

In Figure 18 is shown the 4-photon invariant mass distributions for K_L and K_S decays from the 4-mode data set (20% sample). The over 65,000 CP violating $2\pi^0$ decays appear above a background of residual $3\pi^0$ decays of only about 0.4% with little uncertainty. We have on tape about 300,000 $K_L \rightarrow 2\pi^0$ decays and the corresponding statistical error is about 0.0005 on ϵ'/ϵ. We expect that the systematic error can be reduced to this level and many "special condition" runs have been taken for systematic studies, but of course a very extensive analysis is required before we know for sure. To reach such a precision, it will be necessary to keep all systematic errors below about 0.1%.

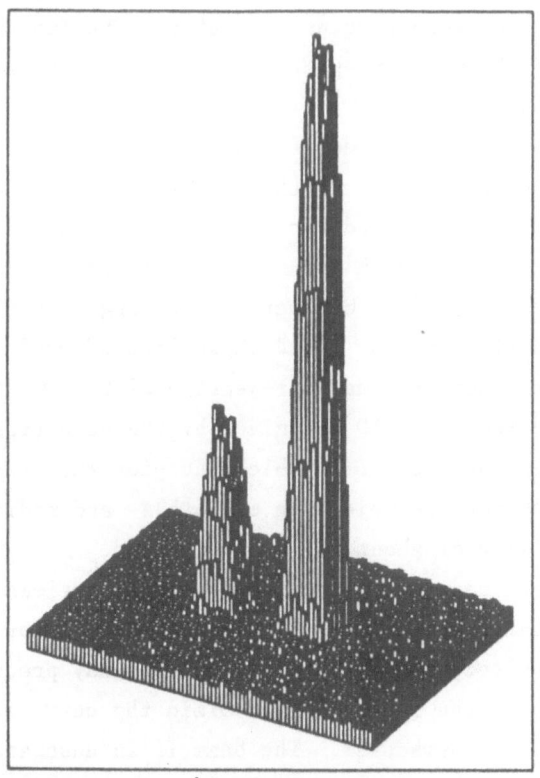

Figure 17 $2\pi^0$ profile at lead-glass.

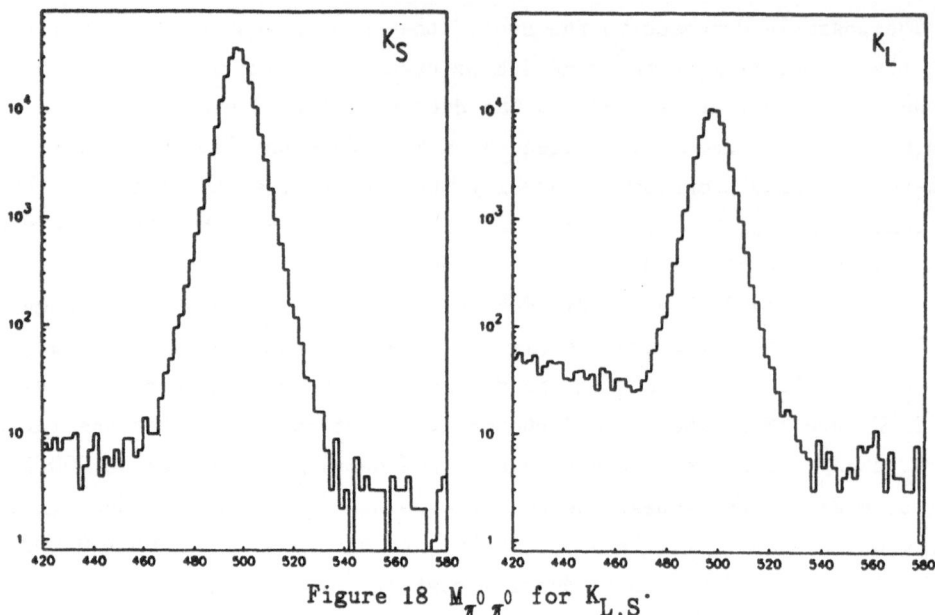

Figure 18 $M_{\pi^0 \pi^0}$ for $K_{L,S}$.

XIII. RARE KAON DECAY EXPERIMENTS

The field of searching for and studying rare decays of the neutral and charged kaon has made good progress recently. The Table shows the present status of most of the current experiments. We will briefly touch on the relevant physics topics.

The $K^+ \rightarrow \pi^+ +$ "nothing" mode is indirectly sensitive to the top quark mass: a signal, corresponding to the $\pi^+ \nu \bar{\nu}$ final state, at the 10^{-10} level per neutrino flavor is expected for a mass of 100 GeV.

The experiment stops the K^+ in an active target as we described earlier. A cylindrical drift chamber in a 1T magnetic field provides about 2% momentum resolution for the π^+ which is "ranged out" in a 40cm "active" range stack. The stack provides a rejection of 2×10^4 against muons. A further rejection factor of 10^5 is given by the detection of the $\pi \rightarrow \mu \rightarrow e$ decay chain within the counter in which the pion was stopped. There are extensive γ veto's for the rejection of both π^0 and radiative γ backgrounds so that a sensitivity of about 10^{-10} is expected.

The $K^+ \rightarrow \pi^+ \mu^+ e^-$ mode, if found, would signal a violation in lepton flavor. Lepto-quarks, particles which carry both flavor and lepton number, have been hypothesized as mediators of such a decay process. Such a mechanism would make the lepton sector mimic the quark sector where the (Cabibbo) mixing is substantial. The beam is an unseparated one with

RARE K DECAY SEARCH STATUS

MODE	EXPERIMENT	RESULT	EXPECTED SENSITIVITY	COMMENTS
$K^+ \rightarrow \pi^+$ + nothing	BNL E787		10^{-10}	Detector works well 10^{-9} next run
$K^+ \rightarrow \pi^+ \mu^+ e^-$	BNL E777	$<1.1 \times 10^{-9}$	1.5×10^{-10}	Limited by beam halo. May pursue $K^+ \rightarrow \pi^+ e^+ e^-$ (357 events in hand) $\pi^0 \rightarrow \mu e < 8 \times 10^{-8}$
$K_L \rightarrow \mu e$ $K_L \rightarrow ee$ $K_L \rightarrow \pi^0 e^+ e^-$	BNL E780	$<1.9 \times 10^{-9}$ $<1.2 \times 10^{-9}$ $<3.2 \times 10^{-7}$		Limited by beam halo. Will pursue $K_L \rightarrow \pi^0 e^+ e^-$ (10^{-10}) in E845
$K_L \rightarrow \mu e$	BNL E791	$<3.0 \times 10^{-10}$	$\sim 2 \times 10^{-11}$	Limited by accidentals from K_L decays. May pursue $K_L \rightarrow \pi^0 e^+ e^-$. Will measure $BR(K_L \rightarrow \mu\mu)$
$K_L \rightarrow \mu e$ $K_L \rightarrow ee$	KEK E137	$\lesssim 3 \times 10^{-9}$ $\lesssim 4 \times 10^{-9}$	$\sim 2 \times 10^{-11}$	Will pursue $K_L \rightarrow \pi^0 e^+ e^-$ (10^{-10}), in E162
$K_S \rightarrow \pi^+ \pi^- \pi^0$	FNAL E621	$\lesssim 1.5 \times 10^{-7}$	$\sim 3 \times 10^{-9}$	Expected rate $\overset{\sim}{-} 1.2 \times 10^{-9}$ Proposal for upgrade
$K_L \rightarrow \pi^0 e^+ e^-$	FNAL E731	$<5.1 \times 10^{-8}$	$\sim 1 \times 10^{-8}$	

Note: Some results are preliminary while others are published.

5×10^6 K^+ and 10^8 π^+ and p per spill. The Kaons decay in a 5m decay volume and the decay products are measured with two magnetic bends. Electrons are identified with two H_2 $\overset{v}{c}$ counters and a Pb-scint shower counter on one side of the apparatus. The misidentification probability was measured to be ~ 2×10^{-6}. On the other side, electrons are rejected with CO_2 $\overset{v}{c}$ counters and muons are detected in a proportional tube-steel range stack. A sensitivity of 10^{-10} is expected.

Searches for $K_L \rightarrow \mu e$ have similar motivation. One crucial background is the semileptonic decay to $\pi e \nu$ with a subsequent pion decay. Thus excellent mass resolution is required. The experiments have already reached the 10^{-10} level and a sensitivity in the 10^{-11} range is expected. The E791 group has a signal of about 100 very clean $K_L \rightarrow 2\mu$ events.

The $K_L \rightarrow \pi^0 e^+ e^-$ mode has attracted a great deal of recent interest. It is expected to be CP violating in lowest order and have a very large "direct" contribution: the ratio analogous to ϵ'/ϵ for this mode may be as large as unity! The expected branching ratio is in the 10^{-11} range while the best limit to date is three orders of magnitude worse. There are currently dedicated efforts at BNL, KEK, and FNAL to study this mode.

XIV. CPT INVARIANCE

We will now briefly look at tests of CPT invariance in the Kaon system. Recall that we assumed CPT symmetry when we analyzed the M and Γ matrices; we required

$$M_{11} = M_{12} \text{ and } \Gamma_{11} = \Gamma_{12}$$

and that led to the eigenstates

$$K_L = (1 + \epsilon)|K^0\rangle + (1 - \epsilon)|\bar{K}^0\rangle$$

$$K_S = (1 + \epsilon)|K^0\rangle - (1 - \epsilon)|\bar{K}^0\rangle$$

with

$$\frac{1 - \epsilon}{1 + \epsilon} = \alpha = \sqrt{\frac{M_{12}^* - i\Gamma_{12}^*/2}{M_{12} - i\Gamma_{12}/2}}$$

If we now relax this requirement, we find

$$K_L = (1 + \epsilon_L)|K^0\rangle + (1 - \epsilon_L)|\bar{K}^0\rangle$$

$$K_S = (1 + \epsilon_S)|K^0\rangle - (1 - \epsilon_S)|\bar{K}^0\rangle$$

where

$$\frac{\epsilon_S - \epsilon_L}{2} \equiv \Delta = \frac{(M_{11} - M_{22} + i(\Gamma_{11} - \Gamma_{12}))}{\Delta\Gamma - 2i\Delta M}$$

Very little is known about ϵ_S directly. It could be measured, in principle, by determining the semileptonic charge asymmetry in a pure K_S beam.

In fact, we know very little about CP violation in K_S decays. Let us here briefly review what is known.

3π Decays

A natural place to look for CP violating decays of the K_S is to study the (possible) 3π decay.

$\pi^0 \pi^0 \pi^0$

If two of the π^0's have any orbital momentum, it must be even by Bose-Statistics. Thus the third π^0's L valve is also even and therefore

$$P(\pi^0 \pi^0 \pi^0) = P^3(\pi^0) = -1$$

Since $C(3\pi^0) = C^3(\pi^0) = +1$, we have

$$CP(3\pi^0) = -1 \text{ (s wave) and therefore}$$

$$K_S \to 3\pi^0 \text{ is CP violating.}$$

$\pi^+ \pi^- \pi^0$

Given the small Q value in the decay, lets assume $L = 0$. Thus:

$CP(\pi^+ \pi^-) = +1$

$CP(\pi^0) = -1$

$CP(\pi^+ \pi^- \pi^0) = -1$

With $l \neq 0$, we can have $CP(\pi^+ \pi^- \pi^0) = +1$ but this is highly suppressed. We thus define the parameters

$$\eta_{+-0} = \frac{\text{amp}(K_S \to \pi^+ \pi^- \pi^0)}{\text{amp}(K_L \to \pi^+ \pi^- \pi^0)}$$

$$\eta_{000} = \frac{\text{amp}(K_S \to \pi^0 \pi^0 \pi^0)}{\text{amp}(K_L \to \pi^0 \pi^0 \pi^0)} \, .$$

We would expect that

$$\eta_{+-0} = \eta_{000} \overset{\sim}{-} \eta_{+-} \, .$$

Since we have seen that direct (i.e. ϵ') effects are very small what do we expect for the rate of $K_S \to 3\pi$?

$$\frac{\text{rate } K_S \to \pi^+\pi^-\pi^0}{\text{rate } K_S \to \pi^+\pi^-} = \frac{\text{rate } K_S \to \pi^+\pi^-\pi^0}{\text{rate } K_L \to \pi^+\pi^-\pi^0} \times \frac{\text{rate } K_L \to \pi^+\pi^-\pi^0}{\text{rate } K_L \to \pi^+\pi^-} \times \frac{\text{rate } K_L \to \pi^+\pi^-}{\text{rate } K_S \to \pi^+\pi^-}$$

$$= |\eta_{+-0}|^2 \times \frac{0.124}{0.0020} \times |\eta_{+-}|^2$$

$$\overset{\sim}{=} 1.7 \times 10^{-9}$$

so that BR $(K_S \to \pi^+\pi^-\pi^0) = 1.2 \times 10^{-9}$ (CP, violation, part) and similarly

$$\text{BR } (K_S \to 3\pi^0) = 2.0 \times 10^{-9} .$$

How can we hope to observe such a small effect? The best bet seems to be an interference experiment.

Consider a beam which is initially pure K^0 :

$$K^0 \sim K_L + K_S$$

so that the 3π decay rate vs time will be given by

$$N_{3\pi} = |\text{amp}(K_L \to 3\pi) + \text{amp}(K_S \to 3\pi) \; e^{i\Delta Mt - t/2}|^2$$

$$\alpha \; |1 + \eta_{+-0} \; e^{i\Delta Mt - t/2}|^2 .$$

Thus at $t = 0$, for example, one expects a small interference term in the decay distribution given by

$$2 \text{ Re } \eta_{+-0} \overset{\sim}{=} 3.3 \times 10^{-3}$$

This experiment is being performed by the E621 group (Rutgers, Michigan, Minnesota collaboration) at Fermilab. With two targets - one far away where little interference is expected - they have collected about $3 \times 10^6 \; \pi^+\pi^-\pi^0$ decays.

With about 10% of the data analyzed they can say that

$$|\eta_{+-0}| = 0.002 \pm 0.016$$

In addition, the group plans an upgrade to be able to achieve an error of 0.0004 and thus have an $\sim 5\sigma$ determination.

We note that the LEAR experiment is in a good position to also perform the measurement. They have the following advantages

1) Good acceptance at very early decay times where the asymmetry is greatest.

2) Simultaneous "pure" K^0 and $\bar{K}^0$ beams where the interference effect is reversed.

These advantages, coupled with less statistics, make for a comparable measurement.

There is little information on η_{000}. A recent Xenon bubble chamber experiment (V.V. Barmin et al., Phys. Lett. 128B, p. 129) with incident K^+ @ 850 MeV/c identified 632 $3\pi^0$ events, obtaining the 90% C.L. limit of

$$|\eta_{000}|^2 < 0.1$$

Thus the prospects for observing CP violation in K_S decay are reasonably good although there is not much chance of any surprises: $\epsilon'_{3\pi}$ is expected to be of order $\epsilon'_{2\pi}$ and clearly we lack the sensitivity to see any direct effect.

The semileptonic charge asymmetry should be determined by the CPLEAR group and they expect a measurement of 2 Re ϵ_S to a precision of a few σ.

Let us consider by how much ϵ_S and ϵ_L could differ. The unitarity relation becomes

$$\langle K_S | K_L \rangle = \frac{1}{2} \left[\left(1 + \epsilon_S^* \right)\left(1 + \epsilon_L \right) - \left(1 - \epsilon_S \right)^* \left(1 - \epsilon_L \right) \right]$$

$$\overset{\sim}{-} \epsilon_S^* + \epsilon_L \begin{cases} = 2 \text{ Re } \epsilon \text{ [CPT conservation]} \\ = 2 \text{ (Re } \epsilon - i \text{ Im}\Delta) \end{cases}$$

where

$$\epsilon = (\epsilon_L + \epsilon_S)/2$$
$$\Delta = (\epsilon_L - \epsilon_S)/2$$

Then our unitarity relation becomes

$$\text{Re } \epsilon - i\text{Im}\Delta \overset{\sim}{-} \frac{\gamma_S}{2i\Delta M + \gamma_S} \left\{ \eta_{+-} \text{ BR}(K_S \to \pi^+\pi^-) + \eta_{00} \text{ BR}(K_S \to \pi^0\pi^0) \right\}$$

We have noted previously that ϕ_{+-} is consistent with

$$\phi_P = \tan^{-1} (2\Delta M/\gamma_S) = 43.7^0 (2)^0$$

However, there is a measurement of ϕ_{00} which is somewhat different:
$$\phi_{00} = 55.7^0 \pm 5.8^0$$

Numerically, we have

$$\text{Re } \epsilon - i\text{Im}\Delta \overset{\sim}{-} 0.723 \, e^{-i\phi_P} \{0.686 \, \eta_{+-} + 0.314 \, \eta_{00}\}$$

$$=0.723 \, e^{-i\phi_P}\{0.686\times2.275(21)\times10^{-3} e^{i44.6(1.2)^0} +0.314\times2.299(36)e^{i54(5)^0}\}$$

$$\overset{\sim}{-} 0.723 \times 2.28 \times 10^{-3} \times \{0.69 \, e^{i0.9(1.2)^0} + 0.31 \, e^{i10(5)^0}\}$$

Thus we have

$$\frac{\text{Im}\Delta}{\text{Re } \epsilon} \overset{\sim}{-} 0.69 \, \sin(0.9^0 \pm 1.2^0) + 0.31 \, \sin(10^0 \pm 5^0)$$

$$= (1.1 \pm 1.3) \times 10^{-2} + (5.4 \pm 2.7) \times 10^{-2}$$

$$= (6.5 \pm 3.0) \times 10^{-2}$$

We see that we can limit the size of CPT violation in the mass matrix to the level of perhaps 10% of the size of CP violation.

Thus we have

scale of second order weak transition = $\Delta M = 7 \times 10^{-12}$ MeV

scale of CP violation $\sim 10^{-3}$ ΔM $= 7 \times 10^{-15}$ MeV

scale of CPT violation $\leq 7 \times 10^{-16}$ MeV.

The measurement on η_{00} can be improved. The NA31 group and the E731 group will eventually both have measurements at the 1^0 level for ϕ_{00} (relative to ϕ_{+-}).

We see that our unitarity relation gives us CPT violation in the mixing matrix if $\phi_{00} \neq \phi_p$.

We noted earlier that

$$\eta_{+-} - \eta_{00} = 3 \epsilon'$$

and that the phase of ϵ' is known through an analysis of $\pi\pi$ scattering in the relevant energy regime. This conclusion is in contradiction with the experiment of Christenson et al. on ϕ_{00} (at the 2σ level). However, that analysis assumed CPT symmetry in the K^0 decays. Specifically, if we write, (See V.V. Barmin et al., Nucl. Phys. B247 (1984))

$$\text{amp}(K^0 \to 2\pi, I) = (A_I + B_I)\, e^{i\delta_I}$$

$$\text{amp}(\bar{K}^0 \to 2\pi, I) = (A_I^* - B_I^*)\, e^{i\delta_I}$$

Then

$$\text{amp}(K_1 \to 2\pi, I) = \sqrt{2}(\text{Re } A_I + i\text{Im } B_I)\, e^{i\delta_I}$$

$$\text{amp}(K_2 \to 2\pi, I) = \sqrt{2}(i \text{ Im } A_I + \text{Re } B_I)\, e^{i\delta_I}$$

Thus by examination we can associate each term with either a conservation or violation of CP and CPT symmetry as is shown below:

	CP	CPT
Re A_I	cons	cons
Im A_I	viol	cons
Re B_I	viol	viol
Im B_I	cons	viol

The "perpendicular" difference between η_{+-} and η_{00} can be evaluated as:

$$\epsilon' \overset{\sim}{=} \frac{1}{3} \eta_{+-} \sin 10^0 \overset{\sim}{=} 1.3(7) \times 10^{-4}$$

and

$$\frac{\epsilon'}{\epsilon} \stackrel{\sim}{=} 0.06(3)$$

Now

$$\epsilon' = \frac{1}{\sqrt{2}} \frac{\text{amp}(K_2 \to 2\pi, I = 2)}{\text{amp}(K_1 \to 2\pi, I = 0)} = \frac{1}{\sqrt{2}} \frac{i \text{ Im } A_2 + \text{Re } B_2 \, e^{i(\delta_2 - \delta_0)}}{\text{Re } A_0 + i \text{ Im } B_0}$$

and if we take the experiment seriously, we would have

$$\frac{\text{Re } B_2}{\text{Re } A_0} \stackrel{\sim}{=} (1.8 \pm 1.0) \times 10^{-4}$$

or

$$\frac{\text{Re } B_2}{\text{Re } A_2} \stackrel{\sim}{=} 4 \times 10^{-3}.$$

Now how would this relate to other Kaon decays. The $K^+ \to \pi^+\pi^0$ decay is essentially pure $\Delta I = 3/2$ so

$$\text{amp}(K^+ \to \pi^+\pi^0) = (A_2 + B_2) \, e^{i\delta_2}$$

$$\text{amp}(K^- \to \pi^-\pi^0) = (A_2^* - B_2^*) \, e^{i\delta_2}$$

$$\Gamma(K^\pm \to \pi^\pm\pi^0) = |A_2|^2 + |B_2|^2 \pm 2 \text{ Re } A_2 B_2^*$$

$$\frac{\Gamma(K^+ \to \pi^+\pi^0) - \Gamma(K^- \to \pi^-\pi^0)}{\Gamma(K^+ \to \pi^+\pi^0)} \simeq \frac{4 \text{ Re } A_2 B_2^*}{|A_2|^2} \simeq \frac{4 \text{ Re } B_2}{\text{Re } A_2}$$

The experimental value for this rate ratio is $(8 \pm 12) \times 10^{-3}$ so that we have

$$\frac{\text{Re } B_2}{\text{Re } A_2} = (2 \pm 3) \times 10^{-3}$$

A non-zero Re B_2 term would result in a K^+, K^- lifetime difference as well. Experiment has

$$\frac{\tau^+ - \tau^-}{\tau^+} = 1.1(9) \times 10^{-3}$$

which would imply

$$\frac{\text{Re } B_2}{\text{Re } A_2} = (1.3 \pm 1.1) \times 10^{-3}$$

We close this section by saying that the future measurements of $\Delta\phi$ with a precision of 0.5^0 will determine Re B_2/Re A_2 with a precision of about 3×10^{-4}.

XV. THE FUTURE: THE NEUTRAL B SYSTEM

We will close with a discussion of the possibility of observing CP nonconserving effects in the B system in the future.

Consider $\Delta\Gamma$ for the weak eigenstates in the B system. We have

$$\Delta\Gamma \sim 2\pi \sum_F \rho_F \langle\bar{B}^0|H_{WK}|F\rangle \langle F|H_{WK}|B^0\rangle$$

where F are all states which both B and $\bar{B}$ can decay into.

From a study of diagrams contributing to both Γ and $\Delta\Gamma$, we can neglect $\Delta\Gamma$ with respect to Γ since

$$\frac{\Delta\Gamma}{\Gamma} \sim \lambda^2 .$$

Let us consider the time evolution of a neutral B state. We have

$$B_H \sim (1 + \epsilon_B)|B^0\rangle + (1 - \epsilon_B)|\bar{B}^0\rangle$$

$$B_L \sim (1 + \epsilon_B)|B^0\rangle - (1 - \epsilon_B)|\bar{B}^0\rangle$$

$$|B^0\rangle \sim \frac{|B_H\rangle + |B_L\rangle}{1 + \epsilon_B}$$

$$|B^0(t)\rangle \sim \{|B_H\rangle\, e^{i\Delta mt} + |B_L\rangle\}\, e^{-\Gamma t/2}$$

$$= [(1 + \epsilon_B)[1 + e^{i\Delta mt}]\, |B_0\rangle + (1 - \epsilon_B)[e^{i\Delta mt} - 1]\, |\bar{B}^0\rangle]\, e^{\Gamma t/2}$$

$$\sim \{(1 + e^{i\Delta mt})\, |B_0\rangle + \frac{1 - \epsilon_B}{1 + \epsilon_B}\, (e^{i\Delta mt} - 1)|\bar{B}^0\rangle\}\, e^{-\Gamma t/2}$$

Now, let us calculate

$$|\langle 1^- q\bar{\nu}_q|B^0(t)\rangle|^2 = 2\, e^{-\Gamma t}\, \{1 + \cos\Delta mt\}$$

$$|\langle 1^+ q\bar{\nu}_q|B^0(t)\rangle|^2 = 2\, \left|\frac{1 - \epsilon_B}{1 + \epsilon_B}\right|^2 e^{-\Gamma t}\, \{1 - \cos\Delta mt\}$$

Now, neglecting CP violation in the mixing, we have for the ratio of wrong to right sign semileptonic decays:

$$\frac{\text{rate } B^0(t) \to e^+}{\text{rate } B^0(t) \to e^-} = \frac{\int e^{-\Gamma t}\, \{1 - \cos\Delta mt\}dt}{\int e^{-\Gamma t}\, \{1 + \cos\Delta mt\}dt}$$

$$= \frac{\dfrac{1}{\Gamma}\left\{1 - \dfrac{1}{1 + (\Delta m/\Gamma)^2}\right\}}{\dfrac{1}{\Gamma}\left\{1 + \dfrac{1}{1 + (\Delta m/\Gamma)^2}\right\}}$$

$$\simeq \frac{x^2}{2 - x^2} \quad (x \equiv \Delta m/\Gamma).$$

From similar measurements, the Argus and CLEO groups have determined that
$$x = 0.76(16) \quad \text{ARGUS}$$
$$= 0.64 \qquad \text{CLEO (Preliminary)}$$
We are justified in neglecting CP violation in the mixing since
$$\left|\frac{1 - \epsilon_B}{1 + \epsilon_B}\right| \simeq 1 + \frac{1}{2} \left|\frac{\Gamma_{12}}{M_{12}}\right| \sin \phi_{\Delta B=2}$$

and we have $\Gamma_{12} \ll M_{12}$.

How are we likely to see CP violation in the B^0 system? The most promising is an asymmetry in an initial B^0 vs a $\bar{B}^0$ decay to a CP eigenstate.

Let us briefly mention an idea to reduce the price of "flavor tagging" as explained in a recent paper by Atwood, Dunietz, and Grosse-Wiesmann.

Consider the reaction

$$e^+e^- \to b\bar{b} \quad \begin{cases} \sigma \simeq 5\text{nb at the } Z^0 \\ \\ \simeq 1\text{nb at the 4S} \end{cases}$$

$$\frac{d\sigma}{d\Omega} \sim 1 + \cos^2\theta + 2 A_{FB}^0 \cos\theta$$

We expect $A_{FB}^0 = 0.87$ for b quarks with polarized electrons. Thus one can effectively tag events from geometrical considerations alone. Let us consider a particular final state f. We would measure

$$A_{CP} = \frac{N(f, \text{ forward}) - N(\bar{f}, \text{ backward})}{N(f, \text{ forward}) + N(\bar{f}, \text{ backward})}$$

A calculation indicates that to measure an asymmetry to 10% in a mode with BR = 5×10^{-5} (e.g. $\bar{p}p$), one would need to produce about 3×10^7 $b\bar{b}$ pairs with 90% polarized electrons.

The assumptions are

$$\sigma(b \to B_d) = 35\% \to 2 \times 10^7 \text{ tagged } B^0 \text{ or } \bar{B}^0 \text{'s}$$

$$BR(\bar{p}p) = 5 \times 10^{-5} \to 1000 \text{ tagged decays}$$

$$30\% \text{ efficiency} \to 300 \text{ events}$$

$$63\% \text{ tagging efficiency} \to 189 \text{ detected events}$$
$$\text{(detector holes)}$$

Thus one can measure this asymmetry with a precision of about 7%.

With 90% polarization, the asymmetry is diluted by a factor of about 0.7.

Consider now the luminosity requirements:

3×10^7 b$\bar{\text{b}}$ pairs at a cross-section of 5nb would require an exposure of 6000 (pb)$^{-1}$.

With a luminosity of 10^{32} cm^{-2} sec^{-1}, this would need 6×10^7 seconds or about 6 years of running. The hope for addressing this kind of physics at e$^+$e$^-$ machines will fall into the area of ever greater luminosity Z factories, or in the simultaneous reconstruction of many modes with large asymmetries.

ACKNOWLEDGEMENTS

I would like to acknowledge the contribution of my colleagues on the E731 experiment and R. Cousins, L. Littenberg and W. Molzon for discussions on the Brookhaven rare K decay program. I especially wish to thank Tom and Barbara Ferbel for their dedication to the Summer School at St. Croix.

POST COLLIDER PHYSICS

G.Giacomelli

Dipartimento di Fisica dell'Universita' di Bologna
INFN, Sezione di Bologna
Bologna, Italy

1.INTRODUCTION

The present standard model of particle physics, which includes the Weinberg-Salam theory of Electroweak Interactions and Quantum Chromodynamics for Strong Interactions, explains quite well all available experimental results. The theory had a surprising confirmation with the discovery of the intermediate vector bosons W^+, W^- and Z^o and with recent precise measurements of electroweak parameters (1).

On the other hand few physicists believe that this is the ultimate theory, because it contains too many free parameters, there is a proliferation of quarks and leptons and because it seems unthinkable that there is no further unification with the strong interaction and eventually with the gravitational interaction (2). Most physicists consider the present standard model of particle physics as a low energy limit of a more fundamental theory, which should reveal itself at higher energies. It is possible that there are at least two energy thresholds.

The first threshold could be associated to supersymmetric particles or to a new level of compositeness, a substructure of quarks and leptons. This first threshold could be around few TeV and could be revealed with the new generation of colliders (for instance UNK, LHC, SSC).

The second threshold could be associated with the Grand Unification of the Electroweak and Strong Interactions and of quarks with leptons. This unification would appear at extremely high energies, $> 10^{14}$ GeV. It is unthinkable to reach these energies with any kind of Earth-based accelerator. These energies were instead readily available in the first instants of our Universe, at a cosmic time before 10^{-35} seconds. It is in this context that non-accelerator physics, in particular underground experiments, may play an important role. One may look for "fossil" particles, left over from the Big Bang. Magnetic monopoles, the lightest supersymmetric particles, nuclearites, etc. are examples of such "fossil" particles. The Grand Unified Interaction violates some conservation laws, like the conservation of Baryon Number and of Leptonic Numbers. These violations would be a common affair at extremely high energies, while they would lead to very rare phenomena in our ordinary world of extremely low energies. Proton decay would be the most explicit example of Grand Unification with Baryon Number violation. Lepton number violation could manifest in neutrinoless double beta decays. Since all these phenomena are very rare, one needs large apparatuses which have to be shielded from cosmic rays and from ambient radioactivity.

Non-accelerator physics is also important for astrophysics. One may say that supernova 1987A opened up the field of Neutrino Astronomy and one may hope to have, in the future, Gravitational Wave Astronomy and Dark Matter Astronomy. There are also important connections between particle physics and cosmology and astrophysics that can be studied with non accelerator experiments. The connections stem from the hypothesis that the Early Universe was a gas of extremely hot particles and that as time progressed the temperature decreased, the radius increased, there were phase transitions with changes of gas constituents and with changes in the basic forces, from unified ones to non-unified ones.

In the following lectures notes we shall discuss large underground experiments, the new Gran Sasso laboratory and other planned facilities. We shall then discuss the present status of searches for proton decay, n-$\bar{n}$ oscillations, neutrinoless double beta decays, magnetic monopoles, gravitational waves, etc. Neutrino physics, neutrino astrophysics and cosmic ray studies are also considered.

2. EXISTING AND PROPOSED LARGE UNDERGROUND DETECTORS

The experimental problems of underground detectors are to a large extent related to the energy range covered by the experiments. One may distinguish underground studies in three categories (3-5):

i) Low energy phenomena, $E \leqslant 20$ MeV, for which the main problem is the radioactivity background.

ii) Study of events of about 1 GeV energy, like nucleon decays and "atmospheric" neutrino interactions. In this case the main feature of the detector is its mass (generally 1-10 Kt) and its capability of identifying neutrino events.

iii) Detection of throughgoing particles, specifically muons, but also monopole candidates. The main feature of these detectors is area. The muons include also upward going muons coming from neutrino interactions in the rock below the detectors.

The existing large underground detectors (Fig.2.1 shows the flux of cosmic ray muons reaching them) can be divided in three groups: i) water Cherenkov detectors (Figs.2.3 - 2.7); ii) tracking calorimeters (Figs.2.8 - 2.9); iii) liquid scintillator detectors (Fig.2.2). Table 2.1 gives the main features of these three types of detectors, with relation to the three kinds of underground experiments defined above.

For the detection of low energy neutrinos the competition is between large mass water Cherenkov detectors and liquid scintillator detectors. The most important parameter is the energy threshold, which can be around (5-7) MeV for scintillators and 7-8 MeV for water Cherenkov counters.

In the search for nucleon decays the competition is between water Cherenkov detectors and tracking calorimeters (3-12). The latter consist of sandwiches of iron plates and ionization detectors. Water Cherenkov and tracking calorimeter techniques are complementary. The Cherenkov technique allows to reach the highest masses, while the tracking calorimeters provide better space resolutions and good identification of electrons, muons and charged kaons. The largest water detectors are the IMB (8000 t; fiducial mass 3300 t) and Kamioka-2 (3000 t; fiducial mass 880 t). The discrimination between nucleon decay candidates and neutrino events requires large amount of light collected by the phototubes in Cherenkov detectors (20% in Kamioka-2; 8% in IMB-3) and a good spatial resolution in tracking calorimeters.

In cosmic ray studies and in monopole searches the detectors should combine a large area with a good spatial resolution. It is also important to identify upward muons, generated by high energy neutrinos interacting in the rock below the detector, from the much more abundant downward flux of cosmic ray muons. Cherenkov detectors are good for determining the direction, but have a poor spatial resolution. Liquid scintillation detectors may identify upward going muons by time-of-flight measurements, but their space resolution is not very good. The combination of scintillators with tracking calorimeters offers both advantages.

Table 2.1 and 2.2 show the main characteristics of present and of some of the planned detectors, respectively. Considering the water Cherenkov technique, the Super-Kamiokande project (9) should have a fiducial mass of about a factor of 10 larger than present detectors; furthermore, a substantial improvement in the amount of collected light should be obtained. The originality of the Sudbury project (10), designed for the study of solar neutrinos, is the use of heavy water as a target. The DUMAND project, dedicated to high energy cosmic rays and neutrino astronomy, uses a large volume of sea water. Liquid scintillation detectors in the Gran Sasso Laboratory (LVD and MACRO) contain almost 10 times more scintillator than present set-ups using the same technique. Furthermore, spatial resolution is achieved by inserting planes of streamer tubes in the apparatus. The ICARUS project, designed as a universal underground detector, will consist of a liquid argon drift chamber surrounded by external muon detectors. It should simultaneously achieve a rather low energy threshold as well as good spatial and energy resolutions (see Sections 3 and 9).

Fig. 2.1: Cosmic-ray muon flux as a function of overburden in meters of water equivalent (mwe) or depth underground (feet in standard rock) for various underground labs in existence around the world.

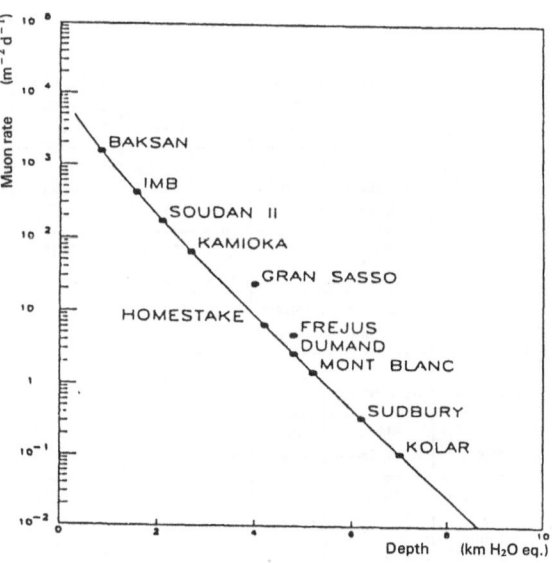

Table 2.1: Main features of present large underground detectors (3). The table does not include the HPW detector, a 0.7 Kt water Cherenkov detector, which was located in Park City, Utah

Technique	Detector	Total mass (t)	Low energy ν — E(e⁻) threshold (MeV)	N decay — Fiducial mass (t)	Light collected (%) / Fe sampl. dist. (cm)	dE/dX	Spatial resolution (cm)	Cosmic-ray muons and multimuons — Area (m²)	Depth (mwe)
Water Cherenkov	IMB I	8000	50	3300	2%		100	550	1570
	IMB III		20		8%				
	Kamioka I	3000	30	880	20%		100	320	2700
	Kamioka II		7.5	780 + active shield					
Liquid scintillator	LSD Mont Blanc	90	5–7				50	50	5200
	LASD Homestake	100	5				15	160	1200
	Baksan	330	12.5				15–35	300	850
Tracking	Kolar Gold Fields I	140	Proportional tubes	60	1.2	Yes	10	30	7000
	Kolar Gold Fields II	260		160	0.6			60	6015
Detectors	Nusex	150	Streamer tubes	113	1	No	1	20	5200
(Fe/Detector planes sandwich)	Frejus	912	Flash tubes + Geigers	550	0.3	No	0.5	100	4800
	Soudan II	(1000)	Drift tubes	~ 650	0.3	Yes	0.5	130	2100

Table 2.2: Main features of planned large underground detectors (3)

Technique	Detector	Total mass (t)	Low energy ν — E(e⁻) threshold (MeV)	N decay — Fiducial mass (t)	Light collected (%)	Spatial resolution (cm)	Cosmic-ray muons and multimuons — Area (m²)	Depth (mwe)
Water Cherenkov	Super Kamioka Japan	32000	7	22000	40	30–50	1900	2700
	Sudbury Canada	1000 D₂O	7	1000 D₂O	40	30–50	300	6200
	Dumand USA (Hawaii)	150000 (sea)					6300	4800
Liquid scintillator	LVD (Gran Sasso)	1800 scint.	7			3	600	4000
+ Tracking	Macro (Gran Sasso)	1000 scint.	10			3	900	4000
Liquid argon drift chamber	Icarus	3500 L argon	7	Liquid argon + magnetic field		0.2	800	4000

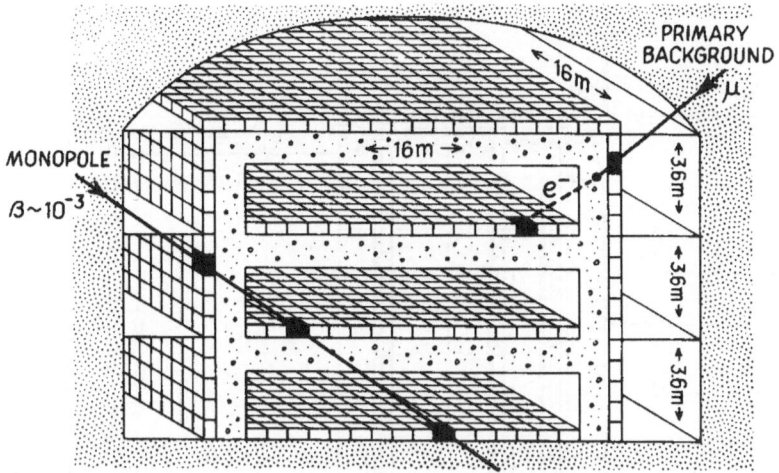

Fig. 2.2: Cross sectional view of the Baksan underground scintillator telescope (6).

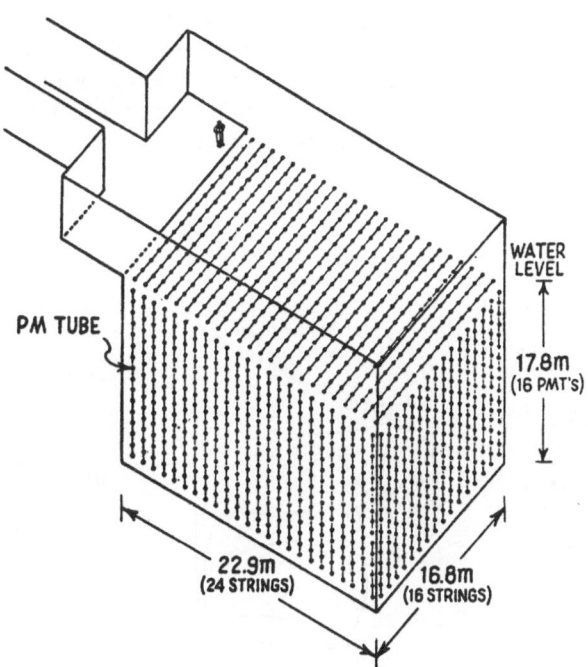

Fig. 2.3: Schematic view of the arrangement of the 2048 photomultiplier tubes used in the IMB water Cherenkov detector (7).

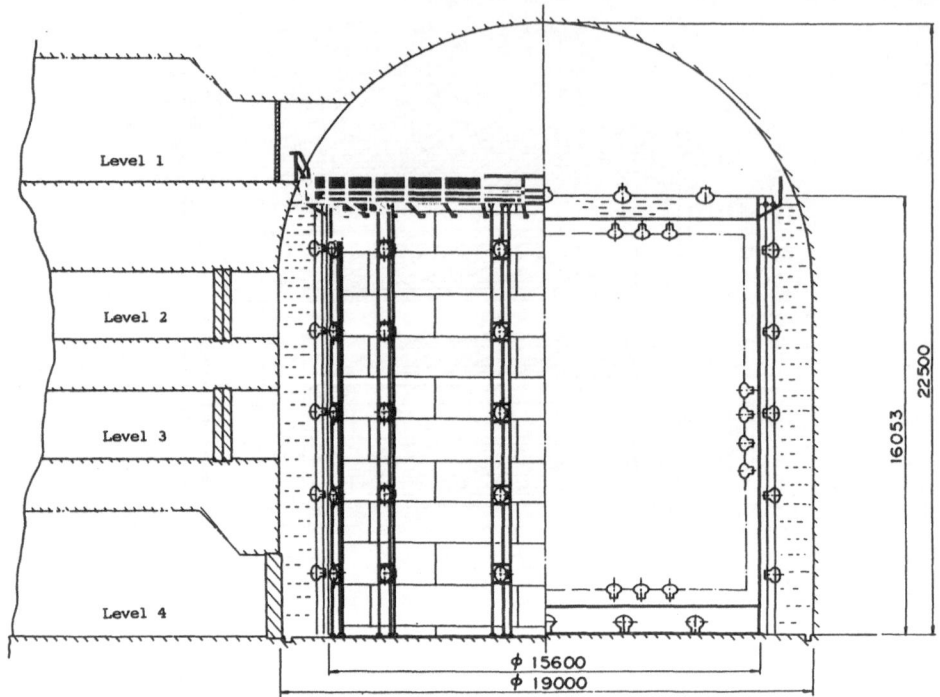

Fig. 2.4: The Kamioka nucleon decay detector, located 1000 m underground in the Kamioka mine, 300 Km west of Tokyo (8). It consists of a steel cylinder, 15.6 m diameter, 16 m high, containing 2900 m^3 of purified water. There are 1050 large photomultipliers (50 cm diameter R1449X) distributed over the inner surfaces of the water tank, 1 PM/m ; the photocathodes cover 20% of the entire surface. There is also an anticoincidence layer, 1.2 m thick, viewed by 123 PM, covering the whole sensitive volume.

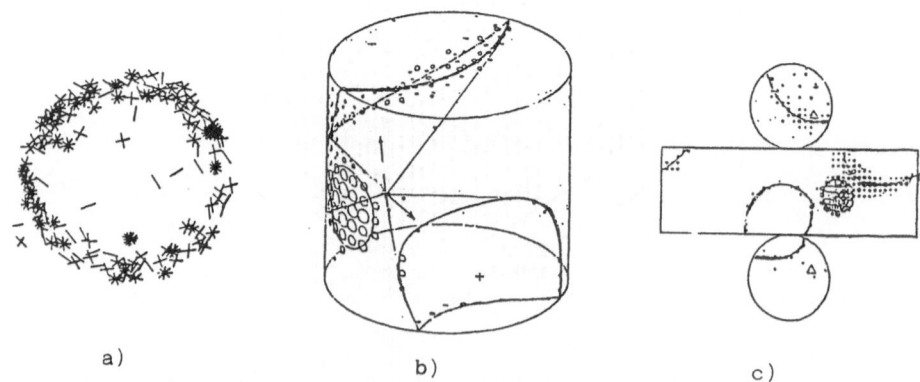

Fig. 2.5: a) Cherenkov light pattern of a stopping muon heading a wall of the IMB detector.b,c) Cherenkov light "rings" for different types of tracks in the Kamioka detector; c) represents the cylinder unfolded.

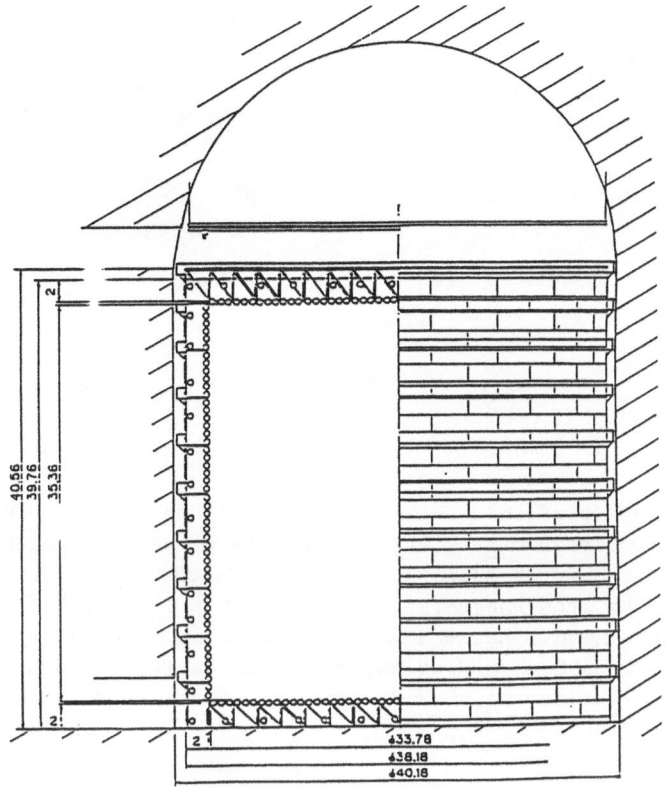

Fig. 2.6: The proposed Super-Kamiokande Detector (9) (units in m).

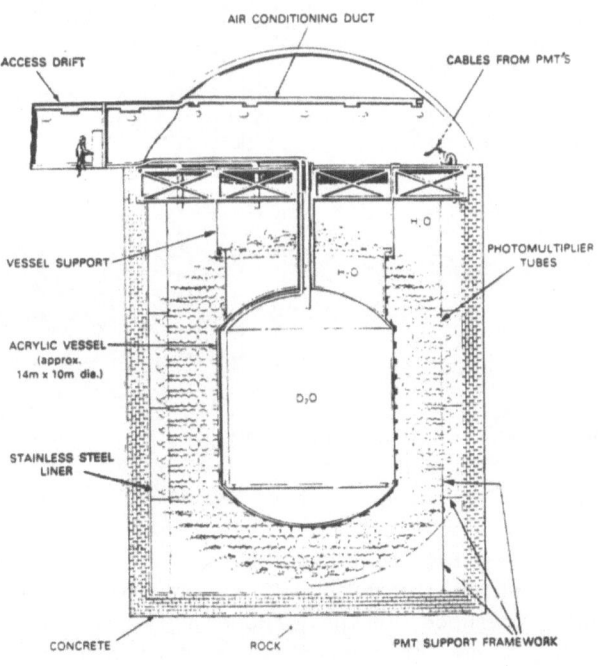

Fig. 2.7: The Proposed Sudbury D_2O Cherenkov Detector (10). The diameter of the cylindrical rock cavity is 20 m.

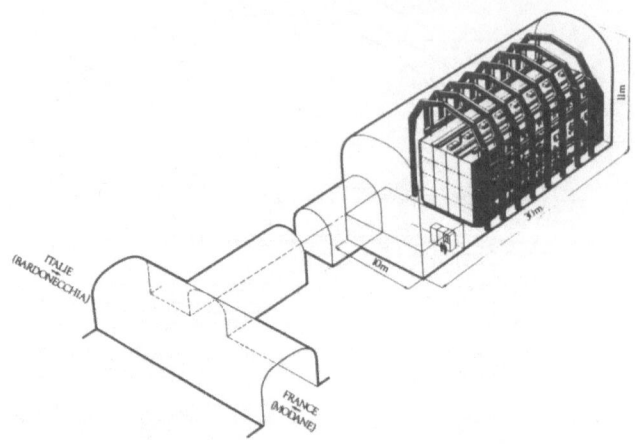

Fig. 2.8: The layout of the Frejus nucleon decay detector, located in the Frejus tunnel, under 4500 mwe (11). It is made of vertical planes, 6x6 m^2 in size, of i) flash chambers (1024 tubes, each of 0.5x0.5 cm^2 cross section), ii) 3 mm iron sheets, iii) Geiger tubes (each 1.5x1.5 cm^2 cross section. One 8 t module has 8 flash chamber planes and 1 Geiger plane; the total mass is 900 t (600 t fiducial mass). Energy resolution: for electrons: $0.15/\sqrt{E}$; for pions and muons of 0.3 GeV/c: 20%; and 3%.

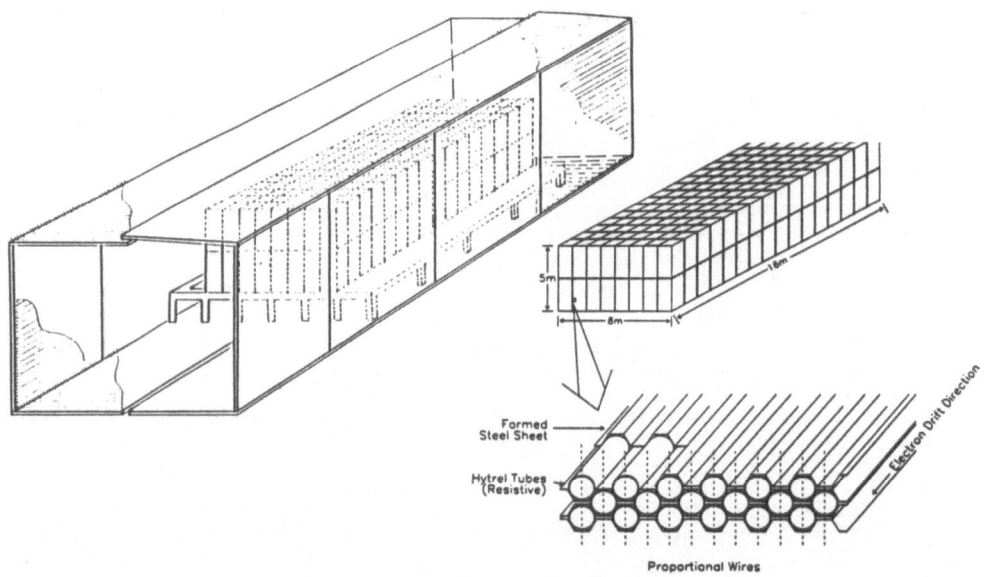

Fig. 2.9: The Soudan-2 detector for nucleon decay will have a total mass of 1100 t, in 256 modules, each 1x1x2.5 m^3 size (12). Each module has formed sheet steel, 1.6 mm thick, stacked vertically, to give exagonal close-packed arrays of holes filled by resistive plastic tubes. Electrical potential applied to tubes gives up to 50 cm drift of ionization electrons, which are detected by vertical proportional wires backed by horizontal cathode strips. Anode wire spacing is 1.5 cm, cathode strip spacing 1 cm. The detector is surrounded by a two-layer array of extruded aluminium proportional tubes (S=1700 m^2).

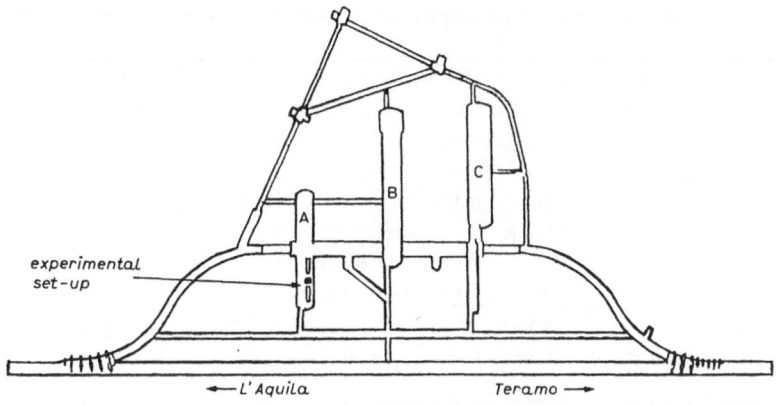

Fig. 3.1: Layout of the Gran Sasso underground lab. The useful cross sections of halls A, B will be 16 m at the base and 16 m high; hall C is larger. Hall A will house LVD and Gallex, hall B MACRO, hall C ICARUS. Laser interferometers are located in the three tunnels at the top.

3. THE GRAN SASSO LAB.

Since this lab will be for the near future the largest underground non-accelerator facility, it is worthwhile to discuss it more fully. The Gran Sasso National Laboratory (LNGS) of the Istituto Nazionale di Fisica Nucleare (INFN) is located on the highway Rome-Teramo, about 120 Km east of Rome (13). The lab consists of a complex of three underground tunnels, each about 100 m long, for a total volume of about 100.000 m^3 (Fig.3.1), and an Extensive Air Shower (EAS) array on top of the mountain at an altitude of 2000 m a.s.l. and about 25° off the vertical from the underground lab.

Table 3.1 gives the basic parameters of the underground lab, as well as the facilities available. The laboratory is at an altitude of 963 meters above sea level, is well shielded from cosmic rays (by about 4000 meters water equivalent (mwe) of rock, see Fig.2.1) and it has a low activity environment (one tenth of Mont Blanc). It is easily accessible by the italian highway network and it is equipped with all the facilities and the safety systems of a typical High Energy Laboratory. The 3 halls are completed and have been equipped with heat insulation, electric power, cranes, safety gates, etc. The installation of experiments has started and some experiments will start data taking at the end of 1988. The physics aims of the 9 approved experiments (LVD, MACRO, ICARUS, GALLEX, Double Beta Decay, Gravitational Wave Antenna, Interferometers, Geomagnetic Field Measurements and the High Altitude Laboratory) may be classified as follows:

1) Detection of particles from external sources.

  a)-Solar neutrinos (E$_{\nu_e}$ < 14 MeV) (GALLEX,ICARUS).   } Neutrino
    -Neutrinos from stellar collapses (E$_{\nu_e}$ < 30 MeV).   } physics,
    (LVD,ICARUS,MACRO)
    -High energy muon neutrinos from point sources(E$_{\nu_\mu}$>1 GeV) } Neutrino
    (MACRO,LVD,ICARUS).   } Astronomy

Table 3.1 Basic parameters and facilities at the LNGS underground lab.

Location:	Latitude $47^{\circ}27'09''$N, Longitude $13^{\circ}34'28''$E, 963 m a.s.l.
Depth:	$\geqslant$1400 m of rock ($CaCO_3$) with $\varrho \simeq 2.8$ g/cm^3, $<Z>\simeq 9.4$
	$\geqslant$3600 m of water equivalent (average of about 4000 mwe)
Activity of rock (Bq/Kg):	^{232}Th=0.25, ^{238}U=5.2, ^{40}K=5.1, ^{114}Bi=4.2
Neutrons:	fast=3×10^{-6} cm^{-2} s^{-1}, thermal 5×10^{-6} cm^{-2} s^{-1}
Muons:	1μ m^{-2}h^{-1}, $<E_\mu> \simeq 240$ GeV
Natural conditions:	6° C, 100% humidity
Operation:	20° C, 50% humidity
Ventilation:	20000 m^3/h (+100000 m^3/h in emergency)
Fast computer data link:	inside $\longleftrightarrow$ outside $\longleftrightarrow$ INFNET.

b)-High energy muons   ⎧ Study of the muon energy spectrum.
  -Muon groups   ⎨ Determination of the energy spectrum and
   (MACRO,LVD,ICARUS) ⎩ composition of primary cosmic rays.

c)-Search for heavy magnetic monopoles and other exotic particles.
   (MACRO,LVD,ICARUS)

d)-Search for gravitational waves (from collapsing stars).

2) Detection of particles from internal sources.

a)-Search for proton decay, in channels which have not been looked for until now (ICARUS).

b)-Search for double beta decay.

3) Geophysics experiments.

a)-Interferometry to measure the time evolution of geological structures.

b)-Measurement of the space and time variations of the geomagnetic field.

3.1 The Approved Experiments

Here follows a brief description of the approved experiments. The three major experiments (LVD, MACRO, ICARUS) are all purpose detectors: each has a primary purpose, but everyone is capable of giving significant information in most of the other physics subjects listed above.

LVD (Large Volume Detector; by an Italian-American-Soviet collaboration; is being installed in Hall A North) (14), Fig.3.2. It has two types of detectors: a) 1600 liquid scintillators, each 1x1x1.5 m^3 and each viewed by 3 photomultipliers, for a total live mass of 1800 tons (LVD/UNO). b) 9 horizontal layers and 5 vertical ones of limited streamer tubes, each 1x1 cm^2 in cross section (LVD/TRACK). The LVD detector is a scaled up version, by a factor of 20, of the Liquid Scintillator Detector (LSD) now operating in the Mont Blanc laboratory.

The main purpose of the LVD experiment is the detection of neutrinos from collapsing stars (supernovae): it should obtain about 900 events in 10 seconds from a star collapse at the distance of our galactic center. The reaction analyzed is $\bar{\nu}_e + p \rightarrow n + e^+$, followed by $e^+ + e^- \rightarrow 2\gamma$ (detection of e^+ and of converting γ's); about $2\mu s$ later one has also $n + p \rightarrow d + \gamma$. The experiment may also study high energy muons, high energy neutrinos, rare phenomena and possibly solar neutrinos.

MACRO (Monopole, Astrophysics and Cosmic Ray Observatory) by an Italian-American collaboration; is being installed in Hall B) (15). MACRO

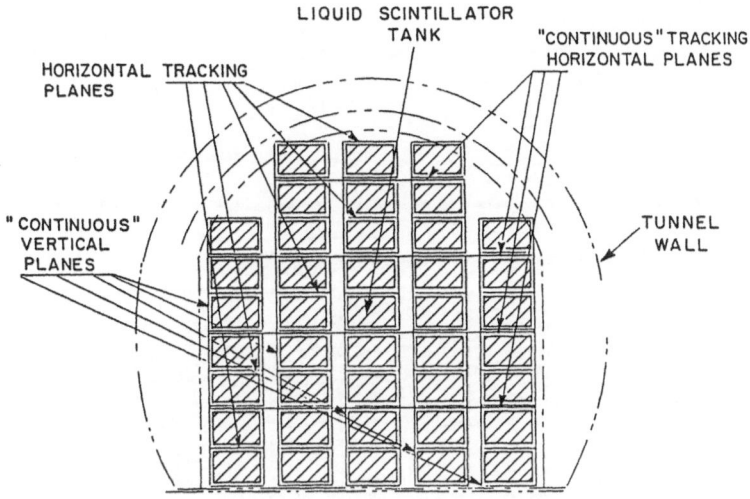

Fig. 3.2: Cross section of the LVD Detector. It consists of liquid scintillator tanks, each 1.5 m^3 in size, of vertical and horizontal planes of limited streamer tubes (14).

(Fig. 3.3) has three types of detectors: a) Three horizontal planes of liquid scintillators, each (0.75x0.25x12) m^3 in size and viewed by two 8'' photomultipliers at each end (the total scintillator mass is about 1000 t); b) 18 horizontal layers of streamer tubes, each 3x3 cm^2 in cross section. Between layers there are slabs of concrete absorbers. c) A track-etch detector consisting of 3 layers of CR39 and 5 layers of lexan.

The sides of the MACRO detector are sealed by one layer of scintillators and 5 layers of streamer tubes. MACRO is basically a large area detector and will have $S \Omega$ = 10.000 m^2 sr.

The main purposes of MACRO are the searches of heavy penetrating particles, such as magnetic monopoles, and the study of high energy muons and high energy muon neutrinos. The detector should be capable of reaching a sensitivity for monopole detection well below the astrophysics bound (Parker bound). The study of rare phenomena is performed with three types of independent detectors in order to achieve redundancy. The detector may also study neutrinos from collapsing stars.

ICARUS (Imaging Cosmic And Rare Underground Signals) will represent a technological step forward, Fig.3.4 (16). It will be a large liquid argon drift chamber, where one should "see" tracks with a space resolution comparable to that of bubble chambers. Moreover one should have many precise dE/dx samplings along the paths of the particles. A small prototype of 2 tons should be ready at the end of this year. A detector of about 300 t will be built at CERN and installed at the Gran Sasso lab where it would mainly be used for solar neutrino studies. The final detector, with a mass of about 3500 t, will be used for proton decay studies and other experiments.

GALLEX (Gallium European Collaboration) is being installed in A-south (17). A large tank, containing 30 tons of gallium, in the Ga Cl$_3$

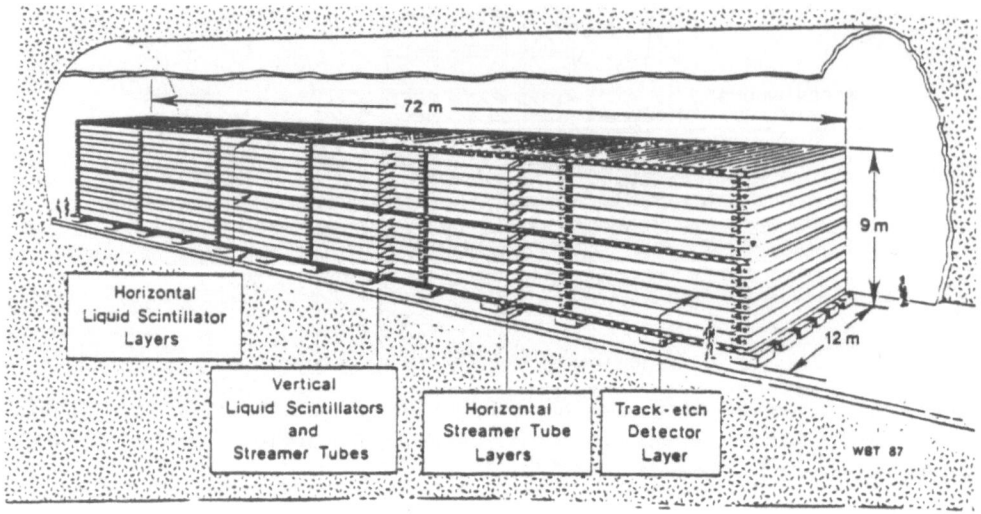

Fig. 3.3: Schematic diagram of the MACRO detector (15). It is a large area apparatus, with 3 layers of liquid scintillators, 18 layers of streamer tubes, absorbers, and one layer of plastic track detectors.

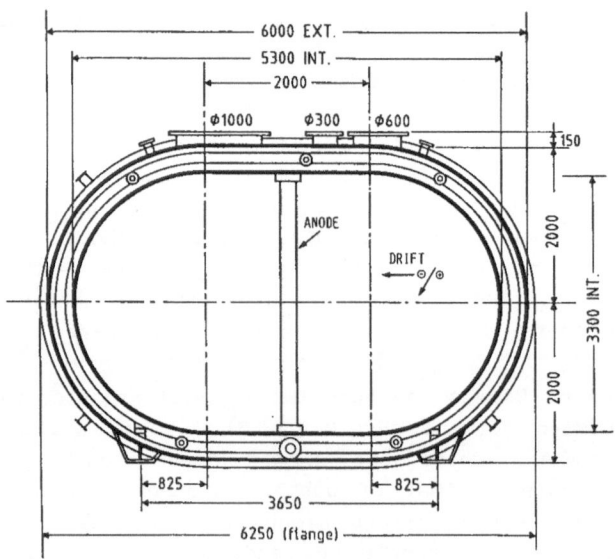

Fig. 3.4: Schematic layout of the 300 t ICARUS prototype experiment (16). It may be considered as two liquid argon drift chambers.

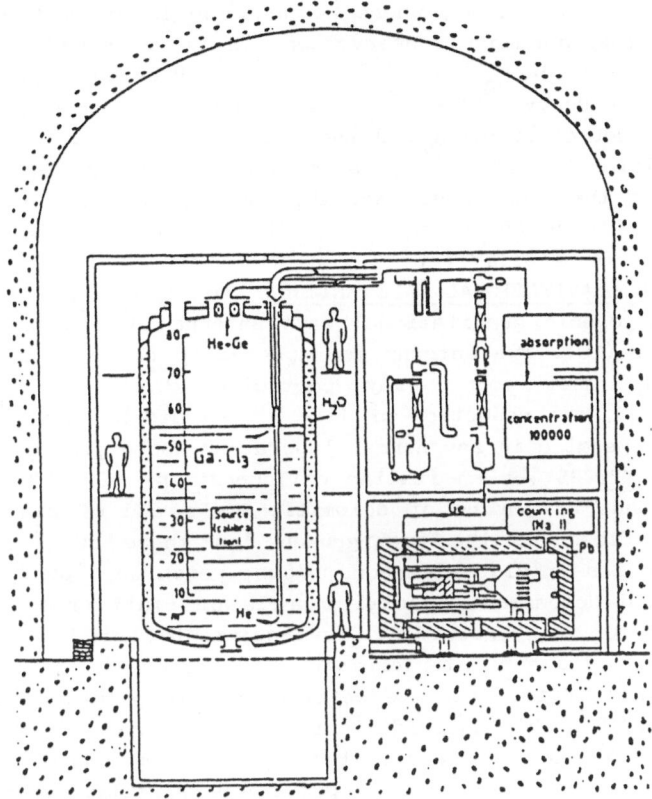

Fig. 3.5: Side view of the Gallex building (17).

form, will be used to study solar neutrinos by radiological techniques (Fig.3.5). Solar electron neutrinos with energies larger than 233 KeV will yield $\nu_e + {}^{71}Ga \rightarrow {}^{71}Ge + e^-$; germanium will promptly form a molecule, $Ge\,Cl_4$, which is a highly volatile gas. Every 10 days the tank will be flushed with helium gas, which will remove the few molecules of germanium chloride; these are transformed into $Ge\,H_4$ and brought to a proportional counter, where one will observe the decay ${}^{71}Ge \rightarrow {}^{71}Ga + \gamma$ (1.2 KeV, 10.4 KeV), which has a decay half life of 11.4 days.

One expects one event per day. The sophisticated procedure will be calibrated with an 800 KCi radioactive source of ${}^{51}Cr$, which yields electron neutrinos of 430 KeV (10%) and 750 KeV (90%). It is worth recalling that the only experiment on solar neutrinos with ${}^{37}Cl$ had an energy threshold of 800 KeV and was thus sensitive only to relatively high energy neutrinos not coming from the main solar nuclear reactions. The gallium experiment will be sensitive to the dominant nuclear reactions.

DOUBLE BETA DECAY, by a European collaboration, will be installed in a special low-activity laboratory (18). This smaller detector will employ multiwire proportional chambers filled with xenon to detect decays of the type $(A,Z) \rightarrow (A, Z+2) + 2 \, e^-$, where no neutrino is emitted and which are forbidden by lepton number conservation. In the approved experiment the decay studied will be $^{136}_{76}Xe \rightarrow \, ^{136}_{76}Ba + 2 \, e^-$. Previous experiments have searched for the decay $^{76}Ge \rightarrow \, ^{76}Se + 2 \, e^-$, where the germanium is at the same time the source of decay and the detecting element. The experiments yielded the lower limit $\tau > 10^{24}$ y for double beta decay. The low activity environment of the Gran Sasso lab, the increase in size of both source and detector and the enrichment of the used isotope ^{136}Xe will allow a large increase in sensitivity.

CRYOGENIC GRAVITATIONAL WAVE ANTENNA An italian group is planning to bring a cryogenic gravitational wave antenna in a special lab under the Gran Sasso (19). The antenna is expected to be so sensitive that it will "detect" most cosmic ray particles. Even if these may be anticoincidized, the reduction of live time would be too severe if the antenna were operated in any normal laboratory above ground.

INTERFEROMETERS (by an italian collaboration) (20). The Gran Sasso mountain is made of layers of dolomitic rocks and of calcareous rocks (limestone). Close to the underground lab there is a fault, which separates the less rigid calcareous rocks, which were bent upward, from the more rigid dolomites. Special tunnels of small cross section have been prepared for this experiment. The equipments, which consist of wire instruments and of laser interferometers of different wavelengths, allow the measurements of the relative position of two points with a relative precision of $1/10^9$. This is sufficient to measure rock movements as well as the effects of the tides and of the free oscillations of the earth. At a later date it is planned to improve the sensitivity of the laser interferometers by two orders of magnitude, reaching the sensitivity required for a laser interferometer detector of gravitational waves.

GEOMAGNETIC FIELD MEASUREMENTS (by an italian collaboration) (21). Portable fluxgate magnetometers will be used to measure microfluctuations in the magnetic field to a fraction of a nanoTesla over a wide range of frequencies. A comparison between measurements made at the Earth's surface and in underground tunnels provide information about the conductivity structure of the interposed earth material. If there is a fault through the Gran Sasso, some electric current might be preferentially channeled in the fault.

EXTENSIVE AIR SHOWER DETECTOR on top of the Gran Sasso mountain (22) (EAS-TOP). This detector is located at an altitude of 2000 m a.s.l., see Fig.3.6. It consists of an array of detectors, for the measurement of the electromagnetic component, and of a muon-hadron detector. The electromagnetic array has 28 modules (each made of 10, 1 m^2 scintillators), separated by 25 m in the central part and by 100 m in the external part. The muon-hadron detector, with a total area of 200 m^2, will have 12 layers of streamer tubes separated by iron and concrete absorbers, for a total thickness of 1 Kg cm^{-2}. The EAS-TOP array will be able to yield: i) electron densities in different points and thus the shower size; ii) the arrival time in different points and thus, by time of flight, the arrival direction; iii) the muon plus hadron content. The apparatus will be operated: i) As an autonomous detector for studies of ultra-high-energy gamma-ray astronomy, cosmic ray anisotropies and primary composition; ii) In coincidence with underground detectors, with the main purpose of measuring the total primary energy of multimuon events.

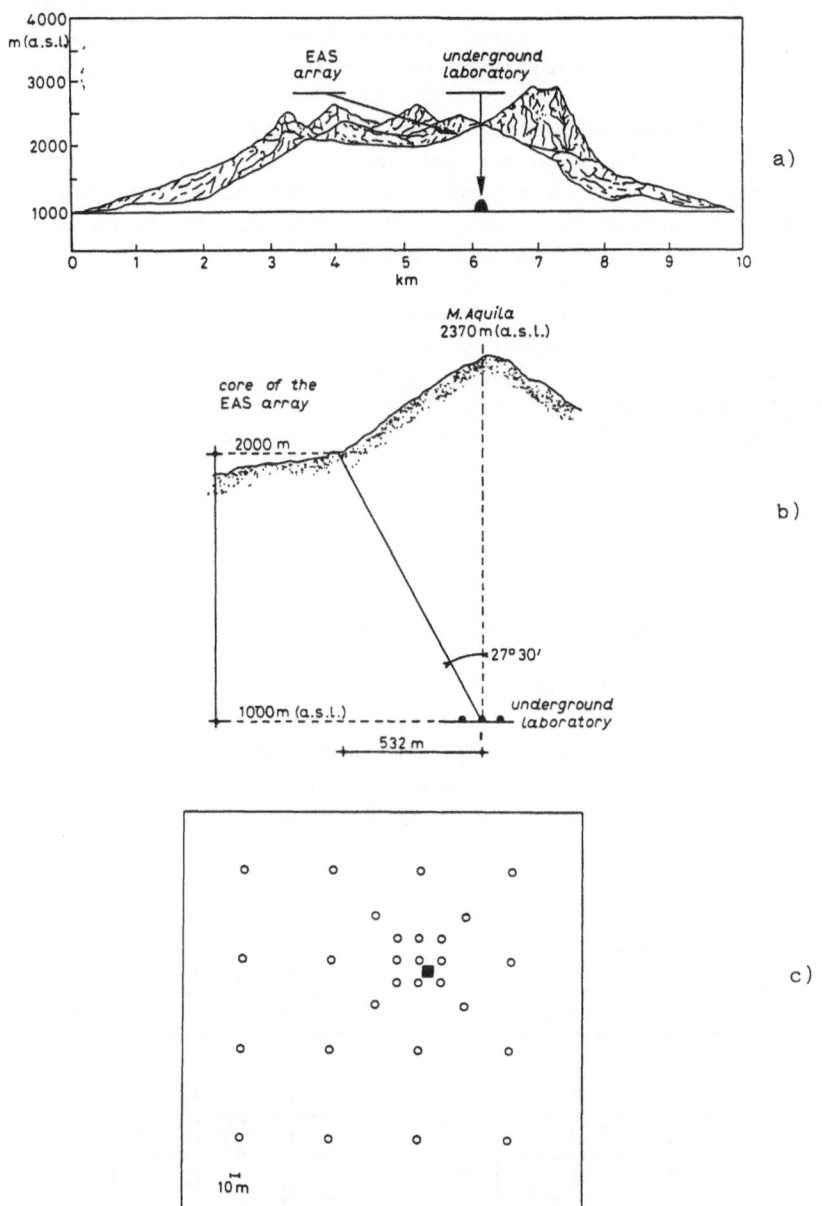

Fig. 3.6. a) b) Profile of the Gran Sasso mountain, location of the underground lab, and of the EAS array. c) Locations of the detectors (o 10 m² modules; ■ hadron-muon detector) (22).

4. PROTON DECAY

GUT theories place quarks and leptons in the same multiplets. Quarks ⟷ leptons transitions are thus possible; they are mediated by supermassive vector bosons X,Y with $m \sim 10^{14}$ GeV, Fig.4.1. A free proton and a bound neutron may decay as $N \rightarrow 1^+ + \text{meson(s)}$ or $N \rightarrow \bar{\nu} + \text{meson(s)}$. The proton lifetime $(\tau \sim m_X^4)$ was first predicted at the level of 10^{31} y; this motivated the construction of large detectors with masses of the order of 1000 t (3).

Present proton decay detectors are of two types: (a) water Cherenkov detectors (IMB, KAMIOKA, HPW) and (b) tracking calorimeters (KGF, Nusex, Frejus, Soudan 2). Water detectors have larger masses, have many free protons and are capable of detecting the sense of the track direction. Tracking calorimeters have a higher spatial resolution and a better π/μ separation at energies of about 200 MeV. Table 2.1 gives a summary of the properties of present large underground detectors (3); Figures 2.3, 2.4 and 2.8 show three of the existing detectors.

The expected number of events per year in a decay channel with branching fraction B is

$$N = f \cdot \underbrace{6 \cdot 10^{23} \cdot 10^9}_{\substack{\text{N.of nucleons} \\ \text{in 1 Kt}}} \cdot \underset{\substack{\text{fiducial} \\ \text{mass (Kt)}}}{M_F} \cdot \underset{\substack{\text{time} \\ \text{(y)}}}{T} \cdot B/\tau \cdot \varepsilon \qquad (4.1)$$

where
f = fraction of relevant nucleon, f=p/(p+n) or f=n/(p+n),
Kt = Kilo-ton,
luminosity = fiducial mass x exposure time=L=M_F T in Kt y,
τ = proton lifetime, in y,
ε = overall detector efficiency.

The main background in large underground detectors comes from atmospheric neutrinos; the background is at the level of about 100 events/Kt y. Most of this background can be eliminated by considering fully contained events and by topological and kinematical constraints.

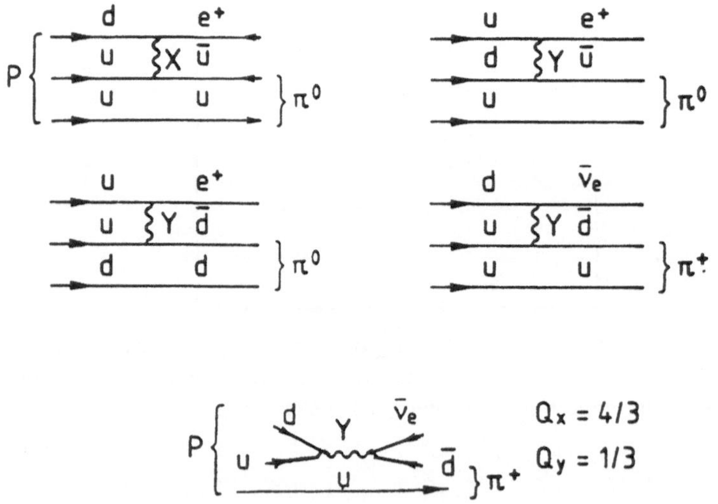

Fig.4.1: Possible diagrams representing nucleon decay ($p \rightarrow e^+ + \pi^o$, $p \rightarrow \bar{\nu}_e + \pi^+$) via the leptoquark bosons X, Y of the Grand Unified Theories.

Table 4.1 Luminosity and τ/B limits for selected background-free decay modes in different proton decay detectors (3,4).

Detectors	IMB	Kamioka	Frejus	All Detectors
L=(Kt y)	4.6	1.3+1.2	1.4	~ 11
τ/B ($p \to e^+ \pi^0$) (10^{32}y)	3.2	1.9	0.5	5.5
τ/B ($p \to e^+ \eta^0$) "	1.0	1.1	0.4	2.6
τ/B ($p \to \mu^+ \pi^0$) "	2.7	1.5	0.5	4.5
Reference	(4)	(5)	(6)	

For the decay modes with a negligible background (f.i. $p \to e^+ + \pi^0$) one has from eq.(4.1)

$$\tau/B > 2.61 \times 10^{32} \; f \cdot L \cdot \varepsilon \qquad (4.2)$$

where L is in Kt y, τ/B in y and the limit is at the 90% confidence level (N=2.3).

For the decay modes with a significant background one has from (4.1)

$$\tau/B > 3.64 \times 10^{32} \; f \cdot \varepsilon \cdot \sqrt{L/R_b} \qquad (4.3)$$

where R_b is the background rate for the considered decay mode when the number of observed candidates is compatible with the expected background rate.

Fig.4.2 shows the luminosity L versus time for most proton decay experiments (3,4). The present combined sensitivity of all experiments is at the level of 11 Kt y. The background-free modes of decay yield lower limits at the level of $\tau/B=$ few 10^{32} y, see Table 4.1 and Fig.4.3a. The limits for other decay modes (with some background) are about 10 times smaller (see Fig.4.3b). Present detectors have almost reached a background rate at the level of 10^{-2} of atmospheric neutrino interactions. In future experiments one needs better rejections and accurate neutrino calibrations at accelerators with low energy (1-5 GeV) neutrino beams.

The present lower limits rule out the simplest SU(5) GUT model. It has been recently enphasized that some superstring models predict $\tau_p \sim 10^{32\pm2}$ y (2).

There are four new proton decay detectors in different state of approval or readiness (see Table 2.2).

The Soudan 2 fine sampling tracking calorimeter with dE/dx measurement is presently being assembled in the Soudan mine (USA), Fig.2.9. It will have a total mass of 1100 t, 166 of which have already been installed. The whole detector may be ready in two years.

The Superkamiokande detector would be a 32000 t water Cherenkov counter (Fig.2.4). It is a direct upgrade of the present Kamioka detector, using large photomultipliers. Superkamioka is a proposal which should be approved at the end of 1988.

Exercise - The number of interactions of 1 GeV atmospheric neutrinos in 1 Kt detector per year is the following:

Rate = cross section x flux x (nucleons/Kt) x time
= 10^{-38} cm^2 x 1 cm^{-2} s^{-1} x 6x10^{32} x 3x10$^7 \simeq$ 180 events/Kt y.

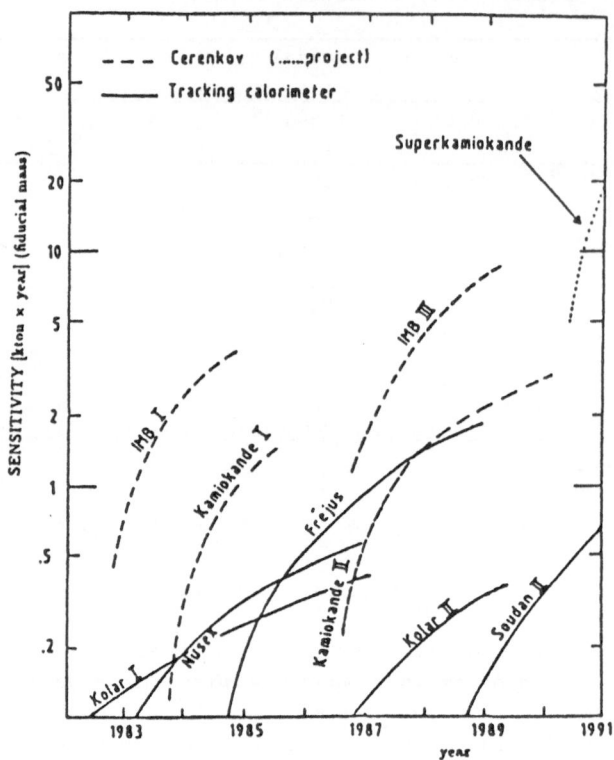

Fig. 4.2: Luminosity L = M_Ft versus time for several proton decay detectors (4).

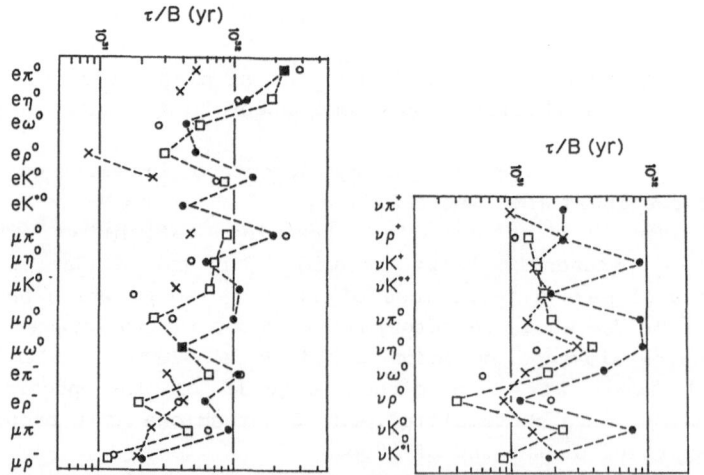

Fig.4.3: 90% CL lower limits for τ/B from several proton decay experiments. (a) N→l$^+$+X; (b) N → $\bar{\nu}$+X (3,4).

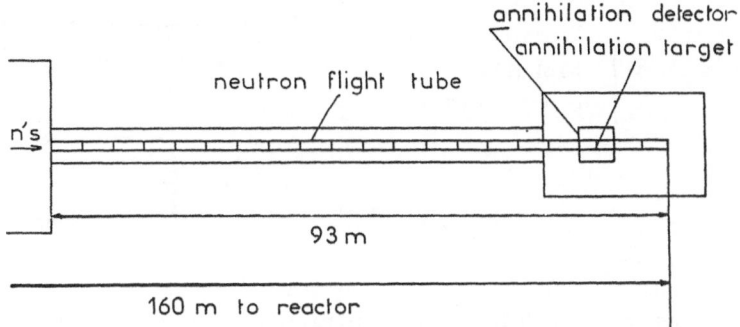

Fig. 5.1: Schematic layout of a n-n̄ oscillation experiment at a high flux neutron beam from a reactor. One uses "very cold" neutrons moving in a zero magnetic field region. If a neutron transforms into an antineutron, it will yield a large signal in the annihilation detector (23).

5. n-n̄ OSCILLATIONS

If $\Delta B=1$ processes are allowed one should also expect $\Delta B=2$ processes, like n$\rightarrow$n̄ transitions. These neutron-antineutron oscillations could be detected in a beam of thermal neutrons impinging in a large calorimeter, where an antineutron could readily be observed, see Fig.5.1.

From the above lower limits on the bound neutron decay discussed in Section 4 one may obtain a lower limit on the n-n̄ oscillation time considering the limit on bound neutron lifetime (τ_n) and a typical rotation time inside a nucleus (T_{rev}):

$$\tau_{n\bar{n}} \simeq \sqrt{\tau_n T_{rev}} \simeq \sqrt{4 \times 10^{31} y \cdot 10^{-23} s} \simeq 10^8 s \qquad (5.1)$$

This limit is now within reach of reactor experiments (23).

6. NEUTRINOLESS DOUBLE BETA DECAY

If for some nucleus (A,Z) the chain of decays
$$(A,Z) \rightarrow (A, Z+1) + e^- + \bar{\nu}_e \qquad (6.1)$$
$$\rightarrow (A,Z+2) + e^- + \bar{\nu}_e$$

is forbidden by energy conservation (see Fig.6.1) or strongly suppressed, the processes may be possible in a single step, with emission of two electron antineutrinos (24,25)
$$(A,Z) \rightarrow (A,Z+2) + 2 e^- + 2 \bar{\nu}_e \qquad (6.2)$$
or with the emission of a majoron
$$(A,Z) \rightarrow (A,Z+2) + 2 e^- + x \qquad (6.3)$$
or simply
$$(A,Z) \rightarrow (A,Z+2) + 2 e^- \qquad (6.4)$$

While decay (6.2) is allowed (but rare), the neutrinoless double beta decay (6.4) is forbidden by lepton number conservation. (6.4) would be allowed if ν_e and $\bar{\nu}_e$ would be identical and if they have a non-zero mass. At the quark level the neutrinoless double beta decay would be interpreted as

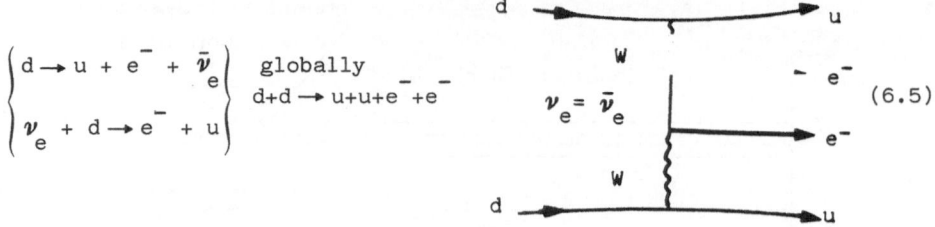

$$\left.\begin{array}{l} d \rightarrow u + e^- + \bar{\nu}_e \\[2mm] \nu_e + d \rightarrow e^- + u \end{array}\right\} \begin{array}{l} \text{globally} \\ d+d \rightarrow u+u+e^-+e^- \end{array} \qquad \nu_e = \bar{\nu}_e \qquad (6.5)$$

Even-even nuclei for which the normal β-decay is energetically forbidden are good candidates for neutrinoless double decay searches (Fig.6.1). The energy spectrum for the sum of the energies of the two electrons, $E = E_1 + E_2$, is different in the three cases (6.2-6.4): a line for process (6.4), a continuum peaked at low E for process (6.2) and a continuum peaked at higher E for (6.3) (Fig.6.2).

Most of the direct searches for neutrinoless double beta decays performed until now use materials which act both as source and as detector, as in $^{76}Ge \rightarrow {}^{76}Se + 2 e^-$ (6.6) Moreover the experiments performed purely calorimetric measurements.

So far the experiments have not found neutrinoless double beta decay. Experiments with ^{76}Ge yielded the limit $\tau > 10^{24}$ y. In some models this corresponds to an effective neutrino mass larger than about (1-3)eV (28).

Typical used germanium masses range from 0.5 to 7 Kg; they include about 10^{24} ^{76}Ge nuclei. It has to be remembered that normal germanium contains 15% ^{76}Ge. In order to increase the number of ^{76}Ge by an order of magnitude a collaboration between several groups plans to use several kilograms of enriched germanium, containing 85% ^{76}Ge.

The allowed $2\bar{\nu}$ double beta decay (6.2) depends only on the relatively unknown nuclear matrix element. The extraction of this process from purely calorimetric experiments, using ^{76}Ge, is very difficult since the total electron energy has a continuum spectrum; thus background subtraction requires a good knowledge of all possible background sources.

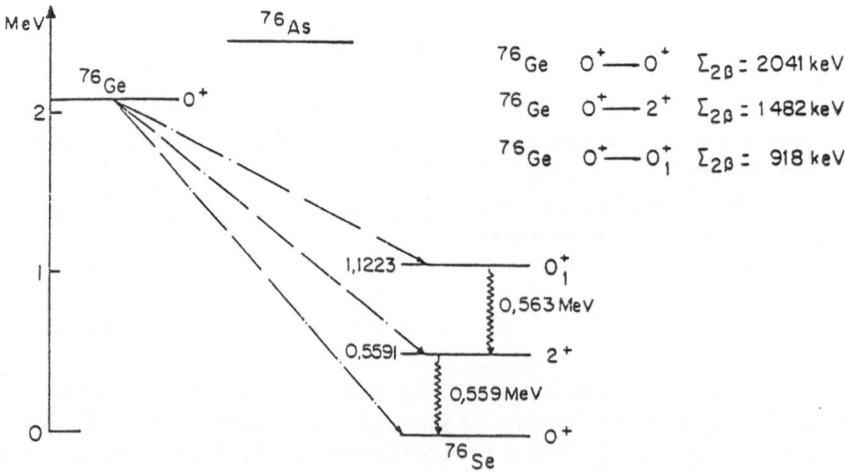

Fig.6.1: Energy levels for ^{76}Ge double β decay (3).

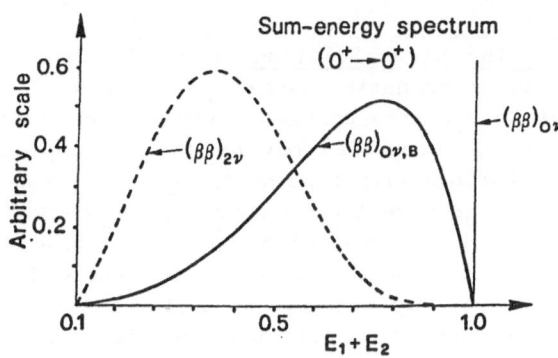

Fig.6.2: Spectra for the sum of the two electron energies for the decays 6.2-6.4.

Last year geochemical experiments have established the existence of the $2\bar{\nu}$ bouble beta decay
for $^{82}Se \rightarrow ^{82}Kr$ with $t_{1/2} = (1.7 \pm 0.3) \times 10^{20}$ y.
More recently this type of decay has also been observed in standard double beta decay experiments (The UC-Irvine group observed $^{82}Se \rightarrow ^{82}Kr$ using a TPC chamber) (24,25).

Several groups are planning to use visual detectors, separating the spatial detection of the two electrons. This may considerably reduce the background due to multi-Compton scattering of γ-rays. The Milano group plans to use several Kg of ^{136}Xe (natural Xe contains 8.9% of ^{136}Xe; one could have enrichment up to 60%). The single beta transition $^{136}Xe \rightarrow ^{136}Cs + e^- + \bar{\nu}$ is forbidden because the atomic mass difference is -69 KeV. Instead double beta decay $^{136}Xe \rightarrow ^{136}Ba$ would be allowed with a favorable transition energy of 2479 KeV. The groups is testing in the Gran Sasso laboratory a multi-elemens proportional chamber using 45 liters of xenon at 10 bars.

7. MAGNETIC MONOPOLES
7.1 Introduction

In 1931 Dirac introduced the magnetic monopole in order to explain the quantization of the electric charge. He also established the basic relation between the elementary electric charge e and the magnetic charge g (eg=n$\hbar$c/2), but had no prediction for the monopole mass. This "classical" pole was searched for at every new accelerator, in particular at the newest colliders where poles could be produced in high energy reactions of the type $e^+ + e^- \rightarrow M + \bar{M}$, $\bar{p} + p \rightarrow M + \bar{M}$ (M=monopole, $\bar{M}$=antimonopole) up to a mass of 800 GeV. This is a mass far too small for the main interests of this school (25).

In 1974 it was realized that Grand Unified Theories of the basic interactions predict the existence of magnetic monopoles, whose mass is related to the mass of the vector boson carrier of the unified interaction; the magnetic charge should be a multiple of the Dirac charge. The flux predictions vary wildly, from too many to too few monopoles.

7.2 Summary of the Properties of Magnetic Monopoles

Properties Based on the Dirac Relation. From the Dirac relation, eg $= n\hbar c/2$, one has the basic magnetic charge $g_D = \hbar c/2e = 137\ e/2 = 3.29\ 10^{-8}$ CGS units (for n=1, e=electron charge; there may be justifications for multiply charged poles). Monopoles may be trapped in bulk paramagnetic and ferromagnetic materials with binding energies of about 200 eV.

Mass. GUT monopoles shold have large masses, $m_M \sim m_x/\alpha \sim 10^{16}$ GeV where α is the GUT coupling constant. In the following we shall use a mass $m_M \simeq 10^{16}$ GeV for the stable pole, though the situation could be more complex.

Electric Charge. Electrically charged poles (dyons) should have a mass somewhat larger than that of a pole without electric charge. A bound (M+p) or (M+Al) system behaves as a dyon.

Size and Structure. The GUT monopole is an extended object with a core radius $r_c = 1/m_x \simeq 10^{-29}$ cm, a region up to $r=10^{-16}$ cm where virtual W^+, W^- and Z^o may be present, a confinement region, $r_{conf}=1$ fm, a fermion-antifermion condensate up to $r_f=1/m_f$. For $r \gtrsim 3$ fm the GUT pole behaves as a point pole, which generates a field $B=g/r^2$.

7.3 Monopoles. Cosmology and Astrophysics

The only place where superheavy poles might have been produced is the early Universe, when the temperature was comparable to the pole mass. Different types of poles might have been produced, at different times in the early Universe; at least the lightest pole should be stable. GUT poles should have been produced at $t=10^{-35}$ seconds as topological defects at the points where several Higgs domains joined, or as pole-antipole pairs in the collisions of very high energy particles ($q+\bar{q} \rightarrow M+\bar{M}$).

If monopoles and baryons were generated at the same time they could have been produced with roughly the same abundances, even if the production mechanisms are different. After production, poles and antipoles had a small chance to annihilate each other because of the rapid expansion of the Universe. But now monopoles are not abundant: may be the number of casual domains in the early Universe was small, may be the Universe cooled down before the GUT phase transition or poles were produced at reheating, and the temperature was low. In the inflation scenario, monopoles were diluted with respect to baryons if inflation took place between the generation of poles and the generation of baryons.

As the Universe expanded and cooled the pole kinetic energy decreased. Poles probably clustered like ordinary matter in galaxies; but galactic magnetic fields acted as monopole accelerators. We thus expect on Earth a flux of poles with velocity $v > 3 \times 10^{-5}$ c, may be with peaks at $v = 10^{-4}$ c (local poles) and at typical galactic velocities (10^{-3} c); there should also be poles with higher velocities.

A bound on pole density and flux is obtained requiring that the pole mass density be smaller than the critical density of the Universe. For a uniform pole density one has for 10^{16} GeV poles: $F < 3 \times 10^{-12} \beta$ poles $cm^{-2} s^{-1} sr^{-1}$. If poles cluster in galaxies the limit would be five orders of magnitude larger.

A limit on monopole flux is obtained as a consequence of the existence of the galactic magnetic field (Parker bound); for $m_M = 10^{16}$ GeV one has: for $\beta \lesssim 3 \times 10^{-3}$: $F \lesssim (10^{-15} n)\ cm^{-2} s^{-1} sr^{-1}$, for $\beta > 3 \times 10^{-3}$: $F \lesssim (10^{-15} \beta n/3 \times 10^{-3})\ cm^{-2} s^{-1} sr^{-1}$. This limit, plotted in Fig.7.1 versus m_M for poles with $\beta = 10^{-3}$, is a reference for pole searches.

Table 7.1. Comparison of monopole search experiments. The last column gives the number of events per year at the Parker limit ($10^{-15}\,cm^{-2}\,s^{-1}\,sr^{-1}$).

Experiment.	Detection Method				$S \cdot \Omega$ ($m^2\,sr$)	Track-ing	Time Res.	Expan-dable	Events /y Parker
	Induc.	Scintil.	Gas	Etch					
IBM III	10 grad.	--	--	--	10	No	Poor	Yes	0.003
Baksan	--	3200	--	--	1850	No	Yes	No	0.6
Frejus	--	--	$Ar-CO_2$	--	880	Yes	Yes	No	0.3
Kamioka	--	--	--	CR39	>10000	No	No	Yes	8
MACRO	--	600	He, C_5H_{10}	CR39	>10000	Yes	Yes	Yes	4

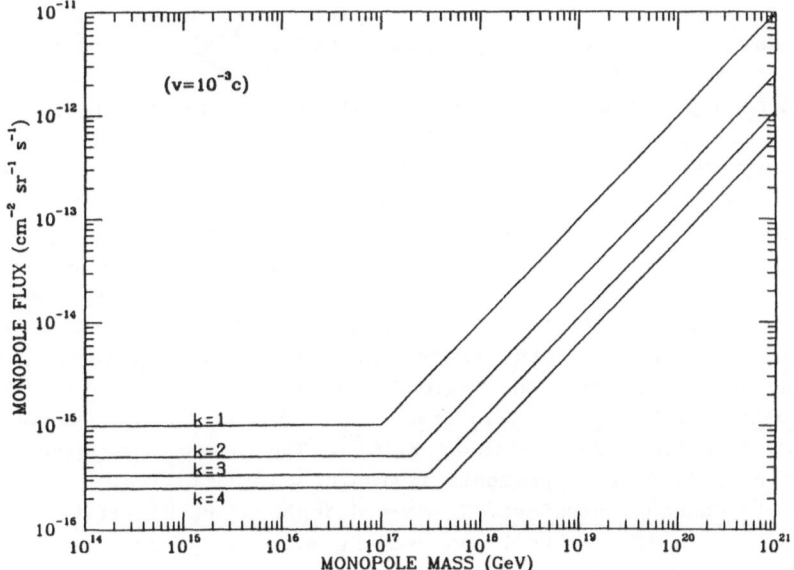

Fig.7.1: The Parker limit for poles with magnetic charge $g=kg_D$, k=1-4.

7.4 Searches for Cosmic Poles

Searches with Induction Devices. Monopole detection with a superconducting ring is based only on the long-range electromagnetic interaction between the pole magnetic charge and the macroscopic quantum state of the superconducting ring. A $g=g_D$ pole of any velocity would be observed as a jump of two flux quanta and with a persistent electric current. Induction detectors have grown in size and complexity. Fig.7.5 shows the combined upper limit from searches with induction devices.

Energy Losses of Monopoles in Matter (Figures 7.2-7.4). For $\beta > 10^{-2}$ a pole with magnetic charge g behaves as an equivalent electric charge $e_{eq} = g\beta$; the energy losses, due to the interaction of the pole with atomic electrons, are proportional to e_{eq}^2 . The energy loss of $g=g_D$ relativistic poles would be 10 GeV $g^{-1} cm^{+2}$, 4700 times higher than minimum ionizing electrically charged particles.

The classical methods used to estimate energy losses of $\beta < 10^{-2}$ protons yield a β threshold for each ionization and excitation level; this results in a threshold for poles in scintillators at the level of $\beta \sim$ 6x10^{-4}. But it was found experimentally that the light yield from very low velocity protons was higher than predicted. This could arise from level mixing effects, from high-velocity tails in the electron momenta distributions, etc. As a consequence there is a stronger belief for a larger light yield of poles in the 3x10^{-4}-6x10^{-4} β-range, Fig.7.3.

For gaseous detectors the threshold is estimated to be around $\beta \sim$ 10^{-3}. On the other hand a passing monopole may displace, in the atoms of the material, atomic energy levels by several eV, leading to mixing and crossing of levels. This effect was found for H, He, Ne and may exist for many more materials. Since the collision time (10^{-15} s) of a pole with an atom is much shorter than the de-excitation time (10^{-8} s) of the atom, the atom would be left in an excited state. The de-excitation energy may be used for ionizing a molecule mixed with the atomic gas. Thus gaseous detectors may detect poles down to $\beta = 10^{-4}$ using an appropriate gas mixture.

For plastic track detectors the relevant energy loss for the "latent" track is the restricted energy loss, that is the energy loss in a cylinder of radius 100 μ around the pole direction. The restricted energy loss is coincident with the total loss for $\beta < 10^{-2}$; it is only a fraction (80-90%) of the total energy loss at higher β-values, where some energetic -rays leave the region of production.

Searches With Scintillation Counters. Scintillation counters should be capable of detecting poles with $\beta > 3x10^{-4}$. The present combined upper limit from scintillator experiments is indicated in Fig.7.5.

Searches With Gaseous Detectors. Gaseous detectors are sensitive to poles with $\beta \gtrsim 10^{-3}$. Use of the Drell and Penning effects brings the threshold down to $\beta = 10^{-4}$. At present the Frejus proton decay detector has the largest SΩ ($\simeq$ 880 m^2 sr). The overall limit is indicated in Fig.7.5.

Searches with Plastic Track Detectors. The passage of heavily ionizing particles may be recorded in insulating materials which range from CR39 plastic sheets to mica. The latent track may be made visible by chemical etching. The most sensitive material is CR39; the sensitivity of CR39 to poles with $g=g_D$, $g=3g_D$ and to bound poles is shown in Fig.7.4. The combined limit with plastic track detectors is shown in Fig.7.5. Indirect searches using ancient muscovite mica samples, rely on the following

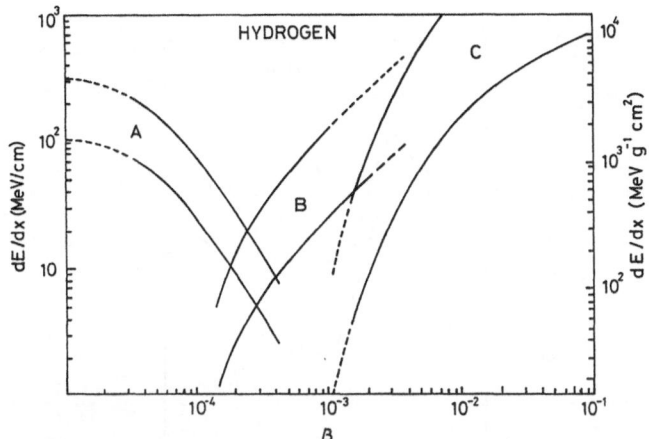

Fig.7.2: Energy losses in atomic hydrogen as function of the pole magnetic charge (Lower curves for $g=g_D$; higher curves for $g=3g_D$). The vertical left scale is in MeV/cm; the right scale is in MeV g^{-1} cm^2. Curves A are due to pole–atom elastic collisions, B to excitation and C to ionization energy losses. Dashed lines indicate velocity regions where the approximations for the calculations may break down.

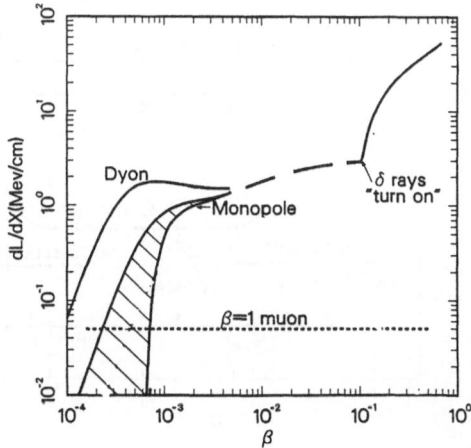

Fig.7.3: Light yield per cm of scintillator for a bare monopole with $g=g_D$ and for a dyon (25).

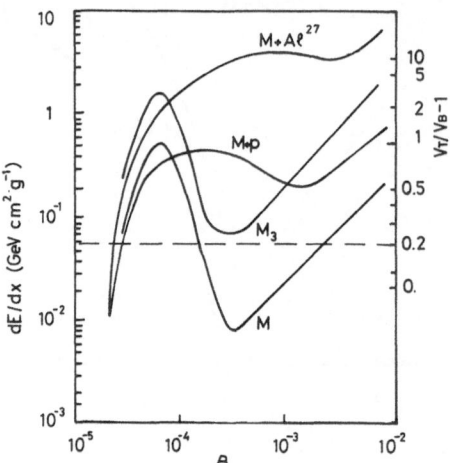

Fig.7.4: Restricted energy losses in CR39 for free and bound poles with $g=g_D$ and $g=3g_D{}^{18}$. The right vertical scale gives the reduced etch rate; the horizontal line represents an energy loss threshold in CR39 (25).

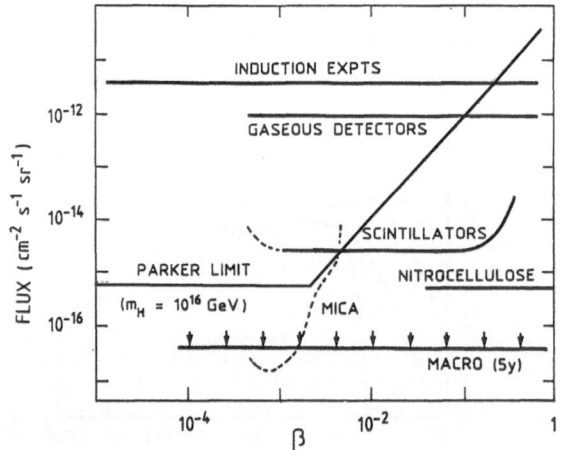

Fig.7.5: Compilation of upper limits obtained using different techniques for a flux of cosmic poles (25).

scenario: A pole with zero or negative electric charge enters the Earth, captures an ^{27}Al nucleus forming a bound state, which passes through a naturally occurring mica, where elastic nuclear collisions yield a trail of lattice defects, that survive as long as the mica remains unheated. Mica samples of ages of 5×10^8 years were cut, cleaved into several $100 \, \mu m$ thick sheets, etched, reassambled in a pair of sheets, scanned in trasmitted light with the microscope focused on the common surfaces. From the no observation of pits the authors estimated an upper limit flux at the level of few 10^{-18} $cm^{-2} s^{-1} sr^{-1}$ for $2 \times 10^{-4} < \beta < 2 \times 10^{-3}$, (Fig.7.5).

7.5 Monopole Catalysis of Proton Decay

A GUT pole may catalyze baryon-number violating processes of the type $p + M \rightarrow M + e^+ + mesons$. If the $\Delta B \neq 0$ cross section were large, a monopole would trigger a chain of baryon decays along its passage through a large detector, such as those which study baryon decay. The Kamioka water cherenkov detector yielded limits at the level of $F \lesssim 1.9 \times 10^{-15}$ $cm^{-2} s^{-1} sr^{-1}$ for $\sigma_c = 100$ mb, $3 \times 10^{-5} < \beta < 10^{-3}$ (globally $F \sigma_c < 10^{-12}$).

The catalysis argument applied to the protons of our sun leads to the possibility that the sun emits electron neutrinos with an energy of 35 MeV. The Kamioka detector estimated an upper limit $F \cdot \sigma(mb) < 4 \times 10^{-22}$ ($\beta /10^{-3}$) $cm^{-2} s^{-1} sr^{-1}$. The number of poles in the sun would be less than 1 pole per 10^{12} g of solar material.

The catalysis of nucleon decay by poles may be a source of energy for astrophysical bodies; it may lead to observable effects for those bodies which do not have an important internal source of energy (planets or stars which have used up most of the nuclear fuel). Estimates of these effects lead to strong constraints; the best limits come from x-ray emission by neutron stars ($F \sigma_c < 10^{-19} - 10^{-25}$). But there are many hypotheses which could vitiate the conclusions. For instance a small catalysis cross section would make this discussion irrelevant. Another example: it was indicated that limits for pulsars do not include the effects of magnetic fields where poles may be accelerated and ejected from the star.

In the cosmological QCD phase transition at the time $t = 10^{-6}$ s the quark-gluon plasma should have been converted into hadrons. It was suggested that more than 80% of the baryon number finished in the form of strange quark matter with any value of mass number from few tens to the values for neutron stars. This strange matter, which could be a candidate for the dark matter in the Universe, should be more stable than ordinary matter if it contains the same number of up, down and strange quarks. However strange matter should be vulnerable to GUT poles, since they should be capable to change the baryon number, with reactions of the type $u_1 + u_2 + M \rightarrow e^+ + d_3 + M$, where 1, 2, 3 are color indices. The study of the effect of a flux of poles on strange matter via this reaction in the early Universe places an upper limit on the density of poles at the QCD transition. Considerations on the gravitational capture of poles by strange matter yield upper bounds on present pole density if strange balls of different sizes are surviving from the quark-gluon $\rightarrow$ hadron phase transition.

8. NEUTRINO PHYSICS

Non-accelerator experiments may yield important information on several aspects of neutrino physics, in particular on neutrino mass and decay, on neutrino oscillations and on neutrinoless double beta decay (24, 26). At present there are no compelling indications for any of these effects: the neutrino seems to be a Dirac left h·anded, massless and stable particle.

Neutrino mass

The best direct upper limits from tritium decay for ν_e (Fig.8.1, Table 8.1) and from accelerator experiments for the other neutrinos are $m(\nu_e) < 18$ eV, $m(\nu_\mu) < 270$ KeV, $m(\nu_\tau) < 70$ MeV (24,26). A limit of about $10-20$ eV may be established on $m(\bar{\nu}_e)$ from the difference in arrival time of antineutrinos of different energies coming from Supernova 1987A.

From "common sense" one expects: $m(\nu_e)$: $m(\nu_\mu)$: $m(\nu_\tau) = m(e)$: $m(\mu)$: $m(\tau)$. Since from astrophysical considerations the sum of the neutrino masses should have a bound at less than 100 eV, one could expect $m(\nu_e) \sim 10^{-2}$ eV, $m(\nu_\mu) \sim 1$ eV, $m(\nu_\tau) \sim 10$ eV. These are values well below the present possibilities of direct measurements.

Table 8.1 <u>Tritium decay experiments</u> (26). The experiments measure the upper part of the electron (positron) energy spectrum and try to establish if the spectrum stops before the end point calculated assuming a massless neutrino (see Fig.8.1).

Investigator	Location	Spectrometer	Source
• Lyubimov	Moscow	Toroidal magnet	Valine–t
• Kündig	Zurich	Toroidal magnet	C–t
• TJB/HR/JFW	Los Alamos	Toroidal magnet	$t_1 t_2$ gas
• Ohshima	Tokyo	$\pi\sqrt{2}$ magnet	Cd–Arachidate–t
Fackler	Livermore	Retarding E-S	Al_2O_3–t, t_2 solid
Clark, Frisch	IBM-Yorktown	Retarding E-S	?
Jelley	Oxford	Cylindr. mirror	Cd–Palmitate–t
Sun	Beijing	Iron-core magnet	Lit, PAD–t
Stoeffl	Livermore	Toroidal magnet	$t_1 t_2$ gas
Lobashev	Moscow	Retarding E-S	$t_1 t_2$ gas
Boyd	Colombus	Retarding E-S	t_2 solid
Wellenstein	Brandeis	Cylindr. mirror	t_2 gas
Kalbfleisch	Oklahoma	?	?
Otten	Mainz	Retarding E-S	t gas
Daniel	Munich	Iron-core magnet	Hf–t
* Simpson	Guelph	Si	Si–t
* Derbin/Popeko	Leningrad	Si	Si–t

• Have results

* Expts. completed

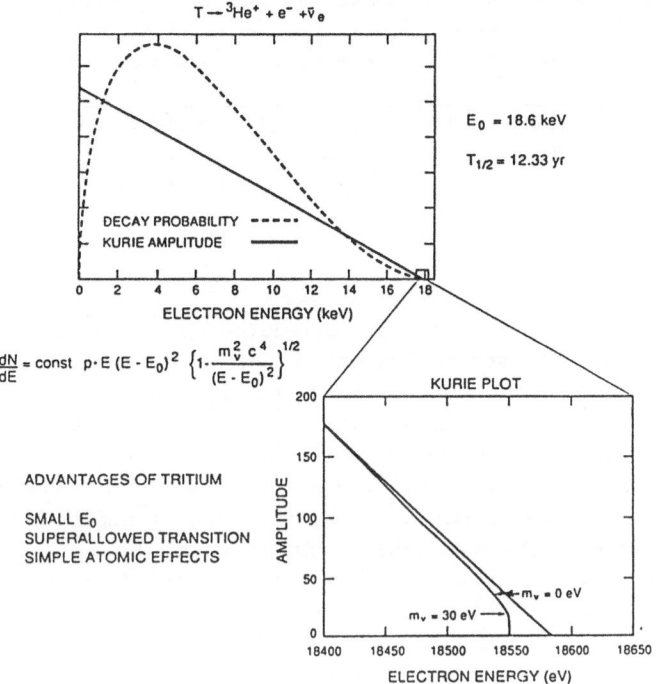

$$\frac{dN}{dE} = \text{const } p \cdot E \, (E - E_0)^2 \left\{ 1 - \frac{m_\nu^2 c^4}{(E - E_0)^2} \right\}^{1/2}$$

Fig. 8.1: Kurie plot for tritium decay (26).

Neutrino decay

The decay $\nu_\mu \to e^+ e^- \nu_e$ was searched for in high energy beams without success. The detection of $\bar{\nu}_e$ from Supernova 1987A, with a flux comparable with expectations proves that the lifetime of $\bar{\nu}_e$ is larger than 10^5 y.

Neutrino oscillations

Neutrinos of one flavor may change (oscillate) into neutrinos of another flavor provided that one or more neutrinos have a non zero rest mass. Neutrino oscillations may be studied either via the disappearance of one type of neutrino or via the appearance of neutrinos of a different type. For $\nu_\mu \to \nu_e$ oscillations in vacuum one writes the probability of a ν_e appearing as (2 flavor case)

$$P(\nu_\mu \to \nu_e) = 0.25 \sin^2 2\vartheta \, \sin^2 \pi D / \lambda_v \qquad (8.1)$$

where ϑ is the mixing angle and D is the distance source-detector; λ_v, the oscillation length in vacuum, may be written as

$$\lambda_v \text{ (Km)} = 2.5 \, E_\nu \text{ (GeV)} / \Delta m^2 \text{ (eV)}^2 \qquad (8.2)$$

where $\Delta m^2 = m_1^2 - m_2^2$. In matter (containing electrons) the ν_e's have an extra interaction term due to $W^\pm$-exchange, which is not present for the other neutrinos (Mikheiev-Smirnov-Wolfenstein effect, ref.27). In this case the oscillation amplitude is enhanced

$$\sin^2 2\vartheta \Longrightarrow \sin^2 2\vartheta / [1 + (\lambda_v / \lambda_o)^2 - 2 \, (\lambda_v / \lambda_o) \cos \vartheta] \qquad (8.3)$$

where λ_o depends on the characteristics of matter,

$$\lambda_o = 1.63 \times 10^7 \text{ m} / [(Z/A) \varrho \text{ (g cm}^{-3})] \qquad (8.4)$$

The Kamioka, IMB and Frejus experimenters have looked for neutrino oscillations using "atmospheric neutrinos", that is neutrinos produced in the atmosphere (3,28). Energetic primary cosmic rays produce pions and kaons which subsequently decay into ν_μ and ν_e (Fig.11.3). The atmospheric neutrino flux at sea level, in the range 1-100 GeV, comes mainly from $\pi \to \mu$ decay, while the neutrino flux for $E_\nu > 100$ GeV comes mainly from $K \to \mu$

Table 8.2. Results on neutrino oscillations using atmospheric neutrinos (28).

Expt	Ratios	e-like	μ-like	e-like/μ-like
Kamioka	esp.data/Montecarlo	1.08 ± 0.11	0.59 ± 0.06	1.79 ± 0.26
IMB	"	--	0.88 ± 0.11	--
Frejus	"	--	--	1.41 ± 0.31
Nusex	"	--	--	0.73 ± 0.37

decay. Neutrino oscillations may be detected using atmospheric muon neutrinos that have traversed the earth, enter the proton decay detectors from below and interact as electron neutrinos inside the detectors The situation is rather complex and one should always compare the event spectrum with the predictions of a "realistic" Montecarlo. Moreover the number of events is rather small (less than 200 per experiment).

The Kamioka group finds a lack of muons (compared to Montecarlo predictions), see Table 8.2. The quantity with the smallest systematic errors is the ratio (e-like events/μ-like events) (that is ($\nu_e + \bar{\nu}_e$)/($\nu_\mu + \bar{\nu}_\mu$)) always referred to a Montecarlo calculation. Kamioka found a three standard deviations effect. The IMB and Frejus data are compatible with no-oscillations, but they cannot rule out the Kamioka result. The Kamioka effect looks more important for interactions with a visible energy smaller than 700 MeV/c; Frejus does not see this effect. Further checks are also needed on the Montecarlo itself. It may thus be concluded as Suzuki (28) said that further data and better Montecarlo calculations are needed.

Fig.8.2 is a compilation of limits on neutrino oscillations (29).

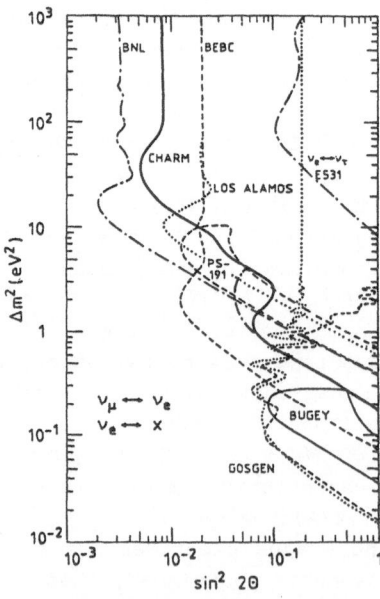

Fig.8.2: Compilation of limits obtained from different neutrino oscillation experiments using neutrinos from accelerators and nuclear reactors (29). The excluded regions are on the upper right.

9. NEUTRINO ASTRONOMY

Underground detectors of large area, of large mass and of high precision may yield important information on the astrophysics of neutrinos. Neutrino astrophysics or, as is now more commonly called, neutrino astronomy, is a new observational window on the Universe, Fig.9.1. Low energy neutrinos of few MeV come continuously from the interior of stars like the sun; slightly higher energy neutrinos, with few tens of MeV, are copiously emitted in supernovae explosions. High energy neutrinos with hundreds of GeV may come from non-thermal powerful point sources. Since neutrinos interact rarely, the observed neutrinos come directly from their sources, without suffering the very many interactions typical of photons. Because of all these sources, the number of neutrinos in the Universe is increasing continuosly. It has also to be remembered that the Universe should be filled with "fossil" low energy neutrinos (of 2×10^{-4} eV) from the Big Bang. Their number should be comparable to that of the microwave electromagnetic radiation. At present there is no possibility of detecting fossil neutrinos. Neutrinos of (1-100) GeV may also come from the sun where annihilations of Weakly Interacting Massive Particles (WIMPS) could take place.

It may be worthwhile to stress advantages and disadvantages of using neutrinos in "Astronomy". Advantages: i) neutrinos are stable, travel at the speed of light and are electrically neutral: thus they do not get deflected in magnetic fields. ii) neutrinos interact weakly; as a consequence they are not absorbed, carry direct informations from their actual source and may be observed day and night. The disadvantages are connected with the fact that they are difficult to detect and that one needs large and expensive detectors.

9.1 Solar neutrinos

According to the standard model of the sun, all its energy is produced in a series of thermonuclear reactions and decays at the center of the sun. This solar "thermonuclear reactor" is very small compared to the size of the sun. The emitted photons suffer an enormous number of interactions (their mean free path is much less than 1 cm) and reach the surface of the sun in about one million years. Visible sunlight comes from a well defined surface, the photosphere. An important fraction of the energy from the sun is emitted in the form of neutrinos of different energies see Figures 9.1, 9.2 and Table 9.1. Since these neutrinos have small energies (0.1 MeV $< E_\nu < 10$ MeV) they have small interaction cross sections; therefore they escape the sun almost undeflected and their observation would give access to the "small thermonuclear reactor" at the center of the sun.

Most of the emitted neutrinos come from the $p+p \rightarrow d+e^+ + \nu_e$ reaction, which yields electron neutrinos with energies $0 < E_\nu < 0.42$ MeV; these have an average interaction cross section of 1.2×10^{-45} cm^2. The highest energy neutrinos come from $B^8 \rightarrow$ decay; these neutrinos have energies $0 < E_\nu < 14.06$ MeV and an average cross section of 2.8×10^{-43} cm^2.

On Earth should arrive about 7×10^{10} neutrinos per cm^2 e per second. The only experiment which has detected solar neutrinos is the chlorine experiment in the USA, using the reaction $\nu_e + {}^{37}Cl \rightarrow {}^{37}Ar + e^-$. (The experiment uses 615 t of C_2Cl_4. The produced gaseous Ar atoms are flushed and counted as was illustrated for Gallex). This reaction has a threshold of 814 KeV and is thus sensitive to ${}^8B$ neutrinos, which is a

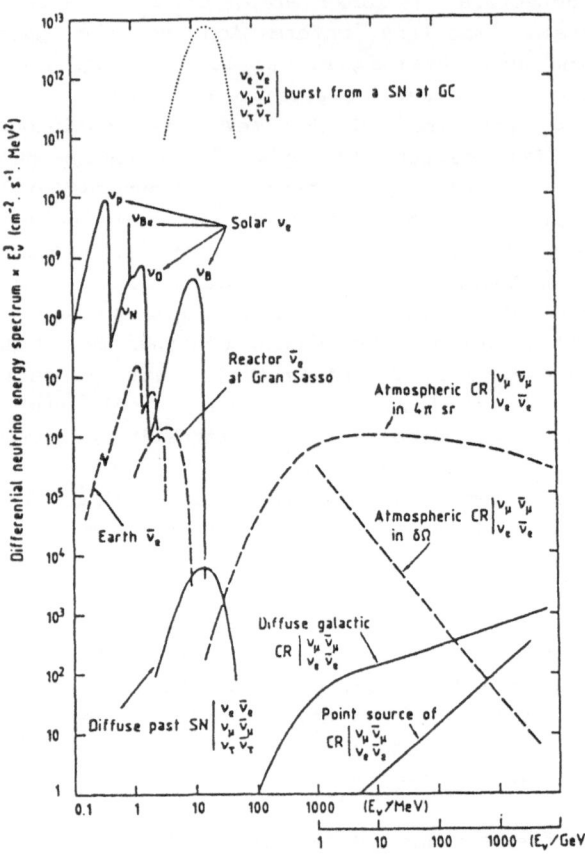

Fig.9.1: Differential energy spectrum of the flux of neutrinos of various origin at the Earth.

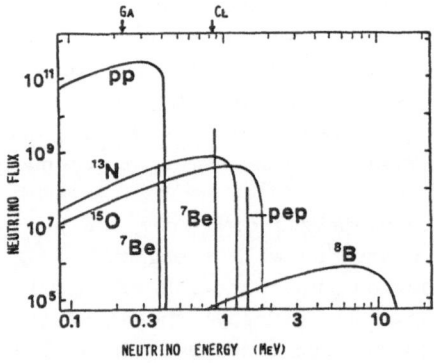

Fig.9.2: Flux of solar neutrinos reaching the Earth versus neutrino energy. (The flux is in neutrinos cm^{-2} s^{-1} for monochromatic neutrinos and neutrinos cm^{-2} s^{-1} MeV^{-1} for non monochromatic neutrinos, given by continuous curves). Arrows indicate the thresholds in Cl and in Ga.

Table 9.1 Hydrogen and Carbon cycles for energy production in the sun. The global equivalent reaction is $4p \rightarrow {}^4He + 2\ e^+ + 2\ \nu_e + 26.7$ MeV.

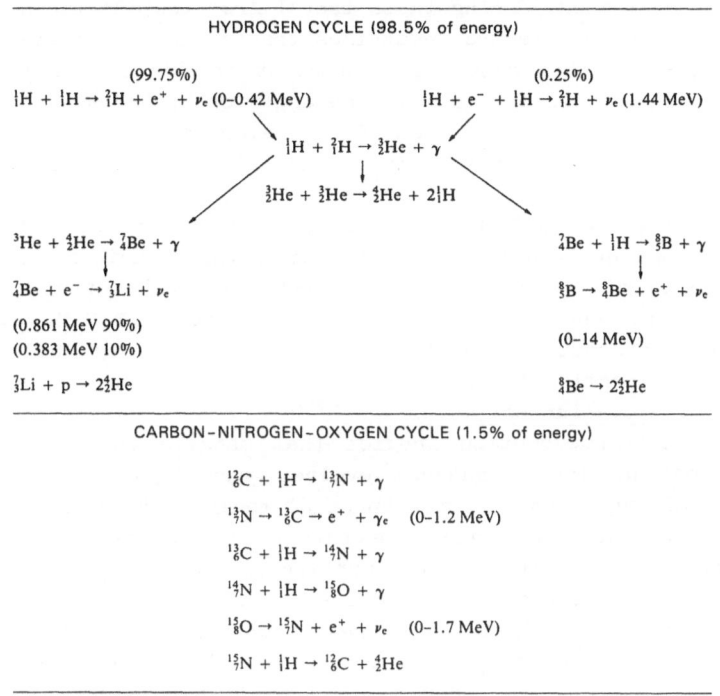

HYDROGEN CYCLE (98.5% of energy)

(99.75%)   (0.25%)
${}^1_1H + {}^1_1H \rightarrow {}^2_1H + e^+ + \nu_e$ (0–0.42 MeV) ${}^1_1H + e^- + {}^1_1H \rightarrow {}^2_1H + \nu_e$ (1.44 MeV)

${}^1_1H + {}^2_1H \rightarrow {}^3_2He + \gamma$

${}^3_2He + {}^3_2He \rightarrow {}^4_2He + 2{}^1_1H$

${}^3He + {}^4_2He \rightarrow {}^7_4Be + \gamma$ ${}^7_4Be + {}^1_1H \rightarrow {}^8_5B + \gamma$

${}^7_4Be + e^- \rightarrow {}^7_3Li + \nu_e$ ${}^8_5B \rightarrow {}^8_4Be + e^+ + \nu_e$

(0.861 MeV 90%)
(0.383 MeV 10%)   (0–14 MeV)

${}^7_3Li + p \rightarrow 2{}^4_2He$ ${}^8_4Be \rightarrow 2{}^4_2He$

CARBON–NITROGEN–OXYGEN CYCLE (1.5% of energy)

$${}^{12}_6C + {}^1_1H \rightarrow {}^{13}_7N + \gamma$$
$${}^{13}_7N \rightarrow {}^{13}_6C \rightarrow e^+ + \gamma_e \quad (0–1.2\ MeV)$$
$${}^{13}_6C + {}^1_1H \rightarrow {}^{14}_7N + \gamma$$
$${}^{14}_7N + {}^1_1H \rightarrow {}^{15}_8O + \gamma$$
$${}^{15}_8O \rightarrow {}^{15}_7N + e^+ + \nu_e \quad (0–1.7\ MeV)$$
$${}^{15}_7N + {}^1_1H \rightarrow {}^{12}_6C + {}^4_2He$$

marginal process among all other nuclear processes in the sun. The experimentally observed flux (2.2 ± 0.3 SNU) is about 1/3 of the expected flux (7.9 ± 2.6 SNU) (26,30). This observation is the basis of the "problem of the solar neutrinos". Recent results from the chlorine experiment show a larger number (5.4 ± 1.1 SNU) of interactions and may indicate a smaller number of missing neutrinos (26), see Fig. 9.3. A recent analysis from the Kamioka detector (31) has shown neutrino interactions in water for $E_\nu > 4$ MeV; they seem to confirm that there are some "missing neutrinos". They have also indication of directionality, which would give a proof that the observed neutrinos really come from the sun.

One clearly needs to confirm the experimental results and study the more abundant neutrinos from the pp reactions. But one may speculate that the origin of the problem could arise either from imperfect knowledge of

Table 9.2 Average cross sections and number of interactions for solar neutrinos. The interactions are in chlorine. One SNU is one capture per second in 10^{36} atoms of Cl.

ν–SOURCE	${}^8B$	${}^7Be$	pp	${}^{13}N$	${}^{15}O$	TOTAL
$\bar\sigma (10^{-46} cm^2)$	10600	2.4	16	1.7	6.6	
SNU	6.1	1.1	0.2	0.1	0.3	(7.9 ± 2.6)

the nuclear reactions in the sun (there may be an anticorrelation with solar activity) or from the behaviour of neutrinos: there could be oscillations of neutrinos, during their travel from the sun to the earth, with a sizable number of electron neutrinos trasforming for instance into muon neutrinos, which then cannot give a count in our detectors. There could also be neutrino oscillations in the interior of the sun.

The Gallex experiment at Gran Sasso aims specifically to the detection of electron neutrinos from the pp reaction. (A similar experiment is being set up in the Soviet Union). Gallex is sensitive to neutrinos with $E_{\nu_e} >$ 233 KeV. In a certain sense it will prove experimentally that neutrinos really come from thermonuclear reactions. ICARUS is also sensitive to neutrinos from the sun, but only to ^{8}B neutrinos, since its energy threshold should be around 4 MeV. Also LVD may give some information on this point. ICARUS may allow some determination of directionality.

9.2 Neutrinos from stellar collapses

Models of stellar evolution predict that massive stars (m 6 solar masses ($m_\odot$), on the main sequence) evolve gradually as increasingly heavier nuclei are produced and then burnt at their centers in a chain of thermonuclear processes, ultimately leading to the formation of a core composed of iron and nickel. Further burning in the shells surrounding the core may make the core mass exceed the so called Chandrasekar limit, $m_{core} > m_{ch} = 5.76\, y_e^2\, m_\odot$, where y_e is the lepton fraction in the core (y_e <0.5). In such case the core implodes in a time only slightly longer than the freefall time (a few milliseconds) and leads to the formation of a neutron star (or of a black hole if the star is very massive). In our

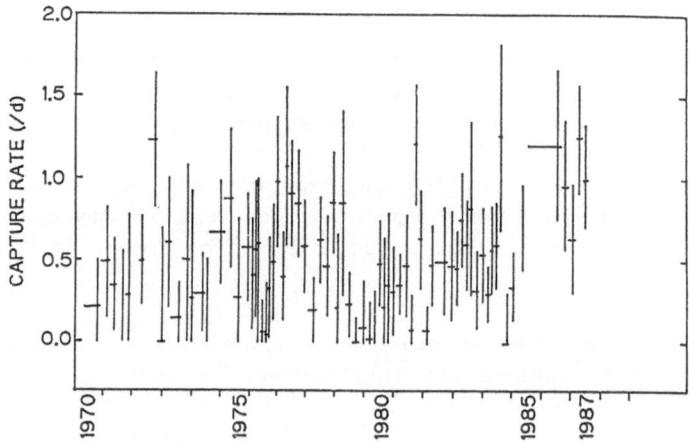

Fig.9.3: Results of the chlorine experiment. The detected reaction is $\nu + {}^{37}Cl \rightarrow e^- + {}^{37}Ar$. Plotted is the number of ^{37}Ar atoms produced per day. This number averaged about 0.4 atoms per day from 1970 till 1985. The average after the 1985 shutdown is larger by about a factor of two. (The right hand scale is in solar neutrino units) (26).

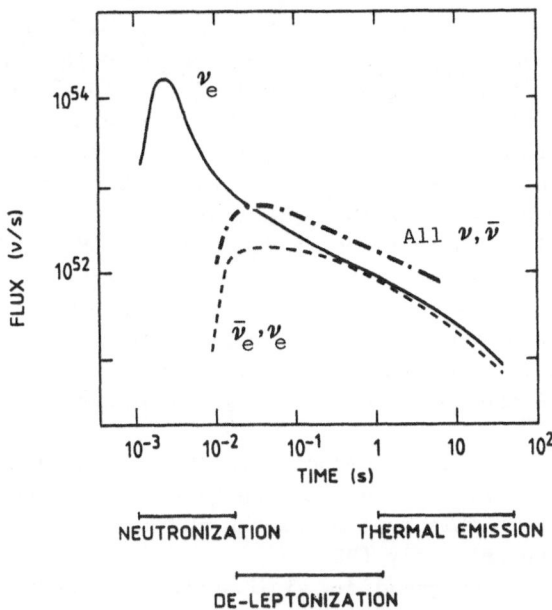

Fig.9.4: Time distribution of the neutrinos from a supernova explosion. The electron neutrino peak arises from the neutrization reaction $e^-+p \rightarrow$ n+ ν_e. The dashed distribution is due to the de-leptonization process $e^+ +e^- \rightarrow \nu_e + \bar{\nu}_e$. The dotted-dashed line is the time distribution of ν_e, $\bar{\nu}_e$, ν_μ, $\bar{\nu}_\mu$, ν_τ, $\bar{\nu}_\tau$ thermally produced at the outer surface of the neutron star (from the "neutrinosphere" at a temperature of about 5 MeV).

galaxy one expects one such event about every 15 years (the optically visible ones are fewer because of dust clouds which limit the visibility of the center of our galaxy).

The total energy released during a stellar collapse will be at least the gravitational binding energy of the residual neutron star, $E = 3 \times 10^{53}$ $(m/m_\odot)^2$ (10 Km/R) ergs. The total energy emitted is typically 10^{53} ergs = 0.1 $m_\odot$, mostly in the form of neutrinos with an average energy of 14 MeV. Thus typically 4×10^{57} neutrinos of each species are emitted. If the collapse occurs at the center of our galaxy, the neutrino flux at the Earth would be about $10^{12}/cm^2$ for each species.

During the collapse, three stages of neutrino emission may be identified (Fig.9.4):

i) <u>Neutronization</u>: $e^-+p \rightarrow$ n+ ν_e, which further reduces the number of leptons in the core. Only ν_e are emitted at this stage, which lasts for a time comparable to the collapse time (few milliseconds) and which leads to a peak luminosity of 10^{59} ν_e/s. A large number of high energy photons is also generated. These, interacting with iron nuclei, give rise to a large number of e^+e^- pairs.

ii) <u>Deleptonization</u>: The reaction $e^+ +e^- \rightarrow \nu_e + \bar{\nu}_e$ leads to ν_e and $\bar{\nu}_e$ which leave the core. The more abundant reactions $e^+ + e^- \rightarrow \gamma + \gamma$ are followed by γ's recreating e^+e^- pairs. This phase lasts about 1 second.

iii) <u>Cooling</u>: The neutron star is very hot and is cooled by escaping neutrinos. Thermal Neutrino emission from the "neutrinosphere" continues for about 10 seconds. The bulk of the neutrino luminosity is emitted during this phase.

All types of neutrinos may be detected via their neutral current interactions with electrons, $\nu_e + e^- \rightarrow \nu_e + e^-$, $\bar{\nu}_e + e^- \rightarrow \bar{\nu}_e + e^-$, etc, with a cross section $\sigma = 1.7 \times 10^{-44} E_\nu$ (MeV cm^2). Since the cross section is small, interactions with electrons lead to a small number of events. It may be worth recalling that the scattered electron "remembers" the direction of the incoming neutrino. The dominant observed reaction is $\bar{\nu}_e + p \rightarrow n + e^+$, which has $\sigma = 7.5 \times 10^{-44} E_\nu^2$ (MeV^2 cm^2). It is energetically possible only with free protons, as in H_2O and in $C_n H_{2n+2}$ detectors. The produced positron annihilates immediately, $e^+ + e^- \rightarrow 2\gamma$, while the neutron is captured after a mean time of about 200 μs (n+p $\rightarrow$ d+γ, with $E_\gamma \sim 2.2$ MeV). Detectors with heavy water, D_2O, could also detect $\nu_e + n \rightarrow p + e^-$. Because of the energy dependence of the neutrino cross section on neutrino energy, the average electron energy is higher, $\langle E_e \rangle = 15$ MeV.

9.3 Supernova 1987A

The observation of the Supernova 1987A in the large Magellanic Cloud, 170.000 light years away, on February 23, 1987 has given a considerable impetus to this type of search and has effectively opened up the field of Neutrino Astronomy (32).

The supernova, as observed in visible light (Fig.9.5) and in the ultraviolet, was not really a typical supernova of type II, but was relatively weak. The burst of neutrinos was observed by the Kamioka (12 events) and IMB (8 events) proton decay detectors (using thousands of tons of water) and by the Baksan (3 events) neutrino telescope (using scintillators), see Fig.9.6. The Mont Blanc detector using 80 t of liquid scintillator observed a probable signal of 5 neutrinos, 4 hours earlier. If both signals are right, it is possible that there was first the production of a neutron star and later, after accretion, the collapse to a black hole. The emitted gas cloud was thin and the neutrinos arrived only hours earlier than visible light, not weeks as anticipated. The Kamioka and IMB neutrinos arrived all within 10 seconds and the flux was approximately the predicted flux. At the last conferences there were lively discussions on the reliability and on the implications of these observations. We shall further discuss these issues in connection with gravitational waves (Section 10). I am afraid that we need another supernova to really solve these problems.

Many new detectors will be capable of yielding important informations on neutrinos from new supernovae (LVD, MACRO, ICARUS, Superkamiokande). Very sensitive detectors, like ICARUS, could in principle look for the general background of 14 MeV neutrinos coming from all previous supernovae.

9.4 Neutrinos from WIMPS annihilating in the sun

Weakly Interacting Massive Particle (WIMP) which could be Dark Matter candidates, can be gravitationally attracted by the sun, trapped in the solar core and thermalized after a few orbits. Then a pair of such particles (W, $\bar{W}$) could annihilate emitting a $\nu\bar{\nu}$ pair (W+$\bar{W}$ $\rightarrow$ ν + $\bar{\nu}$). A candidate for WIMPS is the lightest SUSY particle (the photino ?). The emitted neutrinos would have an average energy of about 1/3 of the WIMP mass, $\bar{E}_\nu \simeq m_W/3$. Thus the neutrinos would have energies between several GeV and several tens of GeV. The IMB and Frejus detectors have looked for such WIMPS: IMB used contained events, Frejus higher energy events. Both found no signal in excess of background in the direction of the sun. The upper limits are 4.9 and 1.6 events per Kt y, for IMB and Frejus respectively.

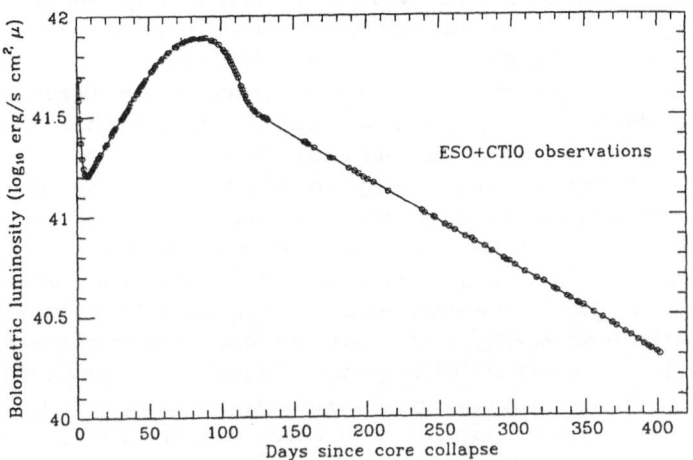

Fig.9.5: The light curve of SN1987A (32).

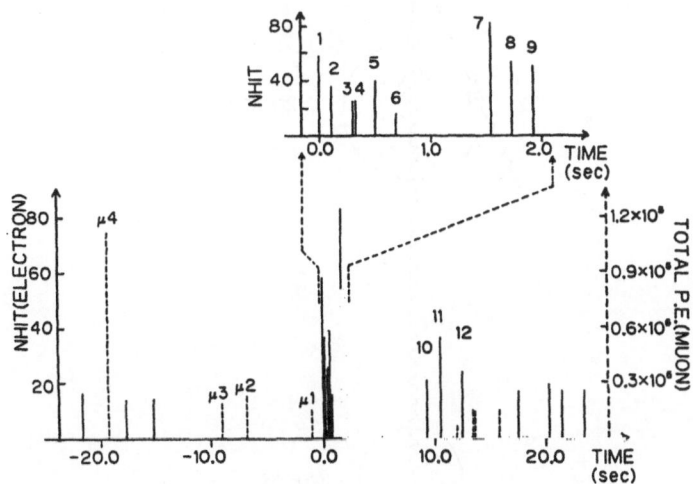

Fig.9.6: The time distribution of the Kamioka neutrino events. Vertical scale is proportional to neutrino energy (32).

9.5 High-energy neutrino astronomy

High energy muon neutrinos can be efficiently detected through their charged current interactions in the rock surrounding the detectors, ν_μ +N →μ+X. Downgoing muons from neutrino interactions will clearly be indistinguishable from the much more abundant cosmic ray muons. Upward going muons from neutrino interactions are seen directly in Cherenkov detectors and may be separated by time of flight from the more abundant downward going muons. These studies are particularly interesting at very high energies where the $\nu - \mu$ angle is smaller and the effective target mass is larger. In order to see point sources one needs an angular resolution of about 1°, comparable or slightly larger than the effective average muon multiple scattering angle in the rock.

In recent years x- and γ-ray astronomy have revealed powerful acceleration mechanisms in many astrophysical bodies; a large flux of energy seems to be radiated by non-thermal processes. In particular multi-TeV gamma-rays with a spectral power E^{-2} have been detected from a number of astrophysical sources, such as Cygnus X-3, Vela X-1 and LMC X-4. These ultra high energy gamma-rays are observed only during two very short intervals, in contrast with softer x-rays, which are modulated more or less following a sinusoidal cycle. From these observations, models of these sources have been constructed in which a companion star eclipses a close by neutron star (Fig.9.7). This last star emits a beam of high

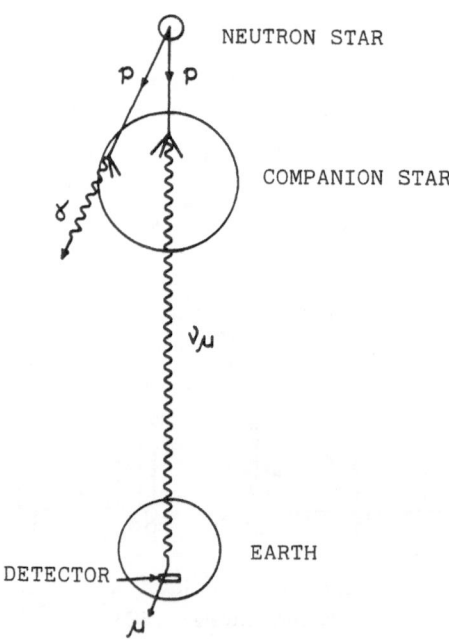

Fig.9.7: A binary x-ray source. The neutron star may accelerate protons up to very high energies. Some of these protons hit the heavy companion star in whose atmosphere interact producing π-mesons and thus, by decay, photons and neutrinos. Photons arrive on Earth only when the line of sight from the Earth grazes the bord of the companion star; neutrinos may arrive for the whole occultation time of the neutron star.

energy protons which collide with the atmosphere of the companion star, producing γ-rays that are heavily absorbed, unless our line of sight grazes the rim of the companion star. If this is true the emission of these gamma-rays is due to π^o-decay. Then muon neutrinos, from the decay of charged pions, should be naturally produced. Because the emitted ν_μ are only marginally affected by the shielding of the companion star, they should be present during most of the eclipse time and the intensity would be correspondingly higher (by a factor 10-30). A neutrino flux with spectral shape E_ν^{-2} would be very efficiently detected, since both the neutrino cross section and the average muon momentum (and hence the "effective target" length) grow linearly with E_ν , thus cancelling the spectral index of -2. Therefore the energy spectrum of the detected neutrinos should be essentially flat.

High energy muon neutrinos are possibly emitted not only by x-ray binaries, but also by expanding supernovae shells (few months or few years after the explosion), from the material surroundig a black hole, a.s.o. Several detectors are now eagarly looking at the remnant of supernova 1987A.

The present larger detectors with surface areas of the order of 100-300 m^2 are somewhat too small. The first detector which can make really full use of the method is the MACRO detector at Gran Sasso, with $S \geqslant 1000$ m^2. In order to perform good studies of high energy neutrinos from point sources larger detectors will be needed, with surface areas at least an order of magnitude larger. Several proposals are being put forward. It is not easy, at the present time, to separate dreams from realities. In any case the proposed detectors are the following:

DUMAND (Deep Underwater Muon and Neutrino Detector) plans to use ocean water as the active Cherenkov medium (33). Stage 1, consisting of a 60 m string with seven photomultipliers at a depth of 4 Km, has been tried in the Pacific, west of Hawai; the detector string was sensitive out to 17 m distance. It used 16-inch Hamamatsu photomultipliers, with hemispherical cathodes, enclosed in oceanographic research glass spheres capable of resisting several hundred atmospheres. Stage 2 DUMAND plans to use nine 330 meters strings, each with 24 sensitive modules (see Fig.9.8). It will be located at a depth of 4.8 Km. It should have a global surface area of about 20000 m^2.

A similar layout is planned for the Baikal lake (34). A soviet group performed a test and a measurement with one string equipped with 6 detecting stations located 50 meters apart at a depth of 1350 meters. Each station consisted of 4 photomultipliers inside oceanographic glass; 2 photomultipliers looked upward and two downward. The experimenters were able to establish an upper limit on a flux of cosmic monopoles via absence of catalysis of baryon decay. The group also measured the transparency of water (absorption coefficient 0.05-0.08 m^{-1} at λ = 480 nm). The Baikal lake has a thick ice crust for two months each year: this makes the deployment of instruments easier. They plan to deploy for 1993 five strings with a total of 200 photomultipliers (Fig.9.9). The layout should have $S\Omega \sim 5 \times 10^4$ m^2 sr and a sensitive volume v = 5×10^5 m^3.

The detector GRANDE (Gamma Ray and Neutrino Detector) would really be very large (S=250x250= 52500 m^2, v= 5×10^6 m^3). It would consist of multiple ring-imaging Cherenkov detectors located in a water filled pit in Arkansas (35). Three neutrino telescopes would look downward (through

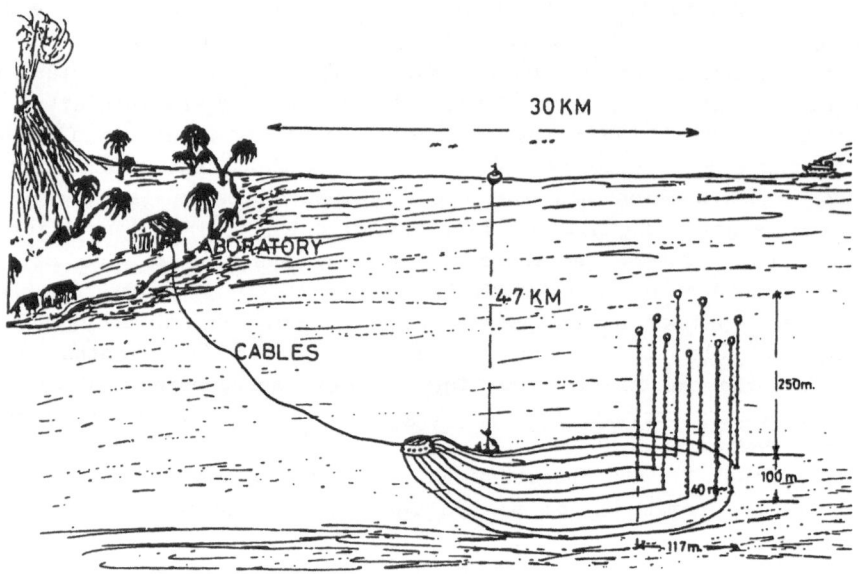

Fig.9.8: Layout of the proposed DUMAND experiment (33).

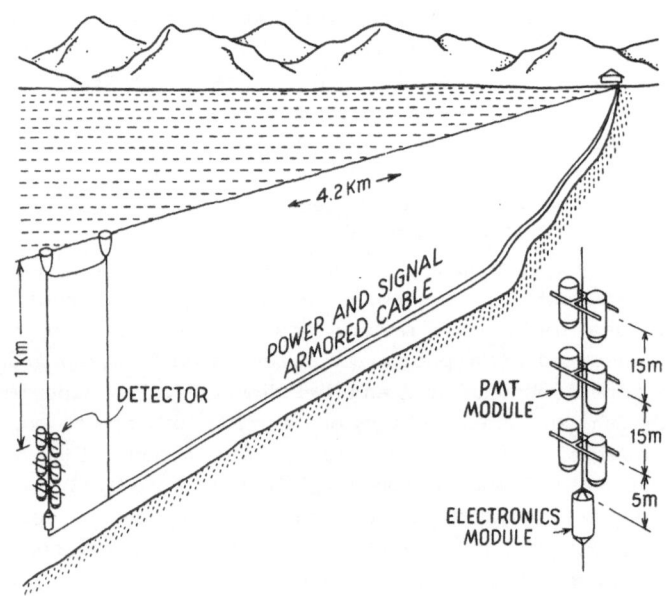

Fig.9.9: Layout the proposed Baikal experiment (34).

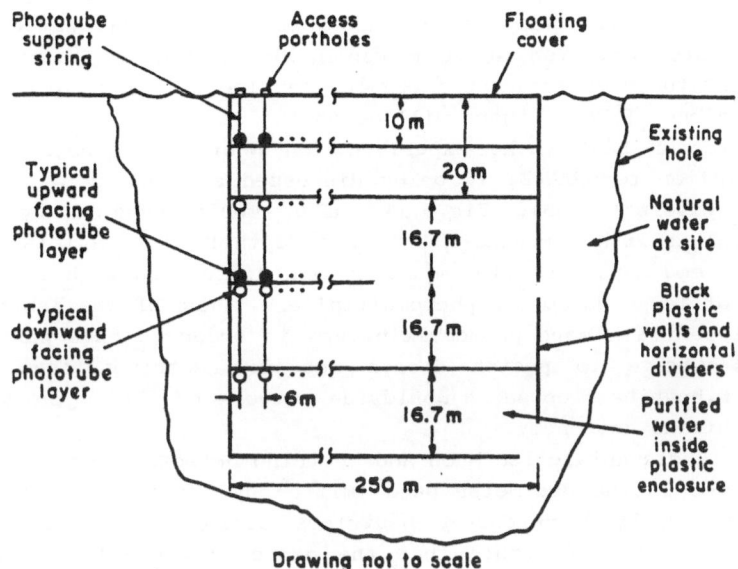

Fig.9.10: Layout of the GRANDE proposal for high energy ν-and γ-astronomy (35).

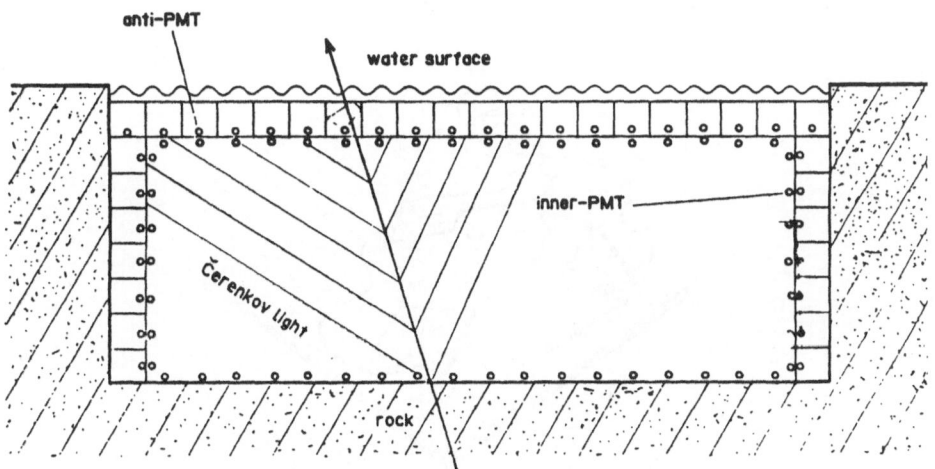

Fig.9.11: Schematic layout of the LENA proposal (36)

the earth), whereas two gamma telescopes would look upward. The aperture of the neutrino and gamma telescopes would be 1.3 sr and 2 sr, respectively. Each telescope would be operated independently. Each should have 1° angular resolution for neutrinos and 0.3° resolution for showers above 10 TeV. The 3 planes of downward facing phototubes are separated by about 17 m; the first plane is situated 20 m below the water surface yielding a total detector of 70 m depth. On each plane there will be 1764, 8" phototubes arranged on a regular lattice with 6 m spacing. Just above each plane of PMT's there will be an optically opaque sheet.

The detector LENA (Lake Experiment on Neutrino Astronomy), which would be similar to GRANDE, is being discussed at CERN by Japanese and European physicists (36), Fig.9.11. LENA would have a cylindrical sensitive volume with a radius R=110 m, a depth D=30 m, a total surface of 38.000 m^2 and a volume v=10^6 m^3. The main difference with GRANDE lies in the use of large Hamamatsu photomultipliers. Part of the inner volume may be equipped with more photomultipliers in order to have part of the detector sensitive to proton decays and to neutrinos from stellar collapses. M.Koshiba proposes a worldwide network of LENA type detectors to study neutrino astronomy.

RAMAND (Radio Antarctica Muon And Neutrino Detector) plans to detect the radiowaves in the decimeter band emitted by the shower of particles produced in ice by high energy neutrinos coming from below, (37,38) Fig.9.12. The authors estimate that the detector would be particularly sensitive to $\nu_e + N \rightarrow e + X$ for $E_e > 10^{18}$ eV for which the effective volume would be larger than 10^8 m^3. Preliminary tests are encouraging. In particular the group has shown that cold ice (at -50°C) has a smaller attenuation to decimeter waves (4–12 db/m). RAMHAND plans to use the same technique with radiowaves produced by neutrinos interacting in the moon, "observed" with radio telescopes on Earth.

Figs. 9.13 and 9.14 illustrate the detection of upward going muons and the field of view of a Gran Sasso experiment.

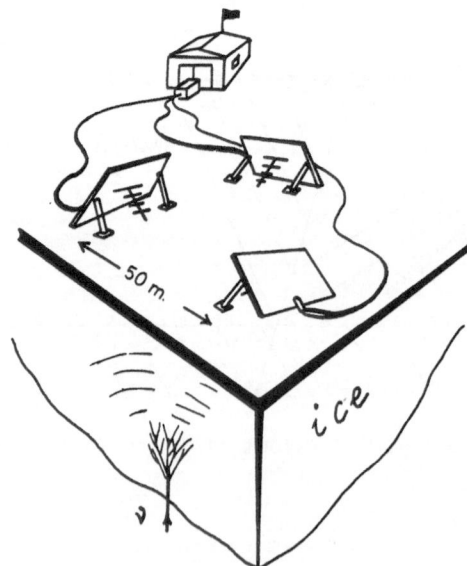

Fig.9.12: Conceptual design of the RAMAND proposal (37,38).

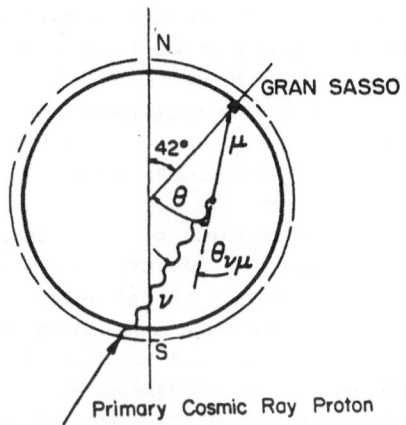

Fig.9.13: Representation of the detection of upward-going muons resulting from high energy neutrino interactions in the rock below a detector at the Gran Sasso Lab (15).

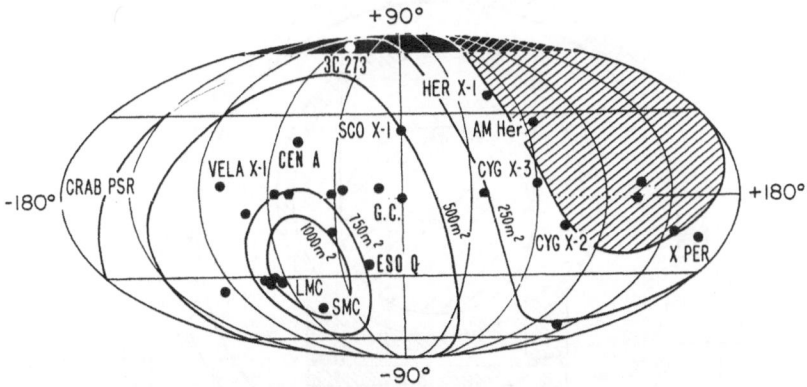

Fig.9.14: The MACRO detector field of view in Galactic coordinates. The contours represent equal time-averaged exposed areas. Locations of a few possible extraterrestrial neutrino sources are indicated. The shaded region in the northern hemisphere is inaccessible to any detector at Gran Sasso due to the background from downward going cosmic ray muons.

10. GRAVITATIONAL WAVES. GRAVITATIONAL WAVE ASTRONOMY

The Earth should be continuously bombarded by gravitational waves produced by far away celestial bodies subject to "strong" gravitational effects. The amplitudes of the gravitational waves emitted by a celestial body should be proportional to its mass, to its acceleration and to the inhomogeneity in its mass distribution. Gravitational waves are emitted when the quadrupole moment of an object of large mass is subject to large and fast variations. Only large celestial bodies subject to unusual accelerations can produce sizable gravitational radiation measurable on Earth. These bodies may be Binary Systems of close by stars; they should yield a periodic emission of gravitational waves, with frequencies from few hundred Hz to 1 MHz. Supernovae explosions may give bursts of gravitational waves, with frequencies of the order of 1 KHz. Also vibrating black holes, star accretion, galaxy formation and the Big Bang should produce or have produced gravitational waves.

A gravitational wave is a transverse wave which travels at the speed of light. A gravitational wave should modify the distances between objects, in the plane perpendicular to the direction of propagation of the wave. These deformations are expected to be extremely small; the largest deformations should be due to gravitational collapses. It has been estimated that a star collapse at the center of our galaxy may produce a variation of the order of 10^{-18} meters per meter of separation of two objects on Earth. The supernova 1987A in the large Magellanic Cloud should have produced a distortion 10 times smaller. A collapse in

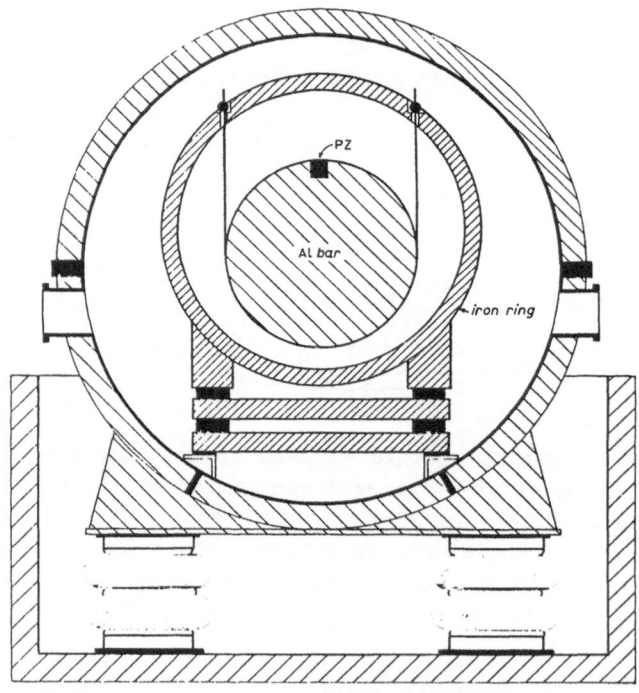

Fig.10.1: The layout of the room temperature Rome gravitational antenna (39). The latest cryogenic antennas are aluminum cylinders, of about 2 tons mass, cooled at 50° mK; vibrations are detected with electromechanical transducers based on d.c. SQUID's.

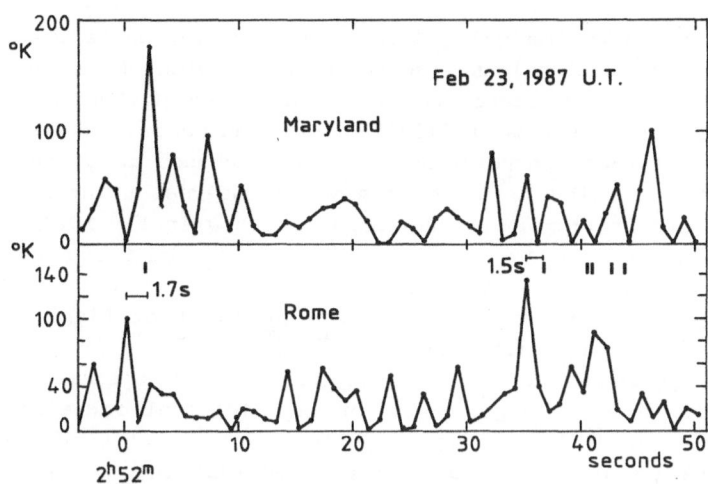

Fig.10.2: Signals from the Rome and Maryland antennas at the time of the neutrino candidates from the Mt.Blanc detector (39). Neutrino candidates seem to come about 1.5 s after gravitational wave above average noise signals.

the Virgo cluster should yield relative variations of 10^{-21}, approximately equal to those due to a binary star.

Very sensitive instruments are needed to observe gravitational waves. The two main lines developed until now are: i) resonating bars at low temperature; ii) laser interferometric systems. Both are almost at the level of being able to detect gravitational star collapses up to the Magellanic Clouds. Several detectors, in coincidence, are needed to ensure that the observed signal is not spurious. A very sensitive resonating bar must be placed underground because it would easily detect cosmic rays, which may be anticoincidized, but which would reduce the effective live time of the antenna.

The detection of gravitational waves would have far reaching consequenses in physics and in astrophysics. In physics it would prove the general theory of relativity, while in astrophysics it would open up a new observational window, related to the detection of violent phenomena in the Universe. One stellar collapse releasing 1 solar mass of energy should give signals corresponding to temperatures of $6°K$, $0.3°K$, $10^{-6}°K$ if it happened at the center of our galaxy, in the Large Magellanic Cloud or in the Virgo Cluster, respectively.

In February 1987 at the time of the Supernova 1987A all cold antennas were shut down. Only two room temperature antennas were operating. The Rome antenna is a cylinder with a mass of 2300 Kg, a resonating frequency of 858 Hz and is oriented at $29°$ to EW, in the SW quadrant (Fig.10.1). A similar antenna was in operation at the University of Maryland (mass=3100 Kg, resonating frequency = 1600 Hz, oriented EW).

The correlations of above average noise signals between the two antennas and of each one with the neutrino events from Mt.Blanc and from (Kamioka + IMB) were analyzed in detail (38). Only weak correlations between the two antennas are found at the supernova time, almost as at any other time, Fig.10.2. There seem to exist correlations during a period of about two hours roughly centered on the 5 neutrino burst of the Mont Blanc detector ($2^h 52^m$), independently between Maryland and Mt.

Blanc and Rome and Mt.Blanc (Fig.10.3). The correlation is observed also in the signal combinations (Rome+Maryland) with Mt.Blanc and in (Rome+Maryland) with Mt.Blanc, where the above average noise signals in the antennas are added or multiplied. The authors estimate that the probability that these correlations be due to chance is of the order of 10^{-6}-10^{-5}. The correlation is found also in analyses using signals only from the two gravitational wave antennas, as show in Fig.10.4. All these effects are mainly due to a dozen of large Maryland and Rome events distributed during the above two hour period. A similar correlation exists for Rome and Kamioka at the time of the Mont Blanc events. No effect is found at the Kamioka time of $7^h 35^m$.

If the signals in the antennas are due to gravitational waves, they would correspond to a large amount of gravitational radiation, may be 1000 times larger than what was expected. This energy problems seems difficult to overcome. It has been speculated that the antennas may have detected a scalar field, a soliton, emitted from a monopole mechanism, not from a quadrupole mechanism. In this case the problem of energy would vanish (40).

It is difficult to draw a definite conclusion from these interesting observations. One clearly needs more data: we welcome another Supernova, possibily in our galaxy, and we need cold antennas as well as more refined neutrino detectors.

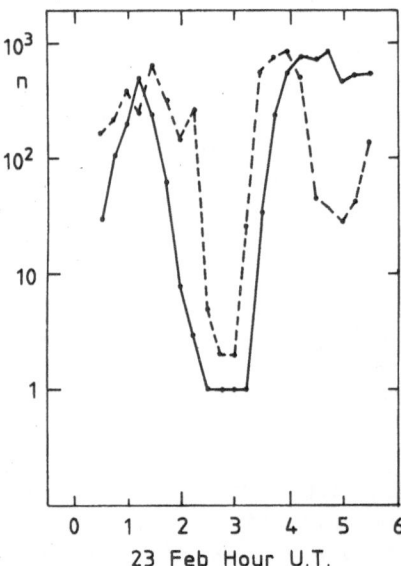

23 Feb Hour U.T.

Fig.10.3: Correlations between the sum of the Rome and Maryland signals with the Mont Blanc signal (continuos line) and of Rome with Kamioka (dashed line) (39).

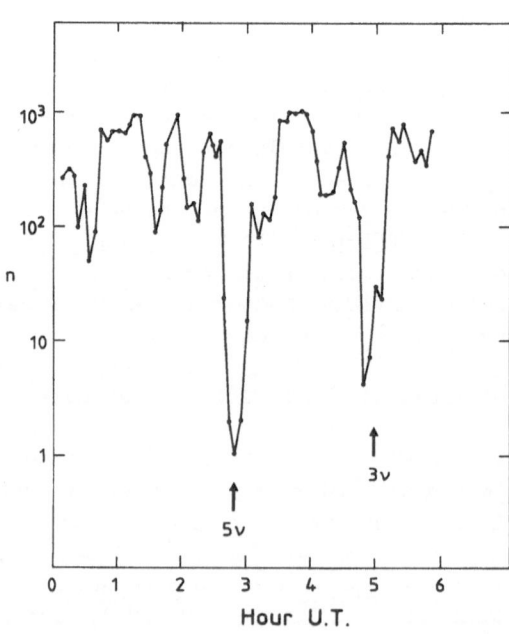

Hour U.T.

Fig.10.4: The quantity n versus time for the product algorithm (Rome)x(Maryland); 5ν indicates the time of the Mont Blanc 5ν burst; 3ν is for a Mont Blanc 3ν burst. The Mont Blanc data are not used in the estimate of n reported in this figure.

274

11. COSMIC RAYS

Cosmic rays are the only sample of matter from outside our Solar System that we can study directly on Earth.

Primary high energy cosmic rays with energies larger than 1 GeV are composed (in number) of protons (92%), helium nuclei (6%), heavier nuclei, lithium to uranium, (1%), electrons (1%), gamma rays (0.1%). At higher energies the percentage contribution of heavier nuclei increases, Fig.11.1. The chemical abundance of the energetic cosmic radiation is similar to the chemical abundance in our galaxy as a whole. This means that the source of cosmic rays cannot be the Big Bang, that is the nucleosynsthesis which happened 200 s after the Big Bang, when essentially only p and He were generated. Cosmic rays must come from a source or several sources with an average galactic composition.

The galactic power output in cosmic rays is estimated to be of the order of 10^{33} - 10^{34} watts (corresponding to about 1 eV/cm^3), to be compared to the 10^{32} watts of radiowaves, 2×10^{32} watts of x-rays, 10^{37} watts of visible light.

Many astrophysicists seem to favor discrete sources as the originators of cosmic rays. The search of discrete sources is thus an important subject, which can be attacked via the study of high energy γ -rays and high energy neutrinos (because they are neutral). It may be that also high energy underground muons carry relevant information, as suggested by some experiments (41). Shock wave acceleration seems to be the mechanism for obtaining the highest energy cosmic rays, while Supernovae shock waves may be the one for the lower energies.

Summarizing: Some of the problems which are being attacked in the field of cosmic rays are: i) the cosmic ray composition at low and high energies, ii) the energy spectrum, iii) searches for space anisotropies (including point sources) and iv) time variations, v) search for antiprotons.

At relatively low energies these problems may be attacked with direct measurements with balloons and satellites. At intermediate energies the large underground experiments may play an important role; for instance the multiplicity of muon groups is sensitive to the primary cosmic ray composition, Fig.11.2-11.6. At still higher energies one may measure electromagnetic showers; the Air Cherenkov detectors first and then the Extensive Air Shower arrays play important roles. The combination of an EAS with an underground detector may yield informations

Table 11.1 - Gamma-ray energy ranges (43)

Energy range(eV)	Classification	Technique	Observation
5.1×10^{5}-1×10^{7}	low/nuclear	NaI crystals	satellite
1×10^{7}-3×10^{7}	medium	Compton telescopes	satellite
3×10^{10}-10^{14}	high (100 MeV)	Spark-chambers	satellite
10^{10}-10^{14}	very high (TeV)	Atmosph.Cerenkov	ground level,mountain
10^{14}-10^{17}	ultra high (PeV)	Air shower array	ground level,mountain
10^{17}-10^{20}	extrem.high(EeV)	Air shower array	ground level,sea level

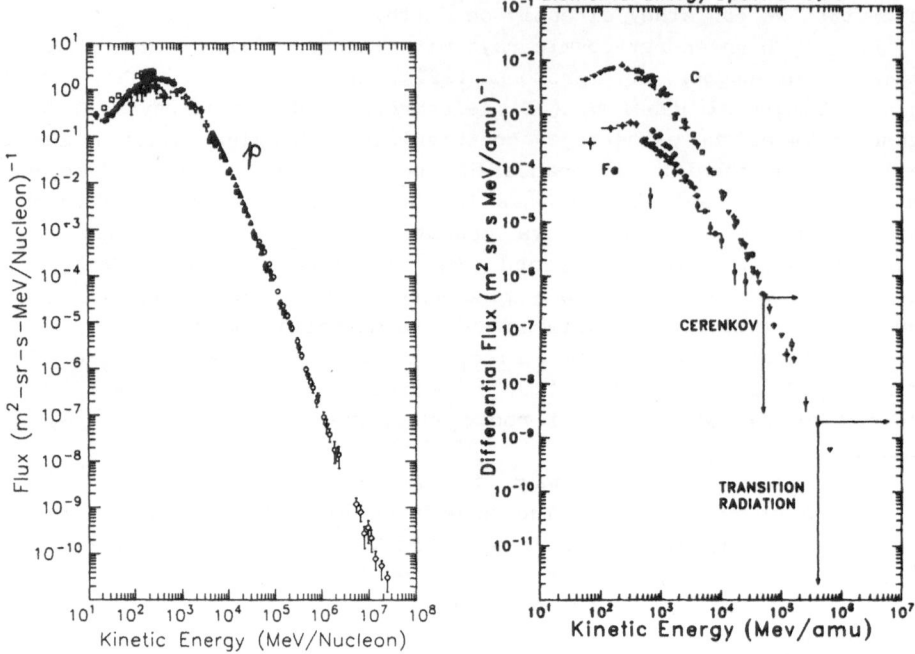

Fig.11.1: Differential flux of nuclei in the cosmic radiation.

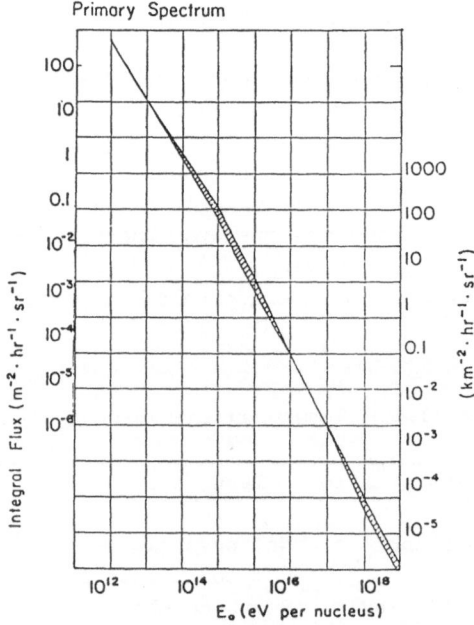

Fig.11.2: The flux of all primary cosmic rays at the top of the atmosphere, shown as an integral spectrum. Experimental uncertainties are represented by cross-hatching.

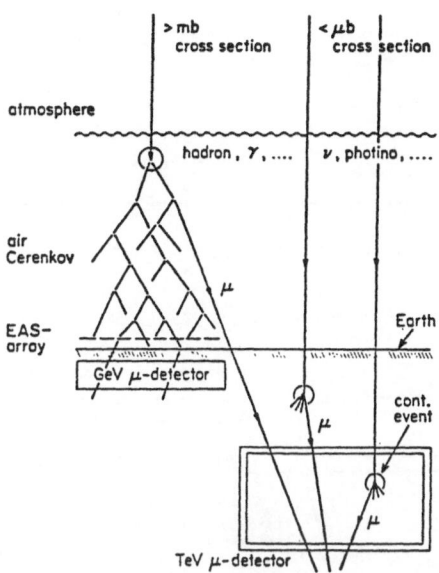

Fig.11.3: Observational techniques for cosmic rays with large and small interaction cross sections with matter (41). Particles with large cross-sections (hadrons, photons) interact near the top of the atmosphere and initiate an electromagnetic shower (either directly or via $\pi^{o} \rightarrow \gamma \gamma$). This shower has a muon component from charged pion production followed by the decay $\pi \rightarrow \mu \nu$. Muons of GeV energy can be detected by detectors shielded by a few metres of earth. TeV muons can reach deep underground detectors, which can reveal the presence of weakly interacting particles, such as neutrinos.

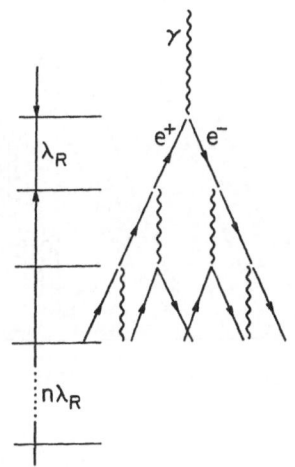

Fig.11.4: γ-ray initiated cascade in the atmosphere (41). Each layer represents a radiation length λ_R. The number of layers is limited by the maximum energy of the incident photon.

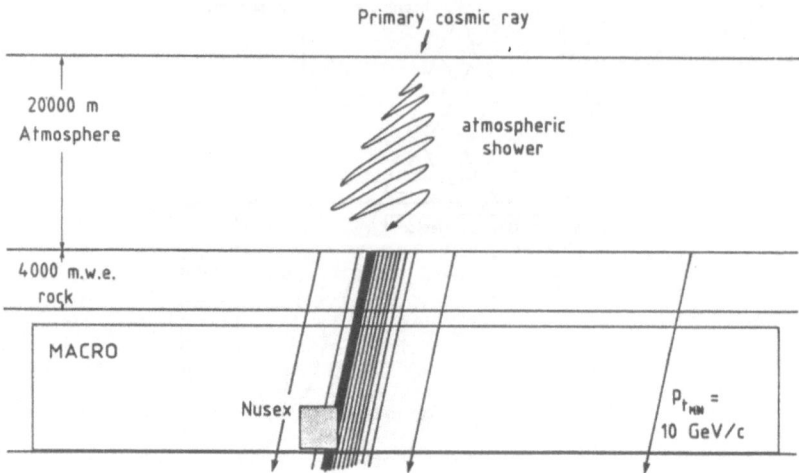

Fig.11.5: Illustration of the detection of a muon group with a small detector (Nusex) and with a large area detector (MACRO). The bulk of muons are from low-p_t phenomena; a single muon from a high-p_t parent particle is indicated.

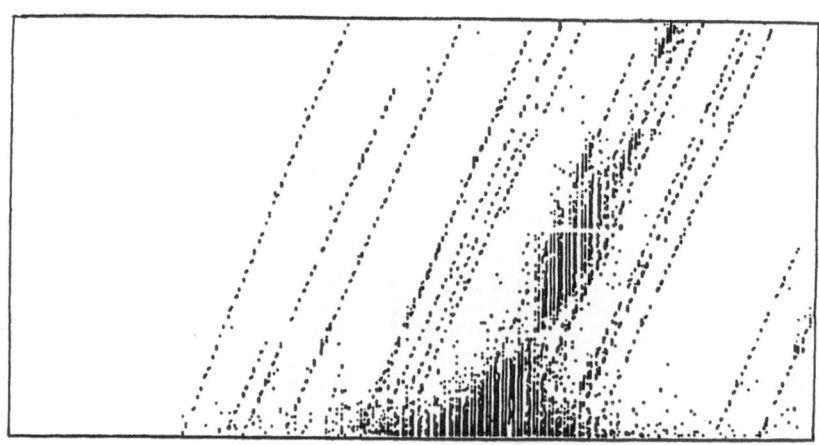

Fig.11.6: A muon bundle observed in the fine-grained Frejus detector.

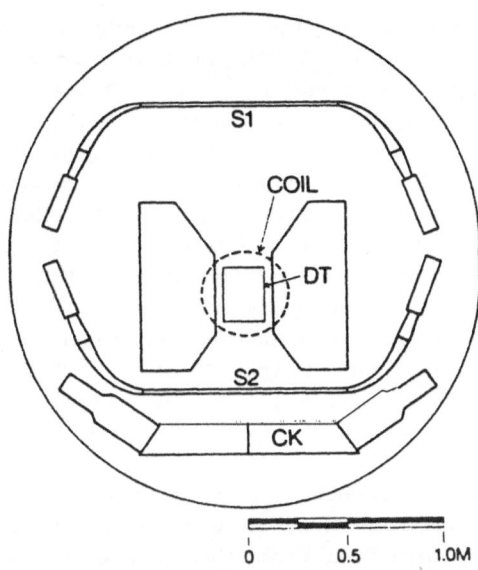

Fig.11.7: Schematic of the PBAR instrument (42).

on energy and multiplicity more directly. Present underground detectors are too small in size to detect the full transverse size of the largest muon groups. MACRO and LVD have a sufficient size to study muon groups in detail. In order to reach the highest energies one needs very large EAS arrays, as discussed below.

Let as biefly discuss the problem of antiprotons and of antimatter. Conventional models of cosmic ray propagation predict the production of secondary antiprotons through collisions of high energy cosmic rays (mainly protons) with the interstellar medium. For energies of several GeV the predicted $\bar{p}/p$ ratio is of the order of 10^{-4} and is expected to decrease with increasing energy as $E^{-0.6}$. For energies smaller than 1 GeV, secondary production is suppressed and the $\bar{p}/p$ ratio is expected to be smaller than 10^{-5}. Thus ratios above the quoted one may indicate novel physics and/or novel astrophysics. In 1981 was reported a high $\bar{p}/p$ ratio of 2×10^{-4} for the energy range of 0.13–0.32 GeV. New results from a ballon experiment yield only an upper limit $\bar{p}/p < 4.6 \times 10^{-5}$ (85% CL) for the energy range 0.2–0.64 GeV (42), Fig.11.7–11.8. From our understanding of cosmology and astrophysics, heavy nuclei such as iron, are produced only inside heavy stars at the end of a chain of thermonuclear processes. It is thus expected that heavy antinuclei (anti–iron) should be produced only inside anti–stars formed of anti–matter. Thus limits on heavy anti–nuclei yield limits on bulk anti–matter in the Universe.

11.1 Extensive air showers arrays

A cosmic accelerator such as Cygnus X–3 has a luminosity of at least one million suns. This may resolve the question of the origin of very high energy cosmis rays: only from Cygnus X3 the flux at Earth is roughly 10^7 photons km^{-2} $year^{-1}$ with energies $1-10^5$ TeV, which is sufficient to explain almost all of the observed very high energy flux.

The study of very high energy γ–ray emitters is important because γ–rays are not deflected by magnetic fields and one may thus locate the source in the sky (42).

The term "gamma ray" is a generic one that includes photons whose energies may stretch from 0.51 MeV (the rest mass of the electron) to the highest particle energies observed (10^{20} eV). Table 11.1 summarizes the current definitions. It may be worth noting that these divisions have no bearing on the emission mechanisms, which could span several regions. The gamma-ray astronomers in the sixties, who operated detectors from balloons and satellites, were confident that at 100 MeV they had reached the highest astrophysically useful energies. They therefore defined the energy region from 30 to 100 MeV as "High Energy" (HE). This is also the region where the neutral pion decay spectrum peaks. The energy range from 0.51 to 10 MeV contains the astrophysically interesting nuclear decay lines and is usually designated the "Low" or "Nuclear" region. The intermediate region, from 10 MeV to 30 MeV, is the most difficult to explore experimentally, since Compton scattering is the dominant interaction mode.

It is not feasible to launch satellites that can efficiently detect gamma rays of energy greater than 10 GeV. This limitation arises from size and weight constraints and from the source spectra, which steepen sharply with energy. Gamma-ray observations at energies greater than 100 GeV have been made using ground-based detection methods.

The atmospheric Cherenkov technique is currently used in the range 10^{11} to 10^{13} eV. This was used to define the "Very High Energy" (VHE) region. Since the technique can be extended down to 10^{10} eV and up to higher energies, the definition was extended to include 10^{10} to 10^{14} eV. This often is also called TeV gamma-ray astronomy.

Air shower particle arrays are operated at energies greater than 10^{14} eV. We use the term "Ultra High Energy" (UHE) to cover the range from 10^{14} to 10^{17} eV. The upper limit here is somewhat arbitrary, but corresponds to the feasible limit for arrays of moderate size. This is also called the region of PeV gamma-ray astronomy (1 PeV=10^{15} eV).

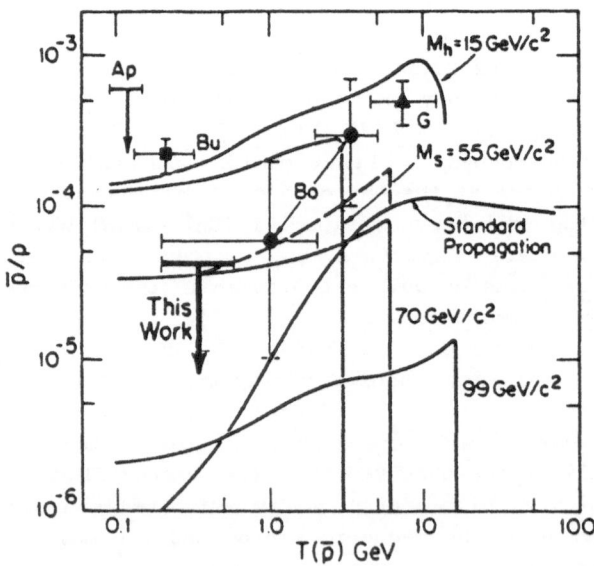

Fig.11.8: Comparison of data on $\bar{p}/p$ with several models (42).

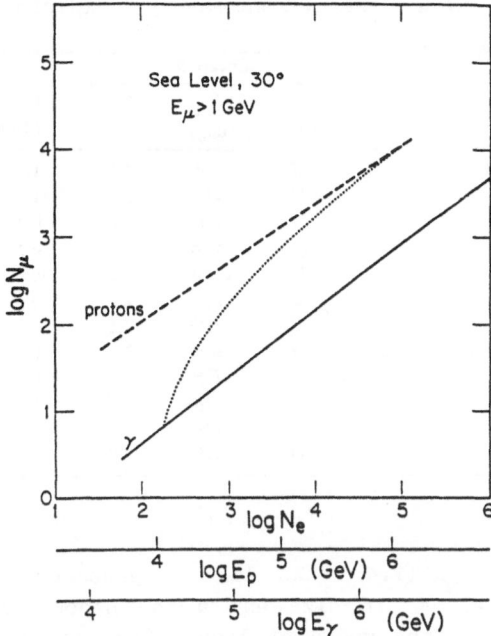

Fig.11.9: Number of muons in excess of 1 GeV (N_μ) versus shower size (i.e. number of electrons N_e) for γ-(solid line) and proton-(dashed line) initiated showers (41). The horizontal energy scales shows the conversion of shower size to initial proton or γ-ray energy. (The dotted line shows the muon signature in a model where the photon becomes strongly interacting at very high energies).

Anticipating the possibility that gamma rays may be a component of the cosmic radiation uo to the highest energies, we define the "Extremely High Energy" (EHE) region from 10^{17} to 10^{20} eV: this is the region of EeV gamma-ray astronomy (1 EeV=10^{18} eV).

As suggested in Fig.11.3 the electromagnetic showers initiated by cosmic ray hadrons are not very different from those initiated by γ-rays. A primary photon converts to an $e^+ e^-$ pair after 1 radiation length λ_R in the atmosphere (which is about 25 radiation lengths thick, with $\lambda_R \simeq 37$ g/cm^2). In subsequent radiation lengths the electromagnetic particles further lose energy by bremsstrahlung $e^\pm \rightarrow \gamma e^\pm$ and pair production $\gamma \rightarrow e^+ e^-$, see Fig.11.4. The prominent feature that can set γ-rays apart from cosmic ray hadrons is their relatively low muon content. The number of muons in a γ-shower is typically a few per cent of that in a hadron shower in which muons are abundantly generated by the decay of the produced $\pi^\pm$. In γ-initiated showers, processes resulting in muons are characterized by small cross - sections; these are i) photoproduction $\gamma +$ nucleon $\rightarrow \pi$, followed by the decay $\pi \rightarrow \mu\nu$, ii) production and subsequent semi-leptonic decay of charm quarks, iii) $\gamma \rightarrow \mu^+\mu^-$ pair production (which is suppressed by a large factor $(m_e/m_\mu)^2$ relative to $\gamma \rightarrow e^+ e^-$). The muon production through these channels is shown in Fig. 11.9. Qualitatively: muons are the discendents of hadrons and the photon is hadronic to $0\,(\alpha)$ $\simeq 10^{-2}$.

Table 11.2 Very high energy gamma-ray experiments (atmospheric Cherenkov)(42).

Country	Organization	Location Lat. (deg)	Long. (deg)	Elev. (km)	Energy (TeV)	Operational
South Africa	Potchefstroom	−27	27E	1.4	1	1985
Australia	Adelaide	−32	143E	0.16	1	1986
Australia	Durham	−31	145E	0.21	0.3	1986
USA	Wisconsin	21	156W	3.0	0.5	1985
USA	Riverside	35	107W	1.5	0.2	1986
USA	Smithsonian	32	111W	2.3	0.3	1983
France	Saclay	43	1W	1.5	0.1	1986
USSR	Crimean Astrophys. Obs.	45	34E	2.1	1	1986
India	Tata	23	78E	1.1	0.5	1987
India	Tata	35	77E	2.7	1	1985
China	Acad. Sinica	40	117E	1.0	1	1987
China	Acad. Sinica	37	97E	3.2	1	1990

The shower of Fig.11.4 looks like a pancake of electromagnetic energy, about $10^2 \sim 10^3$ m^2 in size and a few nanoseconds thick, moving down the atmosphere at the speed of light. Showers in excess of 10 TeV reach the ground, as shown in Fig. 11.3. By recording the arrival time in several detectors of an extensive particle array, the arrival direction of the shower can be reconstructed from the timing sequence. Lower energy showers do not reach the ground. A 100 GeV photon will produce about 100 electrons at 10 Km altitude. Their Cherenkov light does, however, reach earth and can be collected by mirrors viewed by phototubes. The angular spread of the Cherenkov cone is only about 1.5^{o} around the parent photon direction.

Tables 11.2 and 11.3 list the experiments using Cherenkov and Extensive Air Shower Arrays. These last ones use combinations of electron and muon detectors to search for muon-poor showers. Two of the new Arrays are shown in Figs 11.11, 11.12; we have already shown the Gran Sasso one in Fig. 3.6.

The Kiel group, after mapping cosmic rays in the sky for five years, reported a 4σ excess coming from the direction of Cygnus X-3. No other source was found at a significant statistical level. Roughly 20 experiments have by now found evidence for the emission of very high energy γ-rays from the direction and with the caracteristic time stucture of the binary star Cygnus X-3. An incomplete compilation of results is shown in Fig.11.10. Cygnus X-3, seen in radio, infra-red, MeV-γ and X-ray experiments, is observed to emit TeV photons. Most ground-based detectors suggest emission all the way up to 10^5 TeV, while some suggest a cut-off at that energy (open circles in Fig.11.10). The flux can be approximated by

$$F (>E) = 4x\ 10^{-11}/E\ (TeV)\ \text{particles } cm^{-2} s^{-1} \qquad (11.1)$$

for $E \gtrsim 0.1$ TeV and for time-averaged data. Ground arrays yield for certain epochs fluxes higher by more than one order of magnitude. Some TeV experiments have identified periods of increased activity of a few minutes of duration. The "Cygnus beam" has a very rich time structure which has not been deciphered. Periodicities of 4.8 hours, 19 days and

Table 11.3 Ultra high energy gamma-ray experiments (atmospheric Cherenkov and particle arrays) (42).

Country	Organization	Location Lat. (deg)	Long. (deg)	Elev. (km)	Energy (PeV)	Angular resolution (deg)	Operational
Australia	Adelaide	−35	138E	S.L.	1	2.5	.1984
Bolivia	"BASJE"	−16	68W	5.2	0.2	1	1986
Bolivia	SYS Collab.	−16	68W	5.2	0.2	3	1986
UK	Leeds	54	1W	S.L.	1	1	1986
USA	Utah	40	112W	1.5	0.1	0.5	1989
USA	Los Alamos	36	106W	2.1	0.2	1	1986
USA	Dublin	32	111W	2.3	0.1	1	1985
Germany	Kiel	29	18W	2.2	0.1	1	1986
Italy	Turin	42	14E	2.0	0.01	1	1988
Italy	Turin	46	8E	3.5	0.01	5.5	1981
USSR	Erevan	40	44E	3.2	0.1	1	1987
USSR	Lebedev	42	75E	3.3	0.1	3	1974
USSR	Moscow	56	37E	S.L.	1	3	1982
USSR	Nucl. Sci.	43	43E	1.7	0.3	1	1984
Japan	Tokyo	35	138E	0.9	1.0	3	1981
Japan	Tokyo	36	137E	2.8	0.1	1	1988
India	Tata	11	77E	2.2	0.1	2	1984
India	Tata	13	78E	0.9	1	1.5	1984
Antarctica	Bartol	−90	0	2.5	0.1	1	1988

one year have been suggested. Only the 4.8-hour bunching is established; high energy emission occurs very sporadically in these 4.8-hour periods. Other very high energy emitters have been reported: Hercules X-1, the Crab pulsar, 4U0115+63, PSR 1953, LMC X-4, Centaurus A, Vela pulsar, M31, PSR 1802 and the galactic plane. Most findings require confirmation. In Fig. 9.7 we sketched the general picture of a binary source of high energy photons. The system consists of a compact star in orbit around a star that has not yet collapsed. The compact patner somehow accelerates protons, perhaps by a pulsar mechanism or through conversion of energy from accretion of matter from the companion star. The accelerated particles then interact with the companion or the surrounding gas to produce a cascade of secondaries, the stable end products of which are photons, neutrinos, protons, antiprotons, electrons and positrons. The charged particles are injected into the galaxy as cosmic rays, though the electrons and positrons will be much degraded in the source. Some fraction of photons and neutrinos, from $p \rightarrow \pi^{o} \rightarrow \gamma$, $p \rightarrow \pi^{\pm} \rightarrow \nu$, will escape the source, travel in straight paths and eventually reach the Earth. As in any beam dump experiment any new particle for which the beam is above threshold will be produced and beamed to Earth if it sufficiently stable and neutral, e.g. $p \rightarrow \tilde{g} \rightarrow \tilde{\gamma}$.

In most experiments a signal cannot be established without exploiting the fact that on-source showers have a periodic structure superimposed on a continuous (and typically 100 times larger) cosmic ray background. It is customary to define a phase angle from 0 to 1 which traces the binary revolution as shown in Fig.11.13.

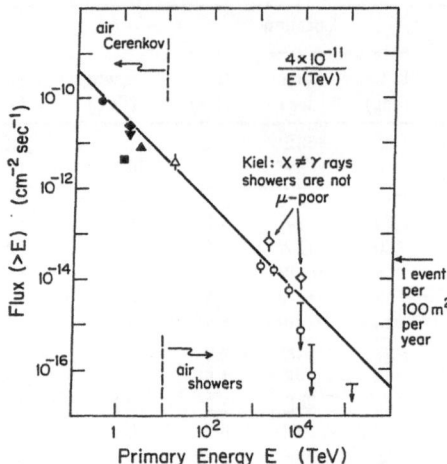

Fig.11.10: Integral flux of very high energy particles from the X-ray binary Cygnus X-3. These particles are called γ-rays although the Kiel experiment challenges this identification. Note the flatter E^{-1} energy dependence compared to the $E^{-1.7}$ fall-off of the cosmic ray flux.

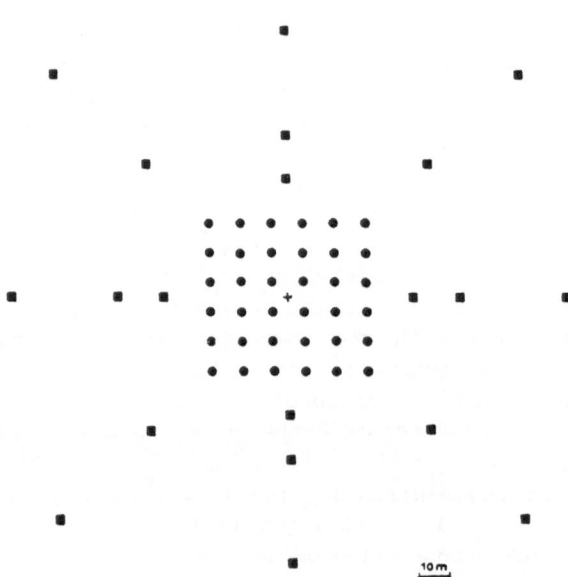

Fig.11.11: The proposed layout of the new University of Kiel gamma-ray telescope. Detectors: 56 scintillation counters, 1 m² each. Squares: 2 detectors with one photomultiplier each for particle density measurements. Circles: 36 detectors with two photomultipliers each for particle density and fast timing measurements.

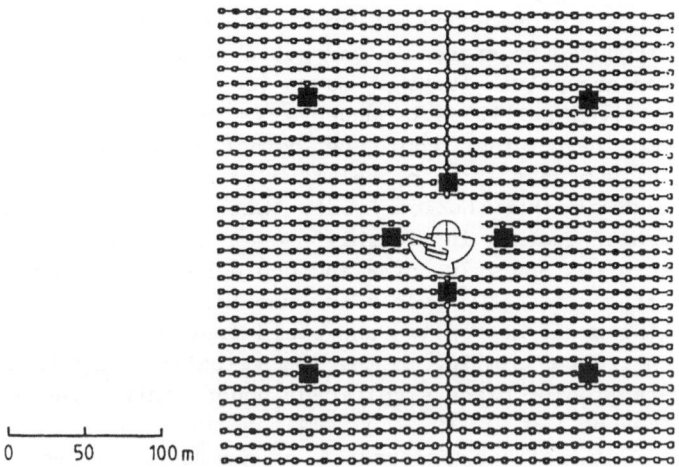

Fig.11.12: Plan of the proposed CASA array showing the location of the 1064 scintillators (43). The "Fly's Eye" air Cherenkov detector is at the centre of the array. Squares show the location of the eight muon detectors being installed by the Michigan group.

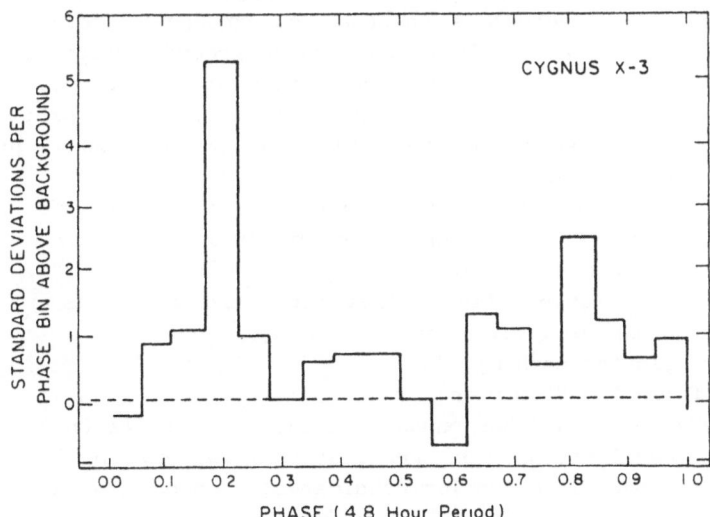

Fig.11.13: Phase distribution of the binary source Cygnus X-3, with a 4.8 h period (43).

12. SEARCHES FOR OTHER EXOTICA

A large area underground detector, such as MACRO, has capabilities for detecting exotic particles via the combined information on particle velocities (via time of flight), dE/dx and time delay of particles relative to muons. Examples of such exotica are:

i) Very massive particles, such as intermediate vector bosons or still heavier ones, produced by extremely energetic cosmic rays; they could be detected in muon groups via large p_t muons (see Fig.11.5).

ii) Heavy long lived particles produced with low velocity by extremely energetic cosmic rays; they could be detected in muon groups via a delayed muon, of somewhat large p_t relative to one or other muons.

iii) Strange quark nuclei (Nuclearites), that is nuclei composed of up, down and strange quarks in roughly equal numbers; they could be absolutely stable for any value of A and could have been produced at the quark-nucleon transition, 10^{-6} sec after the Big Bang. These nuclearites would travel like small meteorites and could provide the missing mass of the Universe. A trigger similar to the low velocity monopole trigger may allow their detection or place significant upper limits over a wide range of nuclearite masses ($10^{-6} - 10^{-12}$ g) (44).

iv) Fast particles, either primordial or produced by very energetic cosmic rays, with anomalously low or high dE/dx.

13. CONCLUSIONS

Non accelerator particle astrophysics has become an interesting and lively interdisciplinary field. We still have good reasons to search for proton decay, for magnetic monopoles, for dark matter candidates, for high energy neutrinos and for other exotica. Esperimental studies on these subjects require both small and very large detectors.

The very large detectors include underground detectors (such as those for proton decay studies) and above ground detectors (such as Extensive Air Shower Arrays). These detectors are still growing in size and are located in new, better equipped laboratories. In particular the Gran Sasso lab represents a considerable step forward and is probably a prototype for future labs.

Small, precise and very refined detectors are needed to search for certain types of dark matter, for gravitational waves and for neutrinoless double beta decay.

Previous experience tells us that there are two kinds of risks in non accelerator experiments:
- the risk of finding nothing;
- the risk of finding "too much".

Glashow (Table 13.1) (45) has enphasized the old reports on free quarks, on neutrino oscillations, on gravitational waves, on proton decay candidates, on magnetic monopoles candidates, on muons from cygnus X3 and others.

To reduce the first risk one needs worthy byproducts (like standard, but good cosmic ray data). As for the second risk one needs caution and redundancy of information, emphasizing, as was done by Suzuky (28), the need of both more data and better calculations.

In any case it would seem that the connection between particle physics, astrophysics and cosmology, enphasized by non accelerator experiments may be one of the most important scientific developments of the second half of the 20th century.

286

Table 13.1 S.L.Glashow CONCLUDING REMARKS at the WOGU7 (1986) in Toyama.

WOGU	Data	Location	Discovery
1	1980	New Hampshire	Neutrino Oscillations
2	1981	Michigan	Proton Decay
3	1982	North Carolina	Magnetic Monopole
4	1983	Pennsylvania	Axions
5	1984	Rhode Island	Sparticles
6	1985	Minnesota	Muons from Cygnus
7	1986	Toyama	The Fifth Force
8	1987	New York	?
9	1988	Aix-les-Bains	?

"We have come to the end of the formal presentations at this joint conference: WOGU*7 and ICOBAN-86. Consider for a moment the past and future of our Workshops:

At each of our meetings, a world-shaking discovery was announced or discussed but never confirmed. Nonetheless, our conferences will go on until, perhaps by the power of persistence alone, we shall find something new. Our patience will be rewarded. Hard work and ingenuity will eventually reveal another hidden facet of nature's masterpiece".

Acknowledgements — I would like to thank many colleagues for informations, discussions and data. I thank all the colleagues of the MACRO collaboration for their cooperation and Ms Gloria Frontali for typing the manuscript.

14. APPENDIX

A1. THE EARLY UNIVERSE

In the so called Standard Model of the Big Bang, the Universe started as a point state of extremely large density and extremely large temperature (46). Time started with the Universe; time zero may be taken for a radius equal to zero, temperature and density equal to infinity. As time progressed the temperature and density decreased, the radius increased. After the Planck time, $t_{pl} = \hbar/(m_{pl} c^2) = 5.4 \times 10^{-44}$ s, one may speak of a Universe composed of particles, very massive and very small particles; in other words we may think of the Early Universe as a gas of particles. The particle composition remained the same for a certain Era. Between two successive Ere there was a phase transition, in general with a change of particle composition.

In terms of Unified forces one had the following Ere and phase transitions (Figs. A.1 and A.2):
- 1st transition: $t = 5.4 \times 10^{-44}$ s, Energy $E = m_{pl} = 1.2 \times 10^{19}$ GeV. Before this time there probably was a unification of all forces; at this time the gravitational interaction separated from the GUT interaction.
- The GUT Era from 5.4×10^{-44} s to 10^{-35} s: in this period one had the Grand Unification (GUT) of electroweak and strong interactions. The total effective number of helicity states of different particles species,

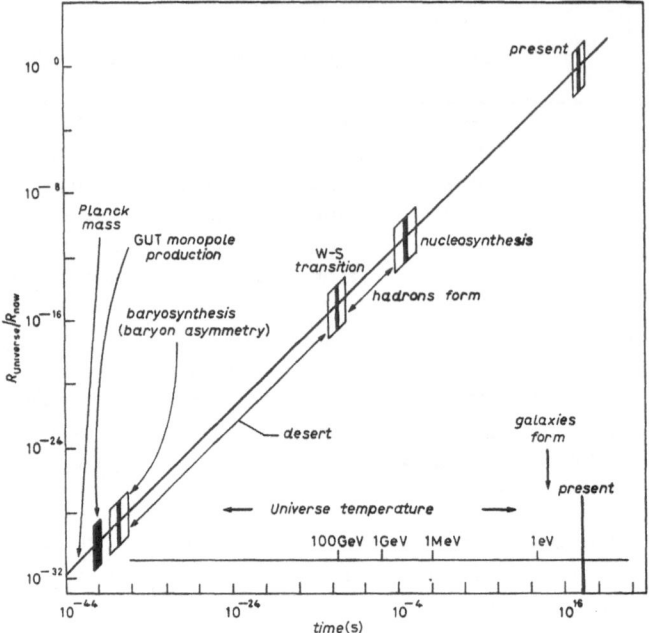

Fig.A.1: Illustration of some phase transitions in the early Universe.

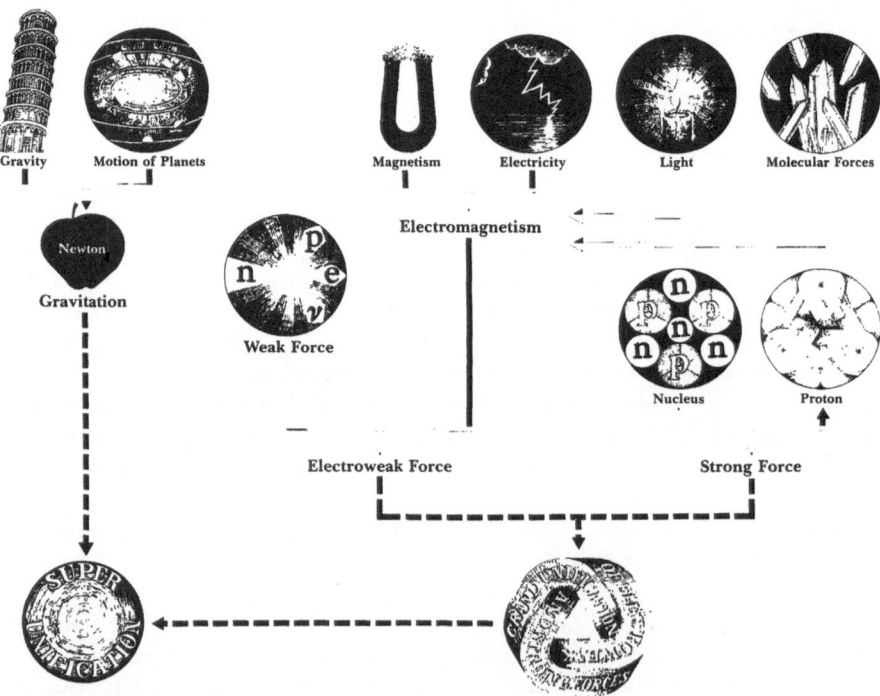

Fig.A.2: Illustration of the Unification of the forces.

that is the number of boson states plus 7/8 of the number of fermion states, was about 160: $N^* = N_b + (7/8) N_f = 160$. In this Era quarks and leptons interacted with each other and interchanged continuously.

- 2nd transition: $t = 10^{-35}$ s, $T = 10^{14}$ GeV. The GUT era ended. There was a phase transition during which magnetic monopoles may have been created as topological defects. Around this transition may have happened an "inflationary phase", when the Universe expanded exponentially. After the transition quarks and leptons are different; strong interactions are different from electroweak interactions. Immediately after this transition was generated a small matter-antimatter asymmetry via a small CP violation in the decay of the X, Y bosons, mediators of the GUT unified interaction.

- Electroweak Era, from 10^{-35} s to 10^{-10} s. During this time the Electroweak interaction was a unified interaction. The relatively short time of this Era corresponds to a very large range of energies. One doesnot know if in this energy range (called a "desert") there are specific energy thresholds, like one connected to substructures of quarks and leptons or to supersymmetric particles.

- 3rd transition: $t = 10^{-10}$ s, $T = 100$ GeV. End of Electroweak Era, that is end of the electroweak unification. After this time the electromagnetic and weak forces are different.

In the standard scenario, the early Universe is assumed to be homogenous, isotropic and radiation dominated. This last statement means that $KT > mc^2$ for any of the particles present. For that period, which extends from $t = 10^{-35}$ s to the epoch of formation of atoms ($t = 10^{10}$ s, $KT = 10$ eV) a few simple formulae connect the cosmic time t, the temperature T, the mass density ϱ, the entropy density $s = S/v$ and the Universe "radius" R (CGS system of units):

$$t \simeq \frac{3.3 \cdot 10^{20}}{\sqrt{N^*} \, T^2} \simeq \frac{2.4}{\sqrt{N^*} \, [KT\,(\text{MeV})]^2}, \qquad (A.1)$$

$$\varrho \simeq \frac{3}{32\pi G t^2} \simeq \frac{4.5 \cdot 10^5}{t^2} \simeq 4.2 \cdot 10^{-36} N^* T^4, \qquad (A.2)$$

$$s \simeq \frac{S}{v} = \frac{2\pi^2}{45} \cdot N^* \cdot \frac{K(KT)^3}{(\hbar c)^3} \simeq 5.0 \cdot 10^{-15} N^* T^3, \qquad (A.3)$$

where K is Boltzmann's constant. In each phase of the Universe during which N^* remained constant there was a state of thermal equilibrium. If the Universe expanded adiabatically, one had $S R^3 = \text{const.}$, $TR = \text{const.}$ The "radius of the Universe" R is the scaling function in the Robertson-Walker metric $ds^2 = c^2 dt^2 - R^2 d\sigma^2$. $T \approx 1/R$ is valid if N^* remained constant; in reality over long times as T decreased also N^* decreased.

During the radiation era one had (CGS units):

$$\left(\frac{\dot{R}}{R}\right)^2 \simeq \frac{8\pi\hbar c}{3 m_{\text{Pl}}^2} \varrho \simeq 5.4 \cdot 10^{-7} \varrho, \qquad (A.4)$$

Other important transitions in the Universe history were the following:

- $t = 10^{-6}$ s, $T = 1$ GeV: Quarks and gluons form hadrons (protons and neutrons)
- $t = 10^{-4}$ s, $T = 200$ MeV: Proton-antiproton annihilation: out of 10^9 annihilations remained one solitary proton.

- t= 14s, T= 0.5 MeV: Positron-electron annihilation: out of 10^9 annihilations remained one solitary electron.
- t= 200s, T= 140 KeV: Nucleosynthesis of helium.
- t= 10^{10} s, T= 10 eV formation of atoms and decoupling of matter (atoms) from the electromagnetic radiation. The cosmic microwave background radiation had its origin at this time.
- t= 1 Gy: Formation of Galaxies, clusters of Galaxies, Voids: the Universe becomes non uniform, at least over scales of hundreds of millions of light years (Fig. A.3) (47).

A2. DARK MATTER

The main evidence for dark matter comes from dynamical studies of spiral galaxies. In these galaxies most of the visible stars are concentrated in a high density nucleus, while in the external regions the stars lie in a plane and the global density is much smaller. One would thus expect that the velocity of the stars, or of any body moving in the galaxy, in similat to that of a planet in our solar system, that is $v \approx r^{-1/2}$.

The observed velocity distribution (measured with Doppler shift methods) is instead constant, independent of r for large r (Figs.A.4 and A.5). Since $v^2 \approx M(r)/r$ one deduces that $M(r) \approx r$ and that matter density decreases as $\varrho(r) \approx r^{-2}$ for large r. This distribution is typical of the distribution a of self-gravitating gaseous matter. From these observations one deduces, if Newton law is correct, that in galaxies is present, besides visible matter, a diffuse halo of dark matter with a total mass 3-10 times that of visible matter. Analyses of globular clusters, elliptical galaxies and clusters of galaxies lead to similar conclusions (48). Of what is composed this dark matter? (49)

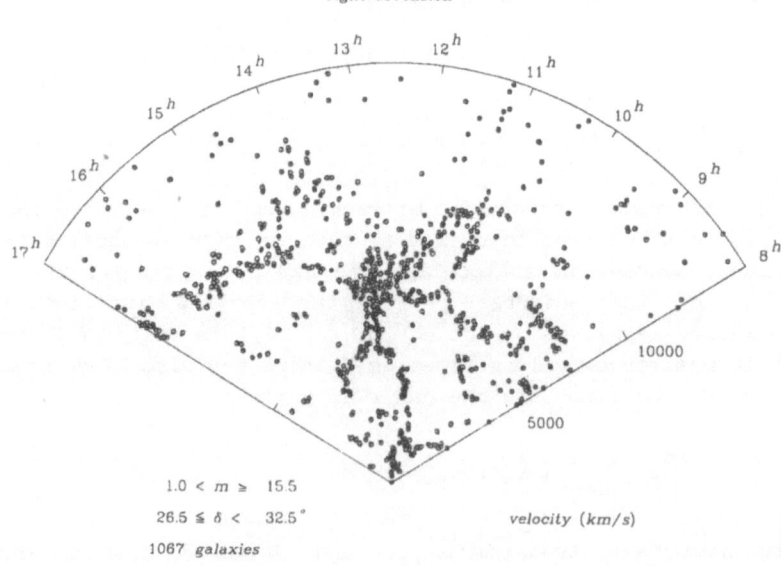

Fig.A.3: Analysis of slices of the sky shows how clusters of galaxies seem to lie in "sheets" and "filaments" (47), at the borders of large "voids" (47). In the figure each point is a galaxy and the velocity is the recession velocity, proportional to the distance from us.

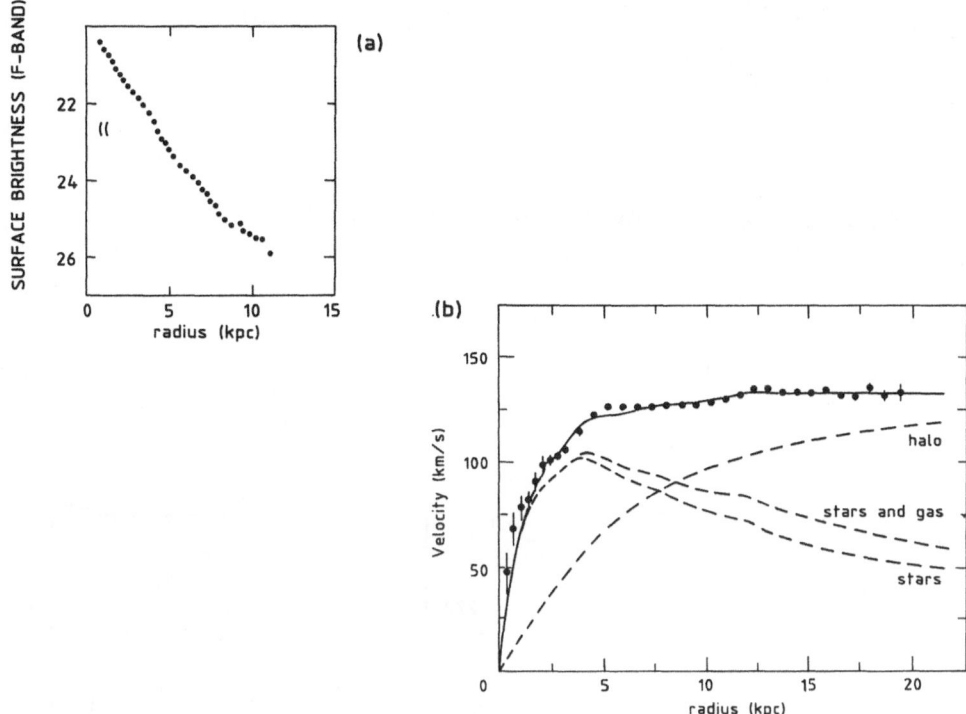

Fig.A.4: (a) Light profile and (b) rotation curve (dots with error bars) for the NGC 2403 galaxy, a spiral galaxy with a core and an extended disc; the solid line is a fit of the rotation curve. Note that the flat part of the rotation curve results from a declining curve for the disc (stars and gas) and a rising one for the halo.

The first hypothesis could be that it is baryonic: large hydrogen clouds, planets (like Jupiter), dead stars (like old white dwarfs and old neutron stars), black holes, etc. But if this type of matter were predominant there is a problem connected with the development of structures in the Universe. The present Universe is highly non homogeneous: stars group in galaxies, galaxies in clusters of galaxies, there are large voids and filamentary structures at larger dimensions. On the other hand the cosmic background radiation is highly uniform. This radiation has a black body spectrum for a temperature $T=2.7^{\circ}$ K, with a temperature variation in all directions of at most $\delta T/T \approx 3\times10^{-5}$. The same could be repeated for the helium/hydrogen ratio. Thus if the Universe was highly uniform at the time when the radiation decoupled from matter (at the time of formation of the atoms, t= few 10^{5} y) there was not enough time for the formation of structures. Thus most physicists and astrophysicists seem to favor a non baryonic dark matter and many candidates have been put forward.

Hot dark matter (HDM) was still in equilibrium with the rest of matter and radiation at T_{QCD}. The neutrinos are examples of HDM. Cold Dark matter (CDM) decoupled much earlier: examples of CDM are Axions, Photinos (or Gravitinos), Magnetic Monopoles, Nuclearites, etc. We are witnessing a large technical development of small and large detectors to try to detect Dark Matter candidates. Most of these detectors concern non-accelerator low energy phenomena; some concern also higher energies.

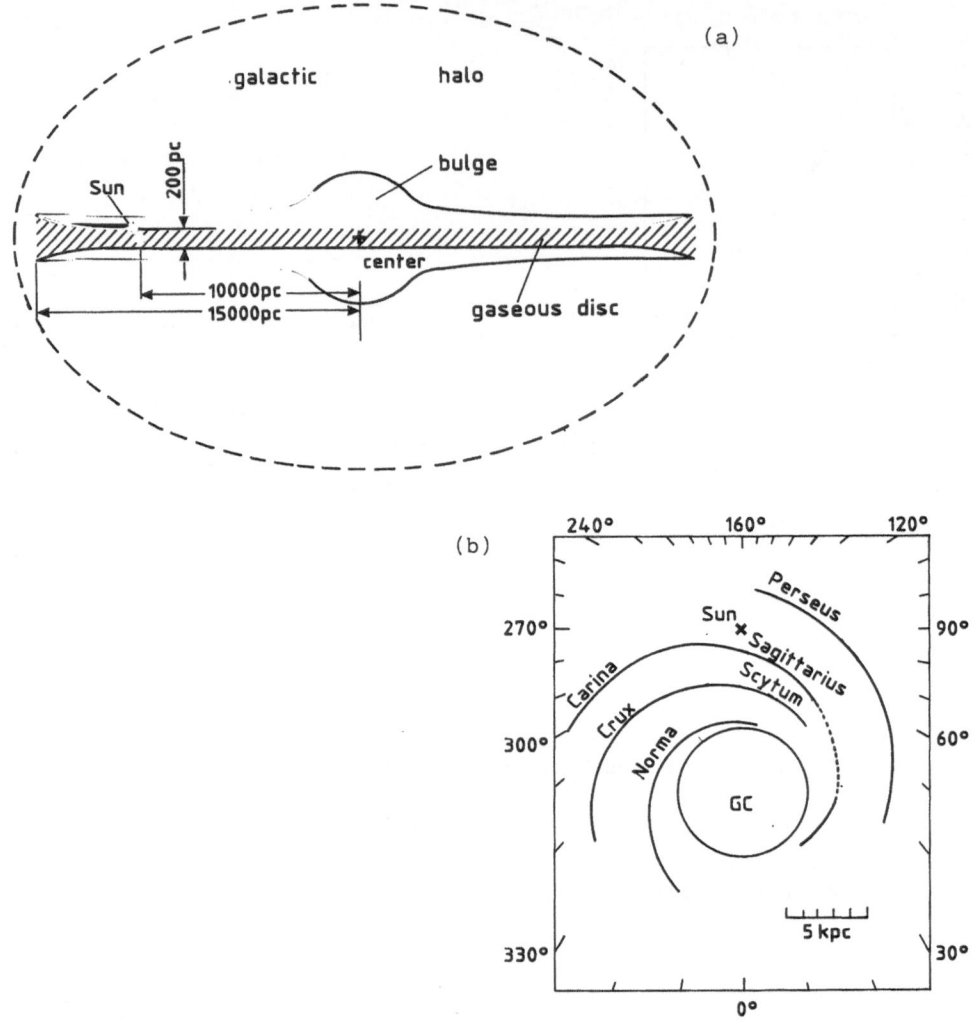

Fig.A.5: Schematic view of our galaxy: (a) edge on (note the bulge and the disc, about 600 light years thick); (b) spiral structure "seen from above" (in reality via radio observation of HII regions).

For instance magnetic monopoles and nuclearites (aggregates of strange matter made of u, d, s quarks) could be detected by large underground detectors. WIMPS with masses of several GeV could be annihilating in the sun and be detected by large underground detectors.

References

1) Costa G., Standard model and beyond, School on non-accelerator physics, Trieste (1988).
2) Nanopoulos D., Superstring implications to particle physics, School on non-accelerator physics, Trieste (1988).
   Beyond the standard model of particle physics, 3rd ESO/CERN Symposium, Bologna (1988).

3) Degrange B., Large existing underground detectors, School on non-accelerator physics, Trieste (1988).

4) Barloutaud R., Review of nucleon decay experiments, Neutrino 88 Conference, Boston (1988).

5) Giacomelli G., Higher energy non-accelerator experiments, Neutrino 88 Conference, Boston (1988).
   Giacomelli G., Underground experiments, Int.Europhysics Conf. Uppsala (1987).

6) Chudakov A.E. et al., Proceedings of the XVIth ICRC, Vol.10, 188, Kyoto (1979). Underground experiments, Int.Europhysics Conf., Uppsala (1987).

7) Bionta R.M. et al., Phys.Lett. 51, 27 (1983);
   Cortez B.G. et al., Phys.Lett. 52, 1092 (1984);
   Haines T.J. et al., Phys.Lett. 57, 1986 (1986).

8) Beier E.W., Proceedings of VIIth Workshop of Grand Unification 79, Toyama (1986).

9) Totsuka Y. Proceedings of the VIIth Workshop of Grand Unification, Toyama (1986).

10) Ewam G.T. et al, Sudbury neutrino observatory, Proposal (1987).

11) Berger C.et al., CEA-Saclay report DPhPE 87-09, submitted to Nucl. Instr.Meth. (1987).

12) Aardasma G. et al., Phys.Lett. 194B, 321 (1987).

13) Giacomelli G., The Gran Sasso Lab, Phenomenology in High Energy Physics, Trieste (1987).

14) Pless I., The Large Volume detector at Gran Sasso, Neutrino 88 Conference, Boston (1988).

15) Calicchio M. et al., The MACRO detector at Gran Sasso, Nucl.Inst.Meth. A264 (1988) 18.

16) Ciocio A.,The status of Icarus,Neutrino 88 Conference, Boston (1988).

17) Kirsten T. et al (Gallium solar neutrino experiment) Proposal for the Gran Sasso Laboratory (1985).

18) Bellotti E. et al (Proposal for an experiment on double beta decay in the Gran Sasso Laboratory) University of Milano report (1985).

19) Pizzella G. et al (A gravitational wave detector at the Gran Sasso Laboratory) (1985).

20) Fiocco G. et al (Interferometria laser per la misura delle deformazioni della crosta terrestre: progetto della stazione del Gran Sasso) Il Nuovo Saggiatore 2/4 (1986) 25.

21) Villante U. et al (A proposal for geomagnetic field measurements in the Gran Sasso Lab) University of L'Aquila Report (1985).

22) Aglietta M. et al (High-energy cosmic-ray physics with an EAS array on the top of the Gran Sasso Laboratory) Nuovo Cimento 9C (1986) 262.

23) Baldo-Ceolin M., n-$\bar{n}$ oscillation experiments, DFPD preprint (1986).
    Fidecaro G. et al., Phys.Lett. 156B, 122 (1985).

24) Fiorini E., Double beta decay, Neutrino 88 Conference, Boston (1988).

25) Giacomelli G., Status of monopole searches, 9th GUT workshop, Aix-Les-Bains (1988), and references therein. Nuovo Cimento Rivista 7 (1984) 1.

26) Kirsten T., Upcoming experiments and plans in low energy neutrino physics, Neutrino 88 Conference, Boston (1988).

27) Mikheyev S.P., Smirnov A. Yu., Sov. J. Nucl. Phys. 42 (1985) 913.
    Wolfenstein L., Phys. Rev. D17 (1978) 2369.

28) Suzuki A., Atmospheric neutrino phenomena, Neutrino 88 Conf.Boston (1988).

29) Bergsma F. et al., A search for neutrino oscillations, CERN/EP 88-51 (1988).

30) Davis R. et al, Science 191 (1976) 191.

31) Hirata H. et al, Search for ^{8}B solar neutrinos at Kamiokande-II, UT-ICEPP-87-04 (1987).

32) Sulak L.R., Review of neutrinos from SN1987A, Neutrino 88 Conference, Boston (1988).
Kamioka: Hirata K.et al., Phys.Rev.Lett. 58 (1987) 1490.
IMB: Bionta R.M. et al., Phys.Rev.Lett. 58 (1987) 1494.
Mt.Blanc: Aglietta M.et al., Europhys.Lett. 3 (1987) 1315.
Baksan: Alexyeev E.H. et al., JETP Lett 45 (1987) 461.
Bratton C.B. et al., Phys.Rev. D37 (1988) 336.1.

33) Stenger V.J., The status of Dumand, Neutrino 88 Conference, Boston (1988).

34) Bezrukov L.B.et al., Modern status of Baikal underwater neutrino experiment, Preprint HE 5.1-8 (1988).

35) Gajeski W. et al., The study of astrophysical sources and high-energy particle interactions with the GRANDE facility, GRANDE. UCI.88-006 (1988).

36) Koshiba M. et al., Resume' of LENA meetings at CERN (1988).

37) Bogomolov A.F. et al, Nucl. Instr. Meth. A248 (1986) 242.

38) Zatsepin G.T., Neutrino Astronomy, Neutrino 88 Conf., Boston (1988).

39) Aglietta M.et al, Analysis of the data recorded by the Mont Blanc neutrino detector and by the Maryland and Rome gravitational wave detectors during SN1987A, To be published (1988).
Amaldi E.et al, Correlation analysis between the Kamioka neutrino detector data and the Maryland and Rome gravitational antenna data, Rome Nota Interna N.918 (1988).
Bronzini F. et al., Nuovo Cimento 8C (1985) 300.

40) De Rujula A., Particle physics from SN1987A, Neutrino 88 Conference, Boston (1988).

41) Marshak M.L. et al, Phys. Rev. Lett. 54 (1985) 2079; 55 (1985) 1965.
Battistoni G. et al, Phys. Lett. 155B (1985) 465.
Berger Ch. et al, Phys. Lett. 174B (1986) 118.
Oyama Y. et al, Phys.Rev.Lett. 56 (1986) 991; UT-ICEPP-87-03 (1987).
Halzen F., On the discovery of very high energy point sources, CERN-TH 4570/86 (1986).

42) Ahlen S.P. et al., Phys.Rev.Lett. 61, 145 (1988).

43) Weekes T.C., Phys.Rep. 160 (1988) 1.

44) Giacomelli G. et al, (Detection of strange quark matter in MACRO) MACRO Int. Rep. 13/86 (1986).

45) Glashow S., Concluding remarks, 7th WOGU, Toyama, Japan (1986).

46) Dressler A., Dynamical parameters of the Universe, Third ESO-CERN Symposium, Bologna (1988).

47) Geller M., Large scale structure of the Universe, Third ESO-CERN Symposium, Bologna (1988).

48) Lynden-Bell D., Evidence for Dark Matter, Third ESO-CERN Symposium, Bologna (1988).

49) Turner M., Candidates for Dark Matter, Third ESO-CERN Symposium, Bologna (1988).

ATTAINING SUPERHIGH ENERGIES WITH e^+e^- COLLISIONS

Ugo Amaldi

CERN, Geneva, Switzerland

1. INTRODUCTION

The purpose of the lectures given at the St. Croix Summer School was to introduce newcomers to the fast developing field of high energy linear colliders. The main emphasis was on the *limitations* posed to the choice of the parameters by the physics of bunch-bunch collisions. Unfortunately very little time was left to discuss the different components of linear colliders, which today are the subject of R&D programs in many laboratories. The interest in this novel accelerator activity is justified by the fact that, in the last twenty years, electron-positron collisions have been producing a continuous stream of first rate physics results. To continue on the same trail it is now necessary to push the centre-of-mass energy to values larger than the $W \simeq 200$ GeV achievable with LEP 200, which should be running in 1993-1994. In 1975 it was shown by B. Richter[1] that, due to synchrotron radiation, the optimal cost of e^+e^- storage rings increases as W^2 and it was quite natural to speculate[2] on the advantages of firing, one against the other, electron and positron bunches accelerated to hundreds of GeV by opposite linacs, since in this case there is no radiation and the cost is obviously proportional to W. A few months after the proposal, which prompted a lot of speculations, I was informed that ten years before M. Tigner had suggested[3] a similar approach to produce electron-electron collisions at few GeV; this proposal had been forgotten, since at few GeV's synchrotron radiation was not yet a problem and storage rings work beautifully. During these ten years, some suggestions had been advanced by G.I. Budker and collaborators in Novosibirsk but not published and were unknown in the West, so that in preparing a review talk for the 1979 Lepton-Photon Conference, I went through the literature and collected all possible precedents. The interested reader can find them described in Ref. 4.

The crossing over point between the cost of circular and linear e^+e^- colliders depends on many assumptions, but everybody agrees that it is around $W \simeq 300$ GeV[4]. Higher energy can be achieved only by linear colliders. At present the designs of 1-2 TeV linear colliders are under study in Europe, Japan, USA and USSR. To transform them into 'projects' will take at least three years, since many problems are still open. Most of them are consequences of the fact that the required luminosities are at least 10 times greater than the largest luminosities obtained at the present electron-positron storage rings: $L \simeq 10^{32}$ cm^{-2} s^{-1}.

2. LUMINOSITY REQUIREMENTS

At $W \simeq 1$ TeV we shall need very large luminosities because the point-like cross-section decreases as $1/W^2$. Fig. 1 is taken from the summary talk I gave in 1987 at the *'La Thuile Workshop on physics at future accelerators'*, where the physics potentials of LHC (a 16 TeV proton-proton collider) and CLIC (a 2 TeV electron-positron collider) were compared[5].

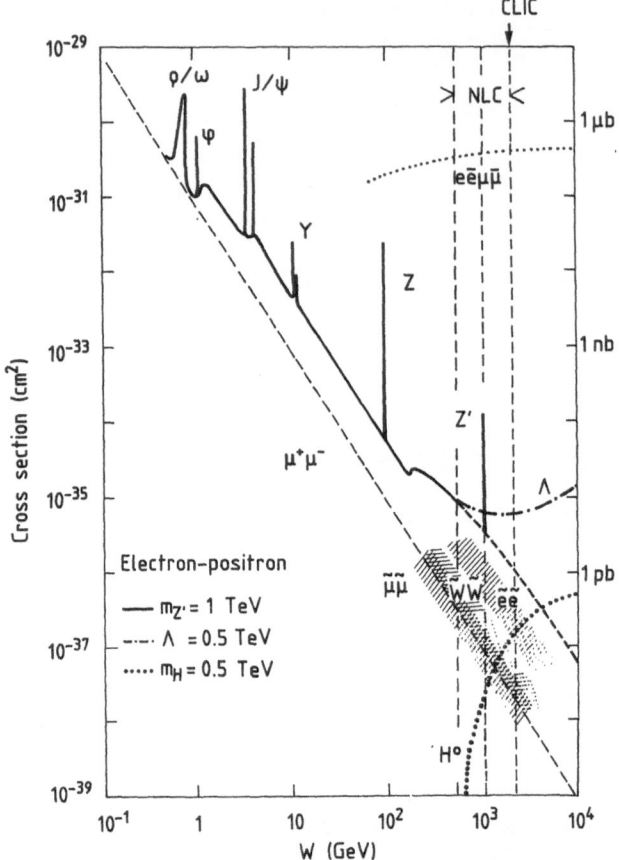

Fig. 1 *Electron-positron cross section as a function of the centre-of-mass energy. The dashed line is the point-like cross section. The continuous line is measured up to about 45 GeV and computed above. The peak due to a 'superstring inspired' Z' is very high. Compositeness with a form-factor scale $\Lambda \simeq 0.5$ TeV would have striking consequences on the total cross section. Sparticles ($\tilde{e}\tilde{e}$, $\tilde{W}\tilde{W}$ and $\tilde{\mu}\tilde{\mu}$) are expected to be produced with cross sections which are of the order of the point-like cross sections (dashed areas). The thick dotted line represents the cross section for the production of neutral higgsons having a mass of 0.5 TeV. (Figure adapted from Ref.5.)*

The continuous curve represents what is known today of the hadronic 'annihilation' cross section, i.e. the sum of the cross sections of all the processes in which the positron and the electron disappear after having annihilated. The figure is based on experimental data up to $W \simeq 45$ GeV and on the predictions of the Standard Model above 45 GeV. Since the overall trend is roughly parallel to the dashed curve, which represents the 'point-like' cross section applicable to $\mu^+\mu^-$ and $\tau^+\tau^-$ production, the figure proves that, in the explored energy range, quarks (and leptons) are point-like particles. The four sets of peaks appearing below about 10 GeV are due to the pair-creation of quark-antiquark pairs, which *together* have the same quantum numbers as the electron-positron pairs and, immediately after production, whirl thousands of times one around the other in a resonance state. The four sets are due to the production of unstable bound states of the quark pairs $u\bar{u}$ and $d\bar{d}$ (ρ and ω), $s\bar{s}$ (ϕ), $c\bar{c}$ (J/Ψ) and $b\bar{b}$ (Y). We would like to complete the figure by plotting the set of resonances due to $t\bar{t}$ bound states, but at present we only know that they must lie above $W \simeq 80$ GeV and below $W \simeq 400$ GeV. There are also indications that the lower limit could be as high as ~ 100 GeV.

The Z peak is not only taller than the others but, according to the Standard Model, it is also of a very different nature, because the Z boson is a *single* elementary particle having the quantum numbers of the initial electron-positron pair and not a composite system. After climbing on the Z^0 peak in 1989, LEP 200 will open the possibility of exploring the much smaller shoulder due to the production of pairs of charged bosons: $e^+ + e^- \rightarrow W^+ + W^-$.

The point-like cross section decreases as W^{-2}, so that at $W = 2$ TeV (CLIC) $\sigma_{\mu\mu} \simeq 2.2 \times 10^{-38}$ cm^2: by running one third of a calendar year at a luminosity $L = 10^{33}$ cm^{-2}s^{-1}, only ~ 200 μ-pairs would be collected. Figure 1 shows that at CLIC the rate of production of W *pairs* in a 4π detector is 50 times larger than for μ-pairs. It is difficult to say whether such a large production rate will be 'useful' for physics or should be considered only as a source of background. Let us now consider the effect of 'new' physics.

In electron-positron annihilations the superstring-inspired Z' appears as an enormous peak, which is drawn in Fig. 1 for $m_{Z'} = 1$ TeV. Clearly at a 1 TeV collider it would not only be seen, but could be studied in detail. The cross section for a higgson of mass $m_H = 0.5$ TeV is large, since it corresponds to $\geq 10^3$ events per year at $W = 2$ TeV and $L = 10^{33}$ cm^{-2}s^{-1}. Compositeness would be signalled by a flattening electron-positron total cross section, as indicated by the dash-dotted line of Fig. 1. For a value of the parameter Λ of the order of 0.5 TeV (i.e. for distances $d \simeq 4 \times 10^{-17}$ cm) the effect shown in the figure is striking, so much so that we are reminded of the surprise of the physicists working in 1969 at Adone when, thanks to the pair-production of point-like quarks, they found a hadron production rate which greatly exceeded the expectations.

In Fig. 1 the shaded areas indicate the cross section ranges of three sparticle channels produced in the annihilation of an electron-positron pair. No unique value can be given because various parameters enter the calculations. In general one can state that selectron pairs are produced about 10 times more abundantly than wino pairs, whose cross section is of the same order as the point-like one (~ 200 events per year for $L = 10^{33}$ cm^{-2}s^{-1}). Smuons are expected to be somewhat rarer, but are very interesting because it was shown at the *La Thuile Workshop* that, if found in the reachable mass range, their bosonic nature (spin $J = 0$) can be *proven* by measuring the angular distribution of the decay muons.

Fig. 1 shows that, around $W \simeq 1$ TeV, the cross sections of new 'expected' phenomena fall in two broad categories: (i) more or less standard Z' production (with $m_{Z'} \simeq W$) and compositeness (with a scale $\Lambda \simeq W$) have $10^{-36} \leq \sigma \leq 10^{-34}$ cm^2 and (ii) SUSY particle production (with $m_{SUSY} \simeq W/2$) and neutral higgson (with $m_H \simeq W/2$) have $10^{-38} \leq \sigma \leq 10^{-36}$ cm^2. In the first case a luminosity $L = (W/\text{TeV})^2 10^{31}$ cm^{-2}s^{-1} gives between 10^2 and 10^4 events/year, where a running year is taken to be equivalent to 10^7 seconds. In the second case, one needs a luminosity $L = (W/\text{TeV})^2 10^{33}$ cm^{-2}s^{-1} to obtain 10^2-10^4 events/year. In short, these are the arguments which at the *La Thuile Workshop* brought physicists to require the 'large' luminosity

$$L \geq (W/\text{TeV})^2 \; 10^{33} \text{ cm}^{-2} \text{ s}^{-1} \tag{1}$$

when entering in the new energy regime opened by TeV linear colliders[5]. Note that this luminosity is needed *for each interaction region*. Since at least two detectors have to be foreseen and, with a linear collider, the two interaction regions can only be served by sharing the available bunch-bunch collision rate, Eq. (1) implies that the collider has to provide a luminosity *twice* larger.

3. DISRUPTION

In the Introduction the opening remarks concerned the poor scaling laws of storage rings when increasing the center-of-mass energy. By now it is well understood that also the scaling laws of linear colliders are *not* very favourable: when in a 'Gedanken' experiment we increase the energy and pass from center-of-mass energies around $W = 100$ GeV (as today achieved, as far as energy is concerned, at the SLAC Linear Collider = SLC) to one TeV and, even more, to many TeV's, one is obliged to change the bunch-bunch 'regime'. At low energies the electrons and positrons radiate classically, when bent by the megagauss magnetic fields produced by the opposite bunch. At many TeV's, to obtain the needed large luminosities ($L > 10^{33}$ cm^{-2} s^{-1}) the dimensions and densities of the bunches have to be modified and the average critical energy of the radiated photons E_c becomes larger than the beam energy E_0; this is the so-called *quantum regime,* which entails bunches which must be few microns long and have transverse dimensions of some tens of Ångstroms. As for storage rings, synchrotron radiation is the cause of the poor scaling laws of linear electron-positron colliders which I want to discuss. More detailed arguments along the same lines can be found in the presentation which I gave at the CERN-USA Accelerator School[6] in 1986.

We call *disruption* the focusing (or defocusing) produced on the particles of one bunch by the electric and magnetic fields of the opposite bunch. Its intensity is described by the *'disruption parameter'* D.

An electron (positron) deflected by the opposite bunch — which is supposed to have a 'round' cross-section of r.m.s. radius σ_x and length σ_z and to contain N particles — radiates electromagnetic energy. A typical particle at a distance σ_x from the axis of the bunch sees a magnetic field proportional to the incoming current ($\propto N/\sigma_z$) and inversely proportional to σ_x: $B \propto N/(\sigma_x \sigma_z)$. The radius of curvature ρ in this field, supposed to be uniform, is proportional to γ/B and thus

$$\rho = \gamma \, \sigma_x \, \sigma_z/(N \, r_e) \, , \qquad (2)$$

where $r_e = 2.82 \; 10^{-13}$ cm is the classical electron radius. The deflection angle is $\theta = \sigma_z/\rho$ and, for small deflections, the focal distance is $F = \sigma_x/\theta = \sigma_x\rho/\sigma_z$. The disruption parameter is defined as $D = \sigma_z/F$, so that it is *small* when the focal distance is *large* with respect to the length of the bunch. By combining $D = \sigma_z^2/\sigma_x\rho$ with Eq. (2) we obtain, for round bunches of r.m.s. radius σ_x and r.m.s length σ_z,

$$D = \frac{r_e \, N\sigma_z}{\gamma\sigma_x^2} \, , \qquad (3)$$

where $\gamma = E_0/mc^2$. When $D \leq 0.5$ the focusing is very little and one can indeed consider the bunch as a lens of focal length F; for $D \geq 0.5$ there is a 'pinch' effect and the effective transverse dimensions of the bunches during the collisions are *smaller* than what one would compute from the properties of the final focus system.

The *natural* transverse area of the bunch is such that

$$\sigma_x^2 = \varepsilon_n \, \beta^*/\gamma \, , \qquad (4)$$

where ε_n is the *'invariant transverse emittance'* of the bunches (which does *not* change in a linear accelerator without wake field and space charge effects) and β^* the β-value of the final focus system at the interaction point. It can be shown[7] that, for a fixed magnetic field at the poletip of the final focusing quadrupoles, the value of β^* increases with the energy of the focused particles as $\gamma^{1/3}$.

The pinch effect reduces the transverse dimensions with respect to σ_x, so that the effective radius can be written as $\sigma_x/\sqrt{H_D}$, where H_D is a function of D. The luminosity increases and takes the form

$$L = \frac{f_r \, N^2}{4\pi \, \sigma_x^2} \, H_D \, , \qquad (5)$$

where f_r is the repetition rate of the bunch-bunch collisions and H_D is the *'enhancement factor'* due to the pinch effect.

For round beams Chen and Yokoya − by using simulation methods[8] − have obtained simple formulae which express H_D as a function both of D and of the ratio $\Lambda = \sigma_z /\beta^*$:

$$
\begin{aligned}
&1 + 2D / 3\sqrt{\pi}, && 0 \le D \le 0.5 , \\
&1 + 2D / 3 \sqrt{\pi} + 0.43 \, [\ln(D/\Lambda)]^2, && 0.5 \le D \le 2 , \\
&1.6 + 0.43 \, [\ln (D/\Lambda)]^2, && 2 \le D \le 100 .
\end{aligned}
\tag{6}
$$

Theses formulae are valid for round beams: $\sigma_x = \sigma_y$. Usually $D \simeq 30$ is considered to be a maximum value for the disruption parameter. For $D \ge 2$ the focusing is so strong the particles oscillate while traversing the opposite bunch. The average number of oscillations performed was computed to be $n \simeq (2D)^{1/2}/2\pi$, so that $n = 1$ for $D \simeq 20$. Chen and Yokoya found that the particles, coming from a narrow ring in the transverse plane, are focused on the beam axis within the oncoming bunch and give rise to the second term in the enhancement factor of Eqs. (6b) and (6c).

For $D = 5$ and $\Lambda = 1/4$ Eq. (6c) gives $H_D = 5.5$, a sizeable effect indeed. Recently Balakin and Solyak[9] contested Eq. (6c) by claiming that above $D \simeq 10$ the numerically calculated enhancement factor *decreases* with D, as initially found by Hollebeek[10]. At $D = 20$ the two results differ by a factor 2. In the following we shall not consider disruption parameters larger than 4, and anyway the discrepancy does not influence our arguments.

By combining Eqs. (3) and (5) with the definition of *beam power*, i.e. the power of one beam,

$$
P = N E_0 f_r
\tag{7}
$$

one gets the *first fundamental relation* of linear colliders

$$
L/(10^{33} \text{ cm}^{-2} \text{ s}^{-1}) \simeq (D \, H_D/29) \; (\text{mm}/\sigma_z) \; (P/MW)
\tag{8}
$$

which express quantitatively the fact that the luminosity is proportional to beam power.

Eq. (8) implies $P > 1$ MW with $\sigma_z \simeq 0.1$ mm and $DH_D \simeq 1.5 \times 3 \simeq 5$, to obtain $L > 10^{33}$ cm^{-2}s^{-1} as needed for *physics reason* at $W > 1$ TeV. This shows why high power beams made of short bunches are a *must* in TeV linear colliders.

4. NATURAL SCALING

Since by now the parameter list of SLC is well known, I reproduce it in Table 1. To scale these project parameters to a $(0.5 + 0.5)$ TeV, which I shall call the *Next Linear Collider* (NLC), it seems at first sight reasonable[6] to keep the same type of accelerator, i.e. the same repetition rate f_r and the same bunch population N.

Luminosity is *proportional* to beam power; and beam power, for N and f_r constant, increases only *linearly* with γ. Since to follow Eq. (1) the luminosity has to be proportional to γ^2, the needed extra factor γ can be introduced in Eq. (8) by keeping $D = $ const and by making $\sigma_z \propto \gamma^{-1}$. From Eq. (3) it is 'natural' to have $\sigma_x \propto \gamma^{-1}$ and scale the various quantities as summarised in Table 2.

In the Table I also assumed that β^* scales as $\gamma^{1/3}$, as shown in Ref. 7, so that $\varepsilon_n \propto \gamma^{-4/3}$ to have the bunch radius of Eq. (4) proportional to γ^{-1}.

The results of the 'natural' scaling laws are collected in the last column of Table 1 which shows that for $W = 1$ TeV one would get $L = 6 \; 10^{32}$ cm^{-2} s^{-1}, close to the goal set by Eq. (1). The last line of Table 2 shows that the *new* parameter δ (to be defined in the next Section) has a frightening γ-dependence and, according to Table 1, it passes from $\simeq 10^{-3}$ at SLC to $\simeq 10$ at the 1 TeV Next Linear Collider we are considering. This is a very serious problem, which we have now to discuss.

Table 1
SLC parameters and scaling to a (0.5 + 0.5) TeV Collider

Parameter	SLC (Project values)	Scaling factor (from Table 2)	Scaled NLC
Beam energy, E_0 (GeV)	50	10	500
c.m. energy, W (GeV)	100	10	1000
Particles/bunch, N	7.2×10^{10}	1	7.2×10^{10}
Bunch radius, $\sigma_x = \sigma_y$ (μm)	1.65	0.1	0.165
Bunch length, σ_z (mm)	1.0	.1	0.1
Repetition rate, f_r (kHz)	0.18	0.18	0.18
Power/beam, P (MW)	0.1	10	1.0
β-value, β^* (mm)	5.0	2.1	10.5
Emittance, ε_n (m)	4×10^{-5}	0.046	$1.9 \ 10^{-6}$
Disruption parameter, D	0.75	1	0.75
Enhancement factor, H_D	2.2	1	2.2
Luminosity, L (cm^{-2}s^{-1})	6×10^{30}	100	6×10^{32}
Beamstrahlung parameter, δ	9×10^{-4}	10^4	9

Table 2
'Natural' scaling laws from SLC to the Next Linear Collider (NLC)

Quantity	γ-dependence	Quantity	γ-dependence
N	1	L	γ^2
f_r	1	P	γ
σ_z	γ^{-1}	$\beta_x^* = \beta_y^*$	$\gamma^{1/3}$
D	1	$\varepsilon_{nx} = \varepsilon_{ny}$	$\gamma^{-4/3}$
$\sigma_x = \sigma_y$	γ^{-1}	δ (defined by Eq. 9)	γ^4

5. BEAMSTRAHLUNG FOR ROUND BUNCHES

As anticipated, the problem is synchrotron radiation in the field of the opposite bunch. According to classical electrodynamics, the energy radiated per unit length by a particle, which moves along a trajectory of radius of curvature ρ, is $P_\ell = 2 \, r_e \, mc^2 \, \gamma^4/(3\rho^2)$. Using for the radius of curvature Eq. (2), the *average fractional energy loss* in a length σ_z for a particle of energy E_0 which is only *slightly* deflected is $\delta \propto P_\ell \, \sigma_z/E_0$ and can be written in the form

$$\delta \simeq 2/9 \ (r_e^3 \, \gamma/\sigma_z) \ (N^2/\sigma_x^2) \, . \tag{9}$$

(The numerical factor, which equals 0.222, has a simple form useful for later simplifications. The exact form computed for gaussian bunches both in the transverse and the longitudinal directions is $8 \, \pi^{1/2}/21 \simeq 0.215$.) The quantity δ has been dubbed *'beamstrahlung parameter'*.

Equation (9) shows that, as anticipated in Table 2, δ increases proportionally to γ^4, if one decides to scale a collider with N = const, f_r = const, P $\propto \gamma$, $\sigma_x = \sigma_y \propto \gamma^{-1}$ and $\sigma_z \propto \gamma^{-1}$, so to have L $\propto \gamma^2$. At SLC $\delta \simeq 10^{-3}$ (Table 1), so that at NLC (W = 1 TeV) with the 'natural' scaling of Table 2 one would have $\delta \simeq 10$, clearly unphysical since a particle cannot loose 10 times the energy it possesses. This is the reason for which, starting from SLC parameters of Table 1, the scaling laws of Table 2 *cannot* be followed above $E_0 \simeq 300$ GeV.

In 1985 Himel and Siegrist showed[11] that at very large energies linear colliders have more favourable scaling laws than the ones implied by Eq. (9). The issue has been discussed and clarified in many recent papers[12] and I limit myself to the presentation of the main results.

In a uniform magnetic field the critical energy of the spectrum of the radiated photons has the form $E_c = (3 \hbar c \gamma^3) / (2\rho)$ so that, using Eq. (2), the *average fractional critical energy* $\overline{Y} = \overline{E}_c/E$ has the form

$$\overline{Y} \simeq 5/(12\alpha) \ (r_e^2 \ \gamma/\sigma_z) \ (N/\sigma_x) \ , \tag{10}$$

where $\alpha = 1/137$ and the numerical factor comes from averaging on gaussian bunches[12]. Himel and Siegrist showed that for large values of $\overline{Y}$ the fractional energy loss is less than the classical estimate given by Eq. (9), but before introducing their result we have to consider flat bunches.

6. DISRUPTION AND BEAMSTRAHLUNG FOR FLAT AND PINCHING BUNCHES

'Flat' bunches have $\sigma_y < \sigma_x$. By introducing the aspect ratio

$$R = \sigma_x / \sigma_y \geq 1 \tag{11}$$

two disruption parameters determine now the focusing in the two planes.

$$D_x = (r_e \ N \ \sigma_z) \ (\gamma\sigma_x \ \sigma_y)^{-1} \ [2 / (1+R)] \ , \tag{12}$$
$$D_y = (r_e \ N \ \sigma_z) \ (\gamma\sigma_x \ \sigma_y)^{-1} \ [2R / (1+R)] = D = RD_x \geq D_x \ . \tag{13}$$

In the following the largest one, D_y, will still be called *'disruption parameter'* and indicated by the symbol D. For flat beams (R > 1) the enhancement factor is reduced with respect to the one given as a function of D in Eq. (6):

$$H \simeq H_D^{\ (1+R)/2 R} \ , \tag{14}$$

and the luminosity is

$$L = f_r \ N^2 \ H / (4\pi \ \sigma_x \ \sigma_y). \tag{15}$$

The *first fundamental relation* (8) becomes

$$L/(10^{33} \ \text{cm}^{-2}\text{s}^{-1}) \simeq [DH/29] \ [(1+R)/2R] \ (\text{mm}/\sigma_z) \ (P/MW) \ . \tag{16}$$

The fractional average critical energy $\overline{Y}$ of Eq. (10) gets a non-trivial modification:

$$\overline{Y} \simeq 5/(12\alpha) \ (r_e^2\gamma/\sigma_z) \ (N/\sqrt{\sigma_x\sigma_y}) \ H_b^{1/2}(D,R) \ , \tag{17}$$

where, following Chen[13], I have introduced a new factor, which depends on D and R

$$H_b \ (D,R) \simeq 4R \ H_D \ [1+R \ H_D^{(R-1)/2R}]^{-2} \ , \tag{18}$$

and which intervenes also in the fractional energy loss when $\overline{Y} << 1$ and there is pinch effect:

$$\delta = 2/9 \ (r_e^3\gamma/\sigma_z) \ (N^2/\sigma_x\sigma_y) \ H_b \ (D,R) \ . \tag{19}$$

For round bunches (R = 1) with pinch effect $H_b \simeq H_D$ and the parameters $\overline{Y}$ and δ are proportional to $H_D^{1/2}$ and H_D respectively. For very flat bunches (R $>>$ 1) Eq. (18) gives $H_b \simeq 4R^{-1}$ and (with $\sigma_x \sigma_y$ = const) $\overline{Y} \propto R^{-1/2}$ while $<\delta> \propto R^{-1}$.

Flat bunches can thus be used to reduce $\overline{Y}$ and δ by large factors, while Eq. (16) is modified *only* by a factor 2.

7. COLLIDER REGIMES

If $\overline{Y} << 1$ the collider runs in the so-called *'classical regime'*. By increasing the fractional critical energy $\overline{Y}$ one enters in the *'quantum regime'*, first studied by Himel and Siegrist. In Ref. 6 I introduced an handy approximation, valid to $\sim 5\%$, which bridges the equations valid in the two regimes by writing:

$$<\varepsilon> \simeq \delta \, F \, (\overline{Y}) , \tag{20}$$

where δ is given by Eq. (19) and

$$F \, (\overline{Y}) \simeq (1 + \overline{Y}^{1/2} + 3 \, \overline{Y}/2)^{-4/3} . \tag{21}$$

The quantity $<\varepsilon>$ is the *average fractional energy loss*, which is equal to δ only when $\overline{Y} << 1$ and can be much smaller than δ of Eq. (19) when $\overline{Y} \geq 1$. It is generally assumed that NLC can accept relatively large values of the fractional loss, as $<\varepsilon> \simeq 0.2$-0.3. This is due to the fact that such a value corresponds typically to center-of-mass spreads $\Delta W/W \simeq 10\%$, which is not too large when running in the "continuum". The relation between the center-of-mass spread due to beamstrahlung $\Delta W/W$ and the fractional energy loss $<\varepsilon>$ is *not* linear. The interested reader can find the relevant arguments in Refs. 6 and 7. The function $F \, (\overline{Y})$, called in SLAC H_γ, is plotted in Fig. 2.

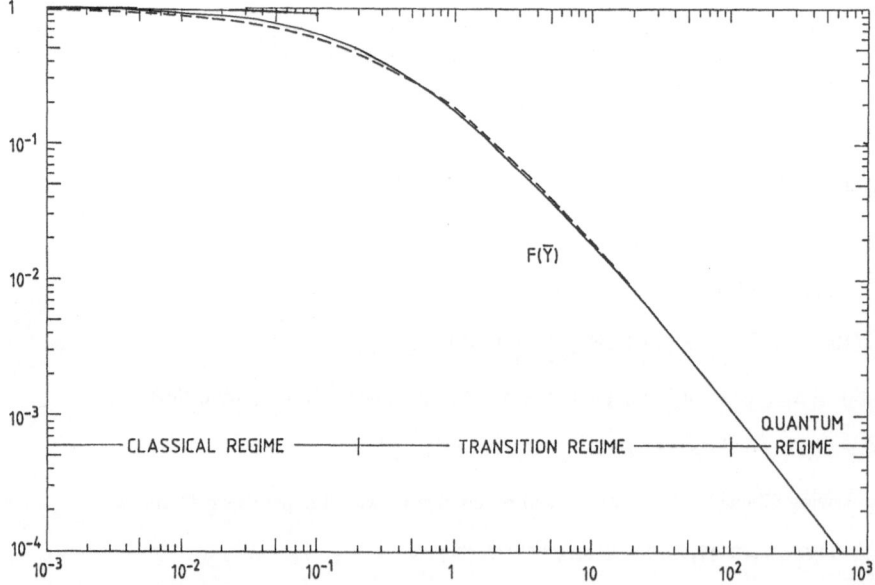

Fig. 2 *The factor $F \, (\overline{Y})$, which multiplies the beamstrahlung parameter δ and gives the fractional energy loss $<\varepsilon>$. The continuous line is the result of Noble numerical calculations[7]; the dashed curve is the simple analytic expression of Eq. (21).*

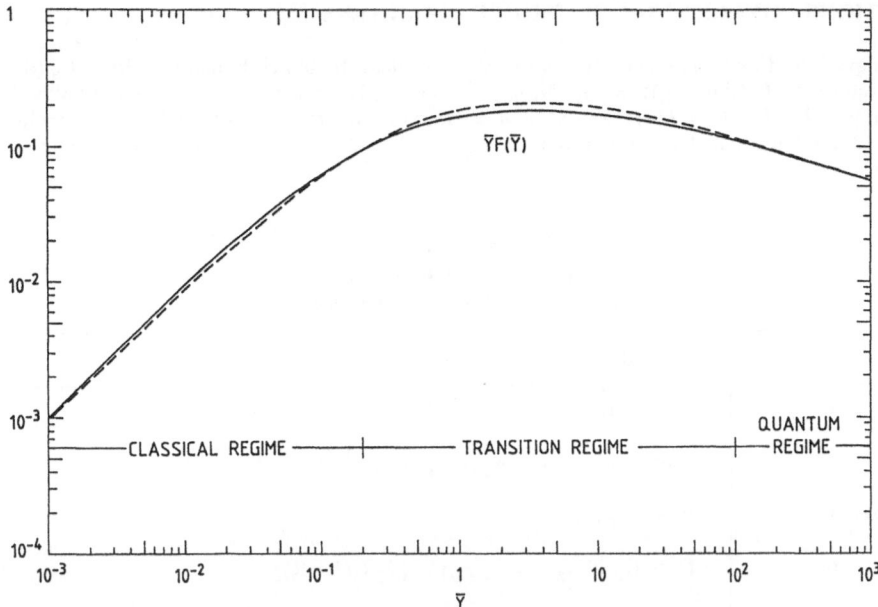

$\bar{Y}F(\bar{Y})$

CLASSICAL REGIME ——————— TRANSITION REGIME ——————— QUANTUM REGIME

Fig. 3 *The function $\bar{Y}\ F(\bar{Y})$, which enters in the second fundamental relation of linear colliders [Eq. (22)], is plotted versus $\bar{Y}$. The continuous curve is due to Noble[12] and the dashed curve represents the simple approximation[6] of Eq. (21): $\bar{Y}\ (1+\bar{Y}^{1}/^{2}+\ 3\bar{Y}/2)^{-4}/^{3}$.*

For the scaled-up version of SLC appearing in Table 1, R = 1 and $H_b = H_D = 2.2$, so that from Eq. (17) $\bar{Y} = 2.9$. Eq. (21) gives $F(\bar{Y}) = 0.075$ and the beamstrahlung parameter $\delta = 9$ of Table 1 implies a fractional energy loss $<\varepsilon> \simeq 0.7$. Such a loss is still *too* large with respect to what is generally considered acceptable ($<\varepsilon> \leq 0.3$): as anticipated, the scaling laws of Table 2 can be applied only up to $E_0 \simeq 300$ GeV.

The fractional energy loss $<\varepsilon>$ of Eq. (20) can be written in a convenient form by expressing δ of Eq. (19) as a function of $\bar{Y}$ and $(L/f_r)^{1}/^{2}$. One obtains the *second fundamental relation* of linear colliders[14]:

$$<\varepsilon> = (16\ \pi^{1}/^{2}\ \alpha/15)\ (H_b/H)^{1}/^{2}\ (r_e^2\ L/f_r)^{1}/^{2}\ \bar{Y}\ F(\bar{Y})\ . \tag{22}$$

Since $<\varepsilon>$ is in practice fixed, the quantity $\bar{Y}\ F(\bar{Y})$ is very important because, together with R and D, it determines L/f_r.

Noble's numerical results[12] for $\bar{Y}\ F(\bar{Y})$ are plotted as a continuous line in Fig. 2, while the dashed line represents the approximation given by the simple form of Eq. (21). Three regimes can be clearly distinguished:

classical regime:		$\bar{Y} < 0.2$
transition regime:	$0.2 <$	$\bar{Y} < 10^2$
quantum regime:	$10^2 <$	$\bar{Y}\ .$

$$\tag{23}$$

In the first regime $<\varepsilon> \propto \bar{Y}$; in the second $<\varepsilon> \simeq$ constant within a factor 2 and $\bar{Y}\ F(\bar{Y}) = 0.15 \pm 0.05$; and in the third $<\varepsilon> \propto \bar{Y}^{-1}/^{3}$. We shall conclude that the Next Linear Collider will run in the intermediate regime, where $\bar{Y}\ F(\bar{Y})$ is about constant. (Note that, since $\bar{Y} = 2.9$, also the scaled version of SLC would have this property.)

8. RELATIONS CONNECTING COLLISION PARAMETERS

We have introduced a large number of quantities to describe bunch-bunch collisions: E_0 (or γ), L, P, R, D (and $H = H(R,D)$; H_b)), σ_z, f_r, N, σ_x, $\overline{Y}$, $<\varepsilon>$. These *eleven* parameters are linked by the *five* equations (7), (13), (15), (17), (22), so that *six* parameters are enough to define the bunch-bunch collisions of any linear collider. To help the reader, the relevant formulae are collected in Table 3.

Table 3
Summary of collider formulae
(the symbol $\simeq$ signifies $5 - 10\%$ accuracy)

Quantity	Formula	Eq.
Energy/beam	$E_0 = \gamma\, mc^2$	
r.m.s. radii	$\sigma_x = (\varepsilon_{nx}\beta_x^*/\gamma)^{1/2}$ $\sigma_y = (\varepsilon_{ny}\beta_y^*/\gamma)^{1/2}$ $[\sigma_x = R\sigma_y]$	(4)
Power/beam	$P = Nf_r\, E_0$	(7)
Disruption par.	$D = (r_e\, N\, \sigma_z)\,(\gamma\,\sigma_x\,\sigma_y)^{-1}\,[2\,R/(1+R)]$	(13)
Pinch factor (R = 1)	$H_D \simeq$ Eqs. (6)	
Pinch factor (R≠1)	$H \simeq H_D^{(1+R)/2R}$	(14)
Luminosity	$L = f_r\, N^2\, H/(4\pi\,\sigma_x\,\sigma_y)$	(15)
Fract.critical energy	$\overline{Y} \simeq (5/12\alpha)\,(r_e^2\gamma/\sigma_z)\,(N/\sqrt{\sigma_x\,\sigma_y})\,H_b^{1/2}(D,R)$	(17)
Beamstrahlung factor	$H_b(D,R) \simeq 4R\,H_D\,[1+RH_D^{(R-1)/2R}]^{-2} \to 4/R$ for R >> 1	(18)
Beamstrahlung par.	$\delta = 2/9\,(r_e^3\gamma/\sigma_z)\,(N^2/\sigma_x\sigma_y)\,H_b(D,R)$	(19)
Fract.energy loss	$<\varepsilon> \simeq \delta\, F(\overline{Y})$	(20)
Quantum factor	$F(\overline{Y}) \simeq (1+\overline{Y}^{1/2}+3\overline{Y}/2)^{-4/3}$	(21)

How do we orientate ourselves in such a complicated multidimensional space? In this Section we first discuss some relations which are simple consequences of the equations of Table 3. But to make full use of them, in the next Section we have to review the scaling laws of linear accelerators.

I address, on the basis of the equations of Table 3, the questions: (i) Will NLC run with sizeable pinch effect? (ii) In which beamstrahlung regime NLC will operate? The answers will be: (i) NLC will utilize the pinch effect to increase luminosity; (ii) NLC will either run in the transition regime or very close to it. The first point is clear when looking at the first fundamental relation of Eq. (16): for a limited beam power the luminosity is proportional to DH!

To discuss the second question we have to write down some new equations. We have already met the two fundamental relations (16) and (22), which are rewritten in practical units in Table 4. The third relation in the Table is obtained by expressing $\overline{Y}$ of Eq. (17) in terms of (L/f_r). The fourth rela-

Table 4
Relations derived from the equations of Table 3 (*)
(practical units are introduced for convenience of use in numerical calculations)

Relation	Eq.
$\sigma_z/mm = [DH/29)] \; [(1+R)/(2R)] \; (10^{33}/L) \; (P/MW)$	(16)
$<\varepsilon> \; \simeq 3.9 \; (H_b/H)^{1/2} \; (L/10^{33})^{1/2} \; (kHz/f_r)^{1/2} \; \bar{Y} \; F(\bar{Y})$	(22)
$\bar{Y} \simeq 0.32 \; (H_b/H)^{1/2} \; (E_0/TeV) \; (L/10^{33})^{1/2} \; (kHz/f_r)^{1/2} \; (mm/\sigma_z)$	(24)
$\bar{Y} \simeq 9.1 (DH)^{-1}(H_b/H)^{1/2}[2R/(1+R)](E_0/TeV) \; (L/10^{33})^{3/2}(MW/P)(kHz/f_r)^{1/2}$	(25)

(*) *In this Table* $10^{33} = 10^{33} \; cm^{-2}s^{-1}$.

tion derives from Eqs. (16) and (24) of Table 4. It relates the parameter $\bar{Y}$ to the *six* most important physical parameters (E_0, L, P, D, R, f_r) which describe *fully* the bunch-bunch collisions. (Note that D and R appear also implicitly in the ratio H_b/H and that $H_b/H = 1$ for R = 1.)

When designing a collider, out of the six parameters (E_0, L, P, D, R, f_r) the energy E_0 and the luminosity L are given by physics, and the present wisdom requires the luminosity given in Eq. (1).

P is fixed by the acceleration technology and the money available, since given a certain efficiency η_{tot} of the energy transfer from wall-plug to the accelerated beam, the power (per linac) is P/η_{tot}. (Today $\eta_{tot} < 1\%$ and in future accelerators one can expect $\eta_{tot} \leq 5\%$.)

The value of D defines the choice between *pinch effect* (D > 0.5) or *no pinch effect* (D < 0.5). As shown by Eq. (4), the aspect ratio R can be varied by using both the transverse emittances ε_{ny} and ε_{nx} and/or the ratio β_x^*/β_y^*.

Eq. (25) of Table 4 shows that, if the luminosity is scaled according to Eq. (1) and with D, R, P and f_r fixed, $\bar{Y} \propto E_0^4$ and at large enough energy the collider **has** to run in the quantum regime ($\bar{Y} > 10^2$).

For a 'standard' NLC (W = 1 TeV and L = 10^{33} cm^{-2}s^{-1}) with P = 1 MW (so that the *total* plug power is ~ 100 MW with $\eta_{tot} \simeq 2\%$) it follows from Eq. (25) that $\bar{Y} = (C/DH) (kHz/f_r)^{1/2}$. It can be easily checked[6] that the numerical factor is $1 < C < 10$ for all values of H and R. A standard NLC, which profits from the pinch effect (D $\geq$ 0.5) *and* runs at the (small) SLC repetition rate (f_r = 0.2 kHz), will thus have $0.2 \leq \bar{Y} \leq 10$ for all reasonable values of R. However, higher rates and beam powers tend to reduce $\bar{Y}$, and one can get $Y < 10^{-1}$ if the aspect ratio R is large. At higher energy (for instance at W = 2 TeV, as for the CERN linear collider CLIC), the choice is again pushed to the transition regime because of the dependence $\bar{Y} \propto E_0^4$ even if f_r is few kHz.

In conclusion, a low repetition rate NLC profiting from the pinch effect will have to run either close or in the *transition regime*. At higher repetition rates and for flat bunches one can choose to sit at the higher end of the classical regime. Anyway, by increasing the energy one is pushed towards the quantum regime, where the scaling laws change. This subject is not discussed here, because such a regime is not interesting for NLC; still in the last Section I shall mention some very recent results which indicate that this regime may not be accessible at all.

9. SCALING LAWS OF COPPER LINACS

It is by now generally accepted that the acceleration techniques of the next generation of linear colliders will be 'reasonable' extrapolations of the ones in use today. The most promising candidates are copper cavities running at high frequency ($10 \leq f \leq 30$ GHz). Superconducting cavities would be an *ideal solution*[15], but at present the achievable gradients are too small and imply long and expensive linacs to reach the TeV energy range[16].

Let us consider a normal conducting linac made of travelling wave sections of length L at frequency f. Given the *stored energy per unit length* W', the *power dissipation per unit length* P_d' determines the decay time of the energy: W'/P_d'. The structure (unloaded) quality factor Q is defined as the ratio between this decay time and the characteristic time of the RF oscillation: $(2\pi f)^{-1} = 1/\omega$:

$$Q = \omega \, W' / P_d' . \tag{26}$$

The quantity τ, proportional to the decay time, fixes the time it takes to build up the field:

$$\tau = W'/(2P_d') = Q/2\omega . \tag{27}$$

(The factor 1/2 corresponds to the usual choice made for high energy colliders[17]. In general $\tau = Q \, \alpha/\omega$ were α is the 'attenuation constant' of the structure.)

Each one of the sections of length L is excited in a resonant mode with an electric field component in the direction of the particle motion; we suppose for simplicity that the excitation is caused by a square power pulse of frequency f, duration τ and peak power P_L. The group velocity v_g, which is the velocity of the energy flow, determines the filling time so that $\tau = L/v_g$. (In the structures to be considered typically one has $v_g/c \simeq 0.05\text{-}0.1$.)

The stored energy *per meter* W' (W/m) is clearly proportional to the square of the accelerating field E (V/m) and to the structure cross-section, i.e. to the square of the wavelength λ. One can thus write:

$$W' = E^2 \, \lambda^2/(2\pi \, cZ) . \tag{28}$$

where Z is an impedance independent of λ for any given geometry of the structure. A typical value is $Z \simeq 300 \, \Omega$, so that at $\lambda = 10$ cm and $E = 17$ MV/m (as for SLC) one needs $W' \simeq 5$ J/m.

The energy is pumped into the structure with a repetition rate f_{rf} in pulses of duration τ. Typically only $\eta_t \simeq 75$ % of the pumped energy is still there (as stored energy W') at the moment in which a bunch of Ne particles crosses the structure, is accelerated by the field E and extracts energy with efficiency

$$\eta = Ne \, E/W' = 2\pi \, cZ \, Ne \, / \, (E\lambda^2) . \tag{29}$$

Of course some power is spent in accelerating the electron and positron beams. When a bunch of charge Ne interacts with the structure it induces a field $E_i \propto$ (Ne) which cancels part of the accelerating field. The average particle of the bunch will thus see the field $(E - E_i/2)$ and this causes a momentum spread. If η is the *fraction* of the stored energy extracted by a bunch, when no particular attention is payed to the problem, the momentum spread is of the order of $\eta/2$. One can do better by choosing the phase of the bunch with respect to the RF wave such that, without beam loading, the particles in the tail of the bunch would see a larger accelerating field.

If b bunches of N particles each are accelerated during a *single* RF pulse ('multibunching'), and extra power is poored in the structure in between bunches to compensate for the energy extracted, the efficiency of the system can be increased without augmenting the momentum spread. Fig. 4 taken from Ref. [18], depicts how the filling of the structure is fitted to the distances between the closely spaced bunches: the 'length' of the accelerating field is longer to compensate for the decrease of the field, so that all particles are given the same amount of energy.

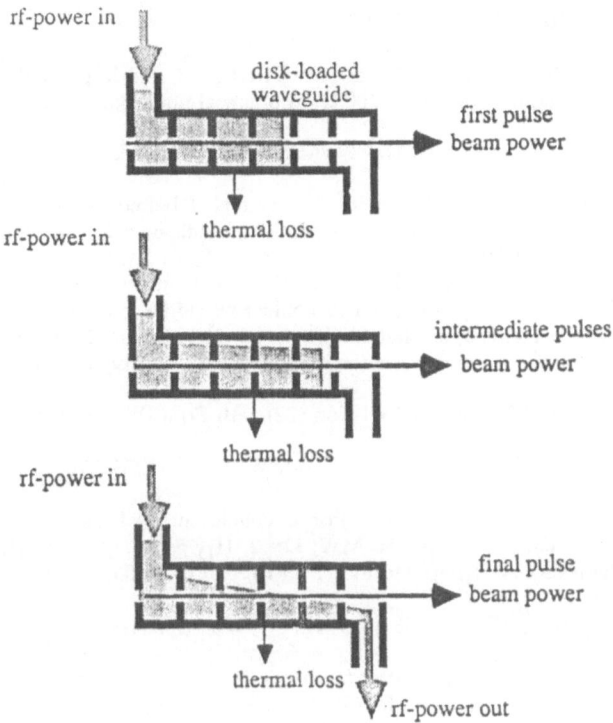

Fig. 4 *In a multibunch scheme the structure is powered in such a way that successive bunches get the same acceleration even in presence of a decreasing field.*

In the case of a multibunch collider the RF power repetition rate f_{rf} is smaller than the collider rate f_r

$$f_r = b \, f_{rf} \, , \tag{30}$$

and the fraction of stored energy which is extracted is roughly

$$\eta \simeq b \, Ne \, E/W' = 2\pi \, cZ \, b \, Ne/(E\lambda^2) \, . \tag{31}$$

Typically for a single bunch $\eta \le 5\%$, because of the requirement on the momentum spread, and one can hope to reach $\eta \simeq 20\%$ in the multibunch compensated scheme.

The total energy transfer efficiency from rf-power to beam is $\eta_{rf} = \eta_t \eta$ and the RF power needed for *two* linacs to obtain two high energy beams of power P *each* is

$$P_{rf} = 2 \, P/(\eta_t \eta) \simeq (2E_0 f_r \, E\lambda^2)/(2\pi \, ceZ\eta_t b) \, , \tag{32}$$

where we have used Eq. (7) and the bunch population N does *not* appear. As already mentioned, in Eq. (32) typically $1/\eta_t \simeq 1.25$, so that with $\eta = 5\%$ ($\eta_{rf} = \eta_t \eta \simeq 4\%$) and P = 1 MW one needs $P_{rf} = 2 \, P/\eta_{rf} \simeq 50$ MW. Note that at the most favorable frequencies the efficiency of producing RF power from plug power is $\sim 70\%$; at the high frequencies needed for the new colliders one can hope to have $\eta_{tot}/\eta_{rf} \simeq 50\%$, so that our choice corresponds to $\simeq 100$ MW of plug power.

10. THE CHOICE OF THE RF FREQUENCY

We can now make contact with the discussion of Section 8, and in particular with the equations of Table 4. By using the quoted value $Z = 300\ \Omega$, in practical units Eq. (32) becomes:

$$P_{rf}/MW \simeq (400/b)\ (E_0/TeV)\ (f_r/kHz)\ (E/MV\ m^{-1})\ (GHz/f)^2\ , \tag{33}$$

which clearly displays the gain in RF power implied by high RF frequencies (i.e. small wavelengths) and by the use of a multibunch scheme ($b > 1$). Eq. (33) could be called the *third fundamental relation* of copper cavity linear colliders. Applied to the NLC of Table 1 with $f = 3$ GHz, $E = 17$ MV/m and $f_r = 0.18$ kHz (as for SLC) it gives $P_{rf} \simeq 72$ MW, a not unreasonable number. However, with such a low gradient the length of a $(0.5 + 0.5)$ TeV collider would be much too large: 2×30 km. If the gradient was increased by a factor 5, so that the length becomes 2×6 km, the power would become $P_{rf} \simeq 350$ MW. This is unreasonable because the plug power P_{ac} has typically to be twice as large.

By combining the second fundamental relation (22) with Eq. (33) one finally gets:

$$b^{1/2}f/GHz \simeq 80(H_b/H)^{1/2}(E_0/TeV)^{1/2}(L/10^{33})\ (MW/P_{rf})^{1/2}\ (E/MeVm^{-1})^{1/2}\overline{Y}F(\overline{Y})/<\varepsilon> \tag{34}$$

which does *not* contain the repetition rate. For a collider defined by the following parameters: $W = 2E_0 = 1$ TeV, $L = 10^{33}cm^{-2}s^{-1}$, $P_{rf} = 50$ MW, $D = 2$, $H_D \simeq 3.5$, $<\varepsilon> = 0.3$, $b = 1$, $E = 100$ MV m^{-1}, which runs in the *transition regime*, (so that $\overline{Y}\ F(\overline{Y}) = 0.15 \pm .05$), Eq. (34) implies

$$f \simeq (40 \pm 13)\ (H_b/H)^{1/2}\ GHz \to (58 \pm 19)\ R^{-1/2}\ GHz\ \ (for\ H_D = 3.5)\ . \tag{35}$$

The first equality shows that for $R = 1$, the frequency is ten times larger than the SLC frequency. The factor $(H_b/H)^{1/2}$, which is equal to 1 for $R = 1$, decreases as $2R^{-1/2}\ H_D^{-1/4} \simeq 1.5\ R^{-1/2}$ for $H_D \simeq 3.5$ and $R \geq 3$, and this gives the second equality in Eq. (35). Flat bunches are thus useful to decrease either the RF frequency or the plug power, since in Eq. (34) $f\ P_{rf}^{1/2} \propto R^{-1/2}$. For $R = 10$ and $H_D = 3.5$ from Eq. (35) the optimal frequency is $f \simeq 18$ GHz.

This argument closes the general discussion of the main scaling laws. We have seen that the Next Linear Collider will have to run with some pinch effect, to increase the luminosity, and in the transition regime. Then the other parameters are practically fixed by the flatness of the bunch and the requirement of not having a too large power consumption. In particular the frequency of the structure *cannot be very different from 20 GHz*. Flat bunches imply a reduction of the frequency which is proportional to $R^{-1/2}$ or, for a fixed frequency, a reduction of the power $\propto R^{-1}$, as shown in Eq. (34).

11. POWER SOURCES

It is easy to write down the peak power $\hat{P}_L$ needed to produce, with a RF square pulse of frequency f and duration τ, the wanted gradient E in a copper structure of length L. As anticipated, the energy ($\hat{P}_L\ \tau$) of each RF pulse has to be larger by a factor $1/\eta_t \sim 1.25$ with respect to the energy (W'L) needed to have an accelerating field E when the bunch crosses the structure. Thus one can write

$$\hat{P}_L\ \tau \simeq 1.25\ W'L\ , \tag{36}$$

and, using Eq. (27),

$$\hat{P}_L/L \simeq 2.5\ \omega\ W'/Q\ . \tag{37}$$

The ratio $\hat{P}_L/L$ is nothing else than the *peak power $\hat{P}'$ per meter* of structure. W' of Eq. (28) can be written in the form $W' = c\ E^2/(2\pi\ Z\ f^2)$, so that

$$\hat{P}' \simeq 2.5\ c\ E^2/(ZfQ)\ . \tag{38}$$

The Q of the structure depends on its design and, for a fixed geometry, scales as $f^{-1/2}$. For a structure of the SLAC design (but having a larger opening so that the group velocity is $v_g \simeq 0.07$ c), the numerical expression

$$Q \simeq 2.2 \; 10^4 \; (GHz/f)^{1/2} \tag{39}$$

can be introduced in Eq. (38) together with the value $Z \simeq 300 \; \Omega$ to obtain

$$\hat{P}'/MW \; m^{-1} \simeq 0.11 \; (E/MV \; m^{-1})^2 \; (GHz/f)^{1/2} \; . \tag{40}$$

For a gradient $E = 100$ MV m^{-1} at the frequency $f = 20$ GHz the needed peak power is $\hat{P}' \simeq$ 250 MW m^{-1} and, for a NLC total length of $5 + 5$ km, the total peak power turns out to be 2.5 terawatts! This number gives a feeling of the difficulties to be encountered in constructing the RF power sources for NLC, especially when compared with the present total SLC peak power (which moreover runs at a much more comfortable frequency): 0.05 terawatts.

Another important factor is the duration τ of the power pulse which, combining Eqs. (27) and (39), is

$$\tau/ns \simeq 1750 \; (GHz/f)^{3/2} \; . \tag{41}$$

The length of structure which has to be powered independently is $L = v_g \; \tau$ i.e., for $v_g \simeq 0.07$ c,

$$L/m \simeq 37 \; (GHz/f)^{3/2} \; . \tag{42}$$

At $f = 20$ GHz, $\tau \simeq 20$ ns and $L \simeq 0.4$ m so that the 10 km collider of our example is made of about 25,000 sections! More than 1500 power sources would be needed, even if each source is connected to 16 accelerating elements.

Power sources of the needed frequency and peak power are not available on the market. Their development is the main subject of any R & D program aiming at the construction of a future linear collider. A review of the ongoing activities would be too long. I limit myself to few short remarks.

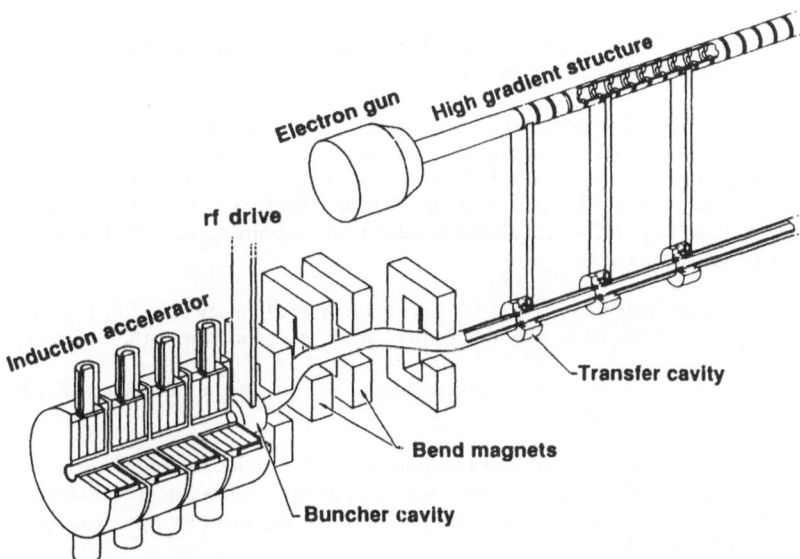

Fig. 5 *The relativistic-klystron linear accelerator considered at present by the Berkeley-Livermore-SLAC collaboration.*

In USSR the Novosibirsk group is working on high frequency klystrons and gyrocons[19]. High frequency klystrons are also under consideration in Japan. In USA the main emphasis has been on the so-called 'relativistic klystron'[20] depicted in Fig. 5. Induction units accelerate an intense beam of a few MeV electrons, which excites one or more cavities which are running at the desired frequency. The same low energy bunches are used more than ones. The frequency chosen for the moment is f = 17 GHz.

At CERN for CLIC the concept is similar, but the electron beam would have a few GeV and would run all along the high gradient accelerating structure[15,17], as initially proposed by Andy Sessler for the Two Beam Accelerator[21]. The intense and highly bunched few GeV beam is accelerated with high efficiency by low frequencies superconducting cavities equal to the one which shall be used for LEP 200. The frequency chosen for the high gradient linac is f = 30 GHz.

The many groups involved in these activities are performing some tests. The only one which already produced concrete results has been performed at Livermore by a Berkeley-Livermore-SLAC collaboration[22]. Using an induction linac and a wiggler, about 2 GeV of peak power were produced for about 20 ns at f ≃ 30 GHz. After a couple of years of uncertainty this Free Electron Laser source has been proposed again as a valid alternative to the sources described above[23].

12. PARAMETERS OF A STANDARD NLC

We can now derive a complete list of parameters for what I have chosen to call a 'standard' Next Linear Collider. The logic of the argument and the numerical values appear in Table 5. The rest of this Section is devoted to comment it.

The input parameters are dictated by physics (E_0 and L), by the desire to reduce beamstrahlung but not to have too flat bunches (R large but not larger than 10), the wish to make a reasonable use of the pinch effect (D ≃ 3.5) and a reasonable power consumption (P_{rf} ≃ 50 MW so that, as discussed in the last Section, P ≃ 1 MW with η = 5%). It is then enough to fix the fractional energy loss at the standard value $<\varepsilon>$ ≃ 0.25 to get all other important parameters by applying the equations mentioned in the Table. The bunch population is reasonably low (N ≃ 10^{10}) and the transverse dimensions of the bunches are *very* small indeed, as in all parameter lists of linear colliders which have beam power of the order of one MW. (For superconducting colliders the power can be more than ten times larger and the cross-section of the bunches are much less demanding[16].)

The Table shows that, as expected, the collider runs at the center of the transition region ($\overline{Y}$ ≃ 0.4) so that the parameter $\overline{Y}$ $F(\overline{Y})$ equals ~ 0.15.

Below the horizontal line appearing in the Table, I list a possible choice of the invariant emittances, which agrees with what is thought possible today with very carefully designed damping rings[6]. At SLC the damping rings give ε_{nx} ≃ 3 10^{-5} m, two times larger than what is needed here, but many recent studies[24,25] have shown that the choice made in the Table (ε_{nx} = 5 10^{-6} m ε_{ny} = 5 10^{-7} m) is possible. Note that, to get R = 10, I have decided to have both $\varepsilon_{ny}/\varepsilon_{nx}$ = 10 and β_y^*/β_x^* = 10; other choices are possible.

The last line in the Table shows that the parameter Λ = β^*/σ appearing in Eq. (6) is larger than ~ 3, as it should not to loose luminosity due to the bunch being too long with respect to the β^* value.

The main parameters of the power sources needed to run such a collider are also fixed. Table 6 collects the formulae and the results.

Few remarks are in order. I choose *one* bunch per RF pulse for safety reason. Without extra plug-power one could roughly double or triple the luminosity with b = 2 or b = 3. This may be possible if one uses structures which have slots to coupled out the higher-order modes excited by the wake field, as proposed by R. Palmer.

Table 5
Construction of a consistent set of parameters for a 'standard' NLC

Input quantities	Computed quantities and/or comment	Numerical value
1. Beam energy E_0		$E_0 = 0.5$ TeV
	c.m.energy $W = 2 E_0$	$W = 1.0$ TeV
2. Luminosity L	physics requirement from Eq. (1)	$L = 10^{33} cm^{-2} s^{-1}$
3. Aspect ratio R	to reduce beamstrahlung	$R = 10$
4. Disruption parameter D	to have some pinch effect	$D = 3.5$
	H_D from Eq. (6) with $A \simeq 4$	$H_D \simeq 4.6$
	enhancement factor from Eq. (14)	$H \simeq 2.3$
	factor H_b from Eq. (18) $\simeq 4/R$	$H_b \simeq 0.42$
5. Beam power P	to consume $P_{rf} \simeq 50$ MW with $\eta_{rf} = \eta_t \eta \simeq 4\%$	$P = 1$ MW
	bunch length from Eq. (16)	$\sigma_z \simeq 0.15$ mm
6. Fractional energy loss $< \varepsilon >$	acceptable to get $\Delta W/W \leq 10\%$	$< \varepsilon > \simeq 0.25$
	f_r from Eq. (22) with $\Upsilon F(\Upsilon) \simeq 0.15$	$f_r \simeq 1.0$ kHz
	Υ from Eq. (24)	$\Upsilon \simeq 0.45$
	$\Upsilon F(\Upsilon)$ from Eq. (21), it checks with above	$\Upsilon F(\Upsilon) = 0.15$
	from Eq.(7): $N = P/(f_r E_0)$	$N \simeq 1.25 \; 10^{10}$
	from Eq. (15): $\sigma_x \sigma_y = H f_r N^2/(4\pi L)$	$\sigma_x \sigma_y \simeq 2800$ nm^2
	$\sigma_y = \sqrt{\sigma_x \sigma_y/R}$	$\sigma_y \simeq 17$ nm
	$\sigma_x = R\sigma_y$	$\sigma_x \simeq 170$ nm
7. Invariant emittances	to get $R = 10$: $\varepsilon_{ny}/\varepsilon_{nx} = \beta_y^*/\beta_x^* = 10$	$\varepsilon_{nx} = 5 \; 10^{-6}$m $\varepsilon_{ny} = 5 \; 10^{-7}$m
	from Eq. (4): $\beta_x^* \beta_y^* = (\gamma \sigma_x \sigma_y)^2 (\varepsilon_{nx} \varepsilon_{ny})^{-1}$	$\beta_x^* \beta_y^* \simeq 3.1$ mm^2
	$\beta_y^* = \sqrt{\beta_x^* \beta_y^*}/R$;	$\beta_y \simeq 0.56$ mm;
	$\beta_x^* = R\beta_y^*$	$\beta_x \simeq 5.6$ mm
	check for Eq. (6): $A = \beta_y^*/\sigma_z$	$A \simeq 4$

Table 6
Parameters of the power sources for a 'standard' NLC

Input quantities	Comment	Numerical value
1. Beam energy		$E_0 = 0.5$ TeV
2. Accelerating field		$E = 100$ MV/m
	collider length $= 2 \times E_0/E$	2×5 km
3. RF total power	to have a plug power < 150 MW	$P_{rf} = 50$ MW
4. Bunches per RF pulse	conservative choice	$b = 1$
	frequency from Eq. (34) and Table 5	$f \simeq 20$ GHz
	wavelength of the structure	$\lambda = 15$ mm
	stored energy/meter from Eq. (28)	$W' \simeq 4.0$ J/m
	extraction efficiency from Eq. (31)	$\eta \simeq 5\%$
	Q-value from Eq. (39)	$Q \simeq 4.9 \ 10^3$
	peak power per m from Eq. (40)	$\hat{P}' \simeq 250$ MW/m
	total peak power	2.5 TW
	duration of the RF pulse from Eq. (41)	$\tau \simeq 20$ ns
	length of the structure (section) from Eq. (42)	$L \simeq 0.4$ m
	total number of sections	$2 \times 12,500$

Once the gradient E and the RF power P_{rf} are chosen, Table 6 shows that the accelerator parameters are frozen. For the 'standard' choices ($E = 100$ MV/m, $P_{rf} = 50$ MW, $\eta_{rf} = 4\%$) we get f $\simeq 20$ GHz with all the consequences listed in the Table. In particular one needs 250 MW/m of peak power. Thus a source delivering ~ 2.5 GW for ~ 20 ns, similar to the one tested at Livermore[22], could power about 10 m, a not unreasonable length which corresponds to 25 sections each 0.4 m long. To power 2.5×10^4 sections one would need 1000 sources.

13. A FINAL COMPARISON

I compare in Table 7 the parameters of my 'standard' NLC with those at present considered at CERN (for CLIC, the CERN Linear Collider) at SLAC (for the TeV Linear Collider, TLC) and at Novosibirsk (for VLEPP, to be built at Serpukhov).

Before comparing the parameter lists, let me underline that at the time of writing we do not have complete parameter lists for TLC and VLEPP. In Table 7 I have used the best information available at the SLAC workshop of November 1988, which I complemented with the missing pieces. To make the comparison more meaningful the luminosity quoted in Table 7 is computed *without* multibunching (b = 1). However, at SLAC the consequences of chosing $b \simeq 10$ (and even $b \simeq 20$) are seriously studied; this would imply automatically either a tenfold increase of the luminosity or a relaxing of the

Table 7
Parameter lists of linear colliders

	'Standard' NLC	CLIC (CERN)	TLC (SLAC)	VLEPP (Novosibirsk)
Energy per beam, E_0 (TeV)	0.5	1.0	0.5	0.5
Gradient, E (MV/m)	100	80	185	100
Frequency, f (GHz)	20	30	17	14
Bunches per RF pulse, b	1	1	1(*)	1
Bunch repetition rate, f_r(kHz)	1.0	1.7	0.36	0.1
RF av. power, P_{rf} (MW), Eq. (33)	50	60	45	10
Particles/bunch, N	1.25×10^{10}	5×10^9	1.4×10^{10}	2×10^{11}
Power/beam, P (MW)	1.0	1.35	1.1	1.6
Disruption, D/H	3.5/2.3	3.3/2.3	5.0/2.2	7.2/2.5
Aspect ratio, R	10	5	130	150
Bunch r.m.s. radius, σ_x (nm)	170	60	390	3000
Bunch r.m.s. radius, σ_y (nm)	17	12	3	20
Bunch r.m.s. length, σ_z (μm)	150	200	70	750
β-value, β_y^* (mm)	0.56	0.28	0.15(**)	1.3(**)
β-value, β_x^* (mm)	5.6	2.3	50	300
Inv. emittance, ε_{nx} (m)	5×10^{-6}	3×10^{-6}	3×10^{-6}	3×10^{-5}
Inv. emittance, ε_{ny} (m)	5×10^{-7}	1×10^{-6}	6×10^{-8}	3×10^{-7}
Beamstrahlung reduction, H_b	0.42	0.84	0.030	0.027
Fractional energy loss, $<\varepsilon>$	0.25	0.33	0.12	0.07
Fractional crit. energy, Υ	0.45	0.77	0.46	0.08
Product Υ F(Υ)	0.15	0.18	0.15	0.05
Luminosity, L (10^{33}cm^{-2}s^{-1})	1.0×10^{33}	1.1×10^{33}	1.0×10^{33}	1.3×10^{33}

(*) *Multibunching (up to b = 20) is seriously considered to increase the luminosity and/or relax some parameters.*
(**) *This number, which I took $\sim 2\sigma_z$, does not appear in the parameter list available to me. Its choice constraints β_x^* and the two transverse emittances.*

very tight vertical dimensions of the bunches. Multibunching is considered a reserve factor for CLIC, while at VLEPP it is most probably excluded by the much greater number of particles per bunch and the consequent larger wake field effects.

The choice of the aspect ratio R distinguishes TLC and VLEPP from the 'standard' NLC and CLIC since, as discussed at length, $H_b \propto 4/R$ and $\bar{Y} \propto R^{-1/2}$. For TLC and VLEPP the chosen value R > 100 implies a large suppression of beamstrahlung radiation. VLEPP even runs in the 'classical regime', since $\bar{Y}$ < 0.2. Which are the consequences of this apparently clever choice? The difficulties have been displaced to two different areas.

By using the parameter lists of TLC and VLEPP as *typical,* we see that *two* routes are opened once R > 100 is chosen. Either — as for TLC — the number of particles is $N \simeq 10^{10}$ and then the bunches are very small vertically ($\sigma_y \simeq 3$ nm), or $N \simeq 10^{11}$ and the bunches transverse dimensions are ten times larger (VLEPP). In the former case the difficulty lies in providing both the very small vertical emittance at the crossing point ($\varepsilon_{ny} \simeq 6 \times 10^{-8}$ m) and the β-value $\beta_y^* \simeq 0.15$ mm. In the second case the problem will be the production of such an intense bunch and the attaining of a *very large* efficiency ($\eta_{rf} \simeq 32\%$). Another problem is hidden in the preservation of the vertical emittance during acceleration of a very intense bunch.

As far as emittances are concerned let me remind that studies made in different laboratories[24] show that well designed and aligned damping rings can produce horizontal emittances $\varepsilon_{nx} \simeq (1 \div 2)$ 10^{-6} m with $N \simeq 10^{10}$. On paper, ratios $\varepsilon_{nx}/\varepsilon_{ny} \simeq 10^2$ are feasible, but to maintain vertical emittances $\varepsilon_{ny} \simeq 10^{-7}$ m during the acceleration in a 5 km long linac one needs[25] alignment accuracies of the order of 10 μm. Note also that the elements of the final focus have to be aligned to a fraction of the final spot size, i.e. to $\sim$ 1 nm (!) vertically for TLC and $\sim$ 5 nm for VLEPP.

The CLIC parameter list is close, not surprisingly, to the 'standard' NLC one, with the added complications that CLIC energy is twice larger. This choice — — which could be reviewed but makes the physics potentialities of CLIC and LIIC equivalent *and* complementary[4] — — would require according to Eq. (1) a luminosity L = 4×10^{33} per interaction point (IP). The parameters considered today and listed in Table 7 miss this goal by a factor 4 (or more, if more IP's are served). The present philosophy is that multibunching will provide it, maybe in a second stage.

The 'standard' NLC and CLIC have R $\leq$ 10 and vertical spot sizes similar to the ones foreseen for VLEPP. The main difference is that the repetition rate is of the order of 1 kHz and *not* 100 Hz. In this case it seems that the main difficulties are hidden in the high repetition rate, which however is not a real problem, since the relativistic klystron solution envisaged for CLIC[17] allows high repetition rates because it uses CW superconducting cavities as power sources. Standard klystrons would not run above the $\sim$ 360 Hz chosen for TLC.

The CLIC and TLC schemes need β-functions at the IP which are at the limit of what can be done. Designs exist for final focus systems which at 0.5 TeV would provide $\beta_y^* \simeq 0.3$ mm and $\beta_x^* \simeq 3$ mm with chromatic corrections for a momentum spread $\Delta p/p$ < 1%. This is more than sufficient for the 'standard' NLC and for VLEPP, but the CLIC parameters (at 1 TeV) and the TLC request (at 0.5 TeV) are at present not achievable by a factor 2 to 3.

Before closing, let me remark that in November 1988 a (until now forgotten) bunch-bunch effect appeared on the scene. This is the materialisation of the beamstrahlung photons in the megagauss magnetic field of the opposite bunch. Then the electrons and the positrons have relatively low energies and are swept by the same field, so to hit the last focusing quadrupole and even the detector. In spite of being discovered after the School, it is important to mention this phenomenon here because it will probably influence the parameter choices of future linear colliders. Pisin Chen reported results of preliminary calculations at the SLAC workshop in November 1988 and concluded that its negative effects may be limited if the fractional critical energy $\bar{Y}$ is chosen to be less than $\sim$ 0.5. Indeed the TLC parameter list of Table 7 is the one prepared to meet such a condition; previously at SLAC values $\bar{Y} \simeq 2$ were considered. For CLIC $\bar{Y} \simeq 0.8$, but little modifications are needed to have $\bar{Y} \simeq 0.5$; VLEPP would run in the classical regime and is not touched by the argument. Of course better calculations are needed. However, for the time being this effect looks very detrimental at energies definitely larger than W $\simeq$ 1 – 2 TeV, since the arguments of Section 8 indicate that to reach them the quantum regime ($\bar{Y} \geq 10^2$) is a must. In conclusion one cannot exclude that this effect will represent an insurmountable difficulty on the way to many TeV linear electron-positron colliders.

314

References

1. B. Richter, Nucl. Instr. and Meth. 136: 47 (1976).

2. U. Amaldi, Phys. Letters 61B: 313 (1976).

3. M. Tigner, Nuovo Cimento 37: 313 (1965).

4. U. Amaldi, *in* "Proc. of the 1979 Int. Symposium on lepton and photon interactions", Fermilab, August 23-29, T.B.W. Kirk and H.D.I. Abarbanel, eds., FNAL, Batavia Ill (1979) p. 314.

5. U. Amaldi, *in:* "Proc. of the Workshop on physics at future accelerators", La Thuile (Italy) and Geneva (Switzerland), 7-13 Jan., 1987, J. Mulvey, ed., CERN, Geneva (1987) Vol. 1, p. 323.

6. U. Amaldi, Introduction to the next generation of linear colliders, CERN-EP 87-169, and *in:* "Frontiers of particle beams", M.Month and S. Turner, eds., Springer-Verlag, Berlin (1988) p. 341.

7. P.B. Wilson, *in:* "Proc. UCLA Workshop on linear collider $B\overline{B}$ factory conceptual design", D.H. Stork, ed., World Scientific, Singapore (1987) p. 373.

8. P. Chen and K. Yokoya, SLAC-PUB-4339 (1987).

9. For a summary see: N.A. Solyak, Collision effects in compensated bunches of linear colliders, Inst. Nucl. Physics, Novosibirsk, 1988, Preprint 88-44.

10. R. Hollebeek, Nucl. Instr. and Meth. 184: 333 (1985).

11. T. Himel and J. Siegrist, SLAC-PUB 3572 (1985), and *in:* "Laser acceleration of particles", C. Joshi and T. Katsouleas, eds., AIP Conf. Proc. 130 (1985).

12. K. Yokoya, Nucl. Instr. and Meth. Λ251: 1 (1986).
R.J. Noble, Nucl. Instr. and Meth. A256: 427 (1987).
R. Blankenbecher and S.D. Drell, Phys. Rev. D 36: 277 (1987).
M. Jacob and T.T. Wu, Phys. Lett. B197: 253 (1987).
M. Bell and J.S. Bell, Part. Accel. 22: 301 (1988).

13. P. Chen, *in:* "Frontiers of particle beams", M. Month and S. Turner, eds., Springer-Verlag, Berlin (1988) p. 495.

14. P.B. Wilson, SLAC-PUB 4310 and *in:* "Proc. of the Part. Acc. Conference", Washington D.C., March 16-19, 1987.

15. K. Johnsen et al., Report of the Advisory Panel on the prospects for e^+e^- linear colliders in the TeV range, CERN 87-12, May 1987.

16. U. Amaldi, H. Lengeler and H. Piel, CERN EF 86-8 and CLIC Note 15.

17. W. Schnell, *in:* "Frontiers of particle beams", M. Month and S. Turner, eds., Springer-Verlag, Berlin (1988) p. 461.

18. W.A. Barletta, High gradient accelerators for linear light sources, UCRL-99268, Rev. 1, Sept. 1988.

19. V.E. Balakin et al., *in* "Proc. of the 1987 ICFA Seminar on future perspectives in high energy physics", P.F. Dahl, ed., Brookhaven National Lab., BNL 52114, Upton, Long Island (1980) p. 244 and references therein.

20. A.M. Sessler and S.S. Yu, Phys. Rev. Lett. 58: 2439 (1987).

21. A.M. Sessler, *in:* "Proc. of the Workshop on laser acceleration of particles", AIP Conf. Proc. 91 (1982) p. 154.

22. M.A. Allen et al., Relativistic klystron research at SLAC and LLNL, SLAC-Pub-4662, June 1988.

23. A.M. Sessler et al., A new version of a free electron laser two-beam accelerator, LBL-25937, Sept. 1988.

24. U. Amaldi, Nucl. Instr. and Meth. A243: 312 (1986).
    R.H. Siemann, *in:* "Advanced accelerator concepts", AIP Conf. Proc. 156 (1987) p. 453.
    C. Pellegrini et al., *in* "New developments in particle acceleration techniques", S. Turner, ed., CERN, Geneva (1987) Vol. II, p. 493.
    M. Bassetti, S. Guiducci and L. Palumbo, *in* "Proc. of the Workshop on heavy-quark factory and nuclear-physics facility with superconducting linacs", Courmayeur (Italy), 14-18 December 1987, E. De Sanctis, M. Greco, M. Piccolo, S. Tazzari, eds., Italian Physical Society, Conf. Proc. Vol. 9, Bologna (1988) p. 169.
    L. Evans and R. Schmidt, SPS/DI/Note/88-1 and CLIC Note 58.
    I.N. Hand and S. Lundgren, CLNS 88/84', May 1988.
    P. Krejcik, CERN/PS/88-47 and CLIC Note 67, *in* "Proc. of the EPAC Conference, Rome, June 1988", in print.

25. R.D. Ruth, *in:* "Frontiers of particle beams", M. Month and S. Turner, eds., Springer-Verlag, Berlin (1988) p. 444.

NONLINEAR DYNAMICS IN THE SSC — EXPERIMENT E778

Stephen Peggs

SSC Central Design Group*
Lawrence Berkeley Laboratory
1 Cyclotron Road
Berkeley, CA 94720

INTRODUCTION

A 1% variation in the cost of an accelerator was not very important forty years ago, when a cyclotron fitted into a single room. Today, when the net cost of an accelerator such as the SSC is measured in billions of dollars, it is much more important to design for an optimum balance between cost and performance. While it is irresponsible to increase the cost of an accelerator more than necessary to make it work "sufficiently" well, it is more irresponsible to construct a machine which almost works, but does not. The problems of large accelerator design lie on the horns of this dilemma. Some aspects of a successful design, such as building in flexibility to enable development in initially unforeseen directions, are almost impossible to quantify. Architectural problems such as these are not addressed here, despite their subtlety and relevance. Neither is the most difficult task addressed — the task of defining what is meant by an accelerator working "sufficiently" well, in terms of needed performance parameters, such as luminosity, lifetime, or linear aperture. Instead, this chapter concentrates on the accelerator physics processes which are expected to limit the performance of the SSC.

There are two broad classes of accelerator physics processes — single particle and collective. Collective effects are caused by the macroscopic electromagnetic fields generated by the numerous circulating charged particles (about 10^{10} particles per bunch). These fields are influenced by the environment, such as the metallic vacuum chamber walls, and act back upon the circulating particles. For example, a single bunch can disrupt itself significantly on one pass through a particular structure in an accelerator. Or, if the fields ring for long enough and have the right frequency, a single bunch can be affected on subsequent turns by the disturbance it laid down on a first turn. Multi-bunch effects occur when a trailing bunch reacts to the ringing fields laid down by preceding bunches.

The performance of the SSC is considered here only in the context of single particle models, in which a test particle circulates a collider for many turns in the presence of static electromagnetic fields. These fields are conceptually divided into linear restoring forces — in which the motion is stable — and nonlinear perturbations. Some sources of nonlinear perturbations are inevitable, in the sense of being designed in (chromatic sextupoles), or of being impossible to design out (beam-beam interactions). Other perturbations are merely random, or accidental, such as imperfections in the magnetic field quality of the

---

*Operated by the Universities Research Association, Inc. for the Department of Energy.

superconducting magnets. The implicit working hypothesis, which will not be justified, is that single particle effects will dominate collective effects in limiting the performance of the SSC. This is not necessarily true for other accelerators, where collective effects are often crucially important.

Accelerator physics is in good company when it considers the problem of single particle stability in response to nonlinear forces. For example, the question of the stability of the solar system is perhaps the best known and longest standing problem in nonlinear dynamics. Here is a system with an age of order 10^{10} periods (years), which, despite the best efforts of generations of mathematicians, has not been proven to be stable. Rigorous mathematical results are hard to come by in even the simplest nontrivial systems, for example, in the three body problem. More valuable than rigorous results, however, are the analytic languages and tools which classical dynamicists have established in their studies of differential systems — systems which are naturally described by differential equations.

The relatively recent advent of powerful computers caused an explosion in the interest paid to nonlinear problems. Computers, by their cyclical iterative nature, tend to make problems look like difference equations. On the other hand, analytic tools tend to make problems look like differential equations, since they are usually much easier to solve than difference equations, using only a pencil and paper. Which representation is truly appropriate depends on the nature of the system involved. For example, it is natural to represent the solar system as a differential system, since gravity acts smoothly and continuously, while an accelerator is naturally a difference system, since the nonlinear perturbations are usually well represented by brief impulses, separated by lengthy sections of linear motion.

At this point a sceptic might argue that numerical methods do not solve physical systems, they merely demonstrate the behavior of their solutions. An appropriate response to this is to point to the important topic of chaotic behavior in both differential and difference systems, a topic that was historically almost completely neglected by classical dynamicists, because of their lack of difference tools. Exact solutions are impossible when the motion is chaotic. Although Poincare recognised chaos as a distinct phenomenon in differential systems in the late 19th century[1], it was the use of computers in simulating chaotic difference systems that led to a broad appreciation of the ubiquitous nature of the phenomenon, and led indirectly to important formal results. According to the common wisdom, "if the only tool you have is a hammer, all your problems look like nails."

Despite all the powerful analytic and numerical tools available, it is still impossible to prove the long time scale stability of single particles in the SSC. At this point a physicist resorts to the traditional defense that pragmatism is more important than rigor. The solar system appears to be comfortably stable for 10^{10} periods. Proton storage rings such as the SPS and the Tevatron, with circulation frequency of about 40 kHz and storage times of about one day, are conservative nonlinear systems which are usefully stable for about 4.10^9 periods. In contrast, the SSC, with a revolution frequency of about 3.5 kHz (the first man-made audio frequency accelerator), needs stability for only about 3.10^8 turns in order to provide collisions for one day. While the time span of the problem has shortened, the time span of the available tools has lengthened — it is no longer uncommon to follow computer simulations of accelerator models for 10^6 turns. Although simulations still fall short of the SSC time scale by about two orders of magnitude, it is reasonable to accept their predictions about the behavior of the SSC, if the simulations agree with the real behavior of existing accelerators operating under relevant nonlinear conditions.

One goal of the E778 nonlinear dynamics experiment is to demonstrate the accuracy of numerical simulations of the Tevatron, when it is put into controlled nonlinear situations which mimic extreme SSC conditions. Another goal is to understand the long and short time scale nonlinear phenomena which are observed. The maximum time scale of the experiment is 10^6 turns if limited data is accumulated on every turn, shorter if more data is taken per turn, or longer if data is taken in periodic bursts. At the time of writing, while anticipating an E778 run in June 1989, it is already possible to say that the short (40,000 turn) time scale behavior is well understood, showing excellent agreement between simulation and experiment[2,3]. Short time scale investigations are now turning to the development of diagnostic and control techniques for the SSC, while the longer time scale investigations are studying phenomenon which, although not critical for SSC performance, have broad interest across the field of nonlinear dynamics[4–7].

The modest goal of this chapter, however, is to give an interested physicist who knows little about accelerators a qualitative description of nonlinear accelerator behavior. Consequently, there is little attempt at rigor, and only some attempt at generality. Models which describe the observed behavior in E778 are emphasised, and some results are shown. Before proceeding to nonlinear discussions, it is necessary to build up a minimum set of accelerator jargon, mostly concerning linear motion. Many references are available if the reader wishes to know more about details of the E778 experiment[8–10], about broader theories of single particle dynamics[11–14], or about the most basic descriptions of accelerator physics[15–22].

CLOSED ORBITS, LINEAR OSCILLATIONS, AND BETATRON FUNCTIONS

Consider launching a bunch of 10^{10} particles from a reference point for one turn around a storage ring. When the particles return to the reference point, they have traced out trajectories which can be pictured as a dense set of fibers in a rope. Each trajectory is described in detail by two functions, $X(s)$ and $Y(s)$, describing the horizontal and vertical displacements from a design orbit down the center of the beam pipe, as a function of s, the azimuthal distance around the ring. A trajectory is uniquely labeled by four initial coordinates — $X(0)$ and $Y(0)$, the initial displacements, and $X'(0) = (dX/ds)(0)$ and $Y'(0) = (dY/ds)(0)$, the initial transverse angles. If the magnetic fields encountered are all static, then it can be shown that there is one and only one trajectory, the "closed orbit," which exactly repeats itself. That is, if the circumference of the collider is C, then

$$\begin{pmatrix} X(C) \\ X'(C) \\ Y(C) \\ Y'(C) \end{pmatrix}_{co} = \begin{pmatrix} X(0) \\ X'(0) \\ Y(0) \\ Y'(0) \end{pmatrix}_{co}$$

[1]

The closed orbit is exactly down the center of the beam pipe, $X_{co}(s) = Y_{co}(s) = 0$, if all the magnets in the storage ring are perfectly aligned, and if they all have ideal fields. In practice, of course, the closed orbit and the design orbit never quite agree, even in the best of circumstances.

The fibers in the bundle of trajectories are tangled. That is, two trajectories can have the same displacements at some azimuth, and cross. However, if the trajectories are represented in four dimensional phase space as $(X(s),X'(s),Y(s),Y'(s))$, then two trajectories may no longer cross, since if they did then the trajectories would become indistinguishable[23]. In almost all of what follows in the rest of this chapter it is possible to ignore the vertical motion, and to consider only purely horizontal motion. The net effect of this simplification here is that the trajectory bundle is now pictured as a set of (X,X') phase space curves smoothly flowing around the machine, instead of tangled (X,Y) displacement curves.

Linear motion in horizontal phase space is described by simple matrices. For example, the phase space coordinates leaving a field free drift of length L are related to the entering coordinates by

$$\begin{pmatrix} X \\ X' \end{pmatrix}_{out} = \begin{pmatrix} 1 & L \\ 0 & 1 \end{pmatrix} \begin{pmatrix} X \\ X' \end{pmatrix}_{in}$$

[2]

Motion across a dipole — a magnet with uniform vertical bending field — is essentially the same as across a drift of the same length, since the coordinate frame rotates with the design orbit. (Note that all beam particles are implicitly assumed to have the nominal energy, so that dispersion in the dipoles may be ignored.) In the last kind of linear magnet, a quadrupole of strength K, motion is described by

$$X'' + K X = 0 \qquad [3]$$

A quadrupole is analogous to a thin lens in light optics, and the coordinate transformation is well approximated by

$$\left(\begin{array}{c} X \\ X' \end{array} \right)_{out} = \left(\begin{array}{cc} 1 & 0 \\ \frac{1}{f} & 1 \end{array} \right) \left(\begin{array}{c} X \\ X' \end{array} \right)_{in} \qquad [4]$$

if its length is much less than f, the focal length of the quadrupole.

Unfortunately, Maxwells law div(B) = 0 leads inevitably to the conclusion that quadrupoles which focus horizontally also defocus vertically, and vice versa. How, then, can a beam be focussed and constrained in two planes simultaneously? The situation is saved by a well known result from light optics, that the net effect of two equal and opposite strength lenses, placed less than their focal length apart, is to focus. Thus a repetitive sequence of FODO cells — Focussing quadrupole, dipole, Defocusing quadrupole, dipole, ... — leads to a net focussing in both planes. The significance of this result was recognised by accelerator physicists in the late 1950's, and was incorporated in the design of the Alternating Gradient Synchrotron, AGS, at Brookhaven, the first "strong focusing" accelerator [24].

The simplest way to describe linear motion is in terms of "normalised" phase space coordinates, (x,x'), which are related to the "physical" coordinates by the transformation

$$\left(\begin{array}{c} x \\ x' \end{array} \right) = \left(\begin{array}{cc} \frac{1}{\sqrt{\beta(s)}} & 0 \\ \frac{\alpha(s)}{\sqrt{\beta(s)}} & \sqrt{\beta(s)} \end{array} \right) \left(\begin{array}{c} X \\ X' \end{array} \right) \quad , \quad \alpha = -\frac{1}{2} \beta' \qquad [5]$$

In this frame a linear trajectory is generally solved by

$$\left(\begin{array}{c} x(s) \\ x'(s) \end{array} \right) = a_0 \left(\begin{array}{c} \sin(\phi(s) - \phi_0) \\ \cos(\phi(s) - \phi_0) \end{array} \right) \qquad [6]$$

That is, motion from one azimuth to another is described by a simple rotation, around a circle of constant radius. Motion in physical phase space amounts to progression around a tilted ellipse, with the transformation from ellipse to circle given by equation [5]. The betatron function $\beta(s)$ which enables this transformation satisfies the differential equation

$$\sqrt{\beta}'' + K(s) \sqrt{\beta} - \beta^{-3/2} = 0 \qquad [7]$$

with periodic boundary conditions. The betatron phase $\phi(s)$ advances smoothly according to

$$\phi' = \frac{1}{\beta} \qquad [8]$$

So, in a normalised phase space description of linear motion the trajectory fibers form a bundle which is circularly symmetric. All the fibers turn around the center of the bundle at the same rate, but the rate varies with the azimuthal position.

The betatron function was introduced above in the classical way, through a differential equation which implicitly assumes that the user is interested in the trajectory as a function of s, the azimuth. It is simpler, and probably more useful, to introduce the beta function in a difference formalism, which assumes that the user is interested in the displacement of a trajectory at a fixed reference point as a function of t, the integer turn number. This is certainly closer to the experimental setup in E778, in which the displacement of a perturbed beam is measured at two fixed neighboring beam position monitors on tens of thousands of successive turns.

In this perspective linear motion for one turn around a machine is described by multiplying successive drift, dipole, and quadrupole matrices together, in order to get the one turn linear map, T, where

$$
\begin{pmatrix} X \\ X' \end{pmatrix}_{t+1} = T \begin{pmatrix} X \\ X' \end{pmatrix}_t \tag{9}
$$

$$
= \begin{pmatrix} \cos(2\pi Q) + \alpha \sin(2\pi Q) & \beta \sin(2\pi Q) \\ -\dfrac{1 + \alpha^2}{\beta} \sin(2\pi Q) & \cos(2\pi Q) - \alpha \sin(2\pi Q) \end{pmatrix} \begin{pmatrix} X \\ X' \end{pmatrix}_t
$$

In normalised coordinates the form of T is even simpler, corresponding merely to a rotation by $2\pi Q$, so that the difference solution is written as

$$
\begin{pmatrix} x_t \\ x'_t \end{pmatrix} = a_0 \begin{pmatrix} \sin(2\pi Q t - \phi_0) \\ \cos(2\pi Q t - \phi_0) \end{pmatrix} \tag{10}
$$

in small but significant contrast with equation [6]. The betatron tune, Q, given by

$$
2\pi Q = \phi(C) - \phi(0) = \int_0^C \frac{ds}{\beta} \tag{11}
$$

is simply the number of twists the trajectory bundle receives in one turn around the accelerator. The solution to the difference equation of motion, equation [10], is only valid for integer t values. So, the graphical representation of this solution plots one (x_t, x'_t) dot per turn. Such a representation is called a Poincare "surface of section." Usually, when many of these dots have been plotted, they appear to join together to make a continuous contour — a circle in the case at hand, assuming Q is irrational.

What happens to this picture when nonlinear perturbations are included? Although everything below is devoted to anwering this question in some detail, it is possible to answer the question in one brief paragraph. Usually, the circular contour is simply distorted away from a circle. More rarely, the tune Q is perturbed to become a rational fraction, say m/n, resulting in the continuous contour being broken up into n distinct smooth contours. And sometimes, the sequence of dots do not eventually form a regular contour, but appear to be randomly placed within a bounded chaotic region of the surface of section.

RESONANCES

Equations [9] and [10] show that only the fractional part of the tune is important for the purposes of discussions with a fixed reference point, because only trigonometric functions of

Q appear. The integer part is dropped from here on, since it is irrelevant. The next refinement is to recognise that the tune is not constant, but is modified by nonlinearities at finite amplitudes, just as the frequencies of near linear oscillators, such as the gravity pendulum, are modified. For example, in one set of E778 conditions which will be referred to below, the tune is approximately

$$Q = Q_0 + k\, a^2 = 0.418 - 7 \cdot 10^{-4}\, a^2 \qquad [12]$$

where Q_0 is the base, or zero amplitude, tune, and the amplitude of the oscillation, a, is in millimetres. The strength of the detuning coefficient, k, depends on the strength of the nonlinearities in the ring.

This detuning means that the rate of twisting in a trajectory bundle changes smoothly with the distance from the center of the bundle. An interesting thing happens when the fractional part Q is equal, or very close, to a rational fraction. According to equation [12], the tune is equal to 2/5 when the amplitude a_5 is about 5.0 millimetres. There are two independent trajectories near this amplitude which exactly repeat themselves after five turns, just as the zero amplitude closed orbit repeats itself exactly after one turn. In the jargon, each of these new trajectories corresponds to five "period five fixed points" on a Poincare surface of section, all with the about the same amplitude a_5, but with different phases, approximately $\phi = \phi_0 + i\, 2\pi/5$, which are visited in turn. For example, if a trajectory is launched at fixed point $i = 1$, after one accelerator turn it returns to point 3, then 5, 2, 4, and then revisits 1 again.

One of these two new trajectories is stable, and the other is unstable. That is, a trajectory launched very close to one of the stable period five fixed points performs linear oscillations around the fixed point, with an amplitude and phase on turn t which are given by

$$\phi_t = \phi_0 + 2\pi \frac{2}{5} t + \delta\phi\, \cos(2\pi\, Q_I\, t) \qquad [13]$$

$$a_t = a_5 + \delta a\, \sin(2\pi\, Q_I\, t)$$

The small oscillation tune Q_I is called the "island tune." Only a limited range of phases, within $\pm\delta\phi$ of the fixed points, are visited by this trajectory — this is what is meant by resonant behavior. A trajectory bundle with resonances included is like a cable wound rope — the strands in each component cable rotate around the center of the cable, and each cable rotates in turn around the center of the rope. Figure 1 shows surface of sections plots for a set of trajectories with different initial amplitudes and phases, taken from a numerical simulation of the E778 experiment, with the realistic values used above. Five resonance "islands" are clearly visible.

Resonances are not expected to be important under normal operating conditions of the SSC. It might correctly be objected that it is impossible to avoid resonances completely, since the number line is dense in rational fractions, and there are resonances everywhere. Fortunately, it turns out that the strength of a resonance drops very quickly with its order, so that normally only an insignificant fraction of trajectories are resonant. It is only necessary to avoid low order resonances, with denominators of less than 10, say. Even in the absence of resonances, however, it is desirable to minimise phase space distortions of the surface of section contours. For example, the distinctive triangular shape in figure 1 leads to unacceptable SSC operating conditions at large amplitudes, according to the design criteria laid down in the Conceptual Design Report of the SSC[25]. The amount of distortion is quantified by the quantity smear, S, where

$$S = \frac{\left(\langle a_t^2 \rangle - \langle a_t \rangle^2\right)^{1/2}}{\langle a_t \rangle} \qquad [14]$$

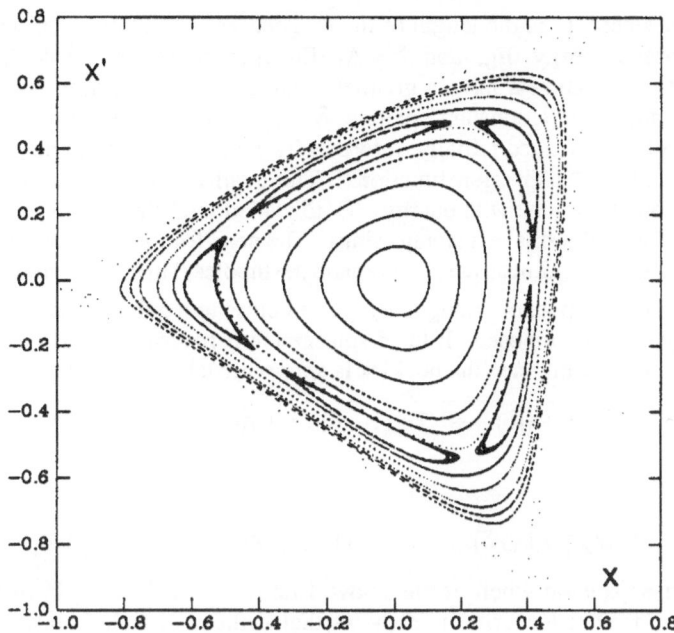

Figure 1. Surface of section plot of several trajectories, from a numerical simulation of the E778 experiment. The value of β is approximately 100 metres, so the five islands at a normalised amplitude of about $0.5\ 10^{-3}\ m^{1/2}$ have a physical amplitude of about 5.0 millimetres.

where angle brackets, <>, denote an average over turn number. That is, the one dimensional smear is the normalised rms variation in amplitude[26].

NONLINEAR SOURCES IN ACCELERATORS

High Order Magnetic Multipoles

The general solution to the Laplace equation for a two dimensional transverse magnetic field is the polynomial

$$B_x + i\,B_y = B \sum_{n=0}^{\infty} (b_n + i\,a_n)(X + i\,Y)^n \qquad [15]$$

where $i = (-1)^{1/2}$, and y is the vertical coordinate. This form is convenient for describing magnets in a separated function accelerator, since then only one of the b_n or a_n is designed to be non-zero. For example, dipoles, quadrupoles, sextupoles and octupoles are described by single b_n values, with $n = 0, 1, 2$, and 3, respectively — a b_n magnet has $2(n + 1)$ poles. Skew magnets, described by a_n values, have no vertical B-field on the horizontal plane. This breaks a design symmetry of most accelerators, and so skew magnets are mainly used for correction purposes.

Returning to a one dimensional analysis again, the effect of a given pure multipole is to deliver a horizontal angular kick

$$\Delta X' = \frac{1}{(1 + \delta)} \frac{(BL)}{(B\rho)} b_n X^n = \frac{1}{(1 + \delta)} k_n X^n \qquad [16]$$

to the trajectory, where L is the length of the magnet (assumed thin), $B\rho$ is the rigidity at the nominal storage energy E_0, and $\delta = \Delta E/E_0$ represents a small deviation from the nominal energy. All magnets have a geometric strength which varies inversely with the energy. A trajectory with a constant positive δ experiences weaker dipoles, and so has a closed orbit which is displaced radially outwards from the center of the magnets by $\eta(s)\,\delta$, where η is called the "dispersion function." The quadrupoles are also weaker, and this leads to a variation of the tune with energy $dQ/d\delta$ called the "chromaticity," which must be compensated in all but the smallest storage rings. The need for this compensation leads to the intentional inclusion of nonlinearities, "chromatic sextupoles," in storage rings.

Consider a thin sextupole of strength k_2 placed close to a quadrupole of strength k_1, so that the two may be superimposed. If the displacement is measured as $Z = X - \eta\delta$, relative to the displaced closed orbit, then the net kick is approximately

$$\Delta X' \approx k_1 (1 - \delta)(Z + \eta\delta) + k_2 (1 - \delta)(Z + \eta\delta)^2$$

or   [17]

$$\Delta Z' \approx [k_1 + (k_2\eta - k_1)\,\delta]\, Z + k_2 [1 - \delta]\, Z^2$$

in a polynomial expansion where terms above first order in δ have been dropped. The coefficient of the first order term in Z shows that if the sextupole is powered with $k_2 = k_1/\eta$, then the net quadrupole effect is constant with respect to first order variations in the energy. If there is a sextupole at every quadrupole, all powered in this way, then the net chromaticity is zero. The price to pay for this correction is the second order term in Z, a deliberate nonlinear perturbation of approximately constant strength for all trajectories. Chromatic sextupoles are the principle source of nonlinearity in most electron storage rings, but not in large superconducting storage rings like the SSC.

Conventional storage rings use "iron dominated" magnets, with fields below the saturation level in iron, about 2 Tesla. The field is shaped by the iron, and excellent field quality is easily guaranteed by stamping the magnet laminations with the right shape — two flat poles for dipoles, four hyperbolic poles for quadrupoles, and so on. The location of the current carrying conductors is of almost no consequence. On the other hand, the field in superconducting magnets is "conductor dominated" — determined almost entirely by the location of the conductors. If the available current density is infinite, the theoretical solution for the current distribution required to create a pure M-pole field is trivial — a circular current shell of $I = I_0 \cos(M\,\theta/2)$, where θ is the angle around the magnet center line. In practice, a significant thickness of superconductor is required to make a dipole field of 6.6 Tesla in the SSC magnets, as shown in Figure 2. It is not possible, even in the ideal design of a two coil layer magnet, to remove all unwanted high order multipole components. This leads to systematic b_n errors in SSC dipoles. It is mechanically much harder to locate conductors accurately than to stamp out magnet laminations, especially when the profile of the cable is not quite uniform, and the magnetic forces are very strong. Manufacturing variances like this lead to random b_n and a_n errors.

Nonetheless, the strongest nonlinear fields in SSC dipoles are due to "persistent currents" on the surface of the superconducting filaments. When Type I superconductors are cooled below their critical temperature, they completely eject any externally imposed magnetic field by generating a compensating surface current. Type II superconductors, such as the Niobium-Titanium commonly used in superconducting magnets, allow partial flux penetration. These persistent currents generate error fields throughout a magnet, with a magnitude which is function, not only of the nominal field strength and distribution, but also of the magnetic history. Persistent current effects are hysteretic.

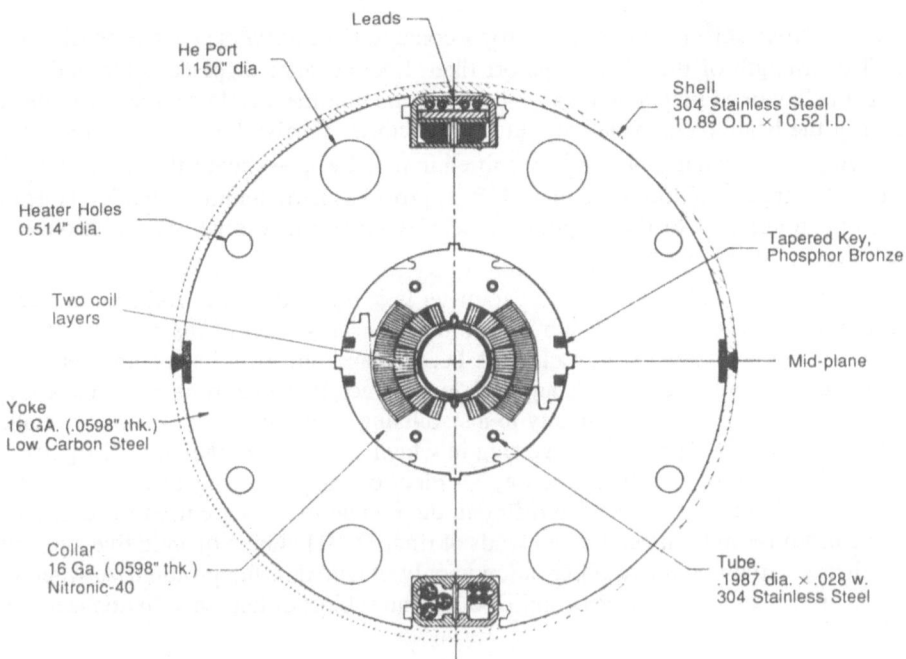

Figure 2. The "cold mass" core of a 6.6 Tesla superconducting SSC dipole. The beam pipe, of 4.0 centimetre inner diameter, is surrounded by a two layer coil which is constrained by a non-magnetic stainless steel collar. The collar, in turn, is constrained by a magnetic steel yoke.

The allowed systematic b_n's of the persistent error field have $n = 2, 4, 6$, et cetera, with the sextupole and the decapole being of most concern — the persistent sextupoles are far stronger than the chromatic sextupoles at the nominal 1 TeV injection energy of the SSC. Fortunately, the perturbation drops rapidly in strength with increasing energy, and is negligible at 20 TeV. Injection energy is the worst time to have field errors present, because then the beam size is largest — trajectories explore more of the bad field region — and the magnetic rigidity is smallest. To make matters worse, Type II persistent currents decay with time. If uncompensated, this decay causes a continuous drift in the chromaticity while beams are being injected, followed by a rapid large jump when the energy ramp is begun. These problems are foremost in discussions which contemplate an increase of the SSC injection energy to 2 TeV.

Beam-Beam

Most contemporary electron and proton storage rings are limited in their performance by the beam-beam effect. Consider a test particle passing through a counter-rotating bunch of particles at a nominal collision point of a storage ring — without a hard collision. The test particle experiences macroscopic electric and magnetic fields which give its trajectory a nonlinear kick. For example, a horizontally displaced proton passing through a round Gaussian bunch of size σ receives an angular kick

$$\Delta X' = -\frac{4\pi\xi}{\beta} \frac{2\sigma^2}{X} \left[1 - \exp\left(\frac{-X^2}{2\sigma^2}\right) \right] \qquad [18]$$

where ξ, the "tune shift parameter," is proportional to the transverse charge density in the bunch. The strength of the kick drops off like $1/X$ at large displacements, unlike the polynomial behavior of magnetic kicks, since now the nonlinear field source is localised at the center of the beam pipe. Small amplitude trajectories receive kicks which are linear in displacement, as in a quadrupole, and are shifted in tune by ξ — hence the name, tune shift parameter. At large amplitudes the tune shift approaches zero, and the situation is usually stable, again in contrast to the magnetic case. Beam-beam resonances are strongest at intermediate amplitudes of a few sigma.

The maximum operational tune shift parameter is of order 0.02 per collision in electron rings, and of order 0.004 per collision in proton rings. This order of magnitude difference is largely due to the difference in transverse beam shape (electron beams are flat, proton beams are round, both are bigaussian) and to the fact that electrons produce a lot of synchrotron radiation, leading indirectly to a stabilising damping of the transverse motion. The SSC will be the first proton storage ring in which synchrotron radiation is significant, with a damping time of less than one day — electron ring damping times are typically measured in milliseconds. Somewhat different theoretical models are used to successfully explain the beam-beam limits in the two kinds of ring[27–31]. Good quantitative agreement between theory, simulation, and observation is only obtained in the proton case when tune modulation affects are taken into account[32–36]. The subject of tune modulation is returned to below.

Contemporary colliders store only a few — less than ten — bunches per beam, with particle and antiparticle beams counter-rotating in the same vacuum chamber. However, the frequency of collision decreases as the machines get larger, leading to a decrease in luminosity, unless there is an increase in the number of bunches or the charge per bunch. Both of these solutions come into violent conflict with the beam-beam limit, loosely defined as the maximum allowable tune shift per turn (not per collision). The SSC resolves this dilemma, and the associated problem of producing copious numbers of antiprotons, by filling two vertically separated storage rings with thousands of bunches of protons, longitudinally spaced by about 5 metres. Collisions between counter-rotating bunches are only allowed where they are useful. Consequently, the beam-beam effect is not expected to be critical in the SSC, although its presence will be noticed.

Radio Frequency Cavities

So far, the longitudinal motion of a test particle relative to the center of its own bunch has been ignored. Only transverse motion has been considered, although sometimes the test particle has had a constant off energy parameter δ, and a displaced closed orbit. For example, a closed orbit trajectory with a large δ of 10^{-3} at a place with a typical dispersion function η of 3 metres is displaced outwards by 3 millimetres, and its single turn path length is about 2 centimetres longer than the design orbit. At the end of each turn a particle following this trajectory lags farther and farther behind the center of its bunch — if the speed of the particle is independent of its energy, a reasonable assumption in the relativistic limit. What, then, keeps a bunch of particles together? The answer is, a small number of short radio frequency cavities, each applying a longitudinal voltage which depends on the test particles longitudinal displacement from the center of its bunch. For the sake of simplicity, suppose that there is only one cavity, with a typical wavelength of about one metre. A nominal particle passes through this cavity when the field is zero, but a particle that arrives late loses energy, and an early particle gains energy.

So, the energy displacement δ is not constant, but oscillates, with a typical period of hundreds of turns in proton storage rings, and tens of turns in electron storage rings. As is shown below, this situation appears at first sight to be analogous to the simple gravity pendulum. However, there is a crucial difference — the radio frequency restoring force is not applied continuously, like gravity, but is only applied as an impulse, once per turn of the accelerator. That is, the gravity pendulum is a differential system, while longitudinal motion in an accelerator is a difference system. Difference systems like this which are analogous to the gravity pendulum are described by the "standard map," which is so named because of its universal importance and frequent occurrence in many different nonlinear manifestations.

Even though nonlinear longitudinal motion does not limit the performance of any accelerator, the standard map is pedagogically well worth studying. Most resonant situations, including the most complex, can be reduced to a standard map by appropriate coordinate transformations (at least in principle). Equivalently, the standard map demonstrates many of the properties of more complex situations, such as chaos, the change of tune with amplitude, and the useful limits of a Hamiltonian description. For these reasons, longitudinal motion is the first nonlinear topic discussed here in detail.

LONGITUDINAL MOTION — THE STANDARD MAP

Suppose that a test particle circulates around a storage ring containing one radio frequency cavity. If the azimuthal reference point at which a Poincare surface of section is to be constructed is just before the cavity, then one turn consists of i) passage through the cavity, followed by ii) traversal of the rest of the machine. Although the RF cavity is typically several wavelengths long, it is reasonable to approximate it as an infinitesimally short impulse by integrating the electric field that the particle experiences into a single voltage. So, if the particle trails behind the center of its bunch by a positive distance of Δs when it passes through the cavity, its off energy parameter on turn $t+1$ is related to that on turn t by

$$\delta_{t+1} = \delta_t - \frac{V_{RF}}{E_0} \sin(\theta_t) \qquad [19]$$

where the RF phase angle

$$\theta_t = 2\pi \frac{\Delta s_t}{\lambda_{RF}} \qquad [20]$$

is a natural longitudinal coordinate. The total path length during one turn varies with the energy according to

$$C = C_0 + 2\pi <\eta> \delta \qquad [21]$$

where $<\eta>$ is the average dispersion function in the bending dipoles. The additional term modifies Δs and θ,

$$\theta_{t+1} = \theta_t + \frac{(2\pi)^2 <\eta>}{\lambda_{RF}} \delta_{t+1} \qquad [22]$$

Equations [20] and [22], taken together, constitute the one turn map for longitudinal motion. Notice that the right hand side of [22] includes terms with both subscripts t and $t+1$.

It is now convenient to make a coordinate transformation, whose physical meaning will soon become apparent. Replacing θ, δ and the physical parameters in [20] and [22] with the quantities

$$q = \theta, \quad p = 2\pi \left(\frac{<\eta>}{\lambda_{RF}} \frac{E_0}{V_{RF}} \right)^{1/2} \delta, \quad \Delta T = 2\pi \left(\frac{<\eta>}{\lambda_{RF}} \frac{V_{RF}}{E_0} \right)^{1/2} \qquad [23]$$

the map becomes

327

$$p_{t+1} = p_t - \Delta T \sin(q_t) \tag{24}$$

$$q_{t+1} = q_t + \Delta T \, p_{t+1}$$

This is the standard map. If q is always small, it has an approximate linear solution of

$$q_t = q_0 \cos(2\pi \, Q_S \, t) \tag{25}$$

where the small oscillation tune, Q_S, is called the synchrotron tune in the particular case of longitudinal oscillations. It is given by

$$\cos(2\pi \, Q_S) = 1 - \frac{\Delta T^2}{2} \tag{26}$$

which shows that even small amplitude motion is unstable under the standard map if ΔT is greater than 2 . In electron storage rings Q_S is typically between 0.01 and 0.1, while in proton storage rings it is typically an order of magnitude smaller.

The physical meaning of the coordinates q and p, and the parameter ΔT, is made clear by taking the standard map to be the numerical representation of a rigid pendulum of unit length, with the acceleration due to gravity set equal to one. In this case q corresponds to the angle the pendulum makes with vertical, p corresponds to the angular velocity of the pendulum dq/dt, and ΔT corresponds to the integration time step size. Since the continuous time T is given in terms of the discrete time t by

$$T = t \, \Delta T \tag{27}$$

and since $2\pi \, Q_S = \Delta T$ for small time steps, the small angle motion of the pendulum is simply

$$q(T) = q_0 \cos(T) \tag{28}$$

It is not surprising that the time step must be much less than the natural period of the system — much less than one — for such a discrete representation of a differential system to be accurate. What is surprising, perhaps, is that the dynamics of analogous differential and difference systems are qualitatively different.

The most compact way to describe the differential pendulum system is by means of a Hamiltonian,

$$H = \frac{1}{2} p^2 - \cos(q) \tag{29}$$

which, by definition, is shorthand for the equations of motion

$$\frac{dq}{dT} = \frac{\partial H}{\partial p} \tag{30}$$

$$\frac{dp}{dT} = -\frac{\partial H}{\partial q}$$

Trajectories of the pendulum system follow contours of the Hamiltonian function, because H is explicitly conserved, since

$$\frac{dH}{dT} = \frac{\partial H}{\partial p}\frac{dp}{dT} + \frac{\partial H}{\partial q}\frac{dq}{dT} = 0 \tag{31}$$

by substitution of the equations of motion[30]. The rate of progress along a contour depends only on the local slope of the Hamiltonian function. These two properties make it easy to picture the behavior of a one dimensional system, if only a Hamiltonian can be constructed from the equations of motion.

Figure 3a shows the contours of the pendulum Hamiltonian. Equivalently, it shows the Poincare surface of section of the longitudinal motion of particles in a storage ring, in the limit that Q_S goes to zero. A trajectory near the center of the plot exhibits stable, limited, oscillations — the pendulum has a maximum absolute angle, or, equivalently, the particle is trapped inside a single RF "bucket." A trajectory near the top or bottom will eventually reach all values of the coordinate q — the pendulum is rotating continuously, or the particle is not associated with any particular bunch. The boundary between these qualitatively different kinds of motion, trapped and untrapped, is called the "separatrix." It takes an infinite amount of time to move once around the separatrix, since the slope of H is zero at the "unstable fixed points" where the pendulum is inverted and motionless. Equivalently, the longitudinal tune of a particle shifts from Q_S at the centre of the RF bucket, to zero at its edges.

From the pendulum point of view, the standard map is merely an approximation of the differential equations of motion, via

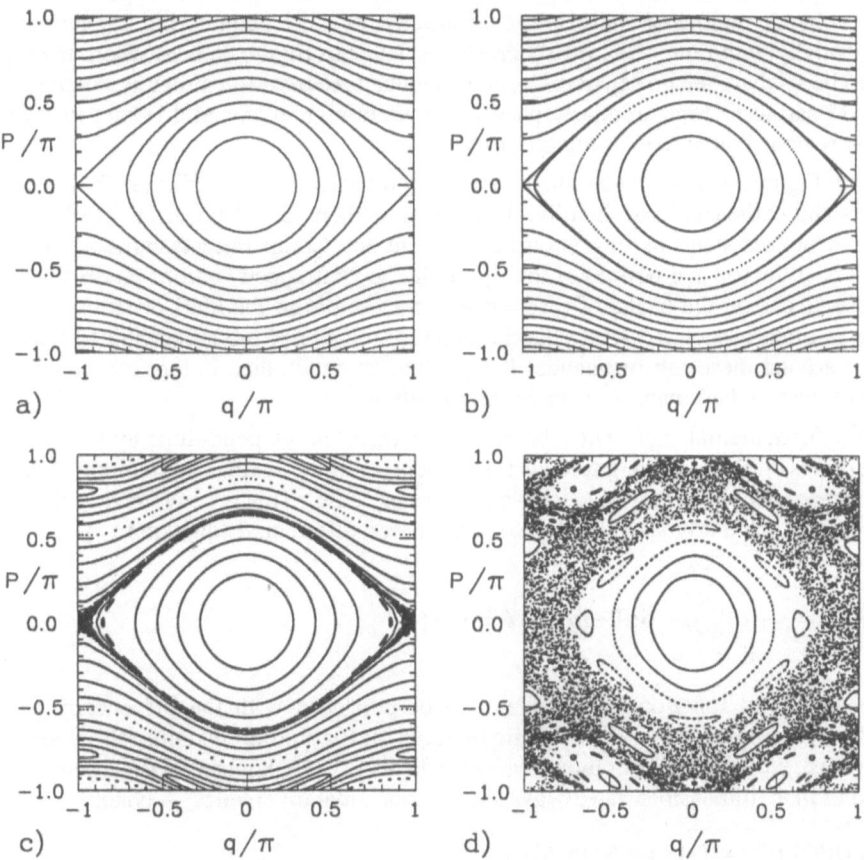

Figure 3. Standard map trajectories, with four different synchrotron tunes, Q_S, corresponding to different integration time steps, ΔT . a) Contours of the Hamiltonian $H = (P^2/2) - \cos(q)$, representing the differential pendulum. b) $Q_S = 0.06$. Almost indistinguishable from a), with no sign of chaos, even close to the "separatrix." c) $Q_S = 0.12$. A narrow chaotic region appears near the "separatrix," and some secondary islands appear. d) $Q_S = 0.18$. Most of phase space is chaotic, surrounding complex island structures.

$$\Delta q = \frac{\partial H}{\partial p} \Delta T \qquad\qquad\qquad [32]$$

$$\Delta p = -\frac{\partial H}{\partial q} \Delta T$$

However, from the RF cavity point of view, the Hamiltonian representation is merely an approximation of the difference equations of motion. It depends upon the physical case in hand whether a continuous or discrete representation is more appropriate. Figure 3b shows how trajectories with a small value of $Q_S = 0.06$ respond to the standard map almost exactly as if the system was continuous. (To keep the plot symmetric, the azimuthal reference point has been moved to the center of the RF cavity, instead of just before it). One dot is drawn per iteration of the map, for many iterations of several different trajectories. Except for one trajectory, these dots appear to form continuous lines — "KAM surfaces"[37–40] — looking like the contours of the continuous Hamiltonian.

In a region of phase space which is regular, where trajectories form KAM contours, two infinitesimally close neighboring trajectories diverge linearly with time. When Q_S is increased to 0.12 and 0.18, in Figures 3c and 3d, chaotic trajectories appear with scattered dots. In chaotic regions of phase space, infinitesimally close neighboring trajectories diverge exponentially with time. Chaos first becomes visible in Figure 3c near where the separatrix used to be — there are no separatrices in difference systems — in a region which is bounded by KAM surfaces. Most of phase space is chaotic in Figure 3d, and it is hard to say whether chaotic regions bound regular regions, or vice versa.

Both Figures 3c and 3d also show secondary resonance island structure in addition to the main island at the center of the plot. For example, there is a chain of 16 small islands near the (now non-existent) main separatrix in Figure 3c. Comparatively large islands are also visible near the top and bottom of the figure, in the "untrapped" part of phase space. These are resonances on the backs of resonances, an example of the kind of recursive structure which is often associated with chaotic behavior. It is not too surprising to learn that, if motion around these sub-resonances is examined in detail, then it, too, can be described in terms of the standard map. And so on, ad infinitum.

The fundamental difference between the differential pendulum and the difference pendulum is that the restoring force is time independent in the first "autonomous" case, and is time independent in the second, "non-autonomous" case. This is conveniently illustrated by rewriting the standard map as a single, second order, differential equation in q

$$\frac{d^2q}{dT^2} = -\sum_{n=-\infty}^{\infty} \delta(T-n\Delta T)\,\Delta T\,\sin(q) \qquad\qquad [33]$$

where the delta function $\delta(\ )$ is not to be confused with the off energy parameter. Neighboring trajectories in one dimensional autonomous systems show only linear divergence, while non-autonomous systems can also show exponential divergence. Systems of two or more dimensions can always show exponential divergence — chaos.

SEXTUPOLES — THE HENON MAP

One of the earliest dynamicists to attempt a general numerical study of nonlinear maps was Henon, an astrophysicist[41]. He found that the map which now bears his name "exhibits all the typical properties of more complicated mappings and dynamical systems." This one-dimensional map is directly relevant to accelerator physics, as it describes an accelerator in which there is a single nonlinearity, a thin sextupole, of unit strength. In normalised coordinates the motion around the linear part of the machine amounts simply to a coordinate rotation, so if the reference point for the Poincare surface of section is just before the sextupole, then the map from turn t to turn t+1 is just

$$\begin{pmatrix} x_{t+1} \\ x'_{t+1} \end{pmatrix} = \begin{pmatrix} C & S \\ -S & C \end{pmatrix} \begin{pmatrix} x_t \\ x'_t + x_t^2 \end{pmatrix} \qquad [34]$$

where $C = \cos(2\pi Q_0)$ and $S = \sin(2\pi Q_0)$, in which Q_0 is the tune of a small amplitude trajectory. Figures 4a through 4d are taken almost directly from a paper by Henon, showing surface of section plots of several trajectories for four values of the control parameter Q_0, near to 1/3, 1/4, 1/5 and 1/6 .

Four different kinds of trajectories can be loosely distinguished. <u>Regular non-resonant</u> trajectories are found close to the origin of each of the figures. The trajectories are regular, but become distorted away from circles at moderate and large amplitudes. As discussed both above and below, the deviation from circularity is conveniently measured by the smear.

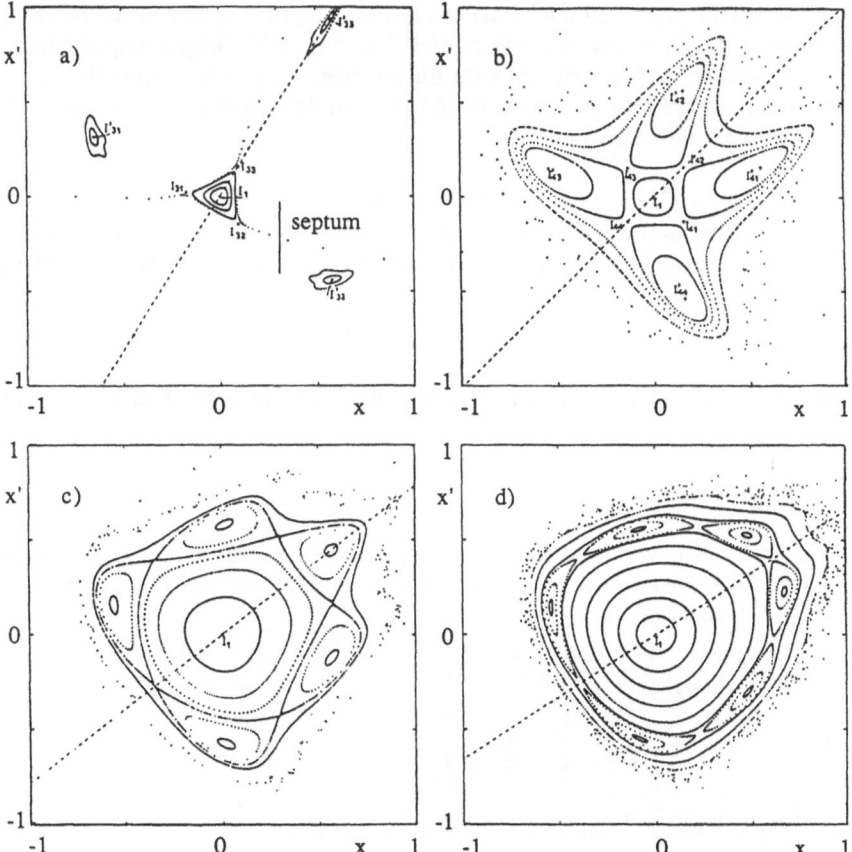

Figure 4. Trajectories obtained by Henon[41] from iterating his map with different base tunes, Q_0 . When $Q_0 \approx 1/n$, n resonance islands appear. a) $Q_0 = 0.324 \approx 1/3$. The stable triangle and the divergent arms are well described by first order theory, but the outlying islands are not. b) $Q_0 = 0.2516 \approx 1/4$. Four big islands at a small amplitude, with Q_0 very close to 1/4. c) $Q_0 = 0.211 \approx 1/5$. Five islands, surrounded by a stable KAM contour, and then chaos. d) $Q_0 = 0.185 \approx 1/6$. The six islands are almost rotationally symmetric—they resemble each other and the standard map structure.

Under normal stable operation of a storage ring the beam fills only the small smear region. The deviations from circularity increase as the amplitude gets larger, until the motion breaks up into <u>regular resonant</u> trajectories, forming a chain of resonance islands. The number of islands corresponds to the denominator of the rational fraction nearest to Q_0. <u>Chaotic trajectories</u> occur at the largest amplitudes in the figures (except in Figure 4a where chaotic points have been removed for the sake of clarity). While some of these trajectories are bounded, as in the case of the standard map, some appear to diverge to infinity. This is because the x^2 nonlinear term in the Henon map is unbounded, unlike the $\sin(q)$ term in the standard map.

<u>Rapidly divergent regular</u> trajectories can be seen in figure 4a, in the form of three arms of widely spaced dots whose amplitude increases rapidly from turn to turn. This behavior is very useful in the controlled extraction of particles from a storage ring, as illustrated by the inclusion of a "septum" in the figure. The septum is a current or charge carrying metallic membrane, arranged so that there is no magnetic or electric field on the inside, while on the outside a particle is deflected into an extraction line. For the extraction efficiency to be high, the septum must be thin compared to the amplitude increase in three successive turns. More and more particles are squeezed out of the stable triangle at the center of the beam pipe by gradually moving the base tune Q_0 closer and closer to 1/3. Despite appearances, these trajectories are really only regular resonant trajectories, since, given enough time and an enormous vacuum chamber, a particle following one of these trajectories would eventually return to the small amplitude region.

Smear, and the First Order Discrete Hamiltonian, H_1

It is relatively straightforward to solve the equations of motion for the distortions which perturb the circular trajectories near the center of the plots — at least to first order in the sextupole strength g, where

$$\Delta x' = gx^2 \qquad [35]$$

First, though, it is convenient to introduce "action-angle" coordinates, J and ϕ, where

$$\begin{pmatrix} x \\ x' \end{pmatrix} = \begin{pmatrix} (2J)^{1/2} \sin(\phi) \\ (2J)^{1/2} \cos(\phi) \end{pmatrix} \qquad [36]$$

That is, the action J behaves very much like the betatron amplitude, while ϕ is explicitly the betatron phase of the trajectory under study. It is easy to show that the motion from turn t to $t+1$ is described to first order in g by

$$\begin{pmatrix} J \\ \phi \end{pmatrix}_{t+1} = \begin{pmatrix} J \\ \phi \end{pmatrix}_t + \begin{pmatrix} -\dfrac{\partial H_1}{\partial \phi} \\[2mm] \dfrac{\partial H_1}{\partial J} \end{pmatrix}_t \qquad [37]$$

where the one turn "discrete" Hamiltonian H_1 is given by

$$H_1 = 2\pi\, Q_0 J + \frac{g}{3\sqrt{2}}\, J^{3/2}\, [\,\sin 3(\psi + \phi) - 3 \sin (\psi + \phi)\,] \qquad [38]$$

in which ψ is the constant phase of the single sextupole, relative to the reference point. The first term in this discrete Hamiltonian corresponds to the linear phase advance of $2\pi Q_0$, as expected.

Not one, but sixteen sextupoles dominate the nonlinear behavior of the Tevatron in the E778 experiment. More generally, then, the Hamiltonian is written as

$$H_1 \;\; = \;\; 2\pi\, Q_0\, J \;+\; \sum_{\{ik\}} V_{ik}\, J^{i/2}\, \sin(k\phi + \phi_{ik}) \qquad\qquad [39]$$

where the sum is over ik pairs

$$\{ik\} \;\; = \;\; \{33, 31\} \qquad\qquad [40]$$

The constants V_{ik} and ϕ_{ik} are obtained from a vector sum of the terms proportional to g in equation [38], over all sextupoles. Equation [40] is not exactly correct, but only describes the motion correctly to first order in sextupole strength, since higher order terms have been ignored. This first order result for H_1, and the results which follow, are easily generalised further. For example, if octupole nonlinearities are also present, the set $\{ik\}$ is extended to become $\{30,31,44,42,40\}$. Or, if two dimensional motion in the presence of sextupoles is to be described, it is expanded to $\{ijkl\} = \{3030,3010,1210,1212,121\text{-}2\}$, where $j/2$ is the exponent of the vertical action, and l is the coefficient of the vertical betatron phase. What is hard is to extend the description to higher order in nonlinear strengths.

The action is a smooth function of the phase, $J(\phi)$, if the motion is regular and non-resonant. Substituting the lowest order solution of phase motion

$$\phi_t \;\; = \;\; 2\pi\, Q_0\, t \;+\; \phi_0 \qquad\qquad [41]$$

into the difference equation of motion for J in [37] gives

$$J(\phi + 2\pi\, Q_0) \;-\; J(\phi) \;\; = \;\; -\,\frac{\partial H_1}{\partial \phi} \qquad\qquad [42]$$

Using the Hamiltonian in equation [39] gives, for perturbations small compared to J_0,

$$J(\phi) \;\; = \;\; J_0 \;-\; \sum_{\{ik\}} \frac{k\, V_{ik}}{2\,\sin(k\pi Q_0)}\, J_0^{i/2}\, \sin(k\phi + \phi_{ik} + k\pi Q_0) \qquad\qquad [43]$$

A resonance denominator appears here, for the first time — if the base tune Q_0 is near an integer or an integer divided by three, then one of the terms in the sum becomes large and can destroy the original assumption that the perturbation is small. The k=3 term causes the characteristic triangular shape seen in Figure 4. In order to describe the 4, 5, or 6 fold structure that leads up to the resonance islands in Figure 4, it is clearly necessary to include higher order terms in the discrete Hamiltonian.

Substituting the phase motion given in [41] into [43] gives the action as a function of turn number,

$$J_t \;\; = \;\; <J> \;-\; \sum_{\{ik\}} \frac{k\, V_{ik}}{2\,\sin(k\pi Q_0)}\, J_0^{i/2}\, \sin(2\pi\, k\, Q_0 t + \phi_{0ik}) \qquad\qquad [44]$$

where $<J> = J_0$ is the average action, and $\phi_{0ik} = k\phi_0 + \phi_{ik} + k\pi Q_0$ is a constant phase. In terms of amplitude rather than action, the motion is

$$a_t \;\; = \;\; <a> \;-\; \sum_{\{ik\}} \frac{k\, V_{ik}}{2^{i/2+1}\,\sin(k\pi Q_0)}\, a_0^{i-1}\, \sin(2\pi\, k\, Q_0 t + \phi_{0ik}) \qquad\qquad [45]$$

According to the definition given in equation [14], the one dimensional smear is

$$S \;\; = \;\; <a>\, \sqrt{\frac{3^2\, V_{33}^2}{2^6\,\sin^2(3\pi Q_0)} \;+\; \frac{V_{31}^2}{2^6\,\sin^2(\pi Q_0)}} \qquad\qquad [46]$$

333

showing that the smear due to sextupoles increases linearly with amplitude, for small amplitudes. The displacement $x(t)$ at the reference point, which can be detected by a beam position monitor (BPM), is

$$x_t = a_t \sin(2\pi Q_0 t + \phi_0) \qquad [47]$$

$$= <a> \sin(2\pi Q_0 t + \phi_0)$$

$$- \sum_{\{ik\}} \frac{k V_{ik}}{2^{i/2+2} \sin(k\pi Q_0)} a_0^{i-1} \Big[\cos(2\pi(k-1)Q_0 t + \phi_{0ik} - \phi_0)$$

$$- \cos(2\pi(k+1)Q_0 t + \phi_{0ik} + \phi_0) \Big] \qquad [48]$$

Since $k = 1$ or 3 for sextupoles, Fourier analysis of a turn-by-turn BPM signal reveals harmonics at $2Q_0$ and $4Q_0$, in addition to the fundamental signal.

Experimental observation of smear

Two analyses are readily available for measuring smear from turn-by-turn position data, corresponding to treatment in the frequency domain and in the time domain. While E778 has so far concentrated on time domain measurements of one-dimensional motion, frequency domain measurements will be essential in the imminent studies of two dimensional oscillations. Both analysis techniques are complicated by the finite size of the beam, as will be seen.

The basic experiment is very simple — kick the beam horizontally on turn 0, inducing an oscillation of amplitude a_{kick}, and observe the ensusing oscillations for at least a hundred turns on two neighboring BPMs. If the two signals on a given turn t are $x_1(t)$ and $x_2(t)$, then the amplitude a_t on that turn is given by

$$a_t^2 = c_{11} x_1^2 + c_{12} x_1 x_2 + c_{22} x_2^2 \qquad [49]$$

where the coefficients $c_{11}, c_{12},$ and c_{12} depend on the beta values at the two BPMs, β_1 and β_2, and on the betatron phase advance $\Delta\phi_{12}$ between them. For example, if $\beta_1=\beta_2$ and $\Delta\phi_{12} = 90$ degrees, then $c_{11} = c_{22} = 1$, and $c_{12} = 0$. Having established the time sequence a_t for a sufficient number of turns, the smear is obtained directly from equation [14]. Practical problems associated with non-zero closed orbit offsets, and β and ϕ errors, are easily overcome.

The data taken and processed in this way in Figure 5 show that, instead of the amplitude being approximately constant (within smear variations), there is an initial gaussian decay of the signal. This decay is due to the finite size of the beam, which implies a distribution of initial amplitudes in a typical range $a_{kick} \pm \sigma$, where σ is the gaussian beam size. The spread in amplitudes leads to a spread in tunes across the beam, of size $\Delta Q = \sigma(dQ/da)|_{a_{kick}}$, causing the signal to decohere with a gaussian time constant of $1/\Delta Q$ turns. It is straightforward to compensate for the decoherence in calculating the smear during one dimensional motion.

The equivalent frequency domain measurement consists of Fourier analysing the signal from either one or both of the BPMs, as described theoretically in [47], and then reconstructing the values of V_{33} and V_{31}, ready for substitution into [46] for evaluation of the smear. Using only one BPM leads to some problems in reconstructing the V values, since the response at $2Q_0$, for example, depends on $V_{33}, V_{31}, \phi_{033},$ and ϕ_{031}. Information from two BPMs is needed to derive the phases ϕ_{033}, and ϕ_{031}, or to construct the amplitude time series for subsequent Fourier analysis. Finite beam size also causes

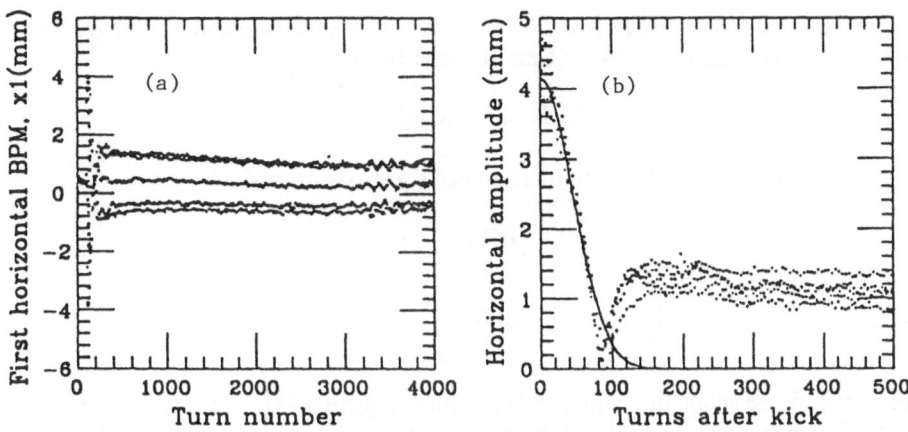

Figure 5. Typical data from the E778 experiment, showing both Gaussian decoherence and a persistent signal. a) Raw turn-by-turn data taken by one of the beam position monitors, over 4,000 turns. The signal strength initially drops very rapidly after a transverse deflection of about 4 millimetres, finally leaving five persistent signal lines with a very slow decay rate. b) The reconstructed amplitude over the first 500 turns, showing that the initial decoherence is well fitted by the solid line Gaussian. About 30% of the beam is trapped in a resonance island at an amplitude of about 4 millimetres.

problems in the frequency domain, by broadening peaks which would otherwise be sharp. So long as the peaks do not overlap, it is relatively straightforward to dreive the single particle equivalent amplitudes and phases.

Five Islands — the Single Resonance Hamiltonian, H_5

The simple solution to the equations of motion given above, [41] and [43], breaks down when the motion is resonant. or nearly resonant. For example, suppose that the coefficient set $\{ik\}$ in equation [43] is somehow extended to include $k=5$ (and hence $i \geq 5$) in trying to describe the phase space distortions close to the five-fold island structure seen in Figure 4c. As the tune Q_0 approaches $1/5$, the shrinking resonance denominator $\sin(5\pi Q_0)$ eventually leads to a violation of the original assumption of small distortions — and predicts infinite distortions when $Q_0=1/5$, on resonance. The source of this error lies in assuming the simple phase advance per turn given by equation [43]. When a trajectory is trapped in a resonance island, the solution is better given by an expression like [13]. The trapped and untrapped approximate solutions for the phase illustrate the topological difference between resonant and non-resonant motion. Given enough time, the phase of a non-resonant trajectory will come arbitrarily close to any given phase. In contrast, the phase of a resonant trajectory has only a limited range of possible values.

There are two important experimental questions to ask about a resonance. First, how wide is it? That is, what amplitude or action range does it span. Second, what is the tune Q_I at the center of the islands? These questions are answered theoretically by concentrating on a single resonance denominator, say five, and by developing a Hamiltonian description of the five-turn motion. That is, whereas so far t has been implicitly integer, soon t will be an integer which is exactly divisible by five. As a starting point, assume that the tune Q_0 is close to $2/5$, as in the resonance investigations of the E778 experiment, and assume that the one turn Hamiltonian H_1 has somehow been developed to include a complete set of coefficients with $i,k \leq 5$, specifically $\{ik\} = \{33,31,44,42,40,55,53,51\}$.

It can be shown that only terms with k=0 and k=5 survive when H_1 is averaged over 5 turns, so that the "single resonance Hamiltonian" becomes

$$H_5 \;\; = \;\; 2\pi\left(Q_0 - \frac{2}{5}\right)J \;\; + \;\; V_{40}J^2 \;\; + \;\; V_{55}\,J^{5/2}\sin(5\phi + \phi_{55}) \qquad [50]$$

This is just shorthand for the five-turn difference equations of motion

$$\begin{pmatrix} J \\ \phi \end{pmatrix}_{t+5} \;\; = \;\; \begin{pmatrix} J \\ \phi \end{pmatrix}_t \;\; + \;\; 5\begin{pmatrix} -\dfrac{\partial H_5}{\partial \phi} \\[2mm] \dfrac{\partial H_5}{\partial J} \end{pmatrix}_t \qquad [51]$$

by analogy with the single turn equations of motion, equation [37]. The meaning of the three terms in H_5 becomes clear when the partial differentiations in [51] are performed. For example, the first term corresponds to a five turn phase advance of $5 \cdot 2\pi\,[Q_0 - (2/5)]$, independent of the action. Subtraction of 2/5 from Q_0 is justified by noting that it leads to an inconsequential subtraction of 4π from the five turn phase advance. The subtraction is motivated by making the coefficient of J a small number. Next, differentiation of $V_{40}\,J^2$ with respect to J leads to a five turn phase advance of $10\,V_{40}\,J$, linearly proportional to the action.

Temporarily ignoring the third term, there is an octupolar tune shift with action or amplitude, given by

$$Q(J) \;\; = \;\; Q_0 \;\; + \;\; \frac{V_{40}}{\pi}\,J \;\; = \;\; Q_0 \;\; + \;\; \frac{V_{40}}{2\pi}\,a^2 \qquad [52]$$

The action J_I at which $Q(J_I) = 2/5$ identifies where the resonance is found. Before examining the behavior of the term in V_{55}, it is convenient to make a coordinate transformation and rewrite H_5 as an expansion around J_I,

$$H_5 \;\; = \;\; \frac{1}{2}U\,I^2 \;\; - \;\; V\cos(5\phi) \qquad [53]$$

where

$$I \;\; = \;\; J - J_I, \quad U \;\; = \;\; 2\,V_{40}, \quad V \;\; = \;\; V_{55}\,J_I^{5/2} \qquad [54]$$

and the value of ϕ_{55} has been conveniently chosen.

Substitution of this Hamiltonian into the equations of motion [51] (with J replaced by I) shows that $(I,\phi) = (0,0)$ is a fixed point — a trajectory launched there is stationary. This is in marked contrast with the usual single turn motion, in which a trajectory always advances by a large phase of about $2\pi Q_0$, even in the absence of nonlinearities. In some region close enough to $I = 0$, then, H_5 may be considered as representing differential equations of motion, continuous in t, which agree well with the difference motion whenever t is an integer multiple of five. In this approximation

$$\begin{pmatrix} \dfrac{dI}{dt} \\[3mm] \dfrac{d\phi}{dt} \end{pmatrix} \;\; = \;\; \begin{pmatrix} -\dfrac{\partial H_5}{\partial \phi} \\[2mm] \dfrac{\partial H_5}{\partial I} \end{pmatrix} \;\; = \;\; \begin{pmatrix} -\,5V\sin(5\phi) \\[2mm] U\,I \end{pmatrix} \qquad [55]$$

which, except for factors of 5, is the familiar case of the pendulum. For small angles, $\delta_\phi \ll 1/5$, the solution of these equations is just

$$\binom{I}{\phi} = \left(\begin{array}{c} 5 \left(\dfrac{V}{U}\right)^{1/2} \sin(2\pi\, Q_I t) \\[2ex] \cos(2\pi\, Q_I t) \end{array} \right) \qquad [56]$$

where it may be assumed that V and U are both positive. The island tune is given by

$$Q_I = \frac{5}{2\pi} (U\,V)^{1/2} \qquad [57]$$

This answers the second of the two key questions about the resonance. Now return to the first question — what is the resonance width?

The approximate representation of the motion by differential equations of motion is valid "close enough" to the center of the islands, and for island tunes Q_I much less than one. A trajectory in this region follows contours of constant H_5 very closely. The shape of the Hamiltonian hillside is a parabolic valley along the I-axis, with a modulation along the ϕ-axis caused by the $\cos(5\phi)$ term which leads to five local minima separated by five saddle points, corresponding to five stable and five unstable fixed points.

The amplitude width of the islands is estimated by assuming that trajectories at least as far as the separatrix follow H_5 contours. (This is explicitly wrong very close to the separatrix, which does not even exist in the difference system.) Since trajectories follow contours of H_5, and since the saddle point (unstable fixed point) is on the boundary between resonant and non-resonant motion, the height of the saddle point, $H_5(0, 2\pi/10)$, is the same as the height $H_5(I_W, 0)$, where I_W is the island half width. This gives

$$I_W = 2\left(\frac{V}{U}\right)^{1/2} \qquad [58]$$

This is readily converted to an amplitude width by dividing by a_I, the resonance amplitude.

Experimental Resonance Observation

In an experiment, the resonance amplitude a_I is adjusted by changing Q_0, so long as it remains inside the dynamic aperture. This has a strong effect on both Q_I and a_W, especially for high order resonances, since Q_I goes like $a_I^{n/2}$, and a_W goes like $a_I^{(n-2)/2}$, where n is the order of the resonance. At first sight measurements of resonances appear to be overconstrained, since there are two parameters in the theoretical model, U and V, while there are three experimental observables, d^2Q/da^2, Q_I and a_W, which are related to each other by equations [52], [57], and [58]. This would provide a stringent test of the model. Unfortunately, life is not that simple, again because of the finite beam size. In practice, d^2Q/da^2 is easily measured to about 10%, but the determination of Q_I and a_W to this accuracy is more difficult.

Figure 5 illustrates typical data obtained by kicking the gaussian beam into a phase space position which partially overlaps fifth order islands. At first the signal undergoes the usual gaussian decoherence. However, there is also a "persistent signal," which has a very small decay rate — it is typically observed for tens of seconds, or millions of turns. This signal is due to particles which do not decohere because they are phase locked within the bounds of a resonance island. If the base tune Q_0 is adjusted to maximise the persistent signal strength, when $a_{kick} \approx a_I$, the persistent amplitude leads directly to the resonance width a_W, through

$$\frac{a_{persistent}}{a_{kick}} = G\,\frac{a_W}{\sigma} \qquad [59]$$

where G is a geometrical factor close to unity which is calculated by numerical simulation[2,3,9]. The beam size σ is assumed to be well known, although in practice it fluctuates from shot to shot. Once measurements of a_W have been made at several values of a_{kick}, the set of data pairs (Q_0, a_{kick}) are analysed to yield an accurate plot of tune versus amplitude.

Measurement of Q_I is not so straightforward. If the beam size is much smaller than the island size, then Fourier analysis of the time series $\{\phi_t - 2\pi (2/5) t\}$ leads to a sharp peak at Q_I, if the phase amplitude $\delta\phi$ in [13] is small. If the phase amplitude is large, then several peaks are seen, at harmonic multiples of a value of a fundamental tune which is smaller than the value Q_I at the center of the island. In practice the beam size is relatively large and Fourier analysis reveals a broad spectrum, due to the spread in Q_I. A better way to measure Q_I experimentally, independent of beam size, is by observing the response of a persistent signal to tune modulation.

TUNE MODULATION

If a set of quadrupoles is perturbed by a small sinusoidal current, the tune of a small amplitude trajectory is modulated according to

$$Q_0 = Q_{00} + q \sin(2\pi Q_M t) \tag{60}$$

where q and Q_M are the tune modulation amplitude and tune. Power supply ripple like this is normally carefully avoided, especially in proton colliders, where any source of noise degrades the storage lifetime of the beam. (This in itself is a good reason for deliberately introducing tune modulation in a controlled experiment.) Noisy quadrupoles are particularly troublesome during the slow extraction of protons, when the smooth approach of the tune to a low order resonance is necessary to ensure a steady spill rate. Special fast quadrupoles are used during slow extraction in the Tevatron, responding to the difference between measured tune and requested tune, to compensate for such noise. It is these quadrupoles which E778 uses in its investigation of resonance behavior in the (q, Q_M) parameter space. As Figure 6 shows, the (q, Q_M) plane is rich in dynamical features. The dotted line in the figure shows the region accessible to the experiment, with maximum q and Q_M values of about 0.01 .

Tune modulation is included in the resonance Hamiltonian near a fifth order resonance by adding a single term to equation [53], to give

$$H_5 = 2\pi q \sin(2\pi Q_M t) I + \frac{1}{2} U I^2 - V \cos(5\phi) \tag{61}$$

This Hamiltonian is still shorthand for two differential equations, not difference equations, because of the very small net motion in five turns. Unfortunately, H_5 is now time dependent, and so is no longer conserved. The two first order equations of motion are now

$$\begin{pmatrix} \dfrac{d I}{dt} \\[2mm] \dfrac{d\phi}{dt} \end{pmatrix} = \begin{pmatrix} -5V \sin(5\phi) \\[2mm] 2\pi q \sin(2\pi Q_M t) + UI \end{pmatrix} \tag{62}$$

or, as a single second order differential equation in ϕ

$$\frac{d^2\phi}{dt^2} + (2\pi Q_I)^2 \frac{\sin(5\phi)}{5} = (2\pi)^2 q Q_M \cos(2\pi Q_M t) \tag{63}$$

This is physically analogous to the motion of a rigid pendulum, of small amplitude natural tune Q_I, which is driven by an external torque. (The factors of 5 could easily be removed by a scale change). Just as longitudinal motion was interesting because of its connection to the universally recurring standard map, the effect of tune modulation on accelerator resonances is interesting as a representation of the driven differential pendulum.

338

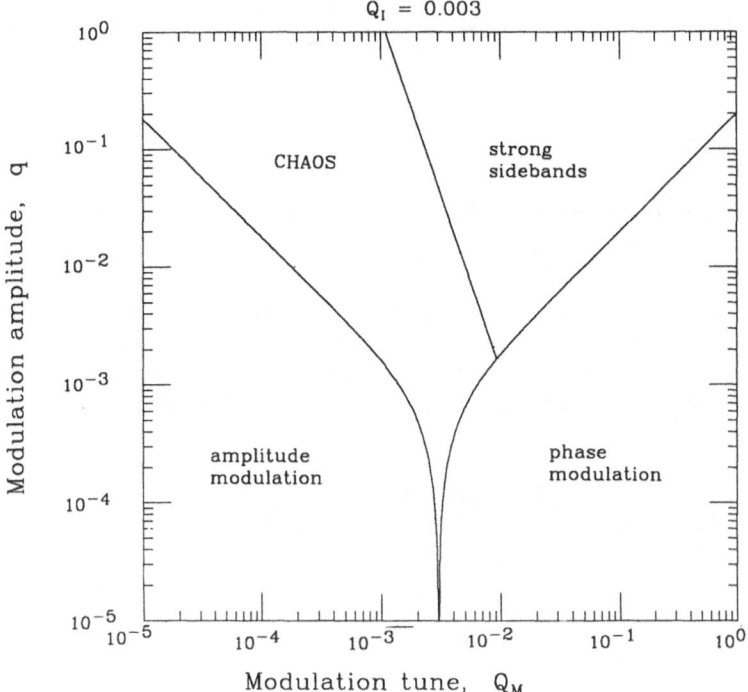

$$Q_I = 0.003$$

Modulation amplitude, q

Modulation tune, Q_M

Figure 6. Dynamical behavior in different regions of the tune modulation parameter space (Q_M, q), for a value of $Q_I = 0.003$. The dotted line shows the region accessible to the E778 experiment, extending beyond the resonance pole at $Q_M = Q_I$ for this particular value of the island tune, which is relatively small.

If the motion is not chaotic, the general form of the solution for the phase of a trapped particle is a double Fourier series expansion in both the driving tune Q_M, and in a free oscillation tune, which is shifted below Q_I at moderate or large amplitudes. However, the experimental observable in E778 is the persistent signal picked up at a BPM, which depends on the average motion of the distribution of trapped particles. It seems reasonable to assume (but is only an approximation) that the center of charge motion which is detected is the same as that of a trajectory with no free oscillation amplitude. Hence we are mostly interested in a solution to the equation of motion which is a single Fourier expansion in the coherent driving tune, Q_M. There is a family of possible periodic solutions, labeled by the integer k,

$$5\phi = k \, 2\pi \, (Q_M t) + \sum_{n=1}^{\infty} c_n \cos(n \, 2\pi \, Q_M t) \qquad [64]$$

where the coefficients c_n are functions of q, Q_M, and Q_I.

The first term in [64], linear in t, corresponds in the pendulum system to gaining or losing exactly k complete turns in one modulation period. In the accelerator system the linear term leads to the possibility of stable resonance islands at a family of sideband tunes, since

$$Q_{solution} = \frac{2}{5} + \frac{1}{2\pi} \left\langle \frac{d\phi}{dt} \right\rangle = \frac{2}{5} + k \frac{Q_M}{5} \qquad [65]$$

Each sideband represents five resonance islands, with centers at an action I_k given by

339

$$Q(I_k) = \frac{2}{5} + \frac{UI_k}{2\pi} = Q_{solution} \qquad [66]$$

so that

$$I_k = k\frac{2\pi Q_M}{5U} \qquad [67]$$

The interesting question is whether the k-th solution is stable. If it is, then it should be possible to observe persistent signals at the corresponding sideband tune, by kicking a beam on top of one of the sideband islands.

Rigorous analytical results for the solutions exist only in the slow and fast modulation limits, when Q_M is much smaller or much larger than Q_I. For large amplitude oscillations in the intermediate region it is necessary to rely on iterative solutions and on simulations. The k=0 solution in the small angle limit 5 $|\phi| \ll 1$ is illuminating. It is given, for all values of Q_M, by

$$\phi = \frac{Q_M{}^2}{Q_I{}^2 - Q_M{}^2} \frac{q}{Q_M} \cos(2\pi Q_M t)$$

and $\qquad\qquad\qquad\qquad\qquad\qquad\qquad\qquad\qquad\qquad\qquad\qquad$ [68]

$$I = -\frac{Q_I{}^2}{Q_I{}^2 - Q_M{}^2} \frac{2\pi q}{U} \sin(2\pi Q_M t)$$

Both expressions include the same resonance denominator, but with different numerators. At constant q, the amplitude of the action oscillation goes to $(2\pi q)/U$ for small Q_M and to zero for large Q_M, while the phase oscillation amplitude goes to zero for slow modulation, and to q/Q_M for fast modulation. This explains the "amplitude modulation" and "phase modulation" labels in Figure 6. The small angle approximation is only appropriate below the boundary line

$$\left| \frac{qQ_M}{Q_I{}^2 - Q_M{}^2} \right| = \frac{1}{5} \qquad [69]$$

which is the solid line in Figure 6 showing the resonance pole at $Q_M = Q_I$.

Rigorous analysis (see below) shows that this is also the boundary of stability for the k=0 solution in the slow modulation limit. Both simulations and a numerical iterative solution to [64] agree that just below the resonance condition, $Q_M \leq Q_I$, this line marks the limit of stability of the k=0 fundamental, but that just above resonance the k=0 solution is stable for all values of q. This shows that the small angle boundary has different physical implications above and below the resonance. Preliminary results from the numerical iterative solution indicate that none of the k≠0 sideband solutions are stable below the resonance[5,7]. In contrast, all of the sideband solutions appear to be stable above resonance, with the possible exception of a small region near the resonance.

Rigorous analysis in the large Q_M limit (also see below) shows that, although the sidebands may be stable, the size of the islands is insignificant below the small angle boundary. If the sideband islands are big enough to overlap with each other and the fundamental chain of islands, there is large scale chaos. Figure 7 shows the appearance of sideband islands when tune modulation with $Q_M > Q_I$ is turned on, in the presence of a single beam-beam interaction with a tune shift parameter just below and just above the critical value required for sideband overlap. The two plots on the left do not include tune modulation, while those on the right do. When the tune shift parameter is increased from

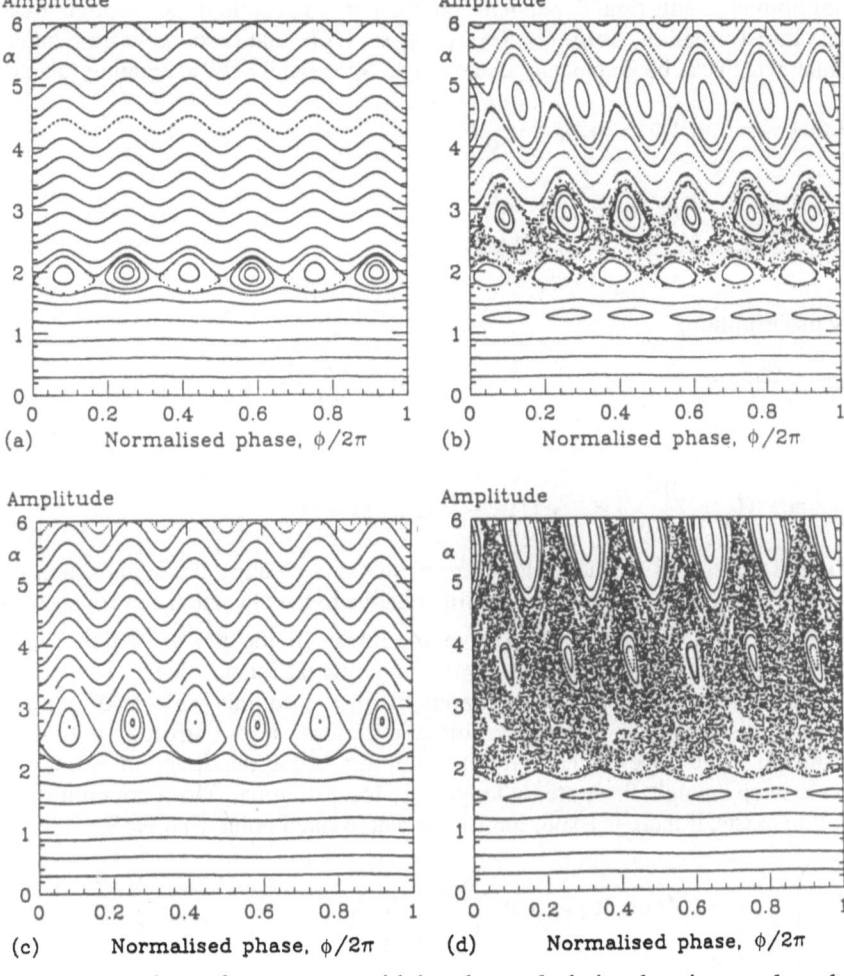

Figure 7. The creation of resonance sidebands, and their chaotic overlap, by tune modulation. A single round beam-beam interaction of strength ξ perturbs the phase space, with a base tune near a sixth order resonance. Plots a) and b), on top, have $\xi = 0.0042$, while the bottom two plots have a slightly stronger value $\xi = 0.006$. Plots a) and c), on the left, have no tune modulation, while those on the right have $(Q_M,q) = (1/194, 0.001)$. Sidebands become visible when the modulation is turned on in b), but the sidebands must overlap for massive chaos to occur, in d). Amplitude α is measured in units of the beam size.

$\xi=0.0042$ in the top two plots to $\xi=0.006$ in the bottom two plots, the sidebands, when they are present, are submerged in a sea of chaos.

<u>Slow modulation — the amplitude modulation region, $Q_M \ll Q_I$</u>

If the tune is changing so slowly that the motion is adiabatic, it is reasonable to approximate the rate of change as constant. As will be seen, the most stringent conditions come when the rate of change is largest, so the most interesting approximation to the Hamiltonian in equation [61] is

$$H_5 \quad = \quad (2\pi)^2\, q Q_M\, t\, I \; + \; \tfrac{1}{2} U\, I^2 \; - \; V\cos(5\phi) \tag{70}$$

This Hamiltonian is still time dependent, but now it is possible to go through a canonical coordinate transformation, from (I,ϕ,H_5) to $(\bar{I},\bar{\phi}, \bar{H}_5)$, that produces a time independent Hamiltonian which can be graphically understood. Specifically, the generating function

$$F_3(I,\bar{\phi},t) \;=\; -\,I\,\bar{\phi} \;-\; \varepsilon t\,\bar{\phi} \;-\; \frac{1}{6}\varepsilon^2\,t^3 \tag{71}$$

with

$$\varepsilon \;=\; \frac{(2\pi)^2\,qQ_M}{U} \;=\; 25\,V\,\frac{q\,Q_M}{Q_I^2} \tag{72}$$

gives, by its definition,

$$\bar{I} \;\equiv\; -\frac{\partial F_3}{\partial\bar{\phi}} \;=\; I + \varepsilon t \;, \qquad\qquad \phi \;\equiv\; -\frac{\partial F_3}{\partial I} \;=\; \bar{\phi} \tag{73}$$

and

$$\bar{H}_5 \;\equiv\; H_5 + \frac{\partial F_3}{\partial t} \;=\; \frac{1}{2}U\bar{I}^2 - V\cos(5\bar{\phi}) - \varepsilon\,\bar{\phi} \tag{74}$$

While the old phase and the new phase are identical, reflecting the suppression of phase modulation in this region, the new action drifts relative to the old action at a constant speed.

The new Hamiltonian has an extra term, linear in the phase, which has serious consequences for the stability of the $k=0$ fundamental island chain. (Note that, as a consequence of linearising the rate of change of tune, solutions with $k\neq0$ are explicitly impossible in this picture). Pictorially, this non-periodic term corresponds to a constant slope of the quadratic valley of Hamiltonian contours, along the direction of the valley. If this slope is steep enough, there are no longer any local minima. There are minima, and the $k=0$ solution exists, if there is a solution for the stable fixed point (I_{FP},ϕ_{FP})

$$\begin{pmatrix} \dfrac{dI}{dt} \\[2mm] \dfrac{d\phi}{dt} \end{pmatrix} = \begin{pmatrix} -\,5V\sin(5\phi_{FP}) + \varepsilon \\[2mm] U\,I_{FP} \end{pmatrix} = \begin{pmatrix} 0 \\[2mm] 0 \end{pmatrix} \tag{75}$$

where the overbars have been dropped. If the $k=0$ islands exist, their centers are at $I_{FP}=0$, with a shifted phase. There are no stable islands at all if $|\varepsilon| > 5V$, that is, if

$$\frac{qQ_M}{Q_I^2} \;>\; \frac{1}{5} \tag{76}$$

This condition corresponds, in this limit, to the small angle boundary in equation [69] .

Figure 8 shows the effect that crossing this boundary has on the measured lifetime of persistent signals observed in the E778 experiment. A set of symbols of a particular kind corresponds to a single constant value of q, at a series of Q_M values. A decay time of 47,000 turns is approximately equivalent to one second in the Tevatron. The decay rate increases dramatically when the stability boundary is crossed, consistent with a fit to the data of $Q_I = 0.0085$. Unfortunately, this method of measuring Q_I is time intensive, since each data point corresponds to a two minute injection cycle of the Tevatron and the analysis is done off-line. It is hoped that in the near future it will be possible to measure Q_I in a single machine cycle, opening up the possibility of a rapid comprehensive scan of resonances across a relatively wide range of tunes.

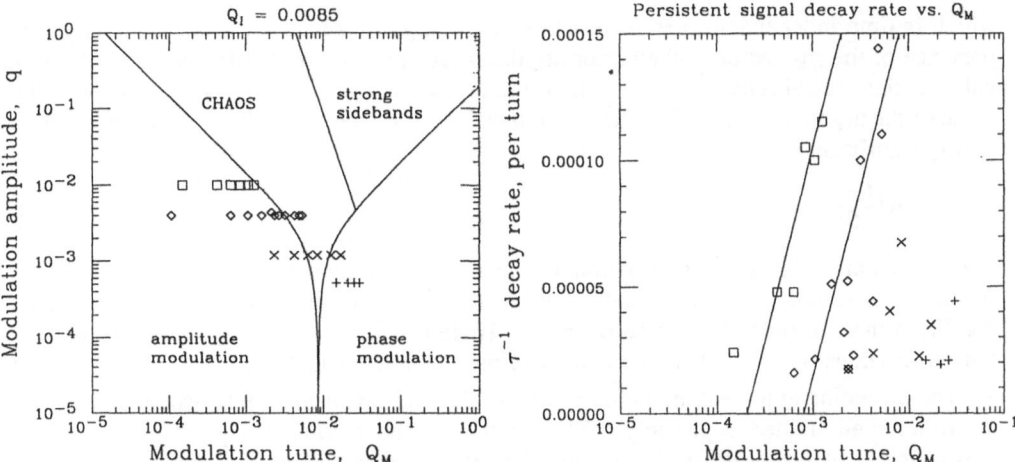

Figure 8. The effect of tune modulation on the decay rate of a persistent signal. Data taken at four values of q reaches from the amplitude modulation region just into the phase modulation region, and into the chaos region. The decay rate of the persistent signal increases significantly as the boundary between amplitude modulation and chaos is crossed.

Rapid modulation — the phase modulation region, $Q_M \gg Q_I$

In this region, instead of approximating the old Hamiltonian and then applying a generating function, a time independent Hamiltonian is found by first applying a generating function and then making an approximation. The appropriate generating function is now

$$F_3(I,\bar{\phi},t) = - I\bar{\phi} - \frac{q}{Q_M}\cos(2\pi\,Q_M t)\,I \qquad [77]$$

which gives, instead of [73] and [74],

$$\bar{I} = I, \qquad \phi = \bar{\phi} + \frac{q}{Q_M}\cos(2\pi\,Q_M t) \qquad [78]$$

and

$$\bar{H}_5 = \frac{1}{2}U\bar{I}^2 - V\cos(5\bar{\phi} + \frac{5q}{Q_M}\cos(2\pi\,Q_M t))$$

$$= \frac{1}{2}U\bar{I}^2 - V\sum_i J_i(\frac{5q}{Q_M})\cos(5\bar{\phi} + i\,2\pi\,Q_M t) \qquad [79]$$

where the J_i are integer order Bessel functions. In this transformation the action remains unchanged, but the phase is modified, appropriate to the phase modulation region. The Hamiltonian is made time independent by concentrating on the vicinity of the k-th sideband, near an action I_k, and then averaging the sum in [79] over one modulation period.

In the limit of large Q_M, not very much happens during one period, and only one turn in the sum survives the averaging. After resynchronising the Hamiltonian to concentrate on the k-th sideband, and dropping the overbars, then

$$H_{5k} = \frac{1}{2}U\,(I - I_k)^2 - V\,J_k\!\left(\frac{5q}{Q_M}\right)\cos(5\phi) \qquad [80]$$

343

which is time independent, and differs from the simple resonant form [53] mainly by the presence of the J_k factor. Whether or not the k-th sideband is significant depends on the value of this Bessel function. As a rule of thumb, J_k is approximately zero if the absolute value of the argument is less than the absolute value of k, the order. That is, the sideband k is only significant if

$$q > |k| \frac{Q_M}{5} \qquad\qquad [81]$$

The right hand side of this equation is the separation of the sideband tune from the fundamental resonance tune. Equation [81] therefore corresponds to the sensible physical condition that, in order for the resonance to be felt at actions near I_k, the tune of such trajectories must be modulated far enough to cross the fundamental.

The preceding argument implicitly presumes that the sidebands can be isolated one from the other, and treated separately. This is true if the sideband separation in action, $(2\pi Q_M)/5U$ according to [67], is larger than the sideband width. If the sidebands are typically wider than they are apart, chaos appears, spanning the action range of the sidebands of significant size. It is easily shown by further approximating the Bessel function, and substituting $J_k V$ for V in [58], that sideband overlap is expected if [81] is true, and if

$$Q_M^{3/4}(5\,q)^{1/4} \;<\; \frac{4}{\pi^{1/4}} Q_I \qquad\qquad [82]$$

This boundary is shown as the second solid line, nearly vertical, in figure 6. Because of the "statistical" approximation of Bessel functions (similar in spirit to approximating a sin function by $1/\sqrt{2}$), this condition is rather qualitative. Depending on the exact phase of the sidebands, some will overlap earlier or later than the condition suggests.

ACKNOWLEDGEMENTS AND DEDICATION

Thanks are due to many members of the E778 collaboration for the inumerable contributions they have made, directly and indirectly, to this chapter. Tong Chen has helped to develop the theoretical analysis of tune modulation, and has suffered in helping to prepare the figures, without the reward of visiting the Virgin Islands.

I dedicate this chapter to Maury Tigner, the director of the Central Design Group of the SSC, who is universally admired and respected by the members of the CDG. Without Maury, the SSC would never have become more than a gleam in a physicists eye.

REFERENCES AND FOOTNOTES

1. Poincare, H., Les methods Nouvelles de la Mechanique Celeste. Paris: Gautier-Villars. 1892
2. Merminga, N., A study of nonlinear dynamics in the Fermilab Tevatron, PhD Thesis, University of Michigan, 1989
3. Edwards and Syphers, An overview of experiment E778, Proc of the ICFA workshop, Lugano, 1988
4. Chao et al., Experimental investigation of nonlinear dynamics in the Fermilab Tevatron, Physical Review Letters p 2752, December 12, 1988.
5. Chen and Peggs, Tune modulation and the driven differential pendulum, to be published in Proc. of the IEEE Particle Accelerator Conference, Chicago, 1989
6. Peggs, Saltmarsh, and Talman, Million revolution accelerator beam instrument for logging and evaluation, SSC-169, Berkeley, 1988
7. Peggs, Hamiltonian theory of the E778 nonlinear dynamics experiment, SSC-175, Berkeley, 1988

8. Chao et al., A progress report on Fermilab experiment E778, SSC-156, SSC-CDG, Berkeley, 1988

9. Merminga et al., An experimental study of the SSC magnet aperture criterion, Proc. of the EPAC, Rome, 1988

10. Peterson et al., Dynamic aperture measurements at the Tevatron, Proc. of the EPAC, Rome, 1988

11. Landau and Lifshitz, Mechanics, Pergamon press, Oxford, 1976

12. Arnold, V.I., Mathematical methods of classical mechanics, Springer-Verlag, New York, 1978

13. Goldstein, H., Classical mechanics, Addison-Wesley, Menlo Park, 1980

14. Lichtenberg and Lieberman, Regular and stochastic motion, Springer-Verlag, New York, 1983

15. Many accelerator school proceedings (AIP, CERN, or joint CERN/US) are broadly circulated. Their eclectic contents are well worth browsing, both for basic introductions and for specialized topics. Perhaps the two most classic accelerator physics references are Courant and Snyder and Sands, below.

16. Courant and Snyder, Theory of the alternating gradient synchrotron, Annals of Physics:3, 1–48, 1958

17. Sands, M., The physics of electron storage rings, an introduction, SLAC-121, Stanford, 1970

18. Edwards, D., An introduction to circular accelerators, AIP Conf. Proc. No. 127, 1985

19. Peggs and Talman, Nonlinear problems in accelerator physics, Annual Reviews of Nuclear Science, 36:287–325, 1986

20. Ruth, R., Single-particle dynamics in circular accelerators, AIP Conf. Proc. No. 127, 1987

21. Wilson, E., Nonlinear resonances, Proc. of the CERN accelerator school, CERN 87–103, 1987

22. Edwards and Syphers, An introduction to the physics of particle accelerators, AIP Conf. Proc. No. 184, 1988

23. This is hinting at conservation of phase space area, as described by Liouvilles theorem. See, for example, Goldstein.

24. Linac designers and constructors claim that they discovered and used strong focussing first, but failed to communicate the knowledge to the accelerator community.

25. SSC Central Design Group, Conceptual design of the Superconducting Super Collider, SSC-SR-2020, Berkeley, 1986

26. Unfortunately there are many slightly different definitions of smear. It is not necessarily possible to convert from one definition to another without making further assumptions about the nature of the motion.

27. Piwinski, A., IEEE Trans. Nucl. Sci. NS-24: 3, 1977

28. Peggs and Talman, Phys Rev D 24: 2379, 1981

29. Myers, S., LEP Note 362, CERN, Geneva, 1982

30. Seeman, J., SLAC-PUB-3182, Palo Alto, Stanford, 1983

31. Keil and Talman, Particle Accelerators 14: 1–2, 109–118, 1983

32. Courant, E., ISABELLE technical note No. 163, Brookhaven, 1980

33. Izrailev, Misnev and Tumaikin, Preprint 77–43, Novosibirsk, 1977

34. Tennyson, J., AIP Conf. Proc. No. 57, New York, 1979

35. Evans and Gareyte, IEEE trans. Nucl. Sci. NS-30: 4, 1982

36. Peggs, Particle Accelerators 17 : 11–50, 1985

37. KAM stands for Kolmogorov, Arnold, and Moser.

38. Moser, J., Nachr. Akad. Wiss. Gottingen, Math. Phys. K1, 1, 1962

39. Siegel and Moser, J. Grund. Math. Wiss. Bd. 187, Springer-Verlag, Berlin, 1971

40. Chirikov, B., Phys. Rep. 52: 265, 1979

41. Henon, M., Appl. Math., No. 3: 291, 1969

DATA ACQUISITION IN HIGH ENERGY PHYSICS

John Harvey

Rutherford Appleton Laboratory
Chilton, UK

INTRODUCTION

The dramatic increase in data volumes generated by modern particle physics experiments poses severe problems for the systems used to collect and process this information. As a result new system architectures have been devised to accommodate the extra processing power needed to correct, reduce and collect the data. Modern systems typically contain hundreds of powerful processing elements interconnected through high speed data acquisition buses and by local area networks. Sophisticated software is used to implement the various system functions.

The data acquisition system has several distinct functions to perform and in the first of these lectures we shall be looking at some of the principles followed in the design of each of its components. The treatment begins with a look at the experimental environment and the requirements placed on the system.

The next step involves choosing an implementation. Here there are many options available and decisions have to be taken concerning the choice of data acquisition bus, processing elements and network connections. In the second lecture some of the recent developments in hardware techniques are mentioned and their use in several experiments described.

Data reduction, formatting, and collection are implemented using software algorithms. Protocols must be defined to control each phase of datataking and to ensure data from each event are assembled correctly. The event data must be continuously monitored to give a rapid indication of malfunction of any part of the equipment. The final lecture illustrates how the problem of developing software for complex distributed processing systems can be allieviated by using technical methods for deriving a design specification. The importance of choosing the most appropriate programming environment is also stressed.

1. SYSTEM ARCHITECTURE

1.1 Experimental Environment

The scope of any system can best be illustrated by looking at the environment in which it has to operate. For a typical High Energy Physics experiment, this can be represented by the diagram shown in Figure 1.

As its name implies, the principal role of the data acquisition system (DAQ) is to process the signals generated in the detector and to write the processed information (ie. "events") onto a data storage device. The reconstruction and analysis of events is normally treated offline and thus this data store serves to delineate the boundary between the DAQ and the subsequent phases in the data processing chain.

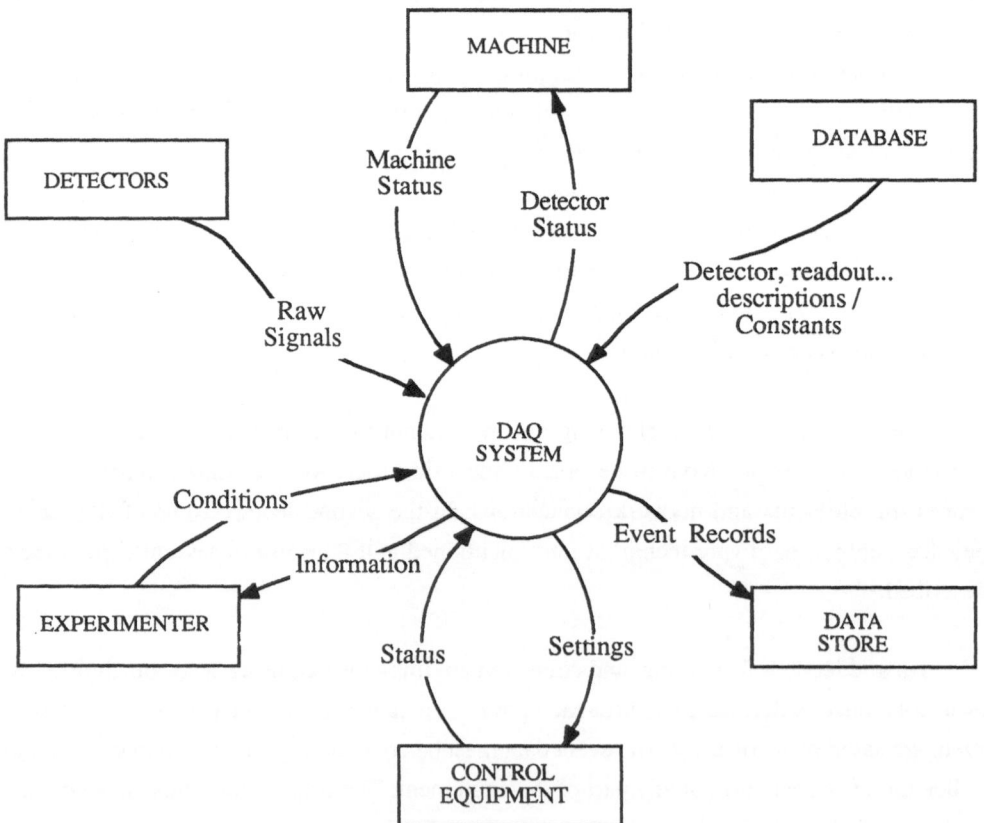

Fig. 1   Diagram showing the boundary between the DAQ system and its environment .

The diagram also illustrates a number of other components with which the DAQ must interact. Firstly it must exchange status information with the system used to control the accelerator. Typically the machine supplies the beam energy, currents and possibly beam polarization information, while the detector supplies information which may be useful for tuning the beam eg. counting rates and currents drawn by the detectors. The coupling between experiment and machine is particularly close for colliding beam machines.

The operator is also shown on the diagram, since he supplies commands which are used to configure the system and to control the datataking state. The system must also provide status information on request, as this provides a means for monitoring the integrity of the data being collected. For complex systems, the provision of a sophisticated operator interface and data monitoring aids has a great influence on the ease with which the experiment can be managed.

The equipment control system is used to set and monitor the operating state of the various detector elements and the readout electronics. This typically involves turning voltage supplies on or off and monitoring temperatures and gas pressures. Nowadays these functions are carried out under computer control and have become one of the major responsibilities of the DAQ. Information accessible through the control system is often essential for an accurate calibration of the detectors and readout electronics and so the regular collection and logging of these quantities is a requirement. Such information may appear as "pseudo-events" on the data storage device.

Also shown on the diagram is a database containing information the system needs access to in order to carry out its functions correctly. In particular, this involves data obtained in an a priori fashion eg. the description of the detector components including survey constants, the description of the readout modules and the detectors they belong to, the different trigger conditions that have been defined and for each trigger which detectors should be used. The database may also contain administrative details, such as a list of members of the collaboration, their particular responsibilities and telephone numbers.

In the above, we have essentially attempted to define a reasonable boundary between the DAQ and its environment. Of course this boundary may vary from experiment to experiment. For example, the event reconstruction may be performed in real-time, in which case the results of the reconstruction will be written together with the raw data onto the storage device.

In addition to a certain functional behaviour, the environment also imposes some performance constraints on the DAQ. It is appropriate to look in more detail how trends in the type of machine, the choice of detector and the physics program have influenced the requirements made on the performance of the DAQ.

1.1.1 The Machines

Physics interest dictates that higher interaction energies must be achieved, which in turn has meant that colliding beams must be employed. As a result, most of the larger experiments are currently running in storage ring machines. In addition, new machines under construction eg. Lep, SLC and HERA employ colliding beams as do those planned for the future eg. SSC, LHC and CLIC. In view of this trend, we shall emphasise the constraints imposed by the storage ring environment on the design of the DAQ.

In storage rings, particle beams are injected into the ring in bunches and are constrained in a circular orbit by a magnetic field. The bunches are then accelerated until the desired energy is reached. Once the bunches are in a stable orbit, they are brought into collision at a number of interaction points around the ring. Experiments are built around these interaction points to study the physics of the collisions. Several collider parameters influence the event rate 'seen' in the detectors. These events come from two sources. Firstly from products of the collisions of particles in the opposing bunches. This sample contains the events of interest which must be measured by the detector and recorded by the DAQ. The sample of interesting events is unfortunately contaminated by so-called background events. One primary role of the DAQ is to distinguish the real events from the background, such that only interesting information is written to the data store.

The event rate resulting from collisions between particles in the opposing bunches is given by the product of the total cross section for the interactions and the luminosity of the beam.

The beam luminosity should be as high as possible to give the best possible event rate. The luminosity is determined by the number of bunches in the machine, the particle density in each bunch and the focusing properties of the beam transport elements. As the beams circulate, particle losses cause the luminosity to decrease and after a time, usually several hours, it becomes necessary to dump the circulating beams and inject new bunches into the machine. The luminosities which can be achieved in practice are indicated in Table 1.

The cross section for interactions is strongly dependent on the type of particles used in the machine. For example, the rate for e^+e^- interactions is given by:

$$R_{tot} \times \sigma_{\mu\mu} \, , \quad \text{where} \quad \sigma_{\mu\mu} = \frac{87 \text{ nb}}{(2 E_{beam})^2}$$

where $\sigma_{\mu\mu}$ is the point like cross section for $e^+e^- \rightarrow \mu^+\mu^-$ and R_{tot} is the ratio of the total annihilation cross section to $\sigma_{\mu\mu}$. R_{tot} varies with energy, indicating the onset of new physics processes. This dependency of the cross section on centre of mass energy is shown in Figure 2 [1]. The basic energy dependence of $\sigma_{\mu\mu}$ results in very low event rates in e^+e^- machines, ranging from <0.1 Hz in the continuum to ~1 Hz on the Z^0 resonance and thus it is practical to record all these events. The rate on the Z^0 peak is dominated by hadronic decays of the Z^0, which are easily identified since the deposited energy and number of tracks seen in the detector are both rather high. However, reactions such as $e^+e^- \rightarrow \mu^+\mu^-, \tau+\tau^-$ and the 2γ processes are characterised by low energy deposit and a small number of tracks, and are therefore more difficult to discriminate from background.

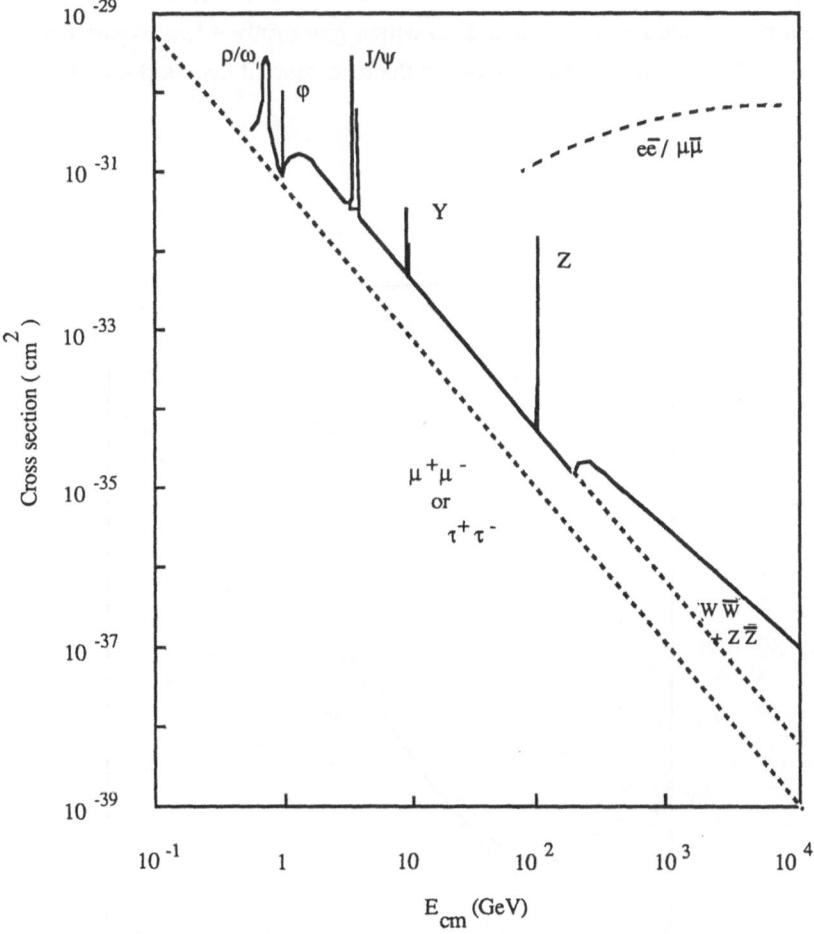

Fig. 2   Electron - positron cross section as a function of the centre of mass energy. The dashed line shows the point-like cross section. The continuous line is measured up to 45 GeV and computed above.

351

The total cross section for proton - proton interactions is shown in Figure 3. Note that the extrapolation to high centre of mass energies is very model dependent [2]. The total cross section consists of three parts: the elastic, the diffractive and the inelastic cross sections. However only the inelastic events contribute particles in the trigger sensitive region of a general purpose detector, since the elastic and diffractive events send all outgoing particles into a narrow cone in the forward and backward directions. Nevertheless the inelastic cross section seen in the UA1 detector, for example, is approximately 40 mb, which at the collider luminosity produces an interaction rate of $>10^4$ Hz. At future colliders this rate becomes $> 10^7$ Hz. These rates imply that there is an interaction for nearly every beam crossing, which is clearly a very different situation than for e^+e^- system. As there is a limit as to the rate at which data can be written to the data recording medium (a few Hz), it is clearly necessary to be highly selective in choosing which events to record. On the other hand, the most interesting physics processes are quite rare. For example, at the $\bar{p}p$ collider at CERN, the cross section for producing Z^0 is about 2 nb which gives only a few events per day. It is important that the mechanism used to "trigger" the recording of events should not be biased

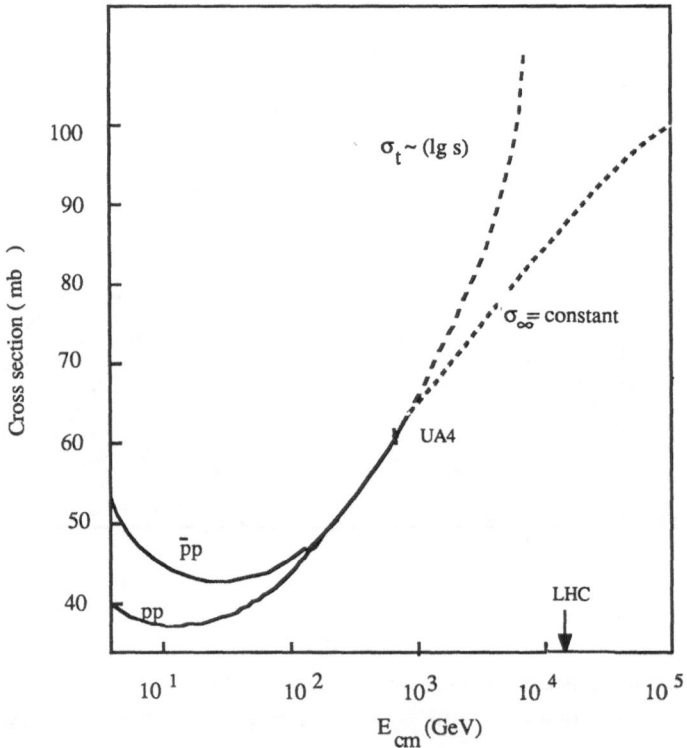

Fig. 3 Compilation of the pp and $\bar{p}$p total cross section data presented together with two
extreme models, $(\ln s)^2$ behaviour and asymtotically constant total cross section.

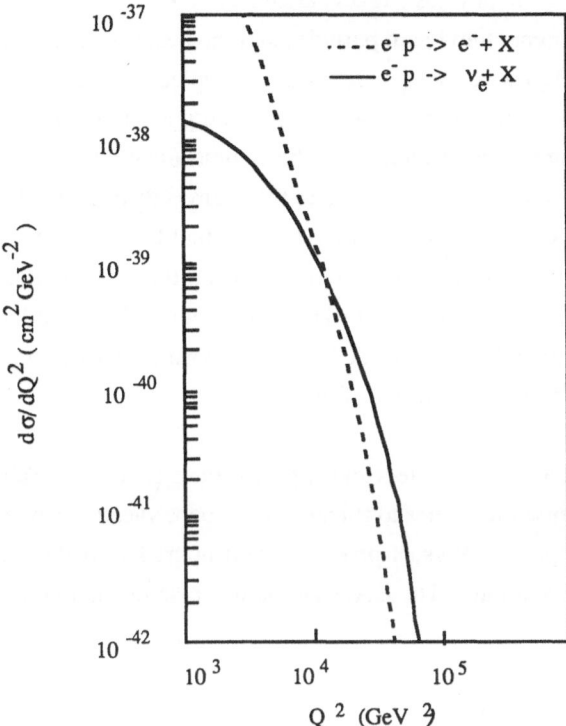

Fig. 4 Differential cross section for neutral current and charged current processes as a function of momentim transfer (Q^2).

such that these interesting events are missed. The inelastic scattering process is dominated by low P_T multiple interactions between constituents of colliding hadrons which results in many low E_T clusters seen in the detector. Interesting events are characterised by hard collisions between constituents and typically give rise to jets with high P_T. The trigger therefore relies heavily on identifying localised energy deposited at high E_T. Thus the problem for hadron colliders is that as the centre of mass energy increases the total inelastic cross section is increasing, whereas the jet jet cross section is decreasing.

The large cross sections at an electron - proton collider come from weak neutral current processes, which result in a scattered electron, and charged current processes, which give an undetected neutrino. Figure 4 shows the differential cross section at HERA ($\sqrt{s}$=314 GeV)[3]. At very low q^2 interactions resulting from almost-real photoproduction will dominate giving a rate of ~100 Hz into the detector. Although interesting in itself, this rate has to be reduced. At medium q^2 the neutral current rate dominates and gives a rate of ~0.1 Hz and at high q^2 the total rate drops to ~10 per day.

The backgrounds to the physics triggers arise from cosmic rays, beam-gas collisions, collisions between off momentum beam particles and the beam pipe, synchrotron radiation and electronic pickup. Rates for these processes are typically in the range from 1 kHz to 1 MHz and they clearly dominate in the case of e^+e^- and e^-p machines. Background rates can be minimised by ensuring a good vacuum in the beam pipe and by the use of collimators and shielding. The topology of background related to the beam is distributed along the beamline and not concentrated at the interaction point. Thus residual background can be effectively reduced by eliminating those triggers not originating from the crossing point of the beams. Cosmic rays are uncorrelated in time with the beam and can be eliminated by gates generated from the timing of the bunch crossing. A sample of cosmic ray events may be used to provide a useful check on the performance of the detector.

A summary on machine parameters and expected rates is given in Table 1. Note that a number of factors can cause these rates to fluctuate eg. a poor vacuum can give rise to higher background rates, unexpected physics processes and upgrades to the machine can also produce unpredictable event rates. The DAQ should be versatile enough to cope with these fluctuations.

1.1.2 The Detectors

A consequence of using colliding beams is that there are few regions where the beams are brought into collision and therefore very few experiments. These experiments are designed as general purpose detectors such that each experiment can study the full range of final states encountered.

Although the variety of physics signatures is large the final state particles give rise to four categories of triggers. A quark or gluon gives rise to a large number of particles confined in a narrow cone ("a jet") whereas a neutrino or photino gives rise to missing transverse energy. An electron or photon can be identified by a localised electromagnetic

Table 1   Machine parameters and primary interaction rates for a number of colliding beam machines

Particle Type	e+ e-		$\bar{p}p$ / pp		ep
Machine Name	Petra	Lep I	S $\bar{p}$pS	LHC	Hera
Bunch Spacing (μsec)	3	22	3.8	0.025	0.096
Centre of Mass Energy (GeV)	40	100	540	16,000	314
Luminosity (cm^{-2} s^{-1})	10^{31}	10^{32}	10^{30}	10^{33}	10^{32}
Interaction Rate (Hz)	< 1	~1	~10^4	~10^8	< 1

Table 2   Data volumes generated by a number of typical experiments

Experiment	Tasso	Aleph	SLD	UA1	SSC
Number of channels	~30,000	~500,000	~200,000	~100,000	850,000
Raw data (Mbyte)	~0.1	~30	~60	~2	?
Event size (Kbyte)	~5	~100	~135	~160	400
Event rate (Hz)	~3	~2	~2	<4	?
Tapes per day	~15	~100	~100	~200	?

shower and a muon by its penetrating properties. Any combination of these trigger signatures can be used to match a particular physics process. A typical experiment will therefore employ a number of different detector elements including:

- Solid state microstrip detectors used to locate decay vertices of short lived particles
- Wire tracking chambers situated in a solenoidal magnetic field for the momentum analysis of charged particles. Typically 100 samples per track are taken involving measurement of drift time and pulse height.
- Calorimetry consisting of an inner layer for the measurement of electromagnetic energy and an outer layer for the measurement of hadronic energy.
- Charged particle detectors for the identification of muons are located on the outside of the experiment. Wire chambers measuring drift time and charge division are typical.

To ensure that the events sampled are unbiased the detector should be hermetic. For practical reasons this is usually achieved by employing a cylindrical geometry with the forward regions close to the beam being closed by "end-caps". An overlap between these components is necessary to ensure that there are no gaps in the acceptance.

Because of the complexity of the events the granularity of these detectors must be very fine in order to give adequate spatial resolution. Silicon detectors have a spatial resolution of several microns and can generate millions of samples of raw information. The cells of the calorimetry are typically arranged in many towers which project towards the interaction region and each tower is divided into several longitudinal segments such that the development of the induced showers can be traced. The size of a tower is chosen to match the size of the shower induced by the particle traversing the detector. This results in detectors with tens of thousands of towers, with each tower giving many samples in depth. In addition, the need for a wide dynamic range and information on pulse shape causes an increase in the data generated by the readout electronics for the calorimetry. In the case of tracking detectors time sliced digitising methods are used to provide multi-hit capability. The net result is that there are enormous volumes of raw data generated on the detector (Table 2).

The channel occupancy is determined by the number of final state particles and this varies considerably with the initial state particles and the collision energy. However, the fraction of the total number of channels activated during an event is still typically less than a few percent, and so the data volumes that have to be readout are significantly less. The reduction factor can be estimated from Table 2 by comparing the raw data rate with the event size.

From the forgoing discussion it is possible to identify several distinct activities that take place as data are collected by the DAQ system:

- signal processing, for collecting the charges produced on the detector channels and for performing digitisation
- trigger processing, for filtering unwanted events and synchronising the readout of data from all detector components in the experiment.
- data processing, for reducing the data volume to include only those detector channels with information; for applying corrections to allow for variations in the response of different channels in the detector and readout electronics (calibration); for gathering together the data from the various detector components; for assembling complete events and writing them onto the storage device.
- control processing, for the setting, regulation and monitoring of the operating state of the equipment

The principles used to establish the way in which these functions should be implemented are discussed in the following sections.

1.2   Signal Processing

As particles pass through the detectors they deposit energy in various ways, eg. through ionization, by radiating or through nuclear absorption. This primary energy deposit may be amplified by the detector and appear as a charge to be readout by the DAQ. Thus the presence or absence of a signal on a detector channel indicates the passage, or otherwise, of a particle in the neighbourhood of the channel. The amount of energy deposited by the particle depends on the type of particle and its energy and thus a precise measurement of the signal shape can be used for particle identification. The precise timing of the signal can be used to correlate the signal with signals appearing on other detector channels.

The signals are typically amplified and shaped at the detector and dispatched through cables to the readout electronics, where they are sampled by digitisers to recover the charge and time. The precision with which this can be done depends on the noise, resolution and dynamic range properties of the electronic components. An absolute calibration can be performed by using particles of known energy in a test beam. In this way corrections can be applied on a channel by channel basis to allow for variations in the response of the detector and electronics. Calibration of the electronics can also be made by injecting a known charge and measuring the response and must be performed regularly to check for drifts in the behaviour of each channel.

1.3 Trigger Processing

The task of the trigger system is to process the information generated by the detector after each bunch crossing and to activate the readout system only for those crossings producing interesting events. The trigger uses the calorimetry to locate energy clusters and tracking devices to identify track segments coming from the interaction point. For triggering on energy the full granularity of the calorimeter is too fine and signals from several adjacent towers must be summed to form one trigger segment. The tracking logic is also designed to use the same trigger segment geometry. The results from each trigger segment are then combined in an attempt to match an acceptable signature. The processing of detector information by the trigger needs only to be sufficiently accurate to achieve an unbiased discrimination between those events that are wanted and those that can be discarded; in the trigger emphasis is placed on getting a correct, unbiased result in the fastest possible time so that precious machine time isn't wasted. For those events to be kept, extra care must be taken to perform a full digitisation with the highest precision.

Each bunch crossing is a potential source of interesting events and so, if the DAQ is to function without introducing deadtime, then the state of the detector must be examined following each crossing. The input rate to the trigger is thus determined by the bunch crossing frequency (~100 kHz), whereas the output rate is limited by the data recording rate (a few Hz). The trigger decision logic is typically organised into a number of levels, which are characterised by decreasing input rates and correspondingly longer processing times. The lower levels are implemented in hardware, due to the short time available for making the decision, whereas subsequent levels can be implemented using software (Figure 5). The software trigger operates on full event information and since there is substantially more time available can make use of more sophisticated event reconstruction algorithms to effect the reduction.

As we shall see, the DAQ architecture is designed to allow the processing at each level to proceed in parallel, such that providing the second, third etc. levels can keep up they do not introduce any dead time into the system. Thus the most acute problem to solve is to provide a deadtimeless first level trigger. The time available for the first level trigger is constrained by the bunch crossing time and can also be influenced by the properties of the detectors used by the experiment. The precise logic used by the trigger depends on the signatures of the interesting events and the background events that must be discriminated against, which as we have seen is strongly dependent on the type of machine.

357

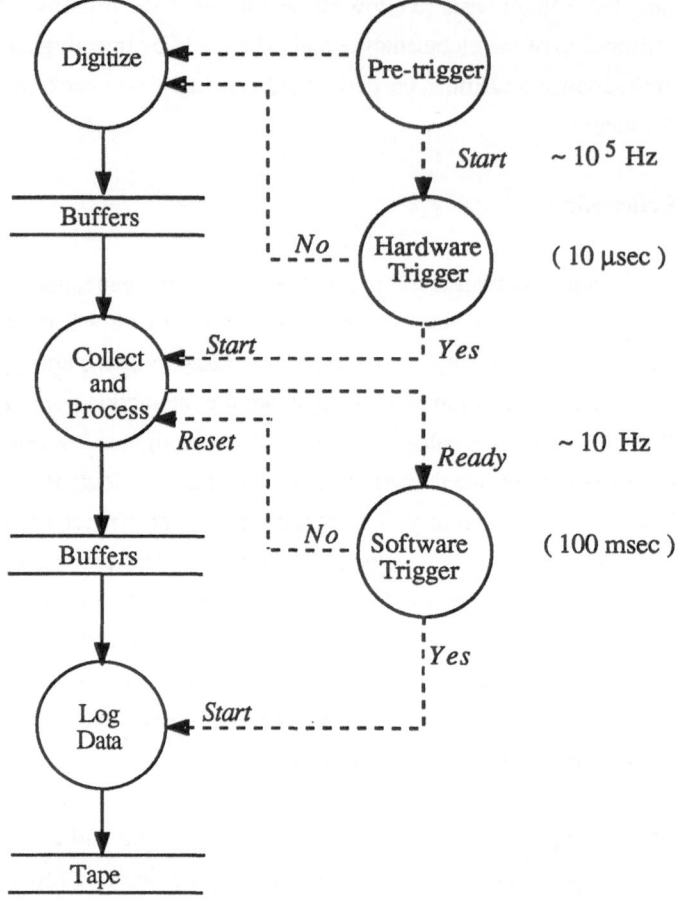

Fig. 5   A logical view of the data readout scheme showing two different
          levels of trigger logic

At LEP the basic triggers used are :

- at least two minimum ionizing tracks
- at least one minimum ionizing track and one energy cluster (low threshold)
- total electromagnetic or hadronic energy (high threshold)

The bunch crossing interval is generous (ie. 23 μsec) and this has been exploited by the
OPAL experiment to implement a deadtimeless trigger according to the scheme shown in
Figure 5.

In the case of Aleph, the inherent deadtime of the TPC created by the long drift time of the primary ionization in the chamber (~40 μsec) means that this detector cannot be used in the first level trigger without introducing a substantial deadtime. A three level trigger system has therefore been chosen. A pre-trigger generated by the bunch crossing signal every 23 μsec starts the digitisation of signals coming from the detector. The first level trigger is used to gate the TPC and therefore must be available quickly (~1.5 μsec) and be limited to less than a few hundred Hz. It uses a cylindrical drift chamber close to the beam pipe, the calorimetry and the muon detector. A level 1 "reject" decision causes the electronics to be reset (this takes ~ 18μsec) and the trigger system is re-enabled to wait for the next beam crossing. Thus no deadtime is introduced. A level 1 "accept" causes the level 2 trigger logic to be invoked. The level 2 trigger uses the TPC information to locate tracks that point to the bunch crossing region; this can be done progressively within the 40 μsec drift time, with the final answer available a few μsec after that. It is expected that this will reduce the trigger rate from ~500 Hz to ~10 Hz. Thus the level 2 trigger causes two crossings to be missed for every level 1 trigger processed and results in a small deadtime of 2%. A third level of triggering is provided to remove any residual background ie. to reach the predicted physics rate (~ 1 Hz).

As we have already seen the problem on hadron colliders is more acute due to the high inelastic cross section. On the CERN $\bar{p}p$ collider bunch crossings occur every 3.8 μsec and nearly half of them produce interactions seen in the detector. In the upgraded UA1 experiment almost half the time between crossings is consumed by shaping and digitisation of the calorimeter signals, cable delays and time needed to reset the electronics. Thus the time available to the first level trigger processor is only ~1.5μsec. The first level trigger must achieve a rejection rate of 10^{-3} and at the same time maintain the highest possible efficiency for triggering on rare physics signals (W, Z, top etc.). It uses a fast digitisation of the calorimeter signals to produce total energy sums, and weighted energy sums to provide a measure of transverse energy and missing transverse energy. Hits in the muon drift tubes are used to identify potential muons. The basic triggers are based on the following conditions:

- an electromagnetic cluster with $E_T > 10$ GeV
- a jet cluster with $E_T > 25$ GeV
- a large total E_T (eg > 120 GeV)
- a muon candidate

These thresholds are required to keep the rates to an acceptable level. However certain physics processes are rather subtle and can only be identified by using combinations of the above with lower thresholds. For example, a W decay to anti-bottom and top followed by top decaying semi-leptonically produces a final state with an electron, two jets, and missing transverse energy. None of these has enough energy to trigger by itself. Thus other common requirements include two electromagnetic clusters above a threshold of 6 GeV, a transverse energy imbalance greater than 17 GeV in coincidence with a jet having $E_T > 15$ GeV, and one muon in coincidence with a jet. The second and third level triggers use drift times from the central detector, drift times from the muon drift tubes, and the full digitisation in granularity and depth of the calorimetry to bring the rate down from ~100 Hz to the data recording rate of a few Hz.

At HERA the trigger problem is dominated by the very high bunch crossing frequency. As it is impossible to construct a first level trigger in the 96 nsec between bunch crossings, the solution adopted has been to clock both trigger information and data into pipelines. Thus in the trigger processor the result of each computation is clocked from one logic step to the next at the HERA frequency (10.4 MHz). The final result emerges after a given length of time determined by the length of the pipeline, which for H1 is ~2.5 µsec and for Zeus is ~5 µsec. A schematic view of the Zeus trigger pipeline is shown in Figure 6. In the case no valid trigger is matched, data in the pipeline are merely overwritten by information from subsequent crossings and the pipeline continues to free-run. Once an "accept" is generated the pipeline is stopped and data are copied into a multi-event buffer. This introduces a small deadtime of ~5%.

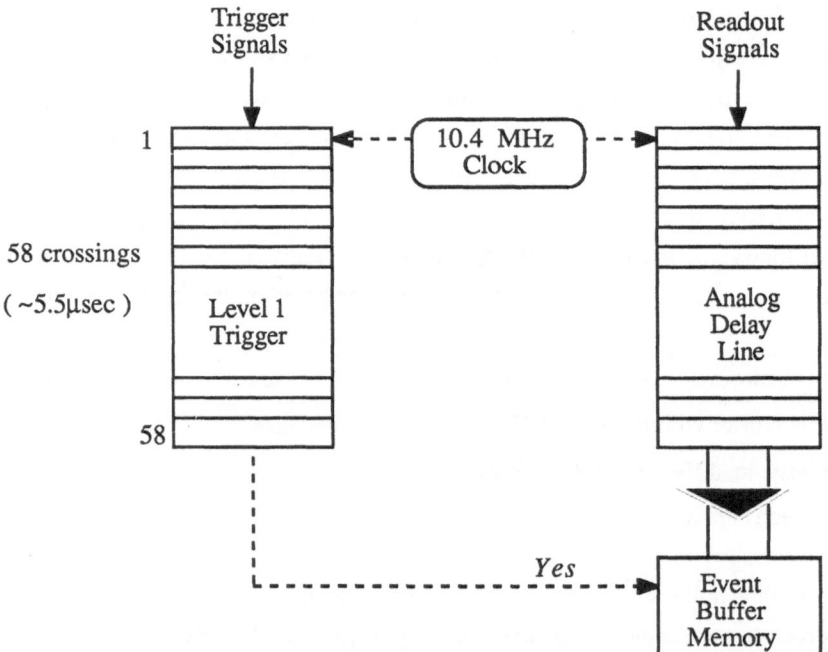

Fig. 6   Block diagram of the Zeus first stage trigger and readout scheme. The depth of the pipeline is sufficient to allow ~5.5 µsec for the first level trigger logic to function .

Once again calorimetry is the key to triggering. Charged current interactions give rise to a hadron jet (large E_T) and a neutrino (missing E_T), whereas neutral current processes generate hadron jets and an isolated electron. Simulations predict that the first level trigger can reduce the rate to < 1kHz. The extra time available to the subsequent trigger levels is used to reanalyse the data in a non-pipelined fashion to further reduce this rate. Once again the final level of triggering is a software trigger for reducing the rate sufficiently to match the data storage writing speed.

It is worth pointing out that on the hadron machines being proposed for the future (LHC and SSC) the triggering problem will be at least an order of magnitude worse than for the situations outlined above. For example at LHC which has a design centre of mass collision energy of 17 TeV the total cross section will be ~100mb, according to model predictions. The design luminosity for this machine is 10^{33} cm^{-2}sec^{-1} and the bunch crossing interval is only 25 nsec. The average number of inelastic events for each crossing is therefore ~ 1.5. Studies indicate that although the trigger represents a serious problem, feasible solutions depending heavily on the calorimetry and pipeline techniques can be found[4].

1.4 Data Processing

In practice, data flow may be viewed as a number of stages, with each stage conforming to the scheme shown in Figure 7. An input data stream is buffered while portions of the data are used to make the trigger decision. Any processing of the data may be performed in parallel with the trigger processing. Once the trigger decision is formed, the data are either read out into the next stage or are simply overwritten by subsequent events, according to whether the trigger result is "accept" or "reject". The filling and emptying of the buffer can proceed asynchronously and providing the buffer never becomes full, no deadtime need necessarily be introduced.

The amount of buffer space needed depends on several quantities and can be estimated using queuing theory[5]. If the input queue is assumed to have a Poissonian time distribution with average λ, the event buffer has depth κ and the processing time distribution is exponential with average η, then the deadtime is given by the formula :

$$\text{Dead time (\%)} = \frac{(1 - \eta) \, \eta^{\kappa}}{1 - \eta^{\kappa+1}}$$

The dead times deduced from this formula by varying κ and η are shown in Figure 7. Note that a complete simulation involves combining the queuing theory equations for adjacent steps and can only be solved numerically.

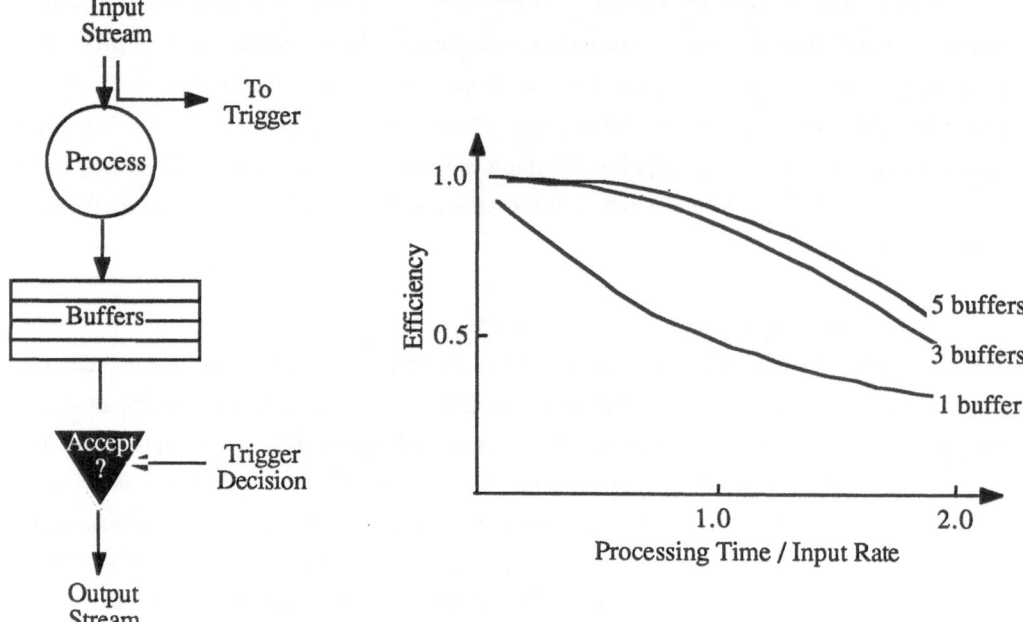

Fig. 7   A schematic view of a single stage in the readout system. The graph gives an indication of how the deadtime can be minimised in this simple case by introducing sufficient buffer space that the dataflow does not become blocked. These results were obtained using simple queuing theory.

Thus a strictly deadtimeless DAQ system can be achieved, providing:

* the first level trigger has been designed to match the bunch crossing frequency
* enough buffer space is available to cope with the variations in data volumes and data processing times at each stage in the data collection

In practice, it may be preferable to omit a stage of buffering if the deadtime introduced is small and a substantial gain in simplicity and expense can be achieved. The first point in the dataflow where buffering can be introduced is after the digitisation ie. in the front-end modules. Here buffers help to de-randomise the data flow into the first stage readout processors. In order to minimise the movement of data the transfer should only be attempted once a good reduction factor has been achieved in the trigger.

In the case of Aleph, a level 2 "accept" invokes the readout of the digitisings and the trigger is not re-enabled until all these digitisings have been collected. The readout of each subdetector proceeds in parallel and therefore the deadtime is determined by the component with the longest readout time (~5msec). This results in an additional 5% deadtime at the nominal level 2 rate of 10 Hz. This deadtime could have been eliminated by introducing an extra layer of buffering in the front-end modules. These buffers are in fact present in the TPC readout, since they are needed anyway to store the flash ADC samples collected over the drift time in the chamber volume. However, since the deadtime is acceptably small it was decided not to include extra buffers for the other detectors and to profit instead from the reduced complexity and expense of the simpler solution.

The downstream stages of the DAQ system are fairly common to most collider experiments (Figure 8). The data flow resembles a tree, with many front-end modules supplying a *readout controller*, which in turn together with other controllers belonging to the same detector supplies another processor, the *local event builder*. This has the responsibility of assembling all the information belonging to its subdetector for each event. The final step in the readout involves assembling the complete event by collecting the data coming from these local event builders. Once the data from a complete event has been assembled it can be processed by the software trigger, following which those events selected can be written to the data store.

The data are processed as they flow through the tree. Zero suppression and channel calibrations are performed by hardware at the front-ends for speed reasons and to ensure that the data volumes transported downstream are kept to a minimum . The down stream readout processors all provide multi-event buffers to de-randomise the data and are responsible for reformatting the data and building valid event data structures.

For the data flow system to work properly strict protocols are needed to synchronize the flow of information. The flow of data is initiated by the trigger system and at the downstream end has to be re-synchronised to build the complete event. Within each subdetector branch the readout can proceed asynchronously. If the event buffer at a particular stage within one of the subdetector branches becomes full, the readout protocol causes the flow of data out of the branch to be temporarily suspended. Thus data starts to pile up in the branch and may result in the readout controllers become blocked. This will cause the trigger to be disabled. Once buffer space becomes available again data starts to flow in the branch and the trigger is re-enabled. The event fragments should carry a label containing the trigger mask and trigger number so that a check can be made when the event is assembled to ensure that the data has been collected together correctly.

This asynchronous pipeline model has been used in the Aleph dataflow model. The effect of varying the buffer size available at each point of the dataflow was estimated by performing a simulation of the data flow system using Monte Carlo techniques[6]. The main inputs to the simulation were the architecture of the system, the trigger rates and their distribution in time, and the event characteristics. For the architecture, the three main subdetectors were modeled and the buffer space available at the various stages in the readout

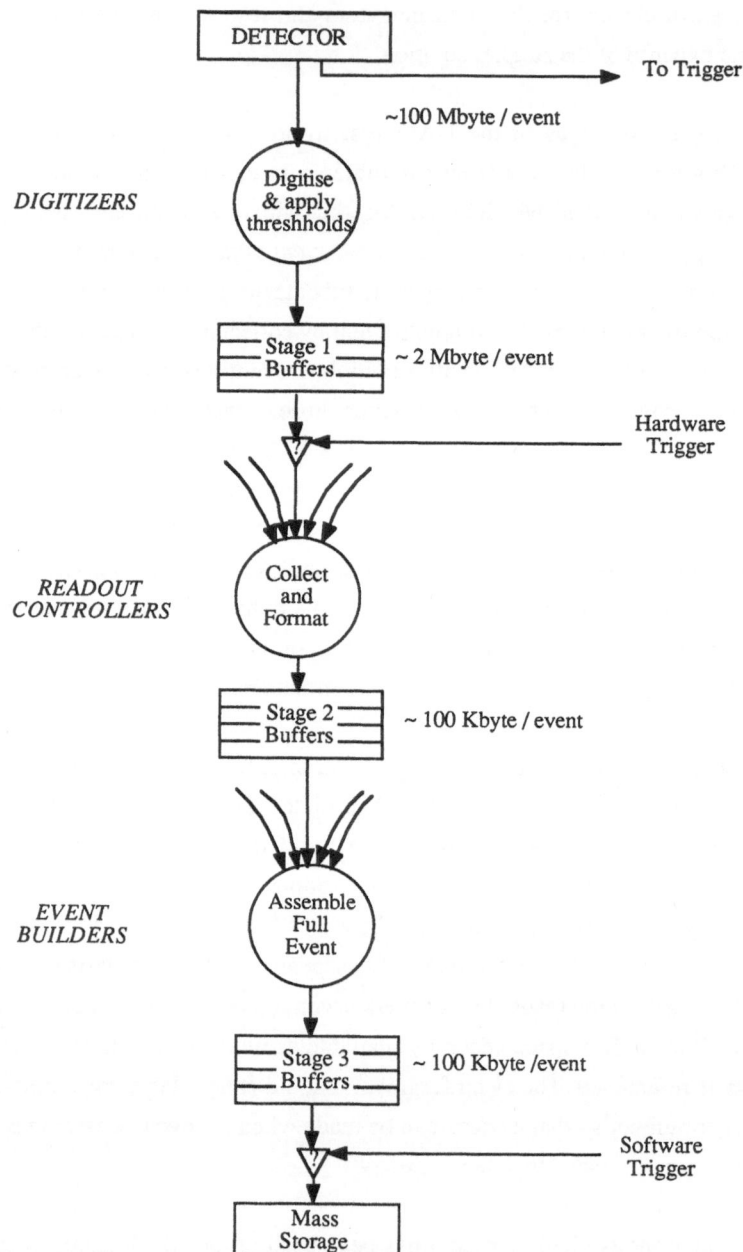

Fig. 8 A readout scheme for assembling complete events in three stages .

was the main variable in the program. Events were generated using the LUND Monte Carlo and then digitised. The simulation was used to determine the average buffer occupancy and also to calculate the overall deadtime as a function of the level 1 rate. The simulation also confirmed that the asynchronous readout in parallel pipelines has the effect of smoothing local peaks and provides the shortest latency in treating input data.

1.5   Control Processing

Techniques exploited in measuring particle trajectories typically involve the use of various gas mixtures, and electric and magnetic fields. An optimum response can be found by studying the behaviour of the detector as various quantities are adjusted eg. gas pressure and mixture, high voltage etc. The response can also be influenced by changes in temperature or atmospheric pressure, or by the presence of gas impurities. To maintain efficient operation of the detector it is important to run the detector under optimum conditions. Any change in these conditions, or in the operational environment, should be noted if it is going to be possible to correctly interpret the information collected from the detector. Other parts of the experiment also need to be controlled, such as the low voltage supply to the electronics, stepping motors etc. Thus a system is needed for setting the operating state of the equipment and for continually monitoring this state.

Some characteristics of modern control systems may be summarised as follows. Firstly on large experiments the number of channels can be substantial. Each detector may be constructed as a number of distinct modules each having its own support system. In addition the readout systems involve substantial electronics at the detector and ~100 crates of electronics for data acquisition. The gas system contains valves, pressure monitors and in certain cases mass spectrometers for detecting changes in gas mixtures or the presence of impurities.

In certain cases stable operating conditions can be maintained by providing a system for automatically regulating the support system. Alternatively, these quantities can be monitored continuously to check that they lie within acceptable limits. Where these limits are exceeded warning messages are automatically generated instructing the operators to take action. Where safety is involved a secure alarm system is required and the system set into a fail safe state.

The data handling problem is modest by data acquisition standards. The total number of channels is typically <10,000. Each channel may be characterised in terms of a few variables, (eg : Channel number, Channel type, Nominal Value, Tolerance, Current Value, Current status). Data needing to be collected and recorded corresponds to only a subset of the total number of channels (~< 1 Kbyte) and the collection frequency is on a timescale of minutes.

It is clearly inconceivable to perform these functions manually. The fast response times required for real-time control and the need for continuously monitoring such a large number of channels indicate the need for a distributed system of simple processing elements. The data rates are more modest , however, and the requirements made on the communication bus used to distribute commands and collect information are not severe.

1.6   Information Processing

The complexity of the experiment raises new problems in the organisation and presentation of so-called secondary data. These data include descriptions of the geometry of the detectors, as well as the relationships between the detector elements and components of the readout system. Many other constants are also generated during the normal operation of a typical experiment.

Initialisation of the data acquisition system proceeds in several stages. Firstly the connectivity and availability of every element needed must be established and each element configured according to the activity requested by the operator. Each activity usually involves selection and initialisation of a particular trigger source and a subset of the readout system.

Calibration often requires triggers to be generated from special sources, such as LEDs, lasers or possibly cosmic rays. The data taken during these periods are used to equalise the response from each detector channel by correcting for variations in the behaviour across the detector and for the response of the electronics. Changes in the calibration with time must also be recorded. The amount of data produced can be enormous and requires sophisticated data management and bookkeeping tools.

A similar situation exists for information collected during monitoring of data collected during normal physics runs. These data are normally generated in histogram form. Nominal behaviour of the apparatus should be accessible to the operator in the form of a set of reference histograms and again it is frequently desirable to keep a history of the performance of the equipment by maintaining copies of sets of histograms with time. Status information, reflecting current datataking conditions, the operating state of the equipment etc. must also be generated.

A major requirement of the data acquisition system is to permit the operator to access this information in a convenient way. Where possible automatic checks should be made to alert an operator to unusual behaviour and diagnostic tools should be available for locating the source of errors after they have been detected. There is clearly a learning process involved in the diagnosis of system faults and it may be advisable to record information on problems and their remedy in a database. It may even be appropriate to use this database as the basis of an expert system for automatically diagnosing system errors.

1.7 System Integration

As already mentioned, a large experiment is usually composed of a many different types of detectors, each requiring its own special "life-support" and readout systems. Thus it is typical for each component on the experiment to be built, commissioned and subsequently managed by a team of specialists. Each team may employ different techniques, either because of the special requirements imposed by the project or because of their own background and expertise. Clearly a special effort should be made during the design phase to minimise duplication of effort by adopting common solutions wherever possible. This problem is exacerbated by the fact that the teams forming the collaboration are often dispersed around the world making communication more difficult.

One consequence of this organization is that during the running of the experiment the various components of the detector may need to be operated independently, which in turn imposes constraints on the design of the data acquisition system. This is typically true during the commissioning phase of the experiment but is also true between normal datataking periods which are normally used for performing calibrations and tests on the equipment. The particular problem to solve is how to permit simultaneous use of the data acquisition system from independent activities without them interfering with oneanother.

This problem has led a number of experiments to introduce the concept of a *partition*.[6,7] In each case the precise definition of what is meant by a partition is slightly different but is broadly defined to be any subset of the data acquisition system which can be configured to function independently of the rest of the system. Each partition consists of a pipeline, where data flows from the front-end electronics to the online computers. More than one partition may exist at any time, thus permitting parallel data streams. Each partition typically has its own set of readout electronics, trigger system and control and monitoring software. The hardware implementation of the data acquisition system often imposes its own constraints on how the system can be partitioned and so the rules for establishing partitions have to be made very carefully.

For partitioning to work properly it is necessary to control access to the resources contained in the partition. The same problem arises in any multi-user system, (eg. any multi-tasking operating system) where access to system wide resources (eg. tape drives) must be controlled to ensure that different users do not interfere with oneanother. In certain circumstances, this may be possible by configuring the hardware to isolate one part of the system from the rest. In many cases it is necessary to control access to resources using a sophisticated resource management scheme implemented in software.

2. SYSTEM REALISATION

The underlying trend towards greater complexity has a number of implications which must be taken into account when making an implementation choice for the data acquisition system.

- the electronics mounted on the detector and the cable plant required to transfer the data to the readout system scales in proportion to the number of detector channels
- the large number of electronic channels implies the use of a large board size to minimise costs incurred due to overheads, such as crates and power supplies
- the use of ECL circuits, FADCs and fast memories results in high power consumption and gives cooling problems
- the large volumes of data generated at the front-ends must be treated and reduced at source
- the bandwidth of the bus must be sufficient to permit the collection of the reduced data efficiently
- it should be possible to permit a partitioning of the readout system to permit parallel readout streams from independent activities
- the use of standards, with good industrial support, is essential, when it is considered that the development of different parts of the system is carried out by many groups from many different institutes

To realise a system many choices have to be made between the technical solutions available. This includes selection of the data acquisition bus, development of electronic circuitry for collecting and processing signals produced by the detectors, and choosing between special purpose processors and commercial microprocessors for data processing applications. More decisions have to be made on the organisation of the control computers, including their interfacing to the DAQ bus and their network connections to eachother and the outside world. These points are treated in the following sections.

2.1 Data Acquisition Buses

It is clear that a versatile multi-purpose bus system is needed to integrate the readout components into a coherent system. In the past, CAMAC has been employed extremely successfully by the majority of high energy physics experiments, but more recently new buses adapted to a distributed processing environment have been introduced. Some of the principal features of these buses will be briefly indicated.

2.1.1 CAMAC

CAMAC systems have been used for several years by the vast majority of particle physics experiments. Its success derived from the fact that it imposed a standard for the instrumentation bus. Through the development of special interfaces, it became possible to connect to host cpus from a variety of manufacturers, such that the instrumentation bus appeared as an extension of the I/O bus of the host. Nevertheless CAMAC has several serious limitations. The main problem comes from the fact that the system was conceived with the host as being the only master on the bus ie. no direct communication between modules in different crates is permitted. This limitation was overcome, to a certain extent, by interfacing CAMAC to private buses through intelligent interfaces. However, when considered together with its other limitations (eg. the data path is only 24 bits wide, the block transfer speed is limited to 0.5 to 2 Mbytes/sec and the board size is too small) it became highly desirable to replace CAMAC with a more modern bus system.

2.1.2 FASTBUS

FASTBUS was conceived within the HEP community as the replacement for CAMAC and has now been accepted as an ANSI/IEEE standard[8]. Figure 9 shows an example of the Fastbus system topology. The basic element is called a segment, which is an autonomous bus connecting master and slave devices. Masters are defined to be devices which are able to access the bus and generate bus cycles, whereas slaves are able to recognize bus cycles and when necessary respond to them. Segments may appear as crate segments and cable segments, and the bus protocols governing the communication between devices are the same

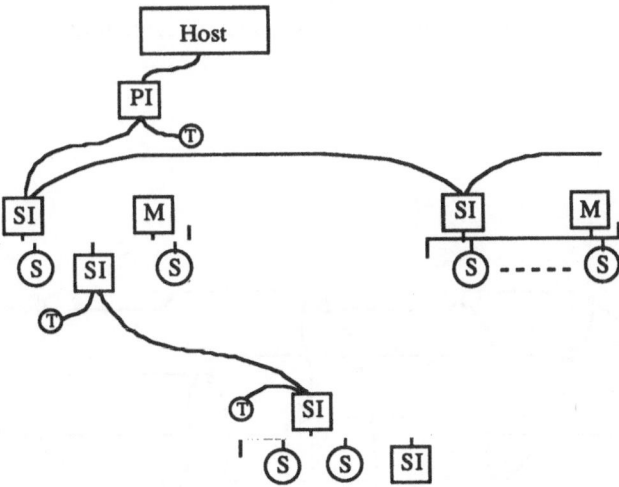

Fig. 9   Schematic showing FASTBUS system topology. The M and S symbols are used to distinguish Masters from Slaves and T represents the termination logic.

for both types of segment. The connection between segments is provided by the Segment Interconnect module or SI. This contains the routing information, which establishes the communication paths through the system and which can be dynamically configured. The Fastbus system is connected to a host computer, which is defined to be the master responsible for initialising the whole system. Some basic features of Fastbus which are exploited when implementing the readout system will be briefly mentioned.

The fastbus protocol uses an address cycle for establishing a lock between a master and a slave, followed by one or more data cycles when data is exchanged. The same 32 A/D lines are used in a time sliced fashion for carrying the address and data information. Three addressing modes are provided:

- geographical addressing, which is determined by the physical location of the module
- logical addressing, used to make application programs independent of a module's physical location
- broadcast addressing, used to send messages to more than one module at a time

There are various control lines available to enable different types of transaction between master and slave. The basic transaction for a handshake read operation is shown in Figure 10. The master places the primary address (module address) on the AD lines followed by the address sync. control line (AS); the slave responds with address acknowledge (AK). The master then asserts the read line (RD) and data sync (DS); the slave responds by placing data on AD and by setting data acknowledge (DK). If no further data cycle is needed, the master will release AS, and the slave AK, thus terminating the transaction.

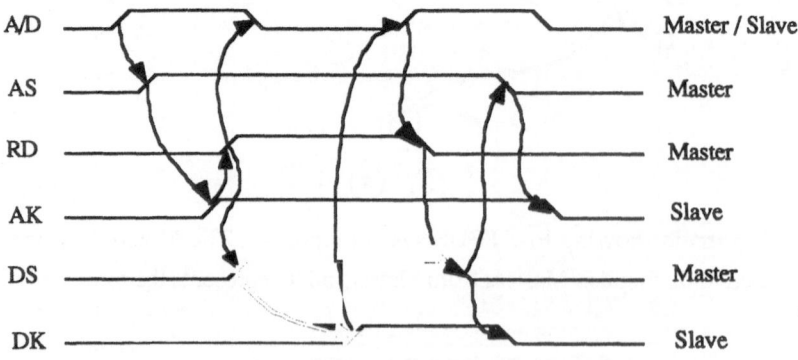

Fig. 10   Basic Fastbus Handshake Read Operation.

A single service request line (SR) is provided on each segment to permit a slave to gain the attention of a master. Each slot in a fastbus crate has a T pin wired onto the address line corresponding to the slot's geographical address eg. the T pin of slot 5 is connected to AD05. The master can determine the source of the SR by issuing a special broadcast called a T pin read.

Thus a sparse data scan, which permits the rapid location of devices containing valid data, can be implemented in Fastbus in the following way :

- a master issues a broadcast to all devices
- slaves with data available set their T-pin line
- the master reads the A/D lines to locate devices with data

In general, systems can exist with multiple masters dealing with multiple slaves on the same segment. An arbitration scheme is provided to prevent contention between competing masters and interrupt messages can be used between masters where needed.

The data flow model described in section 1.4 has been implemented in Fastbus for the Aleph experiment (Figure 11)[6]. In this model data flows from the front end modules along asynchronous pipelines towards a single point of concentration where the complete event is assembled. At the front ends Fastbus is used more as an I/O bus, with almost no multi-processor actions being required. Here the large board size permits a large number of electronic channels to be handled by a single module, which provides a very cost effective solution. Within each branch of the tree, the processors have been implemented as Fastbus modules which are masters towards the upstream modules and slaves towards the down stream modules. Such a scheme has the advantage that it is possible to skip readout stages by connecting the next higher master to a slave; this would not be possible with dual master or dual slave units. For the readout protocol between stages broadcast messages are used heavily, since this avoids a master having to address many slaves sequentially. There is no direct communication between processors at the same level in the hierarchy, since events do not need to be synchronised horizontally. This permits a fully independent parallel readout and therefore does not require modules to wait for slower neighbours.

2.1.3 VME

VME (Versa Modules Europe) is a commercial bus, proposed by Motorola, Mostek, Signetics/Philips and Hitachi[9]. It has emerged as an industry standard, with products being developed by well over 100 companies worldwide. It offers multi-master capability and like Fastbus it has its own addressing, arbitration and interrupt schemes. It also uses an asynchronous protocol. One important difference is that, unlike Fastbus, it was conceived as a single crate system rather than as a general purpose data acquisition system. In addition the VME bus supports 32 address lines and 32 data lines, and private lines are provided for use

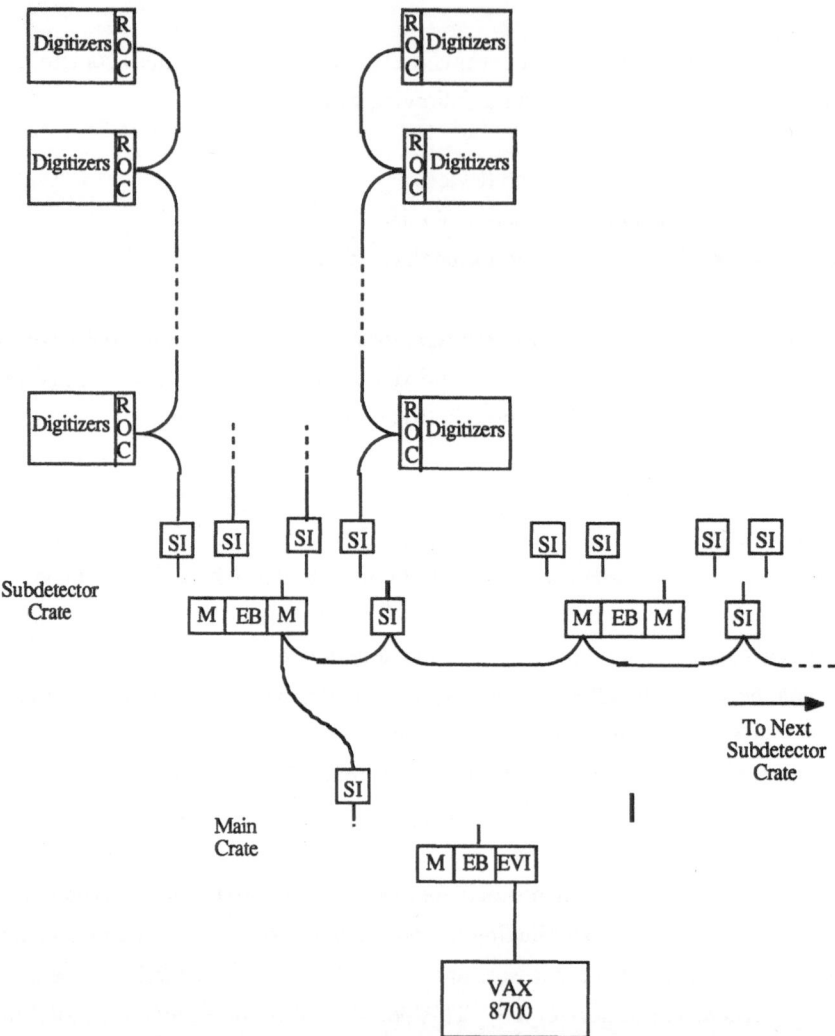

Figure 11 Fastbus implementation of the Aleph readout system. The diagram shows the three readout stages, namely the ReadOut Controller (ROC) which collects data from digitizers, the local Event Builder (EB) which assembles the data from a single subdetector and the main Event Builder which assembles the complete event. The interface between Fastbus and the VAX (EVI) is also shown.

in various extension buses (eg. VMX, VSB). Figure 12 shows the VME/VMX bus architecture.

The use of VME in data acquisition systems has been pioneered by the UA1 experiment, for which techniques have been developed for moving data between crates and modules designed for implementing the readout functions[11]. Many of these developments have been taken up by industry and are now available commercially. An overview of the UA1 readout system is shown in Figure 13 and serves to illustrate how the single crate VME architecture was overcome in a particular experiment. At the front end a parallel readout unit was developed consisting of the readout processor and bus driver module located in one crate and a dual ported VME/VMX memory module located in a second crate. The connection between the crates was made by extending the VMX bus from one to the other. The advantage of separating the modules in this way is that the two global data buses, ie. readout control bus and event data bus, are physically separate. This helps to minimise the contention for the bus by competing masters and thus enables the full bandwidth of the bus to be exploited. Each readout unit is assembling a different part of the same event in parallel with the other readout units. Full events are assembled and processed using a second dual-crate system, the event unit, located further downstream. A special Crate Interconnect bus was developed for connecting the two dual crate systems together.

It is not obvious which of these solutions to adopt, as indicated by the fact that both systems are now in widespread use. Although as a multicrate system, Fastbus is currently better suited to use in data acquisition systems, the complexity of the protocols and the limited industrial support are significant drawbacks. In addition, many of the limitations of VME have been successfully overcome by extending the original specifications. New standards are begining to emerge which will make the VME option increasingly attractive due to its widespread commercial use and support. In practice, both systems are often used in the same experiment, in situations where their relative strengths can be exploited to the full.

2.2   Signal Processing

The information generated by the detectors must be converted from analog to digital form before it can be processed by computers. Each detector channel must be sampled to check its occupancy, and only if it is inactive can it be suppressed from further data processing. Thus the front end electronics represents a substantial fraction of the total cost of the detector.

The first step in the signal processing chain consists of amplifiers and cable drivers for transporting the pulses from the detectors to the electronics racks . At the far end of the cable are line receivers, shaping amplifiers and digitisers. Considerations of space, power dissipation, accessibility and radiation hardness pose severe problems for the electronics

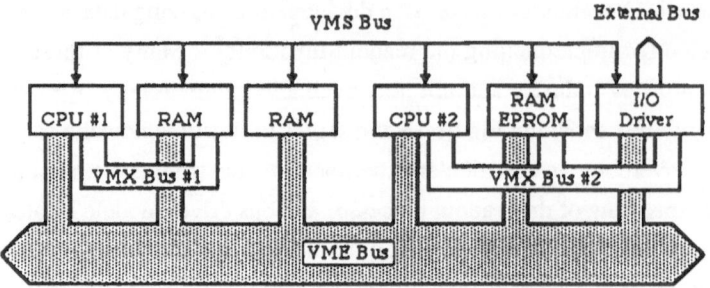

Fig. 12   VME / VMX bus architecture

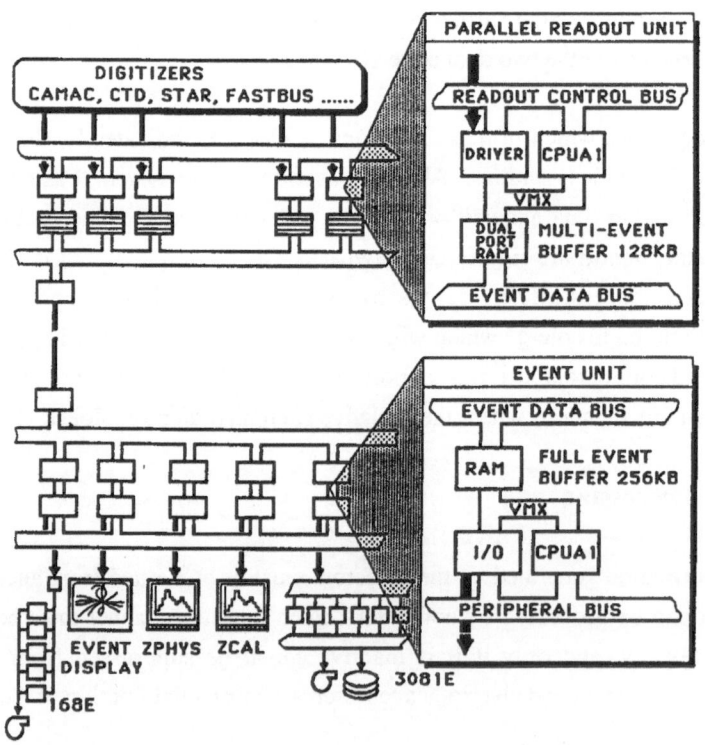

Fig. 13   The UA1 VME readout system [10].

mounted on the detectors. However the advances being made in custom VLSI technology (high density and reliability - low power consumption and fabrication costs), can help to alleviate these problems considerably.

Another problem to be overcome concerns the cabling. The cables themselves occupy large volumes and have to be brought out through the outer detectors. They introduce material which can distort the trajectories and energies of the particles measured in the detector and can also lead to gaps in the acceptance. The connectors are probably the most unreliable link in the whole chain. The cabling can be reduced by multiplexing the signals at the detector. The principle here is to store the pulses from each channel as they are produced and to read them out sequentially on the same cable at some later time when a trigger is formed. Multiplexing results in extra electronics at the detector and a longer readout time; the degree of multiplexing is limited by the time available for performing the readout. However it offers the only realistic solution for certain detectors, such as fine grain calorimeters and micro-vertex detectors, where the number of channels can be enormous. The multiplexing of signals into the same digitisers can result in substantial simplification of the readout and give large cost savings. It is the exploitation of VLSI techniques for packaging the electronics that makes this approach feasible.

On the SLD experiment at SLAC, the low repetition rate of the collider (SLC) has been exploited to permit a high degree of multiplexing in the readout[12]. A reduction in the trigger rate, from 180 Hz to 1-2 Hz, can be achieved in the 5.6 msec between crossings and thus a relatively long time can be allocated for the readout of information from the detector (~50 msec) without introducing substantial deadtime (~10%). Multiplexing at the detector has resulted in a significant simplification of the cable system and at the same reduced the number of acquisition modules needed in the Fastbus readout system. Capacitors are used for storing the charge, which follows the principle used in sample and hold ADCs. The charge can be recovered and sent to the digitisers at a later time on the arrival of a trigger. The number of capacitors needed for each detector channel depends strongly on whether waveform sampling is required. For the drift chambers 512 samples are taken over the full drift time, whereas for the calorimeter the integrated charge on each channel is measured. Thus two special custom VLSI chips have been developed, one for each detector[13].

A block diagram of the drift chamber electronics is shown in Figure 14. On the drift chamber chip (AMU), the input signal from a single detector channel can be switched to 256 sample and hold circuits and sampling rates up to 200 MHz are possible. The input waveform must be interleaved in time over two chips to permit the 512 samples to be taken. During event readout, the data from 128 AMUs (ie. 32,768 time buckets) are multiplexed onto an analog optical fiber at 1.6 μsec per time bucket. This results in a total readout time of 52 msec.

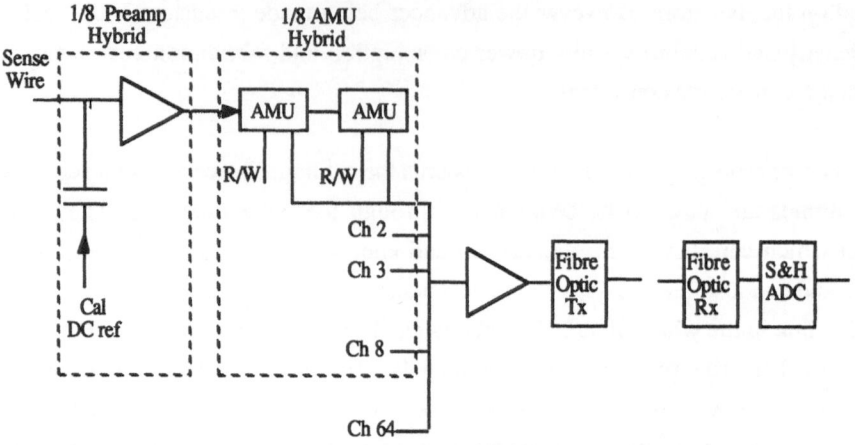

Fig. 14   Drift chamber electronics for the SLD experiment. Data from 64 wires (corresponding to 32,768 time buckets) are multiplexed onto the same optical fibre.

For the calorimeter, a chip with 128 circuits and 32 separate input channels is used (CDU), thus permitting up to four samples of the calorimeter signal. In practice only two samples are taken, one for the baseline and one for the peak. In this case only 192 channels are multiplexed onto the same optical fire giving a readout time of 1.2 msec. This short readout time is required because the calorimeter is completely readout for each beam crossing and used in the trigger.

The noise performance of these circuits corresponds to an 11 bit dynamic range. For the calorimeter the full dynamic range is covered by using two separate CDU channels for each detector channel, one with a gain of one and the other with a gain of 8. To reduce power dissipation, power to the electronics is pulsed only when it is needed ie. on each beam crossing.

It is interesting to note that this scheme corresponds to an analog delay line and thus can be used to implement pipelines on high repetition rate colliders. The Zeus collaboration have adopted a switched capacitor solution[14] for sampling the calorimeter signals and this is shown in Figure 15. The delay line consists of 58 capacitors and is clocked at the HERA bunch crossing frequency (96 nsec). The stored charges are cyclically overwritten until a first level trigger occurs. Again a special VLSI circuit has been developed with four channels on each chip. These components including multiplexors are to be mounted directly on the apparatus.

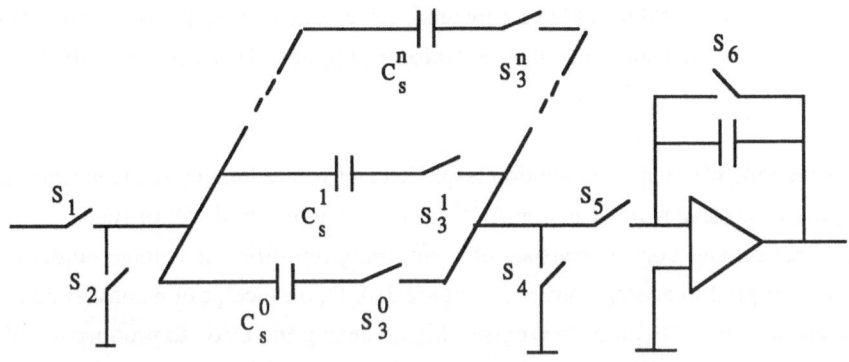

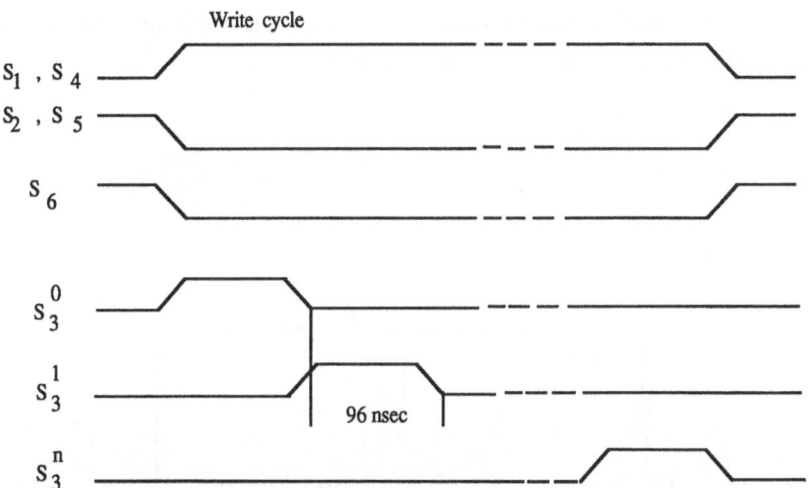

Fig. 15   Principle of the Switched Capacitor delay line for Zeus. In the write phase the switches S_1 and S_4 are closed, and S_2 and S_5 are opened. Only 1 S_3^i is allowed to be closed at a time and, when it opens, capacitor C_s^i will store the charge proportional to the voltage at the time of opening. The switches are closed and opened with the crossing frequency of HERA.

The final step to be performed in processing the signals is the analog to digital conversion (ADC) of the charge contained in the signal. ADCs with high resolution (>12 bits) and long conversion times (>1μsec) have been available for sometime and because of these properties their use has been confined exclusively to the readout stage of the DAQ system. More recently ADCs with somewhat poorer resolution (6-8 bits) but much shorter conversion times (10 - 100 nsec) have appeared, which has opened the possibility of using ADCs for event selection and time-sliced waveform sampling. These are generally known as flash ADCs (FADC).

In situations where high resolution is needed and where long conversion times can be tolerated, the successive approximation ADC provides the best trade off of speed, resolution, power dissipation and cost. It consists of a summing amplifier, a voltage comparator, a register and a digital to analog converter (Figure 16). Upon receipt of a convert command, control logic sets the 2^{n-1} bit of the register high, causing the DAC to produce a half scale input to the summing amplifier in a polarity opposite to that of the input signal. If the input is smaller than half scale, the amplifier output changes sign. This is detected by the comparator, which causes the control logic to reset the 2^{n-1} bit and set the 2^n bit. This cycle is repeated n times at the end of which the register contains the digital representation of the analog input to within 1 part in 2^n. This ADC is generally preceded by a sample and hold circuit as the input voltage must remain constant during the sampling period.

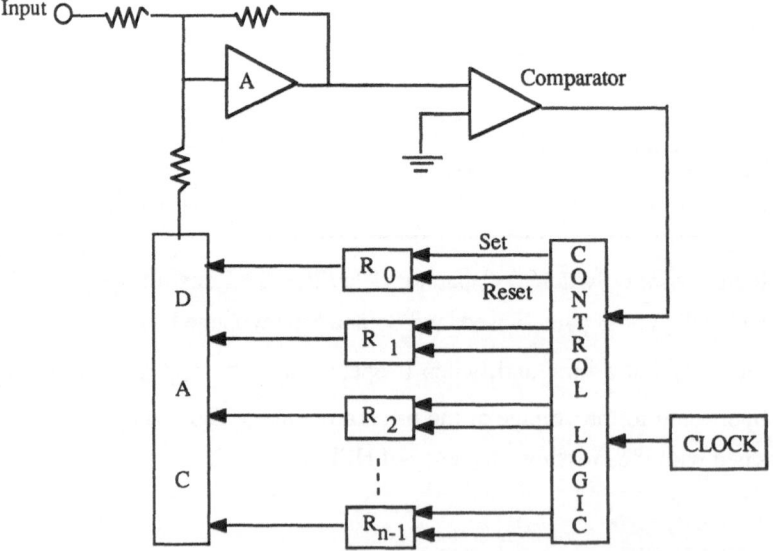

Fig. 16 Analog to Digital conversion by the successive approximation method .

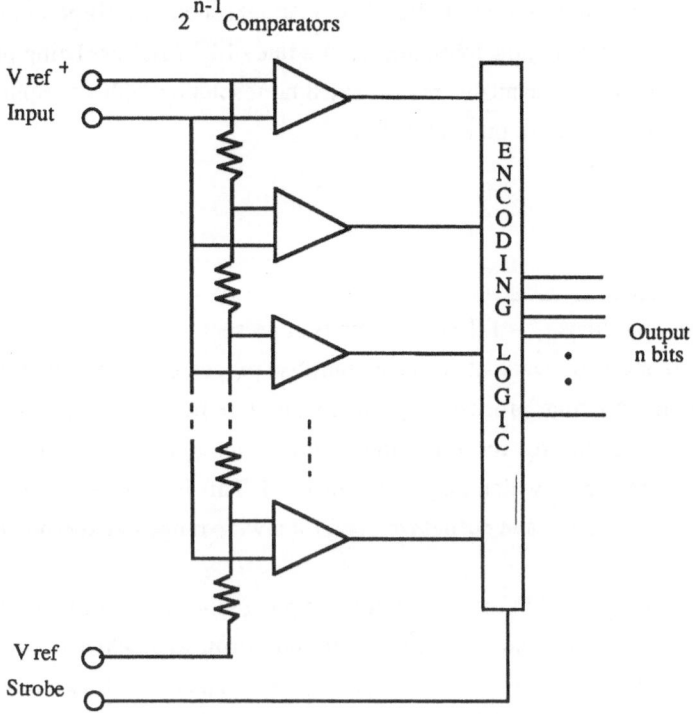

Fig. 17 Principle of the Flash ADC. 2^{n-1} comparator circuits are needed for n bit output resolution.

The flash ADC is a more recent development that owes its existence to the progress made in integrated circuit technology, since it basically comprises a very large number of voltage comparator circuits packed onto a single chip (Figure 17). The input signal is applied continuously to one input of each of a series of 2^{n-1} comparators. The other input is biased by a tap on a reference resistor divider, across which the reference voltage is applied. The comparators continuously monitor the size of the input signal voltage in comparison to the reference voltage. The arrival of a strobe causes the states of the comparators to be latched, and logic circuitry encodes the comparator outputs into binary form. The process is extremely rapid. However, each 1-bit increase in resolution doubles the number of comparators required and also doubles the power dissipation. One problem that is usually encountered is that the resolution of the FADC is inadequate to cover the dynamic range of the signals. One simple scheme for expanding the dynamic range is simply to digitise the signal from each detector channel twice, with amplification factors of order 10 between the two samplings. Another possible way of expanding the dynamic range up to the inherent precision of the ADC is by modulating the reference voltage with the input signal. This gives a non linear relationship between the digital output and the input voltage. The non availability or prohibitive cost of sufficiently fast high resolution FADCs will be a major problem for very high repetition rate machines such as LHC where readout schemes incorporating flash ADCs of 12-16 bits resolution sampling continuously at a rate of 80 MHz are being proposed. The highest performance devices available today have 8 bit resolution with a sampling rate of 100 Mhz, or 10 bit at 40 Mhz, or 12 bit at 20 MHz.

2.3 Processors

2.3.1 Hardwired Processors

For processing times of ~ 1-10 μsec there is at present no real alternative other than to use the hardwired solution. Dedicated special purpose processors offer a very cost effective solution to solving particular processing problems, but are necessarily extremely inflexible and very difficult to adapt for reuse in other applications once their usefulness has been exhausted on the project for which they were designed. This is reflected in the proliferation of dedicated processors that have been developed for a wide range of experiments[15].

The principle followed in the design of these processors is to precalculate functions of arbitrary complexity, store the results in a tabular form, and when needed access the appropriate table entry to obtain a fast real time response. The table entries are typically stored in Random Access Memories (RAMs) or Programmable Array Logic (PALs) and access times of <50 nsec can be achieved. The RAM is functioning as a logic device, behaving simply as a chip that produces a certain output signal (corresponding to the contents of the memory cell) when supplied with a certain set of inputs (used to calculate the address of the cell). In this way mathematical operations, such as computing transverse energy ($E \sin \theta$) can be replaced by a simple memory access. Relationships between quantities such as wire addresses that conform to a valid track segment can be expressed through the memory address and contents of the table. Changing the contents of the RAM therefore changes the trigger conditions and in this way these devices can be considered to be programmable. The problem remains to find a suitable algorithm which can be implemented through the use of look-up tables.

The Aleph level 2 trigger[16] serves as a good example of how look-up tables may be used to build a trigger which relies on the identification of tracks coming from an interaction vertex. Figure 18 shows the r - z projection of a simulated 100 GeV/c $e^{+}e^{-}$ interaction in the Aleph Time Projection Chamber (TPC), which is situated inside a solenoidal magnetic field of strength 1.5 Tesla . The second level trigger looks for tracks with psin $\theta > 1$ GeV/c in the r-z projection as these are essentially straight lines. For a track originating at the interaction point there is a direct correspondence between the drift time t_i at a given radius, r_i, and the track angle θ,

$$cot\ \theta = (\ L - t_i v\ )\ /\ r_i$$

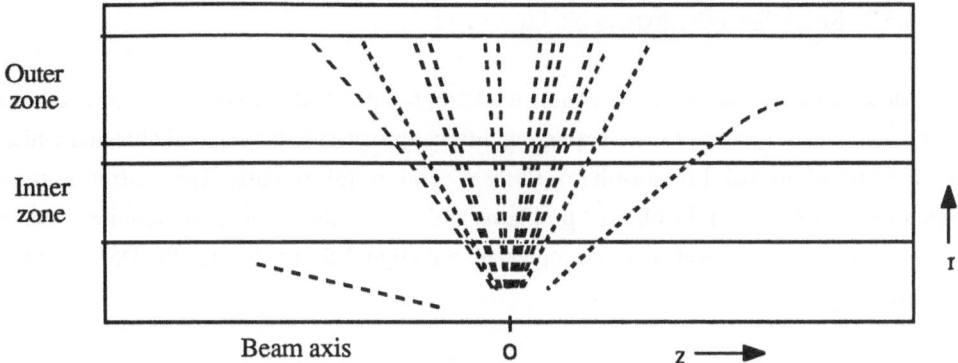

Outer
zone

Inner
zone

r

Beam axis   O   z ⟶

Fig. 18   Simulation of 100 GeV/c e^+e^- interaction in the Aleph TPC. In this r-z
projection the tracks approximate to straight lines pointing to the interaction point.

where L is the total drift length and v the drift velocity. Drift times are measured at several
different radii and these are converted to θ values using the above relation. A track from the
beam intersection point gives the same value of θ for each measurement, whereas a track
originating from up- or down- stream of this point gives a range of θ values. This is
illustrated in Figure 19.

Thus for the purposes of the trigger the r-z plane is divided into θ cells pointing to the
intersection point. The drift time for a hit on a trigger pad is converted to a cell number by
implementing the above relation in a look-up table and the hits within a cell are counted. A
good track is identified when a hit threshold is exceeded, and events with zero tracks or no
tracks from the z region close to the interaction point can be rejected.

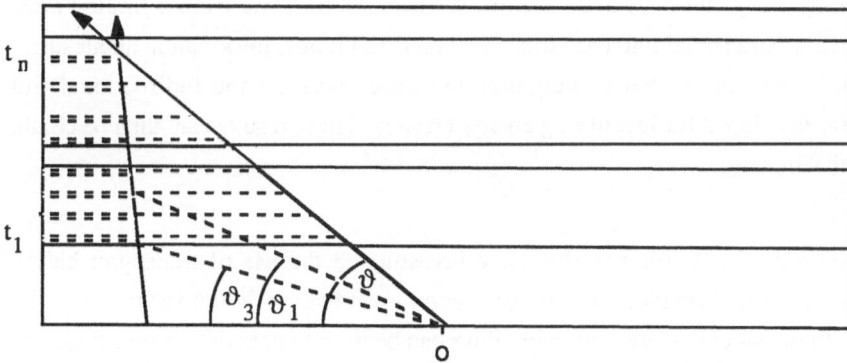

Fig. 19   Tracks from the interaction point have constant ϑ whereas for other tracks ϑ
varies. There is a simple relation between ϑ and the drift time t_n.

381

2.3.2 Specialized Programmable Processors

More recently a new range of commercial programmable devices has become available which use different approaches for making additional processing power available and which can be tailored to solving problems with special requirements. They offer a fully programmable solution to the class of problems which have data processing requirements in the 10 -100 μsec time region. Two examples are the Digital Signal Processor (DSP) and the transputer.

DSPs are widely used in industry for graphics and image processing applications. Their architecture, which is RISC based, has been designed to maximise throughput in data intensive applications, eg. the Motorola DSP 56000 can perform a 24 bit x 24 bit multiplication in 1 machine cycle (ie. 100 nsec). Their wide commercial use has resulted in a low cost per unit, and this, together with their high speed, makes them a very attractive proposition for use in both trigger and readout systems. The DSP is currently being used by Delphi for the implementation of part of their trigger system [17]. The Zeus collaboration are also investigating the use of DSPs for performing data reduction and reformatting[18].

Transputers are also based on a RISC architecture and form a family of processors roughly equivalent in processing power to the MC68000 series. The architecture of the transputer has been optimised for interprocess communication and in particular bidirectional links are provided for connecting cpus. Systems incorporating large numbers of processing elements can therefore offer substantial processing capability providing that the parallel nature of the architecture can be exploited in the application. It is therefore important to look at the concurrency aspects of the various functions of the systems being implemented and to design the architecture of the transputer array accordingly. Two typical arrangements are the pipeline and the two dimensional matrix. For example, the matrix arrangement would be appropriate for locating energy clusters in a calorimeter (Figure 20). Data from a limited area of the calorimeter would be fed into each transputer on a 32 bit bus. Information about surrounding cells could then be passed to neighbouring processors on the bidirectional links and algorithms developed for identifying energy clusters. These results can then be combined to form global quantities.

Again the Zeus collaboration have investigated the use of transputer based VME boards in both the formation of a global second level trigger[19] and for the event building function. In the latter case the transputer links can be used to pass data from subsystem crates to higher levels in the readout thus avoiding the VME crate interconnect problem.

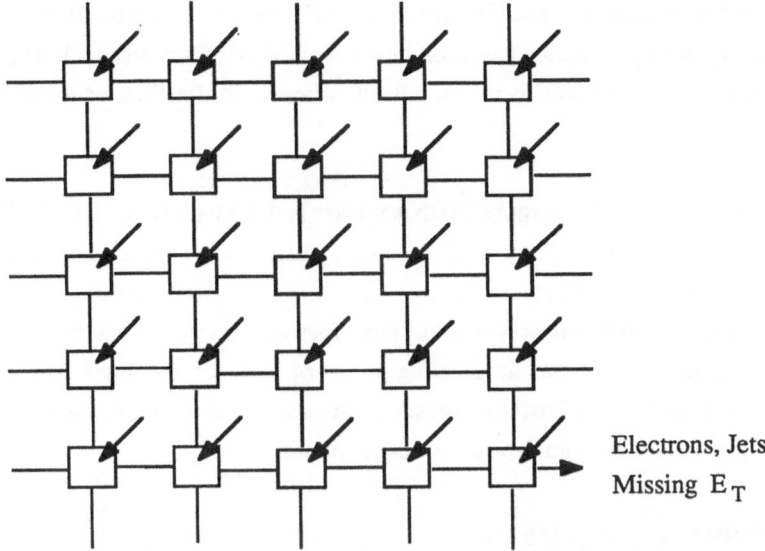

Electrons, Jets
Missing E_T

Fig. 20   A transputer array can be used to look for energy clusters in calorimeter cells. Data from the calorimeter are read into each processor from a given cell on a 32 bit data bus. Processing results can be passed between neighbouring processing elements on the bi-directional links.

2.3.3 Microprocessors

We have already seen from the data flow model described in a previous section the advantages of using a highly distributed processing system for reducing and collecting the data. The availability of powerful microprocessors (32 bit architecture, high speed, good software support) offers a perfect solution for implementing these functions, which have processing times of typically ~ 10 msecs. In addition there are other advantages to be gained from using fully programmable devices. Firstly, sophisticated data structures used by the software analysis packages (eg. typically those used by memory management systems such as ZEBRA and BOS) can be used for formatting the data, such that a unique format is used for all phases of the experiment. Secondly, the microprocessors themselves typically are only occupied for a small fraction of the time in performing their readout functions. The remaining idle time remains a huge resource of cpu power when integrated over the whole system and this can be used for monitoring locally the integrity of the data handled by each device. This reduces further the load on the central online computers.

The programmability of microprocessors offers a very flexible solution to processing problems and data acquisition modules that incorporate microprocessors can usually be used for different functions in the same experiment and even by different experiments. The basic problem in designing the module is to interface the processor to the data acquisition bus. VME being a commercial standard offers certain advantages, in that general purpose processing boards are always available, whereas for Fastbus a substantial design effort is required. In addition new modules based on new chip architectures are made available in VME essentially at the same time as the chip itself, whereas for Fastbus a substantial delay must be foreseen.

An example of a general purpose Fastbus master is the Aleph Event Builder[20], which uses a novel technique for interfacing a 32 bit microcomputer to Fastbus. Complete Fastbus transactions are handled by a Fastbus coprocessor of the M68020 CPU directly rather than via calls to a library. This processor adds new instructions to the standard M68020 instruction set, in the same way as the floating point coprocessor, making Fastbus a native bus of this enhanced cpu. All Fastbus operations are executed as single instructions. For example, a"single word read from dataspace" instruction is:

FBSWRD A_i, A_j, D_k

where A_i and A_j have to be loaded with the primary and secondary addresses and D_k will return the data value. Furthermore, the Event Builder provides at the coprocessor level full support for multi-user operation, creating up to 16 virtual fastbus machines, one per user. This allows the operating system to switch between different tasks using Fastbus with no overhead due to saving or restoring fastbus related registers.

In the block diagram of the unit (Figure 21) the MC68020 and the local area network controller chip for cheapernet (LANCE) are able to access both the processor bus and the memory bus. This dual bus structure has been chosen to allow normal processor operations and network activities while the fastbus coprocessor transfers data from the master port on the crate segment to one or both of the event memories. Access to the memory bus is controlled by an arbitration circuit allowing interleaved access from the cpu and fastbus coprocessor. The event memories are dual ported and accessible from the cable segment slave port. In the most complicated case one could have four concurrent activities taking place on the same memory, one event being read in from the master port, the previous one being treated by the cpu and the one before being read from the cable segment, while a histogram is being transfered via the local area network. The coprocessor contains a microcoded engine which takes care of the communication with the MC68020 and the very complex fastbus protocol, including arbitration, timeouts, error handling, service requests and interrupt messages.

Another interesting development for this board was the replacement of the DMA controller, originally implemented using ~50 discrete MSI chips by two custom integrated circuits. This freed sufficient space on the board to be able to add an extra 1Mbyte of program memory and reduced the overall power consumption at the same time.

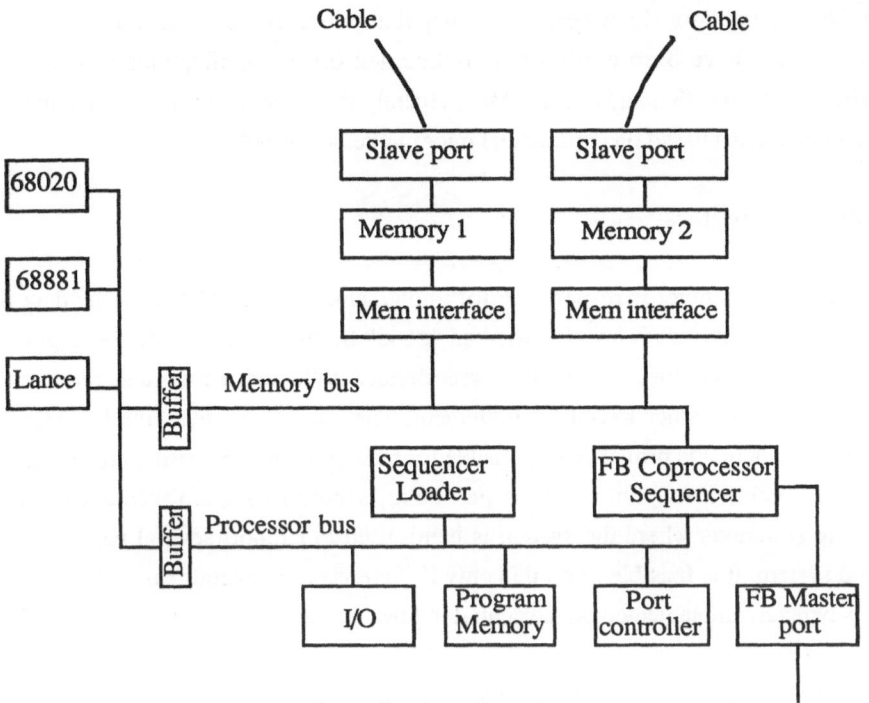

Fig.21   Block diagram of the Aleph Event Builder.

2.4   Processor Farms

As we have already mentioned, the final step in the trigger system normally consists of running some event reconstruction algorithm to perform a more precise analysis of the full event and thereby remove sufficient of the remaining background events to match the tape writing speed. This is a clearly a compute-bound problem requiring significant amounts of cpu power. The event orientated nature of the data means that each event can be treated independently and in parallel with the analysis of other events. These requirements lead to the design of a simple multi processor architecture with no global memory or sophisticated communication between components required. Each processor can run the same algorithm and must have private memory capable of storing at least one complete event. A simple means of passing data and retrieving the results is required together with some control task which keeps track of the status of each cpu and which distributes events accordingly. This scheme is very flexible in that extra cpu power can be added or taken away in a modular way as datataking conditions dictate. The collection of processors connected in this way is usually called a "farm". Processor farms are also used to augment mainframe computers for event processing offline where they again provide a very cost effective solution for compute-bound tasks.

In the past emulators have been a very popular choice as the processing element in a cpu farm[21]. They have been used in both online and offline applications in many HEP laboratories in both the USA and Europe. More recently their dominance in this area has been challenged by systems based on commercial microprocessors[22,23].

2.5   Online Computers

The online computer is responsible for implementing the final step in the data flow model. It must read those events classified as "good" by the final stage in the trigger and write them to the data store. As the data are already collected, formatted, reduced and classified this step is rather trivial to implement. The role of the machine has therefore changed to one where it is primarily responsible for managing the processors now being used for the more routine DAQ functions. In particular, access to system resources must be controlled in situations where the system is being used to support several partitions. In a distributed system, it is feasible to do this only if there is some overall supervisior and this function is typically implemented on the central online computers.

The programming environment on these machines is necessarily quite complex and a powerful, general purpose operating system can offer many facilities for making the programmer's task easier. In many cases robust commercial software packages can be used to supplement the operating system and this is particularly true for data management and operator interface aspects. Software tools for coding, debugging, performance analysis can greatly improve the programmer's efficiency, especially in the development phase of the project.

The need to offer some degree of independence to the various groups responsible for different parts of the detector has led to the use of equipment computers each designated to be used for managing one of the major detectors. Thus it is common to have a number of smaller online machines rather than 1 large machine. This choice can create its own problems when trying to implement the data acquisition functions but it removes the single point of failure problem which tightly couples together different users using a single machine. In general these equipment computers require only access to data coming from a particular detector in the experiment .

A popular choice for the online computer in many HEP experiments is taken from the VAX family. In particular, the option to "cluster" these machines provides an elegant way of implementing a multi-cpu configuration, whilst maintaining centralised system management and permitting the sharing of expensive peripherals. A notable alternative to this approach is found in VME based systems, where the Macintosh is often preferred. The ease with which the Macintosh can be interfaced to VME and the increasing availability of commercial software (eg. database packages and object oriented human interfaces) is making this option increasingly attractive. However centralised control is more difficult to implement using the Macintosh solution.

2.6   Local Area Networks

Good communication services are essential in a system of distributed processors and these can be easily implemented using a commercial Local Area Network (LAN), such as ethernet. The LAN can be used to implement a second communication path (ie. in addition to the data acquisition bus) between the readout processors and the online computers. The availability of commercial LAN controllers, such as the LANCE chip for ethernet, makes the task of connecting a readout processor to the network a relatively easy one.

The LAN is typically used to carry messages to implement the various communication protocols and may be preferred over the DAQ bus for moving small packets of information where bandwidth isn't a problem. It offers a valuable possibility for debugging system errors, when there are suspicions that the data acquisition bus is not working correctly.The LAN may also be used to provide additional communication services, such as remote terminal access and remote file access.

A second use for the LAN is to interface certain peripheral devices, such as terminals and printers, to the computer configuration in a symmetric way. LANs are also used to integrate workstations and can provide on site access to central computing facilities.

3. SOFTWARE FOR DAQ SYSTEMS

As we have seen, data acquisition systems typically consist of hundreds of powerful processing elements communicating through high speed busses and local area networks. Improvements in the commercial fabrication of VLSI circuits have made this possible through the production of a whole new range of electronic devices. It is perhaps fortuitous that these advances in hardware come just at the time that our data processing requirements have become so acute, and it is not surprising that this progress has been eagerly exploited. However, this trend has clearly had a major impact on the degree of sophistication of the software needed to implement the systems. For example, many of the control and acquisition functions that were previously allocated to hardware units are now entrusted to software and so the constraints put upon the reliability and correctness of the software have become much more severe. The long lifetime of the systems also poses problems for the maintenance of software. In particular the reusability of the code will have a major impact on the manpower needed to maintain and modify the system as requirements evolve with time.

In the computing industry much experience has already been gained in developing large software systems. Originally efforts were directed at finding ways of improving the quality of the code, which resulted in a set of programming guidelines being introduced (*Structured Programming* [24,25]). Subsequently new languages with additional redundancy were developed, in which compilers could be used for detecting programming errors. However in large systems the worst errors were traced to a poor understanding of what the system being programmed was supposed to do. Also large projects were found to have specific management problems due to the difficulty of decomposing the system into managable components and establishing effective communication between members of the team of programmers [26,27]. It became apparent that these problems were best overcome by introducing an *engineering like* discipline to software design, in which products are conceived, specified, designed, built, tested and maintained until they are finally discarded. Therefore emphasis is now placed on the earlier phases in the software life-cycle, the motivation being to prevent errors from being introduced or at least to trap them as early as possible. New technical methods have been developed for deriving a more convenient way of specifying the functional requirements of the system and for making a formal system design.

Currently there is great interest in the use of these methods. This is due partly to the increase in scale and complexity of the systems themselves, as well as to the fact that the methods offer a particularly simple and convenient way of formulating the design and can be used effectively in a wide range of applications. A complementary set of management procedures and diagramming techniques, or Methodology, can be used throughout the software life-cycle, ie. from the time it is conceived until it is discarded. In the following sections some of the technical methods that can be used in making a software design for a data acquisition system are described and examples given to illustrate their use.

In addition to the development of design methods a lot of progress has also been made in establishing standards for a wide range of software applications and this has led to many commercial software packages appearing on the market. Various pragmatic decisions therefore have to be made as to what type of operating system to provide, what communication protocols to use and which computer language is the most suitable. Some of the issues involved in choosing the general programming environment are briefly discussed.

3.1 Software Methodologies

A software methodology is a collection of methods, each method being specially adapted to cover some aspect of the software development life-cycle. Broadly speaking, the methods can be sub-divided into four classes:

• Modelling heuristics. These are guidelines for decomposing a system into its components. The result of following these guidelines is to generate a number of models for describing the system's environment, the system's intended behaviour and details of the chosen implementation.

• Modelling Tools. These rely heavily on the use of diagraming techniques as a way of formulating the design. This approach is far preferable to using standard text, which tends to be verbose and often imprecise. The techniques are simple to understand and form a very convenient basis for discussing parts of the system in detail. More detailed descriptions of data items and algorithms are provided using a mathematical style notation and are stored in a dictionary.

• Management procedures. These consist of guidelines for monitoring progress and ensuring that at regular stages in the development cycle the system design is reviewed. Reviewers are normally selected to have a wide range of backgrounds, so that the design can be seen from different perspectives. The status of the project can be more easily assessed, as documentation is always available in a convenient form (ie. models and dictionary).

• Automated Tools. These are themselves software products which can be used to minimise the effort needed to create and maintain the models. They can, in addition, be used to automatically check different parts of the model for completeness and consistency.

Many different methodologies have been developed [28], reflecting the fact that their originators had different philosophies as to the best approach to designing software. Certain techniques are particularly suited to particular applications. However this is still an active area for research and the methods are evolving rapidly as more experience is gained using them. The examples which follow illustrate the Structured Analysis and Structured Design Methodology (SASD)[29,30], which has been used by the Aleph collaboration and by several other groups at CERN and in the United States.

3.1.1 The Modelling Process

The basic distinction drawn when deriving models of the system is to separate clearly what the system should do from how the system is realised. In other words to describe the logical behaviour of the system before decisions are made on how it is to be physically implemented. This helps the designer to avoid mistakes due to preconceptions he may have about the choice of implementation. The precise steps to be followed in deriving the model will depend to a large extent on the nature of the system being modelled, but usually fall within three stages.

The first step is to define the *system context*. This involves making decisions about the true extent of the system ie. what is to be modelled and what is to be left out. It is achieved by defining how the system interacts with its environment and is usually represented by a single function transforming a number of inputs into outputs.

The second step is *system decomposition.* There are at least two ways of decomposing a system into its components. The traditional way is called functional decomposition, which is a strictly top-down approach . At the top level, it involves splitting the system function into its principal sub-functions. Each sub-function may then be used as a starting point for a further level of decomposition. The decomposition continues until, at the lowest level, each sub-function can be described in about half a page of text. This approach has the disadvantage that the only guideline for achieving the decomposition is that the result should reveal minimal interfaces between the various components. An alternative approach, which is particularly appropriate in real-time applications, is to make a list of the stimuli in the system's environment and to model the response of the system to each stimulus. The model is finished by completing the upper layers of the design and, where necessary, by decomposing the event responses into more detailed levels.

The final step is to add details of the *implementation* to the logical model. Where concurrent operation of parts of the system is possible, and even required for performance reasons, the relevant components must be allocated to distinct processing elements. This step also involves allocating components to tasks to be run under the processor's operating system and defining the architecture of the procedures within those tasks. The logical model thus has to be revised to indicate the new partitioning of the system to accommodate the implementation choices. New interfaces between these components have to be defined and the synchronisation of the various elements specified.

3.1.2 Modelling Tools

A model of a software system basically consists of descriptions of the ways in which the system generates a set of outputs from its set of inputs. It also consists of descriptions of the information (ie. data) needed by the system to effect the transformations. A complementary set of tools is needed for describing the various relationships between the data items and transformation processes. In SASD the four techniques shown in Figure 22 are used.

The model is started by drawing Data Flow Diagrams (DFDs), which show the principal data flows in the system (Dn) and the processes which transform them into oneanother (Pn). Sources and sinks of data can also be represented on the diagram(Sn) and data stores may be used to indicate the temporal way data is transferred between processes. The notation can be extended to distinguish data flows, which have real information content, from control flows (Cn), which just serve to trigger some particular action. Decomposition of the system results in a leveled set of DFDs, each diagram representing one level of abstraction of the system design.

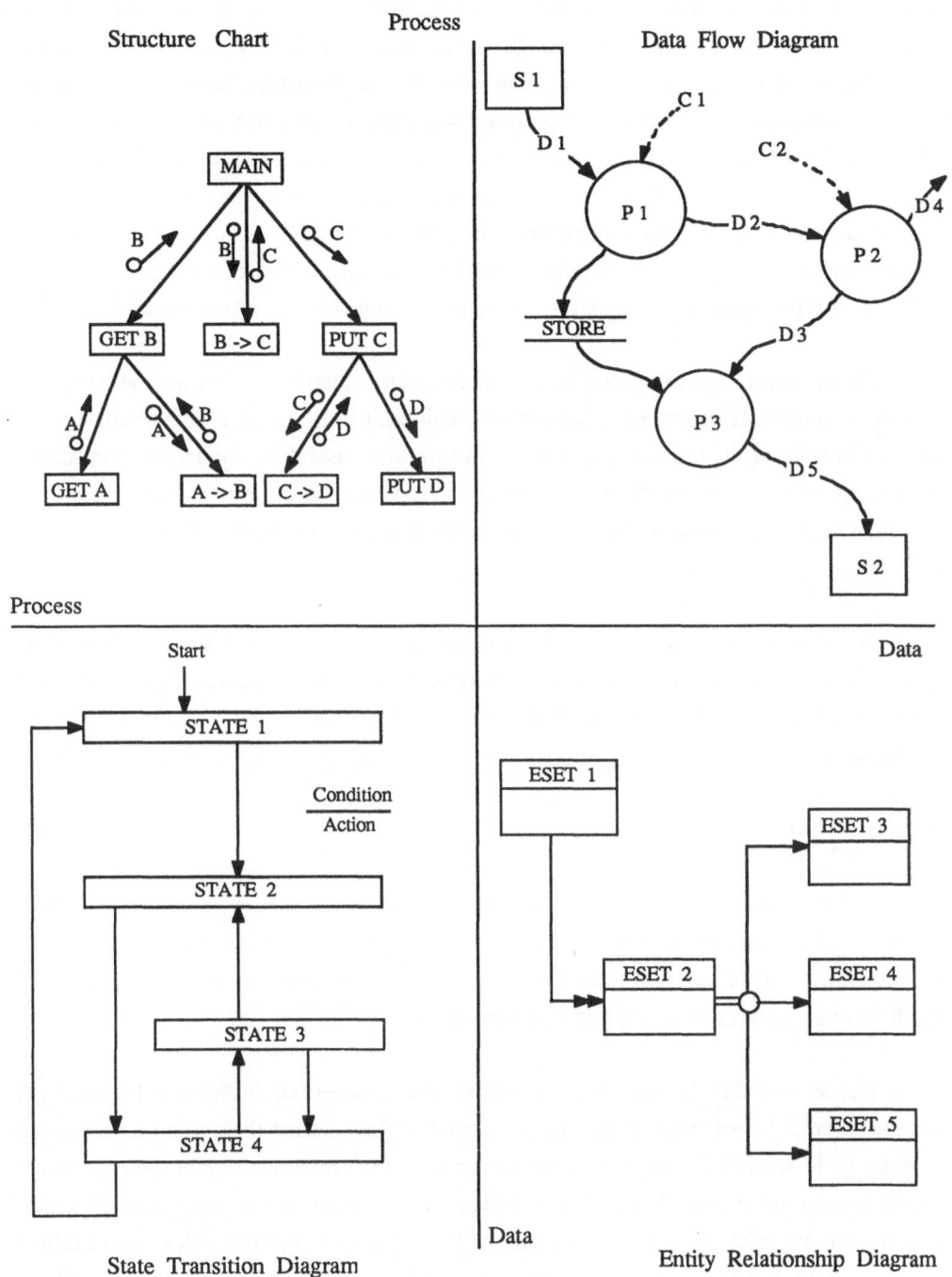

Structure Chart

Process

Data Flow Diagram

Process

Data

State Transition Diagram

Data

Entity Relationship Diagram

Fig 22   A complementary set of diagramming techniques for modelling different aspects of
the software design in terms of the relationships between processes and data items .

In some cases the data flows are composite in nature and certain data stores take on more the character of a database. In these situations, it is difficult to continue with the decomposition of the system until a better understanding of the true content of the data is achieved. The Entity Relationship model provides a tool for describing data in terms of the basic objects, or entities, being manipulated and the relationships between them. Each instance of a particular type of object occupies one entry, or row, in a table, the number of columns being determined by the number of attributes assigned to the object. In principle, adopting such a model avoids the need to duplicate any piece of information and therefore ensures that the data is always consistent. In practice, certain redundant information is usually required to improve the efficiency with which information can be retrieved. This technique is typically used for designing data to be stored in a relational database[31].

The third technique, the State Transition Diagram, views control flows and processes in terms of conditions and actions. The precise relationship between the two is determined by the current state of the system. These diagrams are most useful for modelling transaction processing, where a combination of the state of the system and the transaction requested determines the action to be performed. Note that transaction processing is very common in real-time systems.

The final technique, the Structure Chart, is used for making the detailed design of the system. It shows the system in terms of its software modules and indicates the procedural calls relating them. Parameters passed between program modules may also be indicated on the diagram.

3.2 Practical Examples

In this section an attempt is made to illustrate how the techniques are used in practice, by showing how some features of a typical experiment may be modelled. Emphasis is placed on showing some of the real-time aspects of data acquisition and control, as ways of handling these features have only recently been developed[32,33].

The context of the experiment is described by the context DFD shown in Figure 23. A general feature of most DFDs is that the principal data flow is from the top-left corner of the diagram to the bottom-right corner. Thus, in this case, the prime function of the experiment should be interpreted as the production of new physics results from a measurement of particle interactions. We also see that the boundary of the experiment has been chosen to exclude details of the operation of the accelerator. Physical constants and parameters describing various aspects of the detector (eg.survey constants) are assumed to be available in an a-priori fashion. The experimenter is also seen as part of the environment, since he controls operation of the experiment through commands and monitors system performance through information generated by the system.

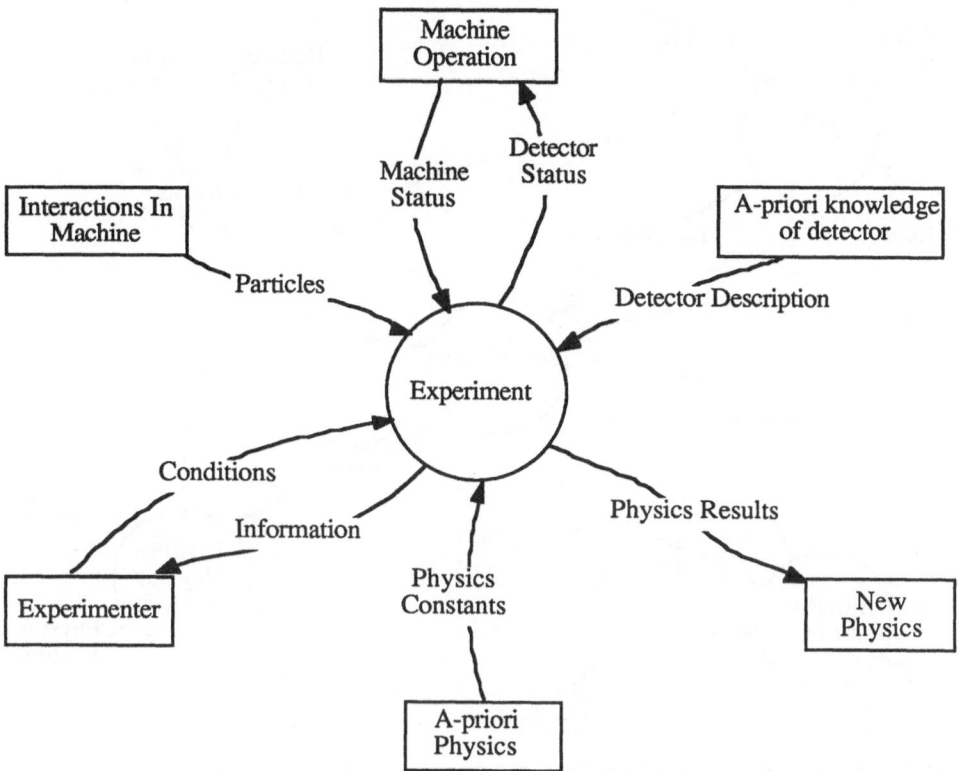

Fig. 23   Context diagram for a typical particle physics experiment.

The first level of decomposition reveals four components of the system (Figure 24). The three major data processing functions are seen to be the assembly of event records from the digitisings produced by the detectors, the reconstruction of the events to identify tracks and compute energy and momenta of their associated particles, and finally analysis of the event topologies to look for interesting features of the physics of the interactions. The same transformations can also be made to process simulated data generated by the fourth component. Comparison of results obtained with the real data and the simulated data help to check whether various aspects of the experiment are understood. This diagram also shows that the four components all access a common database, which is the primary and unique source of all information needed for performing the data processing. Note that, for consistency, the net inputs and outputs on this diagram should correspond exactly to those in the context diagram. This may not be entirely obvious from a cursory glance at the diagram, as some of the flows in the higher levels are actually labelled, for convenience, as composites of several data items. The data dictionary should contain sufficient information, however, to enable this consistency to be verified.

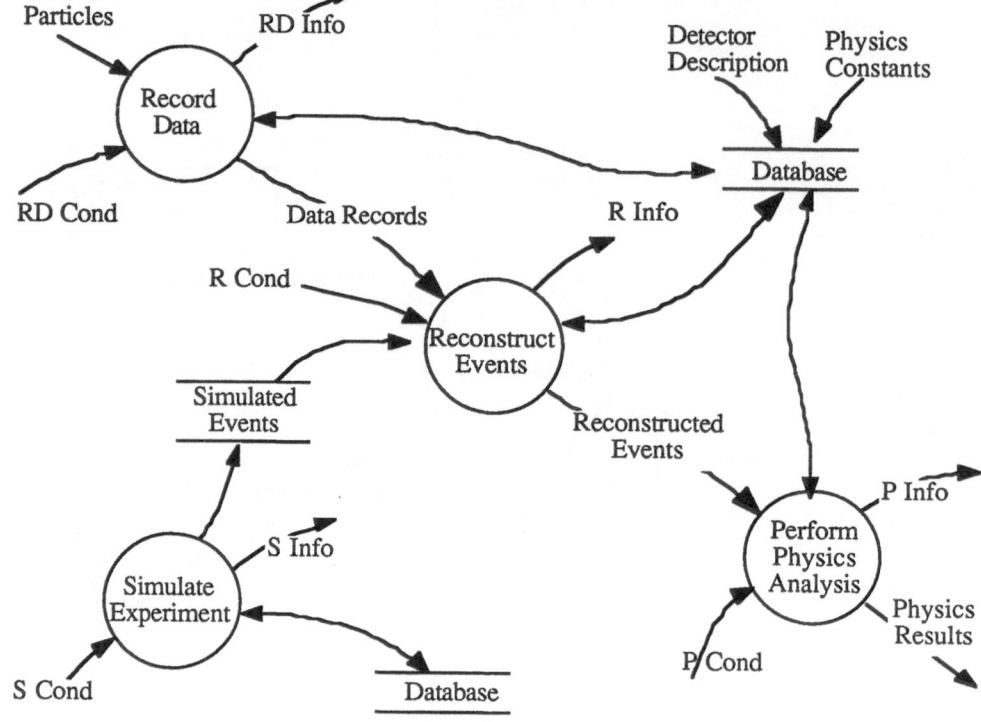

Fig. 24   Data Flow Diagram showing the principle software components resulting
from one level of decomposition below the context diagram.

The two DFDs shown in Figure 25 resulted from the decomposition of the Record
Data transformation and are good examples of the two basic topologies of DFD. The first
diagram shows a data flow which is successively transformed by a number of processes
until it emerges in the required form. In this case, the arrival of data at the beginning of the
pipeline is sufficient to cause the activation of each process in turn. The second diagram
shows a number of processes each of which acts independently of the others. Unlike the
first example, each of these processes is only activated on receipt of a control signal, which
is triggered by the arrival of a stimulus in the system's environment. By keeping track of the
state of the system, the control transformation (dashed circle) can manage the sequencing of
the various actions according to the type of stimulus received.

Data transformations and control transformations are specified using two different
techniques. Data transformations are specified using Structured English, which has a limited
vocabulary comprising terms for representing flow of control and data items appearing on
the diagram. It is most often targeted at a structured programming language familiar to the
designer. The specification for *Build SDEvents from SDDigits* (where **SD** stands for
Sub-Detector) may appear as follows:

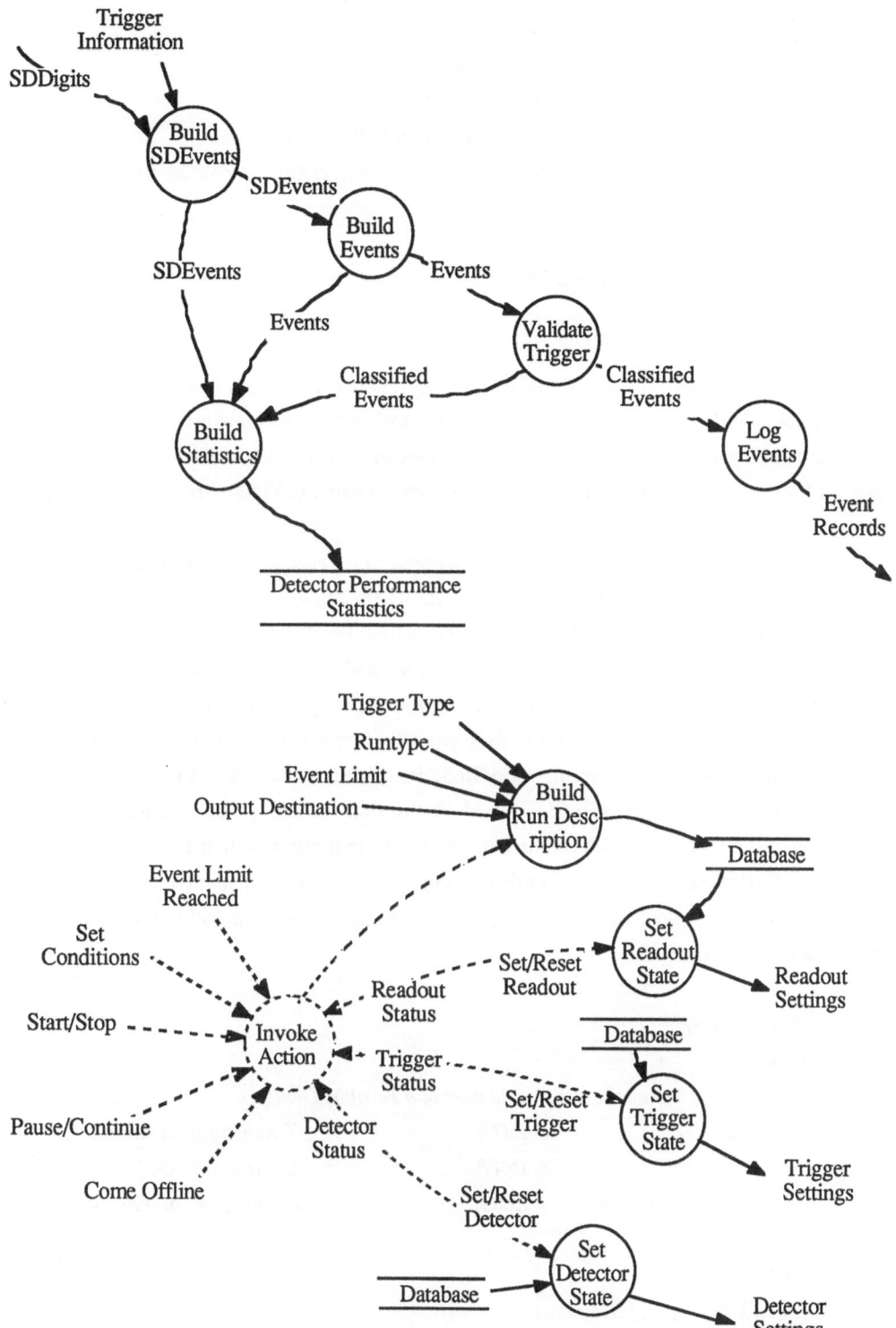

Fig. 25   Two Data Flow Diagrams illustrating the two basic types of topology. The top
diagram shows a dataflow pipeline and the bottom diagram a transaction centre.

Begin

     Get Trigger Mask

     Get Trigger Configuration

     Form Readout List

     Do while Readout List still has entries

          Read SDDigits from next Subdetector on list

          Remove Subdetector from list

     End do

     Fill SDEvent header

End

The control transformation is specified using the State Transition Diagram. The control inputs are the *Conditions* and the processes the control outputs activate are the *Actions*. The state of the system is held in the memory of the control transformation. The State Transition Diagram for the control transformation in Figure 25 is shown in Figure 26.

In the DFD shown in Figure 25, the process *Set Readout State* is responsible for reading information from the database and generating the readout settings needed to prepare the system for datataking. Before this process can be decomposed further, a better understanding of the precise nature of the data is needed. Figure 27 shows a small extract from an Entity-Relationship Diagram describing a readout system implemented in Fastbus. The advantage of such diagrams is that they can be analysed to see if they offer a correct interpretation of the system being modelled, in this case the Fastbus Standard. The correctness of the data structure described in the data model gives confidence in the correctness of the software based on the data model. Data items and their inter-relationships are best specified using a mathematical-style notation. These specifications can be complemented by comments and are managed in a data dictionary. The following are extracts from the Aleph data dictionary[34]:

DEFINE ESET

Port

    : 'Geographical addressable part of a Fastbus Device'

= (GA	*= INTE*	*: 'Geographical Address',*
DA	*= INTE*	*:'Device Address',*
EnLogAdd	*= LOGI*	*:'Enable logical addressing')*

DEFINE RSET

(Port[1,1] -> [1,*] Segment)

    :'A port is identified by a geographical address on a specific segment'

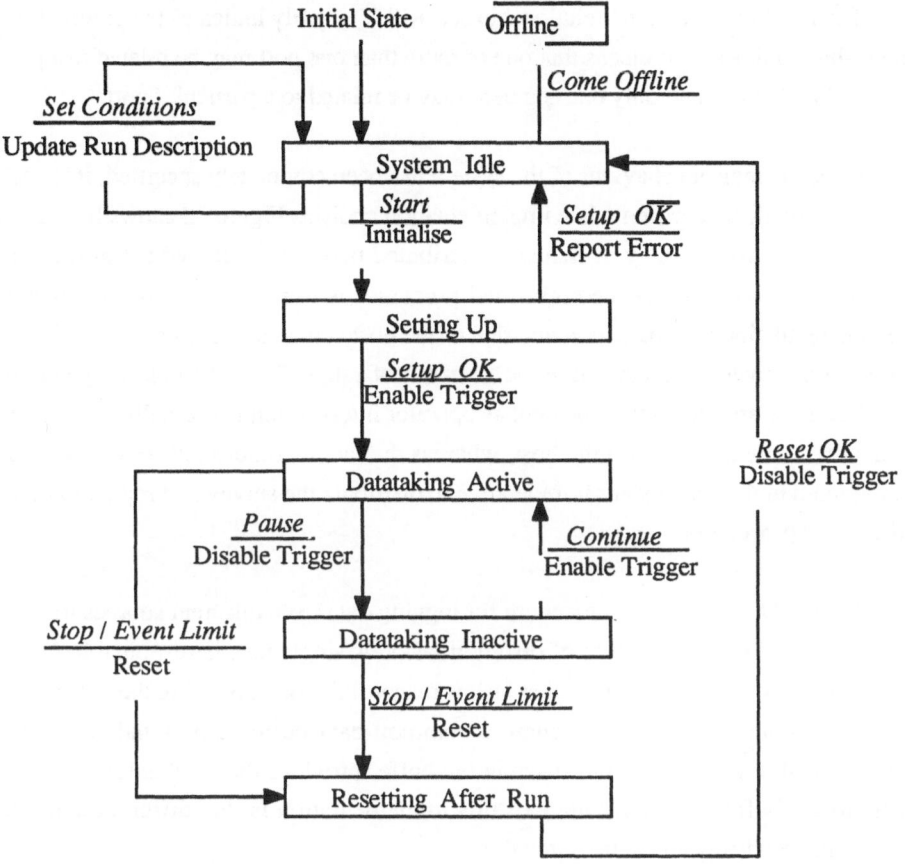

Fig.26 A State Transition Diagram for specifying the control transformation in Fig. 25.

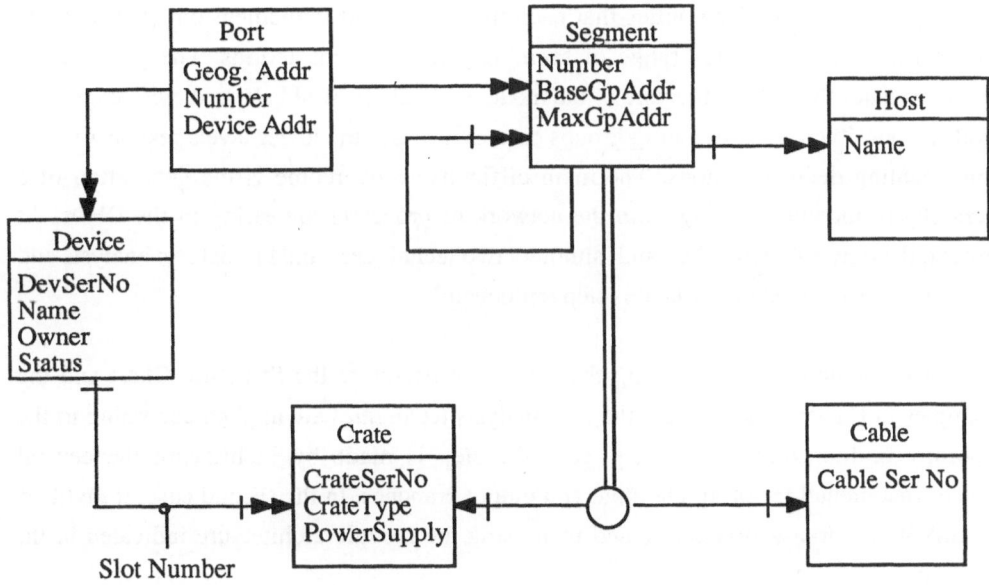

Fig. 27   Extract from the Entity Relationship diagram used to model a Fastbus system.

The notation used in the relationship set (RSET) merely indicates the cardinality of the relationship. In this case it means that one or more than one port may be related to a particular segment, but that one and only one segment may be related to a particular port.

Once the logical behaviour of the system has been completely specified, it is necessary to adapt the model to accommodate implementation choices. Figure 28 shows how a function for regulating a voltage may be actually distributed between an embedded processor and a host processor. The role of the embedded processor is to read the current voltage and to apply the regulation automatically, and that of the host is to determine the nominal value and to report errors when the current value goes out of range. This example is typical of how control systems are often implemented, as operator intervention is normally through the host and databases are managed on the host, whereas the fact that thousands of channels need to be monitored in this way makes it imperative to distribute the survey and regulation functions to dedicated processors.

Figure 29 shows how a procedure for logging data from different sources may actually be implemented using a number of different tasks. A single task is responsible for reading data from each data source and another task is responsible for consuming the data and writing them to the data store. All tasks share a common data buffer and signals indicating the availability of new data and free space in the buffer provides the mechanism for managing access to the buffer. The advantage of this implementation is that different activities can proceed independently and at the same time.

The final stage in deriving the implementation model is to allocate the program modules. The two major principles to be followed in this stage of the design are modularity and hierarchy. Modularity implies that each module should implement one well-defined function and not a complex range of seemingly independent actions. The principle of hierarchy states that the architecture of the modules should be well balanced, with the higher levels responsible for coordinating groups of activities and the lower levels responsible for implementing detailed actions. The main difficulty to overcome is the generation of a hierarchy of modules starting from the network of processes appearing in the DFDs. At present, the only solution is by hand, although two techniques called transform analysis and transaction analysis help to make this step reproducible.

The method for representing the procedure design is the Structure Chart and the examples in Figure 30 show how the two analysis techniques are applied according to the topology of the DFD. In the first case the step is made by identifying the central transformation and factoring the input and output branches. In the second case, it involves identifying the transaction centre and then using the module architecture indicated in the diagram.

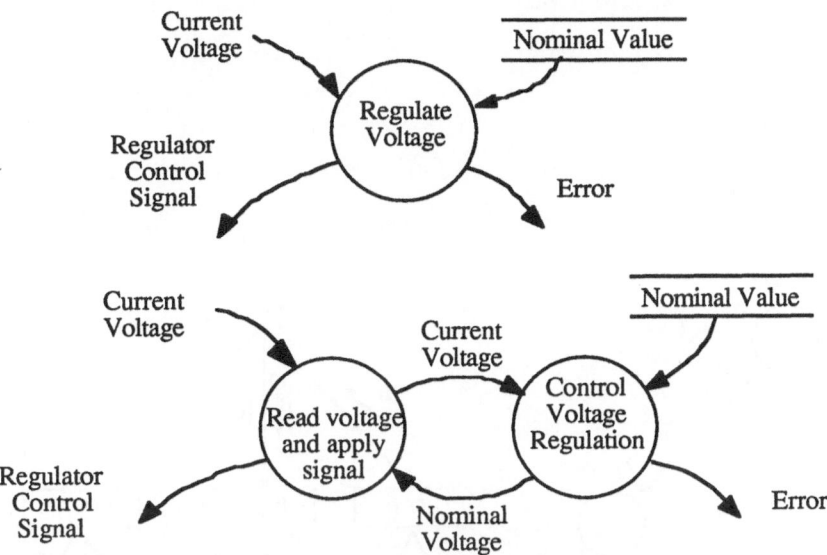

Fig. 28   Deriving an implementation model by allocating the voltage regulation function to two processors .

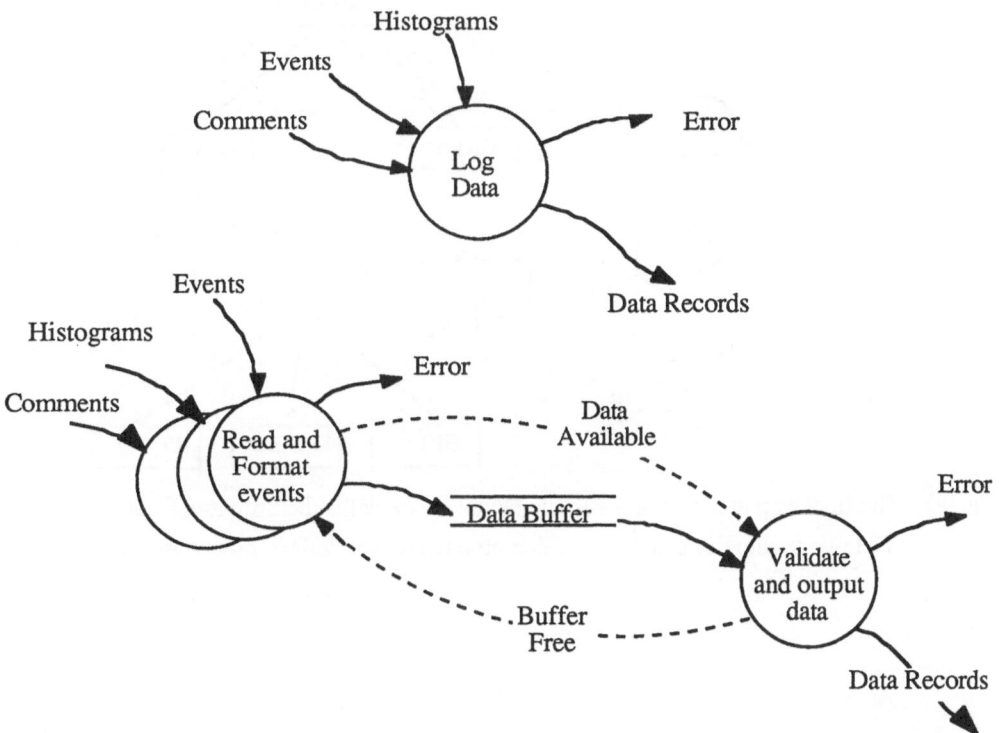

Fig. 29   Deriving an implementation model by allocating different processes to fulfull various aspects of the Log Data function .

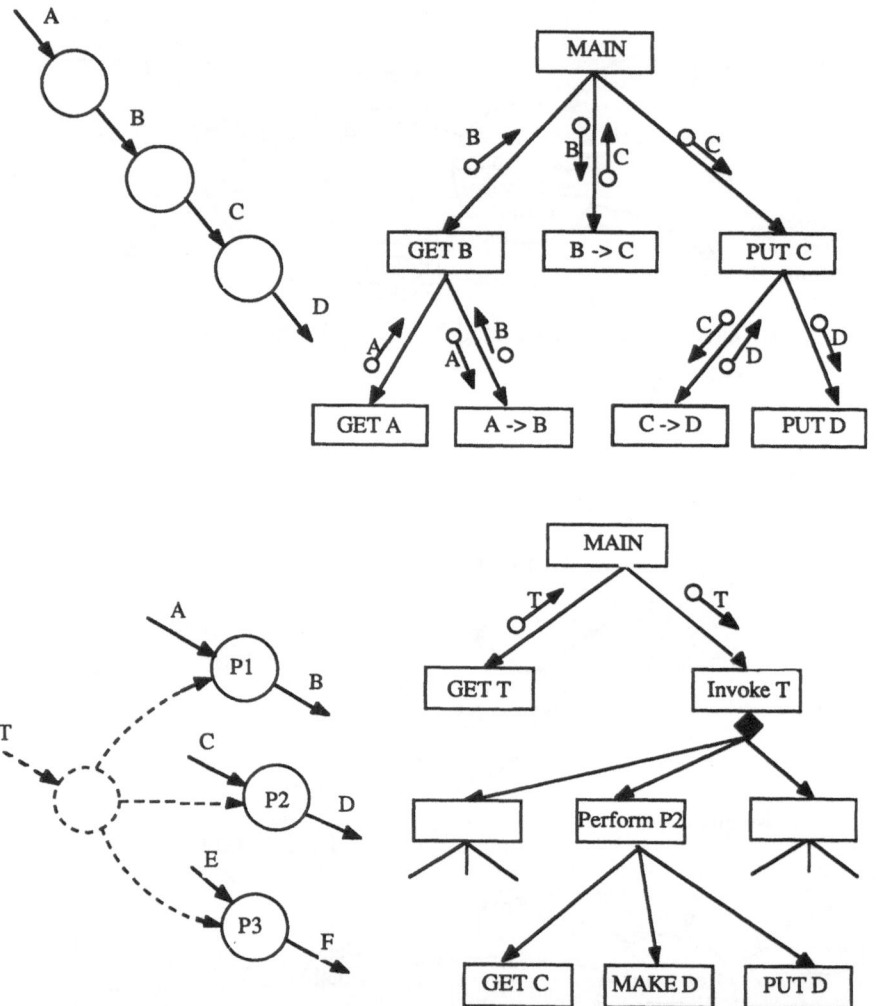

Fig. 30   The final step in deriving the implementation model is the Structure Chart.
The two topologies of DFDs studied give two characteristic procedure structures .

3.3 Evaluating the Model

A big advantage of modelling the system using graphical techniques is that the correctness of the design can be very easily checked by a careful study of the diagrams. Examples of common mistakes, which are easily detected, include processes which create output from data they don't have access to and write-only files.

The content and topology of the diagrams also give an indication of the quality of the design. For example, poorly named dataflows and processes usually indicate that the designer does not have a good understanding of the relevant components. Complex interfaces between components should be avoided as much as possible, as systems with a high degree of coupling between components are extremely difficult to understand and validate, and almost impossible to maintain. The presence of complex interfaces probably indicates that the system decomposition has not been performed correctly.

One of the ways in which program modules can couple is through access to common data areas. In this situation any change to the data structure in these areas can have a major impact on the system, as all modules accessing the data areas directly are immediately affected. This situation can be avoided if these data areas are hidden from the bulk of the system by a set of access procedures. Any change to the data structure only necessitates modifications to this set of routines, which can be built and maintained by the person responsible for maintaining the data. This is known as the principle of information hiding[35].

The architecture of the program modules, as represented in the Structure Chart, may be changed to improve system performance by rearranging modules such that unnecessary procedure calls are eliminated. Thus the modularity of the system is maintained and the basic design is not compromised.

3.4 Automated Tools

The process of entering information into the model and checking the consistency of the different parts of the model is a highly labour-intensive activity and the importance of automated tools for improving the efficiency with which this can be achieved cannot be stressed enough. The development of automated tools has necessarily followed the development of the methods and it is not until recently that really comprehensive automated tool packages have become available[36].

Tools available for making the system design include graphics editors for entering the diagrams and syntax directed editors for entering information into the data dictionary. These tools also provide automatic checking of the diagrams and data dictionary for consistency. In some cases they will also provide options for generating structure charts automatically from dataflow diagrams. Many of the tools were developed for use on PCs to exploit the user friendly interface available. However in large projects the design encapsulated in the database must be available to all those working on the project, which often implies that the database should be accessible over wide area networks. Other tools have been developed for use on workstations and though these tend to be more expensive they do not usually suffer from the limited accessibility of the PC.

Tools to support the coding and testing phases of software development are now made available from computer manufacturers and already are in widespread use. These include syntax directed editors, code convention checkers, code management systems, symbolic debuggers and performance analyzers. They significantly increase the efficiency with which the design can be coded and tested, particularly when used on workstations with window facilities.

In the future, it is likely that these packages will be able to generate code automatically from well specified designs. However, at present, the transition between steps in the software development lifecycle still needs to be made by hand.

3.5 SASD in practice

It is important not to take the guidelines too literally but to adapt them to the project in hand. For example, building prototypes of systems can be a very worthwhile exercise in terms of improving one's understanding of system requirements and of providing input to the design phase. Similarly, evaluation and analysis of existing systems can often provide useful material and in certain cases may be an appropriate point at which to begin the project. Some of the benefits of this approach to software design may be summarised as follows:

- it forces the programmer to think hard about the design and experience has shown that this leads to the production of better software.
- in large projects, a method for partitioning the system into components with well-defined interfaces is essential.
- documentation is available early, ie. before the code, and in a convenient form, and permits frequent review and continuous verification of the design.
- management of software projects becomes a realistic endeavour

3.6 The Software Environment

The software for a system of distributed processors consists of a number of tasks, with at least one task running on each processing element. These tasks will typically need to communicate with oneanother to exchange information and in addition their actions will need to be synchronized. In systems with only 1 processor, or in systems with tightly coupled processors, this communication can be implemented using shared physical memory. In systems where the processing elements are loosely coupled, ie. through LANs or serial links, the communication must be made through the exchange of messages. The programming environment can offer support to the programmer for implementing these functions eg. through system calls to the operating system. In some cases the programming language may already provide these facilities making system calls unnecessary. Those facilities not provided by the programming environment have to be written by the programmer himself.

For the embedded processors the first decision to be made is whether to use no operating system, a system with very basic functionality (ie. realtime kernel) or a system with extensive support facilities. Here the real-time performance of the software is particularly critical and in cases where the distributed functions are very simple it may be preferable to avoid using anything too sophisticated. It is perhaps worthwhile mentioning the sort of facilities a more general operating system can offer to help put these choices into perspective. Firstly it provides a set of utilities for task creation, deletion and scheduling, as well as mechansims for controlling synchronization between running tasks ie. semaphores, mailboxes etc. Utilities are also provided for handling exceptions and processing interrupts. These are quite complex pieces of code and are usually needed in any software system implementing more than a single function. Secondly it provides a framework for adding new utilities, such as new system calls and new device drivers, in an orderly way. Finally it can provide a convenient programming environment through the support of resident editing and compilation facilities as well as symbolic debugging aids.

For the online computer the real-time performance is perhaps less crucial whereas the need for many facilities and general programming tools is more obvious. In this case the options for the operating system are limited to a choice between the manufacturers specific system or portable systems such as UNIX. Although systems, such as VAX/VMS are very popular, a lot of commercial application software is now being developed for use with UNIX, due to its general availability on many manufacturers equipment.

As already mentioned many modern languages support the concept of multitasking by providing facilities for handling task to task communication and task synchronization. In Ada[37], for example, the language supports the concept of shared variables and tasks can be synchronised through the *rendez-vous* mechansim. A special language Occam[38] has been developed for use with transputer arrays. Using Occam interacting concurrent processes can be defined and tasks communicate over the transputer links by a message passing mechanism.

Modern design goals and principles are causing the specifications for programming languages to be revised and new languages to be defined. In general many languages have a rich set of constructs for implementing algorithms but are less able to cope with data structures. In particular, as we have seen, objects need to be defined in terms of their attributes and interrelationships and also by the valid operations that may be performed on them. Older languages, such as FORTRAN, are particularly poor in their management of data items.

4. CONCLUDING REMARKS

We have seen that data acquisitions systems are becoming increasingly complex as the scale of experiments increases. Fortunately the advances being made in the techniques used are just about sufficient to keep pace with functional requirements. The main problems to overcome are conceptual in nature, due to the difficulty of dealing with the complexity of the systems. This is true for the design of VLSI circuits, where physical limits have still not been reached, as well as for the design of software. The computing industry is faced with the same challenge and is developing techniques and software tools for coping with these problems. It is essential that we as a community keep up-to-date with the progress being made and make the appropriate investments in money, time and manpower.

References

1. U.Amaldi,   TeV electron - positron collisions,   Nucl. Instr. and Methods   A243 312:322   (1986)
2. Z.Kunst, Report from working group on large cross section processes, in Proc. of workshop on physics at future accelerators, CERN 87-07 Vol. I, Geneva (1987)
3. G.Altarelli,   Physics of ep collisions in the TeV energy range, in Proc. of the ECFA-CERN workshop, CERN 84-10, Geneva (1984)
4. J.R.Hansen, Report from the working group on triggering and data acquisition, in Proc. of workshop on physics at future accelerators, CERN 87-07 Vol. I, Geneva (1987)
5. A.O.Allen, Probability, statistics and queuing theory, Academic Press, London (1978)
6. Aleph dataflow group, Aleph data acquisition system hardware functional specifications, ALEPH/DATACQ 85-21, CERN, Geneva (1985)
7. D.R.Quarrie, The CDF data acquisition system, IEEE Transactions on Nuclear Science, Vol NS-32, No.4 (1985)
8. The Institute of Electrical and Electronic Engineers (IEEE), IEEE standard Fastbus modular high speed data acquisition and control system, ANSI/IEEE Std 960 (1986)
9. The VME bus specification manual, Revision C.1, Motorola (1985)
10. S.Cittolin, private communication
11. S.Cittolin, The UA1 VME data acquisition system, in Proc. of the 1986 CERN school of computing, C.Verkerk, ed., CERN 87-04, Geneva (1987)

12. D.J.Sherden, The data acquisition system for SLD, IEEE Trans. on Nuclear Science, Vol. NS-34, No. 1, 142:150 (1987)

13. J.T.Walker, S.Chae, S.Shapiro, and R.S.Larsen, Microstore - The Stanford Analog Memory Unit, IEEE Trans. on Nuclear Science, Vol NS-32, No. 1, 616 (1985)

14. Zeus Collaboration, Calorimeter Electronics, in: The Zeus detector status report 1987, J.Krüger, ed., PRC 87-02, DESY, Hamburg (1987)

15. Proceedings of the topical conference on the application of microprocessors to high energy physics experiments, CERN 81-07, Geneva (1981)

16. T.Metcalf, M.R.Saich, and J.A.Strong, The Aleph second level trigger processor, ALEPH 88-127, CERN, Geneva (1988)

17. D.Crosetto, FDDP Fast Digital Data Processor - A modular system for parallel digital processing algorithms, CERN -EP 87-151, Geneva (1987)

18. Zeus Collaboration, Drift Chamber Readout Electronics, in: The Zeus detector status report 1987, J.Krüger, ed., PRC 87-02, DESY, Hamburg (1987)

19. Zeus Collaboration, Second Level Trigger, in: The Zeus detector status report 1987, J.Krüger, ed., PRC 87-02, DESY, Hamburg (1987)

20. R.Benetta, A.Marchioro, G.McPherson, and W.von Rüden, The ALEPH Event Builder, in: Computing in High Energy Physics, L.O. Hertzberger and W.Hoogland, ed., North-Holland, Amsterdam (1986)

21. A.Fucci and K.M.Storr, Using 3081/E emulators in on-line and off-line environments, in : Three day in-depth review on the impact of sprecialized processors in elementary particle physics, Padua (1983)

22. I.Gaines, H.Areti, J.Biel, S.Bracker, G.Case, M.Fischler, D.Husby, and T.Nash, The Fermilab ACP multi-microprocessor project, Proc. of the symposium on recent developments in computing, processor and software research for high-energy physics, R.Donaldson and M.N.Kreisler, ed., Guanajuato, Mexico (1984)

23. B.Jost, private communication

24. E.W.Dijkstra, GOTO considered harmful, Comm. of the ACM, Vol 11, No 3 (1966)

25. C.Bohm and G.Jacopini, Flow diagrams Turing machines and languages with only two formulation rules, Comm. of the ACM, Vol 9, No 5 (May 1966) 366-371

26. B.W.Boehm, Software and its impact : A quantitative assessment, Datamation, May (1973)

27. F.P.Brooks Jr., The mythical man-month, Datamation, Dec (1974)

28. P.Freeman and A.I. Wasserman, Ada Methodologies : Concepts and requirements, ACM Sigsoft Software Engineering Notes, Vol 8, No 1 (1983)

29. T.DeMarco, Structured Analysis and system specification, YOURDON Press, New York (1978)

30. M.Page-Jones, The practical guide to structured systems design, YOURDON Press New York (1980)

31. P.P.Chen, The Entity Relationship model - Toward a unified view of data, ACM Trans. on database systems, Vol 1, no 1, (1976)

32. P.T. Ward and S.J. Mellor, Structured development for real-time systems, Vol 1-3, YOURDON Press, New York (1985)

33. D.J.Hatley, The use of structured methods in the development of largesoftware-based avionics systems, Proc. DASC, Baltimore, (1984)

34. R.Brazzioli, S.M.Fisher, P.Palazzi, and W.R.Zhao, The ADAMO system, ADAMO notes, P.Palazzi, ed., CERN, Geneva (1986)

35. D.L.Parnas, On the criteria to be used in decomposing systems into modules, Comm. of the ACM, Vol 15 No.12, 1053:1058 (Dec 1972)

36. ProMod - GEI, D-5100 Aachen

    SATools and SDTools - Tektronix, Beaverton, Oregon 97075

    StructSoft - 24 Homer St., Parsippany, NJ 07054

    Teamwork - Cadre Tech. Inc., 222 Richmond St., Providence, RI 02903

37. A.Burns, Concurrent programming in Ada, Cambridge University Press, (1985)

38. INMOS Ltd., Occam Programming Manual, Prentice-Hall International, (1984)

RECONSTRUCTION OF CHARGED TRACKS

M. Regler and R. Frühwirth

Institut für Hochenergiephysik der
Österreichischen Akademie der Wissenschaften
Vienna, Austria

To the 65th birthday of my venerated teacher and friend Otto Hittmair.
Meinhard Regler

PREFACE

In these lectures we are concerned with the reconstruction of charged tracks in *experimental particle physics* or, as it is commonly called, *experimental high energy physics*.

By investigating the interactions of elementary particles, experimental particle physics provides the data which are required to verify and develop physical theories providing mathematical descriptions of these phenomena. On a macroscopic time scale the result of an interaction of elementary particles is again particles, which are characterized by their mass, momentum and charge. There are of course additional quantities describing the physical state of a particle, like the angular momentum, and quantum numbers other than the charge. However, the knowledge of *mass, momentum and charge* of all particles produced is deemed sufficient in the reconstruction of an interaction, on the first level of off-line analysis – the *track reconstruction*.

There are two essential ingredients to an experiment:

- the *detector*, an apparatus which is capable of registering particles produced in an interaction;
- the *data analysis software*, i.e.the program which computes the physical quantities mentioned above from the signals produced by the detector.

While neutral particles can be measured only in a *destructive way*, charged particles are detected by measuring the trail of *ionization* they leave behind in a suitably designed detector.

This ionization results in only a small disturbance to the trajectory of the particle, so that – after amplification and conversion into a digital signal by the read-out electronics – it is possible to reconstruct the entire path of the particle through the detector system. Track reconstruction of charged particles will be the main topic of these lectures.

In the first stage of track reconstruction, the raw data are inspected in order to understand the track pattern. For events with a large number of tracks passing several detector modules, this requires the processing of hundreds or thousands of coordinates, and the number of possibilities for grouping them into track candidates is far too large to allow a purely combinatorial approach. Therefore more sophisticated techniques are required in order to offer a unique track pattern to the next stage, the track fit, whenever possible.

The aim of track fitting is to obtain the ultimate geometrical resolution from the measured coordinates, and to submit the hypothesis that the track candidate corresponds to a real physical track to its final confidence test. It has turned out that in most of the cases the least-squares method (LSM) meets best the requirements of track fitting.

In the last stage of track reconstruction the tracks are associated to interaction vertices, either to the primary vertex only, or to an eventual secondary decay vertex. This should result in a complete geometrical description of the interaction or at least of all charged particles produced by it.

It should by now be obvious that the reconstruction of charged tracks and of interaction vertices is an indispensible step on the long way which leads from the raw`data to meaningful physical quantities, which can then be compared to theoretical predictions. In these lectures, first some general aspects will be discussed (Chapter 1). Chapter 2 recalls some basic principles of *pattern recognition,* and then describes different methods of *track finding* in particle physics. In the first part of chapter 3 it is shown how the ultimate geometrical resolution can be obtained from the available information by the *track fit,* and how the track hypothesis is put to its final confidence test. Although primarily the application of the LSM is considered, also results of recent studies of more robust estimators and tests are discussed for the appropriate treatment of outliers and kinks (Frühwirth 1988). Finally, in the last two sections of chapter 3, the reconstruction of interaction *vertices* is treated, first for estimating the position of the interaction point and the momentum vectors of the tracks emerging from the vertex, and then for checking the association of tracks to the primary or to an eventual secondary vertex.

The aim of these lectures can only be to give a minimum basic knowledge of the field, and to facilitate further reading of the literature, an extensive collection of which is provided in (Regler 1989). Therefore only a few references are given here. Readers wishing a more general introduction into the usage of computers in high energy physics are referred to a comprehensive paper by (Metcalf 1986).

1. THE TASK OF TRACK RECONSTRUCTION

1.1 The Experimental Scenario

Since the days of Wilson's cloud chamber, tracks of sub-atomic particles had to be recognized and measured, and starting in about 1960, computers were used to assist in the measurement processes of the photographs.

However, access to high energy physics events with high multiplicities and high event rates became possible only by using the so-called "electronic" detectors, which have nowadays almost completely replaced the cloud and bubble chambers.

In order to provide as much information as possible on final states emerging from particle collisions, huge detectors are now the rule. Hundred of thousands of electronic channels perform the analogue-to-digital conversion or the time digitization of the numerous signals coming out of the detector from several tens of tracks. Several hundreds of Kbytes per event on average have to be stored, and the number of events recorded by an experiment is of the order of 10^7 in e^+e^- colliders, and 10^8 or more in $p\bar{p}$ collisions. About 30 seconds (IBM 370/168 units) CPU-time are needed to process a complex event in the DELPHI detector from the raw input data to the Master Data Summary Tape.

In practice the analysis begins already while an event is being recorded: Fast hardwired processors are able to execute selection algorithms within a few microseconds, and microprocessors allow complex decisions to be taken within a few milliseconds. Thus, by sophisticated multilevel triggering, the amount of data to be recorded on mass storage devices is reduced to an acceptable level. This *on-line* data reduction is of great importance for modern experiments, but it will not be extensively discussed here. However, many algorithms, still restricted to the *off-line* analysis some years ago, can now be executed on-line due to the most recent technical achievements in the field of electronics and computers, and the borderline of what can be done on-line is still moving. This borderline between on-line and off-line analysis is usually drawn at the point where the data are recorded on a mass storage device *and can thus be "replayed"*.

Particle physics experiments can be divided into two groups: *fixed target experiments,* and *collider experiments*. In the first case, a beam of highly energetic particles is directed at a target. In this type of experiment, the center-of-mass energy increases only with the square root of the energy of the incident particle. In collider experiments, two counter-circulating beams of highly energetic particles are made to collide at certain intersection points, and the total energy, which is twice the full beam energy, is available for the interaction itself.

In both types of experiments the detector has to provide the maximum knowledge of all particles emerging from the collision process (origin, direction, momentum, charge and particle type). The art of the experimenter consists to a large extent in designing the detector in such a way that the physically relevant information can be obtained with high probability and high precision. The aim of the reconstruction program is then to translate the tremendous amount of electronic signals into the desired quantities mentioned above. In this lecture only the reconstruction of charged particles will be discussed, and the identification of the particle type will only be mentioned.

The momentum of a charged particle is determined from a measurement of the curvature of the trajectory of the particle (its "*track*") under the influence of a known stationary magnetic field. Therefore a detector system usually contains a *magnet,* sometimes superconducting, which is capable of supplying a field of sufficient strength. The trajectory of the charged particle can be determined from the trail it leaves in devices designed specially to this purpose, the so-called *tracking detectors*. These are capable of detecting a charged track by means of secondary processes (mainly ionization) in the sensitive volume of the detector, which is filled with an appropriate gas. (For high precision vertex detectors also thin silicon wafers are used.) In order to compute the curvature, the position of the trajectory has to be measured in several space points along the track. Thus a *detector system* contains usually several tracking detectors. It has to be kept in mind, however, that a charged particle necessarily interacts with the material the detectors are made of. The resulting disturbance of the track has to be kept as small as possible by designing the innermost detectors, in particular the central tracking detectors, as light and "transparent" as their required mechanical strength permits. The combination of a deflecting magnet with one or several tracking detectors is sometimes called a "*magnetic spectrometer*".

The two types of experiments call for two types of experimental set-ups:

- In fixed target experiments one is confronted with energetic particles in the forward direction and particles with very low energy in the "backward hemisphere". Therefore, long magnetic spectrometers with high precision tracking detectors are required, in order to measure the momenta of the outgoing charged particles with sufficient precision from their deflection in a magnetic field (Fig. 1.1). As most of the particles are going into a relatively small cone of solid angle, good two particle separation is needed. For the low energetic particles in the backward region (target system), a compact but transparent tracking device is needed.

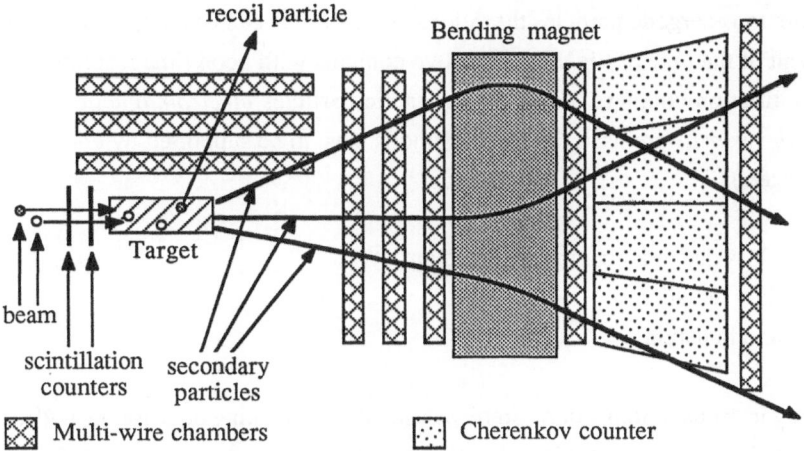

Fig. 1.1   A fixed target experiment. The incident beam particles interact with the target
particles at rest. The momenta of the energetic particles in the forward direction
are measured by a long two arm spectrometer, while the target system must
handle the less energetic particles. (From: M. Regler, "Poster Project")

- In symmetric collider experiments no particular direction is privileged by kinematics. Therefore a compact detector covering the full solid angle is needed (Fig. 1.2). As the detector cannot be very large, due to limited space (and also money), a high magnetic field and precise tracking detectors are required. Again good two-particle resolution is a crucial feature, because of the high multiplicities produced in high energy collisions. At the "Hadron-Elektron-Ring-Anlage" (HERA) at DESY the kinematics of proton-electron collisions is very asymmetric, so that the experiments lie somewhere between collider and fixed target geometry. (In the low $-x$ region the asymmetry is the other way round!)

Colliders are limited in the types of interactions which can be investigated, for instance pp (or $p\bar{p}$), e^-e^- (or $e^- e^+$), or $e^- p$, whereas conventional accelerators allow several kinds of beams and targets. Therefore both kinds of experiments have to be carried out as they fulfill complementary tasks. And although the basic concepts of track reconstruction are the same, they have to be adapted according to the type of experiment under consideration.

1.2 Particle Identification

Particles can be identified in several ways . For charged particles the most important one is the measurement of the velocity of the particle. If the momentum is known in addition, the mass of the particle can be computed and the type can be inferred with a certain probability depending on the precision of the mass determination.

For not too energetic particles the velocity of a particle can be computed directly from a measurement of the *time of flight* between two counters with good time resolution; these are usually scintillation counters. For more energetic particles *electromagnetic effects which depend on the velocity* are used (all these methods have to be supported by knowledge of the track parameters):

- Energy loss due to ionization of a medium
- Cherenkov radiation
- Transition radiation

Energy loss due to ionization can be measured by a tracking detector, provided it works in proportional mode and is equipped with the necessary electronics. Cherenkov and transition radiation are detected by devices specially designed to this end. Information about the particle type can also be drawn from the properties of a shower in a calorimeter. If the particle is unstable and decays inside the detector, its type can also be inferred from an analysis of its decay products. In calorimeters particle identification can be supported by the shape of the shower.

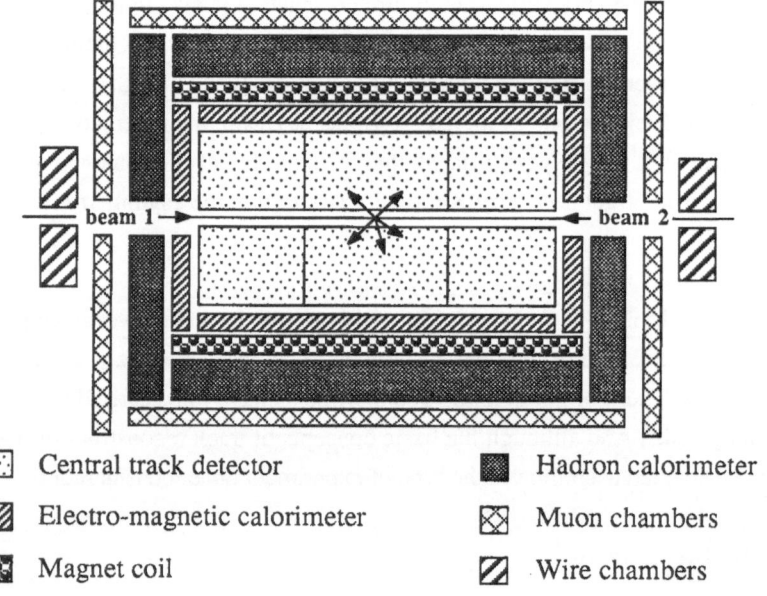

⋮	Central track detector	■	Hadron calorimeter
▨	Electro-magnetic calorimeter	⊠	Muon chambers
▨	Magnet coil	▨	Wire chambers

Fig. 1.2   In collider experiments, no particular direction is priviledged by kinematics. The basic structure of a storage ring detector consists of a "transparent" central tracking device (e.g. microvertex detector and time projection chamber), the electromagnetic calorimeter, the hadron calorimeter (whose iron acts also as return yoke for the magnetic field and as muon filter) and the muon chambers. (From: M. Regler, "Poster Project")

A special behaviour, as far as particle identification is concerned, is shown by muons. They distinguish themselves from all other kinds of charged particles by the fact that they are capable of penetrating several meters of a dense material, e.g. iron or lead. Hence the common technique of muon identification is the following one: A set of tracking detectors ("muon chambers") is put behind an absorber of sufficient thickness ("*muon filter*"), so that only muons penetrate the absorber and are registered by the muon chambers. In a collider experiment the material of the hadron calorimeter is normally used as the muon filter. Since in this case the muon chambers have to cover a large area of several hundred square meters, they are usually designed as drift chambers (Section 1.3).

1.3 Tracking Devices

The most common tracking devices are the *scintillation counters, wire chambers* and *microstrip detectors*.

1. Scintillation counters

In the early times of electronics experiments the role of scintillation counters was in two ways essential:

- Because of their fast response they were (and are still) used for the fastest trigger level (limited mainly by the length of the cables), in order to define a minimal track pattern for an event before being recorded on a mass storage device (this was also the signal to switch on the high voltage for the "spark chambers"); their capability to satisfy complex geometrical requirements is a further advantage, and their reliability, together with a short dead time, is essential to avoid a trigger bias.
- The properties mentioned above made the signals from the scintillation counters also appropriate to be used as "fixed stars" in the complicated task of pattern recognition.

An important application is the use of scintillators in calorimeters with some tracking capability.

2. Multiwire chambers

Since they were invented, multiwire proportional chambers have replaced scintillators in various types of applications, with reduced speed, but with a better resolution. (Nevertheless the role of scintillation counters is still essential in detector systems.) Only a few basic types of multiwire chambers can be described here:

a) The multiwire proportional chamber (MWPC)

The simplest form of a MWPC consists of a thin gas volume confined between two

413

cathode planes. Midway between the cathode planes is an array of parallel signal wires which form the anode (Fig. 1.3). The chamber is put in such a position that most of the tracks cross it more or less perpendicularly.

A track passing the chamber ionizes the molecules of the gas and sets free electrons which are attracted by the nearest anode wire. In the vicinity of the thin wire the electric field is sufficiently high to accelerate the electrons to such energies that they in turn produce secondary ionization electrons. Finally an electron avalanche develops and produces a detectable electric signal on the wire. This signal is then amplified and registered by the electronics attached to the wire. (In fact, in the proportional mode the signal is mainly produced by the movement of the positive ions!)

Under the assumption that only the nearest sense wire gives a signal the uncertainty of the measured position of the track is distributed uniformly between $-d/2$ and $d/2$, where d is the wire spacing, in practice 0.5mm - 3mm. This uncertainty is in principle a deterministic function of the true track position, which is, however, itself random. Therefore the uncertainty of the measured coordinate may be regarded as a random variable. The standard deviation of the measurement error of a single coordinate measurement is then given by the standard deviation of a uniform distribution:

$$\sigma = d/\sqrt{12} \ .$$

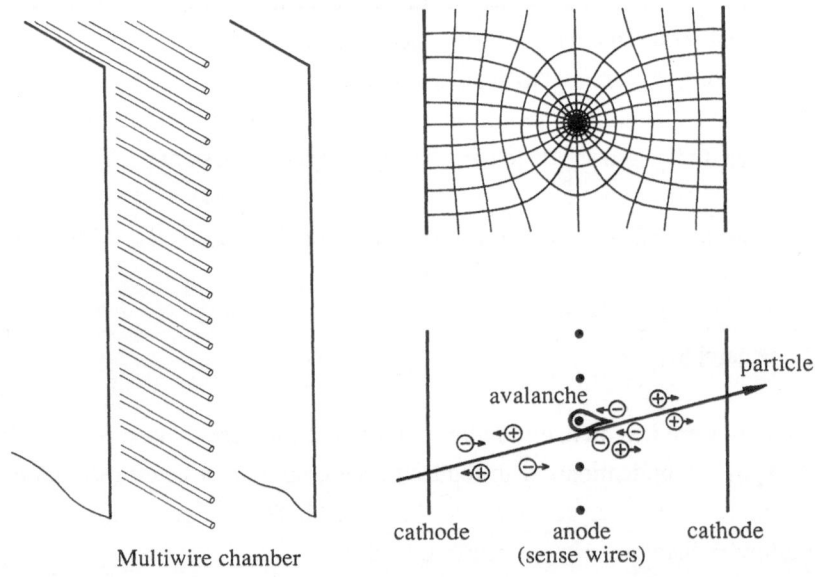

Multiwire chamber

Fig. 1.3   A multiwire proportional chamber (MPWC) with the field lines and the equipotential lines from the high voltage. Charge multiplication occurs in the region around the sense wire where the field strength grows like $1/r$. (From: M. Regler, "Poster Project")

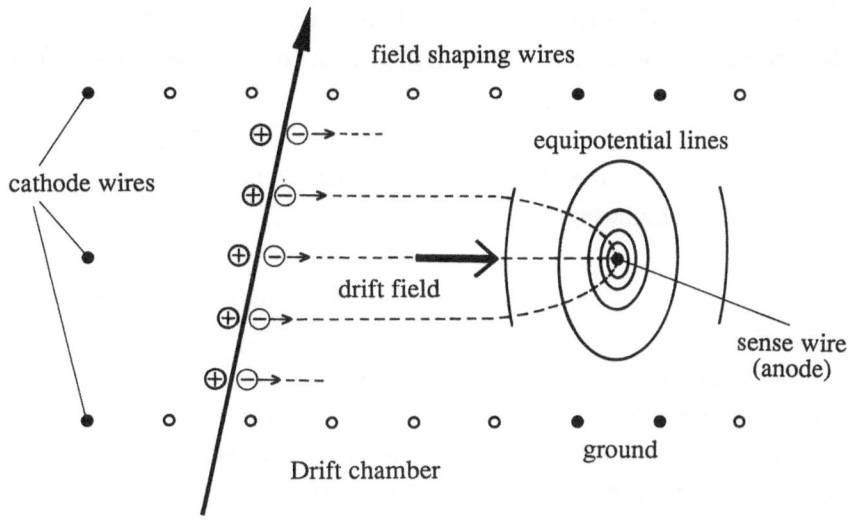

Fig. 1.4   A drift chamber. A fairly constant field is created over most of the drift distance by so-called "field shaping wires". In the region of the sense wire some non-linear corrections must be applied. Because of the diffusion the resolution depends strongly on the drift distance. (From: M. Regler, "Poster Project")

The restriction that a single MWPC measures only a single coordinate for each track can be overcome by picking up the signals which are induced in the cathode by the motion of the ions. By a suitable subdivision of the cathode into strips orthogonal to the wires or into small pads two coordinates of the crossing point can be measured, the third being the position of the chamber plane.

b) The drift chamber

In MWPCs the attainable precision is limited by the wire spacing. Therefore a large MWPC which is to achieve good accuracy has a very large number of wires, each of which is equipped with its amplifier and other electronic devices. This leads to an unacceptably high power dissipation and also to tremendous costs. In order to overcome these constraints, the chamber is modified in the following way: The wire spacing is increased to a few centimeters, and the time lapse between the impact time of the particle and the arrival of the electron avalanche at the sense wire is measured to a precision of a few nanoseconds. Such a chamber is called a drift chamber (Fig. 1.4).

Since in a suitable gas mixture the electron cloud drifts at a constant speed of some 0.03-0.06 mm/nsec towards the anode wire, the measured drift time is directly proportional to the distance of the impact point from the wire. There is, however, an intrinsic left-right ambiguity, since there is no information on which side of the wire the track has passed. The impact time is normally determined by independent scintillation counters or by the beam crossing in bunched colliders. Drift chambers do not stand as high particle rates as do

MWPCs. Due to the relatively long drift times (in the microsecond range) they cannot be used easily in the fast decision trigger. If the drift velocity is not saturated a permanent calibration procedure is necessary.

The measurement error of a drift chamber is generated by several effects. The most important of these are the following:

- The random character of the primary ionization electrons, as far as their number and their spatial distribution is concerned.
- Diffusion along the path of the drifting electrons.
- Different path lengths of different primary electrons due to the inclination of the track and an inhomogeneity of the electric field ("curved isochrones").
- The discretization error of the time-to-digital converter.
- Variation of the drift velocity due to changes in temperature and pressure.
- Locally non-saturated drift velocity and other boundary effects.
- Mechanical tolerances of the chamber.
- Gravitational sagging and electrostatic deflection of the sense wires.

The superposition of these independent effects leads to a measurement error which is Gaussian distributed to a good approximation. The actual variance of the measurement error may depend on the point of impact in the chamber. It has to be determined in a calibration experiment and constantly checked with real tracks. Standard deviations of 0.05mm have been achieved for small drift distances; typical values are in the order of 0.1-0.2mm.

c) The time projection chamber (TPC)

A TPC is in some sense a special type of drift chamber, or rather of a dense stack of drift chambers (Fig. 1.5). The main differences to a conventional drift chamber are the following ones:

- The drift space extends only to one side of the sense wire plane. Thus there is no left-right ambiguity.
- A magnetic field parallel to the drift direction leads to an additional confinement of the drifting electron cloud, preventing excessive transverse diffusion. This allows drift lengths in the order of 1m and above.
- Opposite the sense wire plane there is a two-dimensional array of small pads. These pads receive pulses induced from the pulses in the sense wires. The properly weighted barycenter of the charges distributed over the pads gives the position of the electron avalanche along the wire.

Thus the TPC delivers genuine space points (wire position, barycenter of charges in the pads, drift distance) without ambiguities. The measurement errors are again normally distributed to a good approximation.

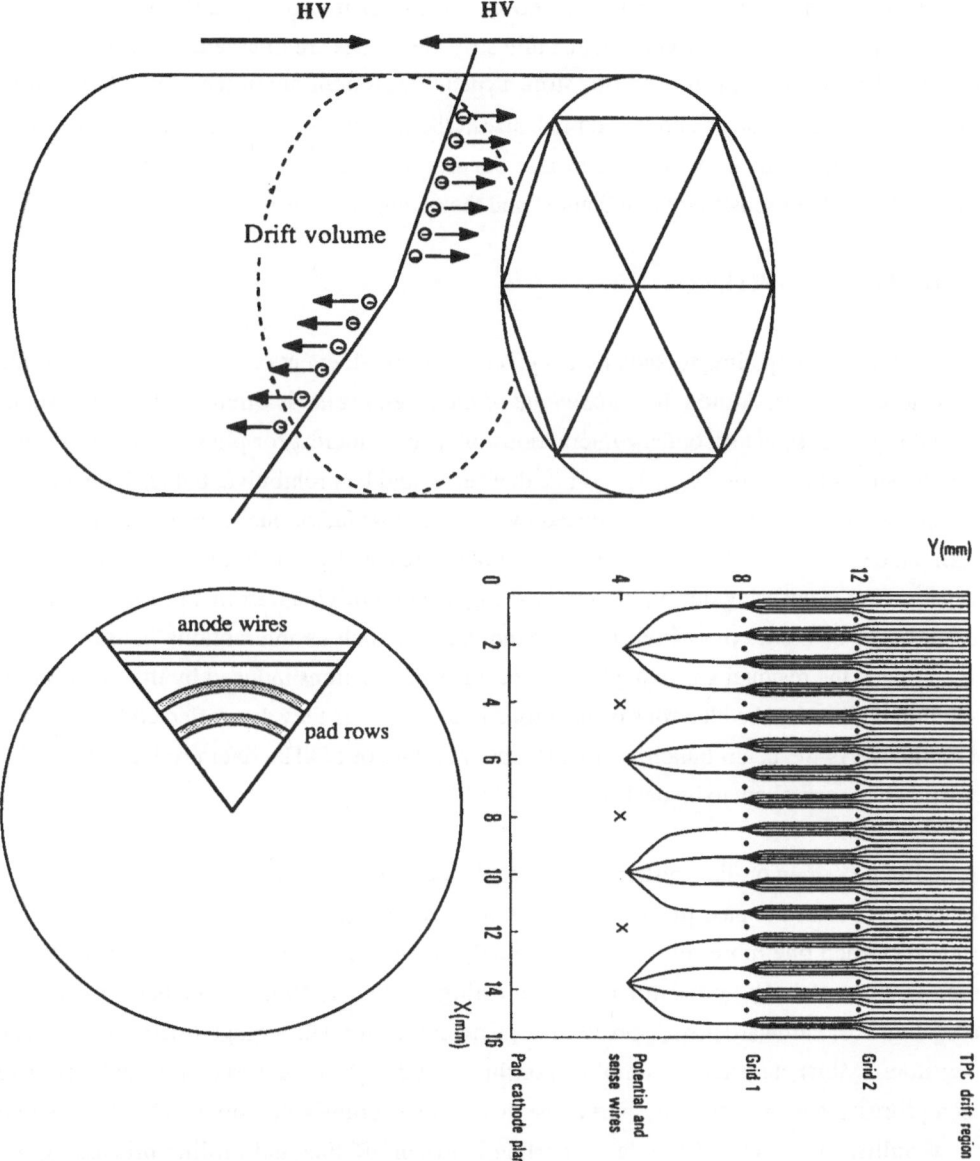

Fig. 1.5   A time projection chamber (TPC). The electrons created by a charged particle drift
over distances of 1m or more. At the endcaps the charge is multiplied in the
strong electric field around the anode wire. Protection grids avoid secondary
activities induced from photons created in the multiplication region. $R\phi$ and z are
measured at all 16 pad-rows with $R_1, .., R_{16}$.

417

3. Microstrip detectors

For high precision vertex detectors which are capable of separating the decay vertex of a short lived particle from the primary vertex, the use of microstrip detectors is the most widely employed in present experiments: thin strip electrodes are mounted on large-size area silicon wafers, with a typical pitch of 25μm. Typical resolutions are of the order of 5μm with a two particle separation of about two pitch sizes; the transit time is of the order of tens of ns leading to a high rate capability; and last but not least they are nearly 100% efficient. Drawbacks are their sensitivity to radiation, and the complicated readout electronics.

1.4 Higher Level Trigger

In high rate experiments such as at a $p\bar{p}$ collider not all events can be transferred to the mass storage device. Although a *subsample of unbiased events* is often needed, the overall data taking must select secondary interactions of special interest for physics. Otherwise not only the amount of events on mass storage devices would be prohibitive, but also the number of important rare events would be reduced by a certain loss factor due to deadtime (recording time). At e^-e^+ colliders like LEP, where event rates are usually smaller, it turns out that only very few events occuring during a beam crossing actually originate from a beam-beam interaction; most of them come from background sources such as interactions between beam particles and the residual gas in the beam pipe, or from reactions induced by the synchrotron radiation emitted by the electrons or positrons in the beam as they are deflected by the magnetic fields. As the beam bunches in LEP cross at a rate of 50kHz, data produced by an interaction have somehow to be dealt with in only 20μsec.

The task of an on-line system is, therefore, not only to perform the usual control of the equipment, but also to pass the raw data at high speed through various levels of the so-called *trigger(s)*, each one more refined and inevitably slower than the previous one, such that a large amount of background (or of uninteresting events) is filtered out before the final recording. Fast hardware processors with large memories for data storage (memory chips can now store 4Mbit) can execute simple algorithms within a few microseconds, and powerful microprocessors with extremely short cycle times allow complex decisions to be taken within a few milliseconds. Therefore the borderline between off-line and on-line processing becomes increasingly fuzzy, and many algorithms which were restricted to the analysis of recorded data, now run on-line. However, it is essential to realize that events rejected by the trigger cannot be "replayed".

As an example a two stage trigger with a continuous event source (no bunches) is discussed: If the *readout and recording* of the data causes a *dead-time* τ_r, it follows from Poisson statistics that the rate of recorded events is (with a scintillator trigger rate f_0):

$$f_r^0 = f_0 \, / \, (1 + f_0 \, \tau_r),$$

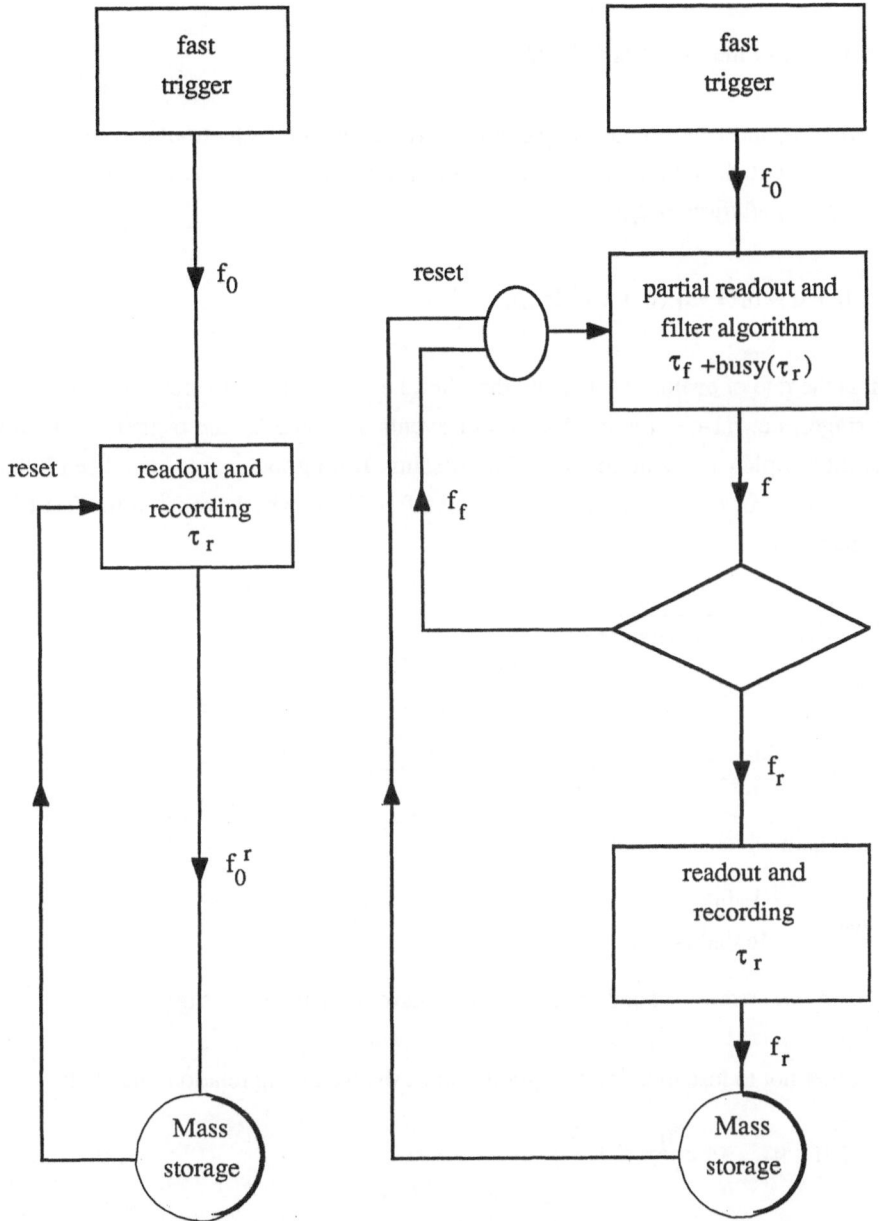

Fig. 1.6.   Data rate flow in a two level trigger logic. If an event is not rejected by the filter, the deadtime remains as long as the experiment is busy with full readout and recording.

with the loss factor $(1+f_0 \tau_r)$, or

$$f_r^0 = f_0 (1 - f_r^0 \tau_r),$$

with the inverse of the loss factor $(1 - f_r^0 \tau_r)$.

With a second level trigger only events arriving when the electronics is not busy with filtering are accepted, and the rate is (with a time τ_f for *partial fast read out, processing, and the reset* and *hopefully* $\tau_f \ll \tau_r$):

$$f = f_r + f_f = f_0 (1 - f_r (\tau_r + \tau_f) - f_f \tau_f),$$

where f_r is the rate of events accepted by the filter, i.e. $f \cdot q$, and f_f is the rate of events rejected by the trigger, i.e. $f(1-q)$; q is the fraction of events accepted by the trigger. This formula holds in the simple case with no parallel processing. If no good events have been killed by the trigger (no error of first kind), and if p is the fraction of good events in the fast trigger, the gain factor G is

$$G = \frac{f_r \, p/q}{f_r^0 p} = \frac{1+f_0 \tau_r}{1+f_0 (q\tau_r + \tau_f)},$$

with

$$f_r = \frac{q f_0}{1+f_0 (q\tau_r + \tau_f)}.$$

The largest gain which can be obtained is

$$G_{max} = \frac{1+f_0 \tau_r}{1+f_0 \, q_{min}\tau_r},$$

with $q_{min} = p$ (no error of second kind, i.e. no background passes the trigger).

In order not to lose in the rate of good events, the following relation must hold:

$$\tau_r \geq \tau_f + q\tau_r, \quad \text{or} \quad \frac{\tau_f}{\tau_r} \leq 1 - q.$$

If the equality holds, no enrichment in good events is achieved, but the amount of mass storage occupied by the data is reduced by a factor of q. In general, however, one can say, that in order to achieve a maximum rejection power the second level trigger should be performed with a minimum of partial fast readout and a minimum of time for processing summing up to τ_f. In bunched colliders τ_f of the second level trigger should obviously be smaller than the time between two beam crossings, $1/f_{cross}$. All these numbers can still be improved by appropriate buffering of events and optimized queuing (Regler 1989).

1.5 Experimental Design

The choice of algorithms to be used in data analysis depends very much on the detector. Therefore great skill is necessary for the design of a suitable apparatus which is expected to collect optimal data from the type of reaction to be studied, but with the constraint of being able to give the information in such a way that it can be treated efficiently by fast computation algorithms (Regler 1981). This can only be achieved by close coordination between hardware and software design during the planning phase of an experiment. This concerns on- and off-line reconstruction, with the additional constraint that for on-line reconstruction the partial readout of the information needed for the trigger algorithm should be particularly fast.

For the off-line track reconstruction, it turns out that a two level strategy is the most adequate, and this should also guide the design of the detector layout. It is the first level, the *pattern recognition,* the performance of which is very much related to the detector layout; the time spent in the second step, the *track fit*, does not depend much on the particular detector, provided that the track fit is appropriatly conceived and coded, and that the pattern recognition works efficiently. There are several reasons for the sensitivity of the pattern recognition to the experimental design. The main reason is the *combinatorial aspect of pattern recognition* which has to associate thousands of often ambiguous coordinates to tracks. Therefore a maximum of data reduction must be achieved, and the number of possible combinations must be reduced at an early stage, e.g. either by associating coordinates to space points before looking for the tracks themselves, or by reducing the dimensionality by operating in projections or in appropriate feature spaces.

In complex modular detectors one may also look for complete track elements in each module rather than only for space points. This part is now called *"local pattern recognition"*, while the proper association of track elements into complete tracks is called*"track search"*. In practice one may need a second iteration of local pattern recognition, since the information from other modules may be helpful in solving some remaining uncertainties from the first stage of local pattern recognition. The recovered information will in turn be used to complete the track search.

An important point is the fact that losses of tracks in the pattern recognition are virtually always correlated with the shape of the event, and therefore result in a *bias* in the sample of found tracks. In particular, a pattern recognition algorithm should not rely too much on the proper working of a single detector module but should be able to cope with the worst possible case, and the designer of the detector has to know which is the minimum of useful information which is absolutely required from his detector.

Finally, it has to be decided in which part of the reconstruction program the main effort should be invested, and whether the gain in execution speed is worth the time spent in the optimization of the program. The CPU-time spent for an accepted event is given by

$$t_{CPU} = r_n \, (t_n + r_{n-1} \, (t_{n-1} \ldots + r_1 \, t_1) \ldots),$$

where r_i is the "reduction factor" of step i ($r_i > 1$), and t_i is the time spent in a reconstruction step averaged over all (accepted or rejected) tracks, similar to (Fig.1.6). From this formula it is obvious that it pays more to speed up the initial steps which are executed more often.

1.6 Software Design

In the design of a large reconstruction program several problems have to be addressed:

- Program structure and development
- Data management
- Data flow
- Graphical representation of the data
- Testing and tuning

In the following only the aspects of software design directly connected to track reconstruction will be discussed. For more details see e.g. (Metcalf 1986, Regler 1989).

1. Program structure and development

Large high energy physics experiments are designed, built and operated by hundreds of physicists. Similarly, the data analysis software is designed, written and used by a large number of people. This fact imposes certain constraints on the way this software is developed. Until recently, the impact of software engineering techniques on high energy physics was practically nil, although there is now a growing consciousness among physicists that for the new generation of experiments clear requirements and design specifications are indispensable. For instance, one of the LEP collaborations uses the SASD (Structured Analysis/Structured Design) approach, both for on-line and off-line programs, as does ZEUS at HERA.

Obviously the modular structure of a large detector system is reflected in the modular structure of the data analysis program. In fact, it is common practice that a group building a particular detector component also supplies the software module which does the local pattern recognition in their detector. The difficulty starts when these modules, written by people with different background and different ways of thinking, have to be put together. Beside the technical problems to be solved in this process, there is an inherent difficulty to be overcome: In order to have completed software available when the experiment starts to take data, one is forced to make assumptions about the hardware, and therefore it is difficult to assess in advance the final performance of the local pattern recognition. This in turn makes the design of the track search a very delicate task, as a too optimistic concept may easily collapse when it is confronted with the actual data. The lesson to be drawn from this is that the design of a reconstruction program should be as flexible as possible and tolerant towards an eventual malfunctioning of the detector, or part of it.

2. Data management

Putting together 10 or 20 modules written by different people into a single program requires a considerable effort of cooperation and coordination. Since nobody can be expected to understand the whole large program, it is natural to think of each module as a *black box*, the internal structure of which is known only by some experts. Each module formally operates on a certain *input data structure* and *transforms* it into an *output structure*, and the module is formally specified by this transformation. The reconstruction programs used in high energy physics are still written exclusively in FORTRAN. In FORTRAN there are only two rudimentary data structures, the homogeneous array of elements and the COMMON block of non-homogeneous entities. It is known since long that these primitive structures are completely insufficient for the formal specification of interfaces between modules. Therefore powerful *data management packages* have been developed, which serve a two-fold purpose: They allow the data to be structured in a way which reflects the modular structure of the detector and the track-vertex structure of the event, and they allow a formal specification of interfaces between modules. Some of these packages, like HYDRA, ZEBRA and BOS77, have become a de-facto standard in the high energy physics community. Since they are all available in FORTRAN77, they run on many different machines. They use a heap storage mechanism whereby storage for an individual bank of data can be requested, and links of various types can be established between it and other banks.

Some experiments have come to the conclusion that a second level of data management is desirable. Such a package operates on a well-defined data structure containing detector data, track and vertex information, and provides a coherent way of retrieving, modifying, creating or dropping data at all levels of the analysis, hiding the technical details of the lower-level package to the user. In addition, it should provide an interface to the interactive analysis system. It also should provide a reasonable protection of the data. It is, however, essential that the application software can access the data without too much overhead, especially if it operates in a global context, e.g. if it looks for overlaps between tracks or searches for a vertex.

3. Data flow

As indicated above, the information from a complex modular detector is processed in several steps. If possible, a *sequential processing* is desirable; sometimes loops are unavoidable in practice.

In a first step (after the calibration) the different signals of the individual detector modules are translated into physical or geometrical quantities. For charged tracks this means to generate first approximation coordinates. For the final definition of the often ambiguous coordinates and for the ultimate translation of drift times more information might be needed.

Finally as much data reduction as possible will be performed in each module. Then the global track search is performed. After a second iteration, track candidates are submitted to the track fit, for the ultimate decision on ambiguities and outliers, and to obtain optimal information on the track parameters. Then the search for primary and secondary vertices is performed, with the final goal of doing physics, often after having performed a fit with the kinematical constraints.

4. Graphics

An important technique in modern data analysis is the *graphical representation of the data*. The ability of modern graphics devices to visualize complex patterns of data in three dimensions allows the operator to recognize unforeseen relations between different subsets of the data. This has had a dramatic impact on track reconstruction.

The graphical display of detector or track data is utilized in several steps of track reconstruction. It is an important tool in the debugging and tuning of the local pattern recognition modules and of the global track search program. The program designer can check that the signals have been associated correctly to the trajectories, one sees where wrong associations have been made or where a trajectory has not been identified at all. This is particularly important for the tuning of the program whith real data, as the true event structure is not known in this case. In later stages of track reconstruction the operator may interactively fit competing or conflicting track candidates or resolve ambiguities by information drawn from a global view of the event, or by reasoning processes which are difficult to build into the program. So data handling can be kept "simple" without the risk of losing rare special topologies. Also the search for secondary vertices and the final selection of interesting events can be facilitated by event viewing. From all this it is obvious that *interaction* with the reconstruction program must be built into the design right from the beginning.

5. Testing and tuning

The testing of the reconstruction algorithms is one of the most important parts of the program development. Although no rigorous formal test procedures are available, a systematic approach is very much advisable. After the purely formal testing of the code, functional testing is best carried out in two steps.

The first step is *module testing*. Individual modules are tested with all sorts of data, valid and invalid. The results are checked against the specifications, the speed and efficiency of the module are closely examined. The results of this examination may result in a partial rewriting of code or in the tuning of some parameters. Module testing should be done in parallel with the development of the detector.

As soon as it is guaranteed that all modules yield an output in accordance with the specifications, one proceeds to the second step, *integration testing*. The complete program is tested, with the emphasis on the proper communication between modules. This should be completed before data taking starts.

Obviously this first round of testing and tuning is done with *Monte-Carlo data*; more testing and tuning will be required when *real data* arrive. Testing with Monte-Carlo data has several advantages: First of all the true values of all physical quantities are known at any stage. Secondly, the complexity of the data can be increased step by step. One starts with an ideal detector behaviour, then adds *inefficiencies* and *outliers*, and finally *background*. The track fit can first be tested without multiple scattering and measurement errors, then with Gaussian errors and multiple scattering; finally the realistic output from local pattern recognition and track search, obtained from a *full realistic simulation*, must be digested. This includes inefficiencies, outliers (see section 3.5 for an exact definition), ambiguities and background.

A well designed *debug output* is essential for efficient testing. Often a debug output with several levels of complexity is needed. In practice it will still develop during real data taking, when the *final tuning* and *testing* must be performed.

1.7 Classification of Results

If one participates in a meeting where the performance of a track reconstruction program is discussed, one gets often the impression of a "Babylonian confusion of tongues". Local coordinate efficiency, single-track efficiency, multi-track efficiency or vertex efficiency are mingled with first and second stage performance, and garnished with words like missing coordinates, missing tracks, faked tracks, ghosts, outliers, significance level, reduction factor, background rejection, bias and losses; the list could still be continued for a while.

The attempt to suggest a common language for pattern recognition specialists would be a hopeless illusion, but a few hints shall be given, how at least each programmer could give his own understandable definition.

The general rule is simple: The *basic set* of definitions must be a *partition*, i. e. a classification of sets which are mutually *exclusive* and *exhaustive*. Two basic definitions should be unable to be fulfilled at the same time, but each situation which can occur in practice must be covered by (exactly) one basic definition. Starting from these basic definitions, larger classes can be defined. In practice often several levels will be necessary. The same basic definitions

TABLE 1
Classification of basically real tracks (single track performance)

	Complete (ε_c)	Incomplete (ε_i)	Lost ($1-\varepsilon_i-\varepsilon_c$)
with extraneous coordinates	• probability p_i of i extraneous coordinates, i=1, . . . • average number $\Sigma\, ip_i$	➡ idem idem ⬆	
without extraneous coordinates	perfect tracks	• probability w_i of i lost coordinates, i=1, . . . • average number $\Sigma\, iw_i$ • loss factor : $\Sigma\, iw_i$ / aver. number of coordinates per track	

can also be arrived at the other way round, by a repeated refinement of the classification. Table 1 gives a possible classification of the performance of single track reconstruction.

This process will now be illustrated on the example of a multilevel program for single-track finding and fitting. First the track candidates must be divided into two different subsets: *real tracks*, i.e. tracks which have something to do with particles, and *faked tracks*, i.e. tracks which are invented by the reconstruction program, with or without the help of hardware background signals. The word "*efficiency*" ε will be reserved for the probability of finding real tracks which are in principle seen by the detector (i.e. the confidence level or the probability of no error of first kind) while the probability of rejecting a track e.g. faked by an earlier stage of the program or by the trigger will be called the *rejection probability* ρ_{rej} i.e. the power of the test (no error of second kind). Tracks faked by the preceding steps will be called background, and tracks faked by the reconstruction module under consideration will be called ghosts. Note that the ghosts of the actual step will *enlarge* the basic set and act as background for the next step. The enlargement is specified completely by the probabilities of generating n ghosts, $\{p_{g,n}, n=0,1,2, ...\}$.

Note that when starting with the track search, two basic reference samples must be distinguished: the number of tracks which could have been seen by the detector, and the number of tracks which are in principle reconstructable from the recorded amount of signals (e.g. at least 3 space points can be found). Pattern recognition should refer to this latter level of performance, while the limitations imposed by the detector performance should be quoted separately whenever possible.

One may still define a *reduction factor* for each module:

$$\left[\frac{\text{number of found tracks (real tracks + remaining background + ghosts)}}{\text{number of input tracks (real tracks + background)}}\right]^{-1}$$

For *multitrack efficiency* an "efficiency matrix" must be given, consisting of the *probabilities to reconstruct l tracks out of k (l<k)*, $\overset{m}{\varepsilon}_{kl}$. For the event type with k tracks, the multitrack efficiency is given by

$$\overset{m}{\varepsilon}_{kl} = \sum_{l=0}^{k} l \; \varepsilon_{kl}/k \; ,$$

and the overall multitrack efficiency can be checked against the single track efficiency (with a probability q_k of the event type with k tracks):

$$\overset{s}{<\varepsilon>} = \left(\sum_{k=0}^{\infty} \sum_{l=0}^{k} q_{kl} \overset{m}{\varepsilon}_{kl} \right) / \sum_{k=0}^{\infty} k \; q_k \; .$$

Note that the reconstruction efficiency depends on the track parameters and the cuts:

$$\varepsilon = \varepsilon \; (p_1, \; ..., \; p_k, \; \text{cuts}).$$

Therefore not only the distortion of the phase space population introduced by the detector geometry must be evaluated, but also the losses caused by the reconstruction algorithms must be considered. Correcting this influence by MC is at least one order of magnitude more time consuming than to correct only for the detector geometry, as a detailed full simulation of the detector behaviour is required for each event type. Sufficient redundancy helps to reduce the distortion and the effort to correct for it. For "efficient" optimization of the computing time a formula is given at the end of section 1.5.

2. PATTERN RECOGNITION AND TRACK SEARCH

2.1 Introduction

Automatic pattern recognition is in principle a problem of artificial intelligence. But although artificial intelligence is a most promising area of research the extremely large gap between the performance of present-day programs and the biological perception system makes it impossible to simulate biological perception, and the only way for the time being for the solution of practical recognition problems is to take a completely independent approach.

There are only few problems where the machine has superior capabilities than the brain. One of them is the capacity for the *quantitative evaluation* of parameters. In contrast to the brain which is a powerful but very general perception system, the "perception technique" of a program can be tailored to a particular problem. This relationship can be *optimized by the interplay of hardware and software design* during the experimental planning.

The basic problem of pattern recognition is to "isolate a class of objects in a noisy field" ("noise" can be real noise or just coordinates generated by other tracks). The job of the pattern recognition program can be facilitated by a simple presentation of the problem, by the correct matching of the measurement resolution with the properties of the object to be found, and by some redundancy.

An interesting link between the biological perception and the computer capability is offered by the availability of powerful graphic packages, for program development and tuning as well as for ultimate decisions on rare event types. The weakness of the human eye in recognizing three-dimensional objects on a two-dimensional screen (note that for two-dimensional objects the human perception is quite powerful) is compensated by the possibility to rotate even very complex tracks or events on a screen, and multicolor pictures give an additional support to the physicist's perception. In the design of an event viewing program the psychological aspect should certainly not be neglected, and it depends very much on the presentation of the event on the screen, whether the man - machine communication is successful.

The software for automatic processing of the bulk of the data taken in a high energy physics experiment is of crucial importance for the success of an experiment, to the extent that – to say it once more – modern detectors have to be designed to take into account the track finding methods that will be employed. The methods used are usual the results of intuition and experience rather than being based on genuine mathematical theory, although some mathematics will also be used in this chapter.

It is a feature of high-energy physics code that the imperfections in the devices used and the widely varying nature of the data result in a large number of branches, which inhibit vectorization of the code so far. However, we believe that the vector processing capabilities of modern machines will finally also influence the pattern recognition methods, of course together with the whole experimental design.

Nowadays one often distinguishes between *local pattern recognition* and *global track search*. The local pattern recognition belongs to a specific detector module and will usually be conceived by the institute responsible for this part of the detector, while the global track search is a common task of the software group. This not only facilitates the planning and co-ordination, but gives also a suitable input for graphics at a very early level. However, this distinction is somehow artificial:

- On the one side, the local pattern recognition may need help from the global track search, in order to solve problems like ambiguities or parallax effects, requiring a second iteration;

- On the other side, some detector modules, e.g. a TPC, are powerful enough to give the almost final track for a large part of the phase space.

Therefore many methods can be used for both local pattern recognition and for track search, and most of the time no further distinction will be made in this chapter between these two levels of track reconstruction.

In this chapter we will first describe the objects of our search, the trajectories of charged tracks. Then the combinatorial aspect will be discussed, followed by local methods, global methods and feature extraction. Finally a few words about vertex search will be said.

2.2 Charged Tracks as Elements of Pattern Recognition

Tracks of charged particles in a static magnetic field are the *classes* we are interested in. Each class in general is composed of a *variable number of objects*, the (straightforward or generalized) coordinates generated by a particle passing through the detector module. Tracks form *distinct classes*, which – beside some exceptions like signals from large area scintillation counters – do not share objects.

The definition of this classification is to a large extent known: a charged particle moving in a static magnetic field obeys to the *equations of motion*, which describe the action of the Lorentz force produced by a magnetic field on a moving charge:

$d^2x/ds^2 = (kq/P) \cdot dx/ds \otimes B(x(s))$,

with:

- s .... path length,
- k .... constant of proportionality,
- q .... charge of the particle (signed),
- P .... absolute value of the momentum,
- $B(x)$ ...... static magnetic field.

A track in vacuum is then represented by the five independent initial (boundary) conditions. These "*track parameters*" *can not be measured directly,* but what is obtained from our detector system is a measurement vector m of dimension $n \leq n_{max}$.

All possible combinations of coordinates (objects) which can occur in an event select the *pattern space* **P** of dimension n_{max} (Fig. 2.1). An ideal track is then a point on a five dimensional manifold **S** ("constraint surface"), according to the constraints given by the equations of motion. Here the two most essential problems enter into the game:

- The magnetic field, which is needed to bend the tracks in order to allow the fifth track parameter (the curvature which is related to the momentum and to the charge) to be es-

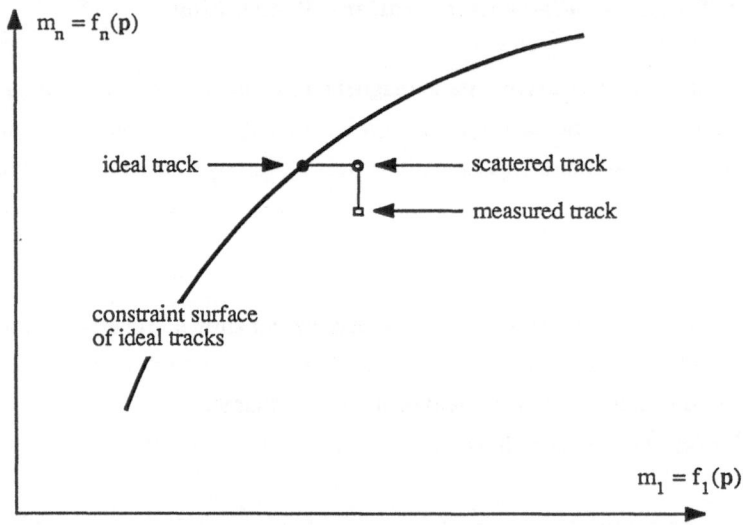

Fig. 2.1   The five dimensional hypersurface which should contain any set of coordinates of an undisturbed track and without measurement errors. Multiple scattering and measurement errors create a certain "halo" by scattering track points out of the five dimensional set of track points into the n-dimensional measurement space.

timated, makes this manifold non-linear;

- The dimension of the measurement vector **m**, n, varies according to the number of the detector modules hit and their response with variable performance, such that only the projection of the populated region of this manifold onto a variable subspace of $\mathcal{P}$ will be known.

In addition, a trajectory is disturbed when a particle traverses the material the detectors are built of, leading to a *stochastic disturbance* of the particle's path; the measurement errors create additional confusion for the final classification. (Fortunately these disturbances are often rather small and create no unsurmountable problems at this level of track reconstruction.) Each track corresponds now to a point in a *"cloud" surrounding the constraint surface*. This deviation from the ideal point will be discussed in detail in chapter 3. Other problems are the left-right ambiguities in drift chambers, and sometimes a minimal knowledge of the particle's path is needed to definitively evaluate the coordinates.

In practice one is looking for one or several *decision functions* **d** such that

$$\mathbf{s} = \mathbf{d}\,(\mathbf{c})$$

defines whether **c** belongs to the class of tracks, where **c** is the set of coordinates of a track candidate.

If pattern recognition methods are chosen such that they rely on direct *"feature extraction"* from the constraint surface $\mathbf{S} \subset \mathcal{P}$, it is important to limit the number of possible subspaces of $\mathcal{P}$ which must be considered. The usual way of doing this is to require a well defined response from each detector (a coordinate, a space point, a space point and the direction, or even also the curvature), irrespectively of how this information has been obtained by the local pattern recogniton of this module; incomplete information will be treated by pragmatic methods like interpolation in a second iteration process only. This requires of course that the track was found at all during the first iteration. A classical procedure is the creation of *space points from a set of at least three coordinates* measured by detectors very close to each other, allowing the application of the triangular condition:

$$x + u + v = \text{const.}$$

The constant can be made zero by an appropriate choice of the origin of the coordinate system. While for tracks normal to the detector planes this formula should be strictly satisfied, inclined tracks suffer from a *parallax error* due to the finite distance of the detector planes. With a minimum of prior knowledge of this track, e.g. that it comes from a common vertex region, the parallax errors can be reduced to a tolerable level. Anyhow modern wire chambers are usually designed with two coordinate readout.

Examples of purely local pattern recognition methods will be given in the next section.

In spectrometers with a simple configuration of the magnetic field it is sometimes possible to give simple explicit functions $\mathbf{f(p)}$ for a reasonable approximation of the manifold $\mathbf{S}$, e.g. the intersection of a helix with the detector planes in a quasi-homogeneous magnetic field. In this case, much more flexibility is available to display the feature pattern, quite independent of the detector performance. Also pragmatic power series have been successfully applied in several experiments.

Four types of data are normally needed to perform the pattern recognition in an experiment:

- The "survey data", defining the exact positions of the different active and passive elements of a detector;
- The "calibration constants" which are needed to convert the "raw data" into physical units, taking into account all properties of the detector devices;
- A description of the magnetic field;
- The proper "raw data".

For the track fit (Chapter 3) also the precise detector resolution is needed.

2.3 Local Pattern Recognition and Global Track Search

It has already been mentioned that the distinction between local pattern recognition and global track search is somewhat fuzzy, and that this distinction is governed not only by the applied algorithms, but also by the integrative role of each module in the total flow of information. And although the construction of space points as objects to be classified by the global track search is a typical example for local pattern recognition, a minimal knowledge of the track is needed to correct for parallax effects. Another illustrative example is an array of streamer tubes with drift time measurements in a LEP detector: this chamber, placed in the forward region of the DELPHI detector, consists of three double layers of drift tubes, rotated by 0, 120 and 240 degrees, respectively (x,u,v). Local pattern recognition proceeds in two steps: In the first stage one looks for *clusters* in each double layer, e.g. for groups of tubes with drift time measurements which may belong to the same track. The drift times in a cluster must be compatible with a straight track passing through the double layer; additional information is drawn from the induced signals in the cathode. Since there is little redundancy in a double layer, a cluster may have several possible solutions (Fig. 2.2).

In the second stage the clusters of all three double layers are combined to space points. All possible combinations of three clusters are subject to a series of tests, each one more selective than the previous one. The first test requires the sum of the three cluster coordinates to be close to zero (see Section 2.2):

$$| x + u + v | < \varepsilon_1.$$

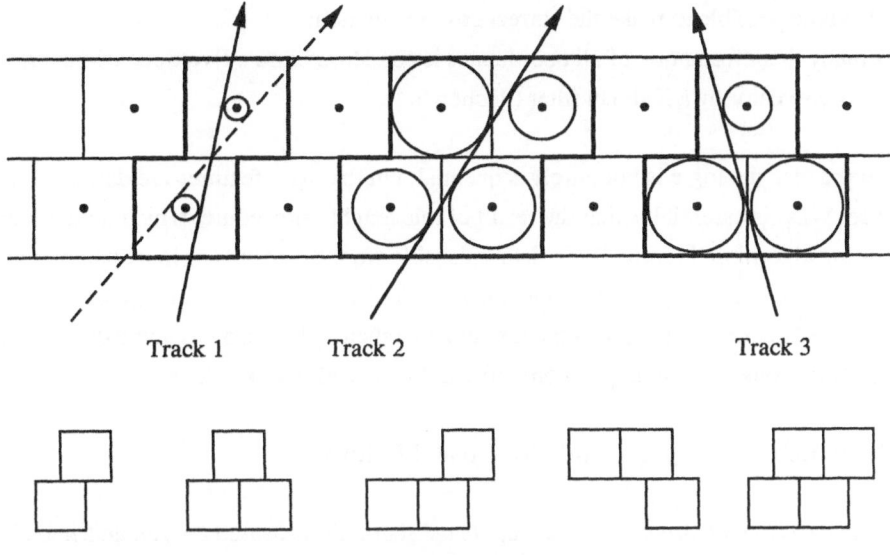

Possible cluster types

Fig. 2.2   Cluster finding and resolution of drift time ambiguities in a matrix of tubes with drift time recording (DELPHI-detector at LEP). Before combining the three co-ordinates, clusters of hit drift tubes are sought in each coordinate (double layer).

The second test requires the sum of the directions found in the clusters to be close to zero (t_x, t_u, t_v denote the slope in the (x,z)-,(u,z)- and (v,z)-plane, respectively):

$$| t_x + t_u + t_v | < \varepsilon_2.$$

Note that $t_x + t_u + t_v$ is exactly 0 for all straight tracks, although the z-positions of the three layers are slightly different. Finally all surviving combinations are subject to a LS-fit and a χ^2-cut.

The interplay of local pattern recognition and global track search shall be illustrated by describing the process of creating a track element from a powerful detector module, the central tracking device in the same storage ring experiment, i.e. a time projection chamber (Fig. 1.5):

- At the beginning a first guess of coordinates $(R\Phi)_i$, z_i, i=1, .., 16 is done from the analysis of the pulse shapes measured by the fast analogue to digital converters;
- The tracks are selected with the help of polar inversion (see Section 2.4);
- Next the measurements are refined by using the pad and wire pulses with an appropriate weight;
- A correction of the focussing or defocussing of the avalanche at the anode wire by the magnetic field is applied;

- It is now possible to make the correct error evaluation;
- Finally the information of all coordinates is condensed into a five-dimensional track parameter vector by a Kalman filter (Section 3.4).

In practice the logic is not purely sequential, but the main features are described by this sequence. What is essential is that the result of this lenghty procedure, which includes even a track fitting algorithm, is formally indistinguishable from the results of local pattern recognition from any other less powerful detector module. This has the advantage of keeping the formal data structure for the global track search symmetric, but also has some drawbacks, the most important one being to make more difficult the detection of kinks.

2.4 Combinatorial Aspects and the Road Method

The *purely combinatorial method* splits the set of all measured coordinates into all possible subsets. Then a track model is compared to each subset (track candidate), in order to find the largest subsets of coordinates fitting the model. In the ideal case, tracks should not share corodinates, and the "rest class" (coordinates left over) should consist only of real background. Although this method is appropriate for very simple cases, it is too time-consuming for most practical applications. If we have to examine the correspondence of the track model with all possible subsets of coordinates, the computing time becomes prohibitive even if the track model is as simple as a straight line (e.g. in the detector telescope in front of and behind the bending magnet of a fixed target forward spectrometer (Fig. 1.2)). The number of possible subsets of n coordinates is:

$$\sum_{k=5}^{n}\binom{n}{k} = \sum_{k=5}^{n} \frac{n!}{k!\,(n-k)!}\,.$$

For n=100 it is larger than 10^{30}! Even if we need only a μs for each test, the processing will last much longer than the existence of the universe. Therefore the purely combinatorial method is prohibitive.

In reality many modifications make this method still applicable for simple cases:

- The subset should only contain coordinates from different detector planes,
- often pattern recognition will be performed in projections, and
- reconstruction of space points reduces the numbers of objects by another important factor.

In addition, *"master points"* or the knowledge of the interaction region might help to reduce the detector volume out of which the coordinates have to be compared with the track model. Here, *the existence of a reliable subset of "reference detectors" is required, which are hit by all tracks, are fully efficient* (or redundant to recover from inefficiencies), and allow all the tracks to be well separated one from each other. Because of their good time resolution, hits onto scintillator arrays ("hodoscopes") or groups of wires from MWPCs can give a valuable contribution to track selection even when one has to deal with high event rates.

Together with some prior knowledge (e.g. the interaction region) the combination of master points as first guess for the track candidates should now be reduced to a reasonable number, increasing with the track multiplicity only by the power of two or three.

Having got track candidates, a "mark" is set to define a region where additional coordinates could ly: the "*road method*" is the most frequent way to complete track finding. Using a simple track model for the initial track candidate (straight line, parabola, spline curve, etc.), a road is defined inside which additional points are looked for. Its width may depend on the quality of the model, the detector resolution, the particle's momentum, multiple scattering, or the inhomogeneity of the magnetic field. Appropriate formulae for road widths must be found empirically from MC-simulation and from real data.

In spite of the limited number of track candidates it is still essential that the track model is simple enough that the decision of adding coordinates from the remaining detectors can be taken after a reasonable amount of computation spent only for the interpolation.

In principle, the better the model, the narrower the road can be, which avoids the pick up of background or coordinates from another track. It is therefore *essential to plan the detector layout such that simple models can be used for the track search*, possibly followed by a second iteration with a more refined track model for almost final track candidates only.

The simplest model would be a straight line, but this is definitively not applicable in the presence of a magnetic field, or at least not in the plane perpendicular to the main field component. Therefore suitable transformations should be looked for, in order to *transform the classification problem* to its simplest case, *the classification of straight lines*.

As in many storage ring experiments we have a central detector in a quite homogeneous field measuring coordinates $(R\phi)_i$ at constant radii R_i, a "polar inversion" is frequently used as the appropriate transformation.

In a homogeneous magnetic field parallel to the z-axis the particle trajectory is a helix, and its projection onto the (x-y) plane a circle. If the trajectory comes from the interaction region, one point of the circle is (x=0, y=0), and the equation of the circle takes the simple form:

$$x^2 - 2xx_0 + y^2 - 2yy_0 = 0,$$

with:

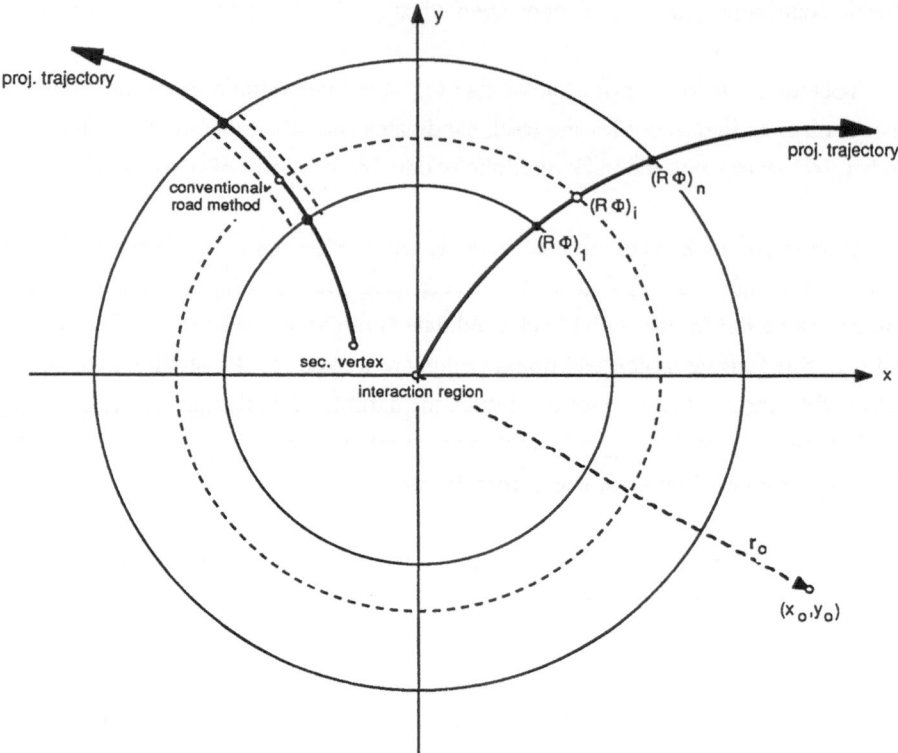

Fig. 2.3. Cylindrical coaxial detectors with the axis parallel to the magnetic field. The road method can be applied either with circles, or with straight lines after a suitable polar inversion (see also fig. 2.5). The innermost and the outermost detector are built such as to give a reliable master point.

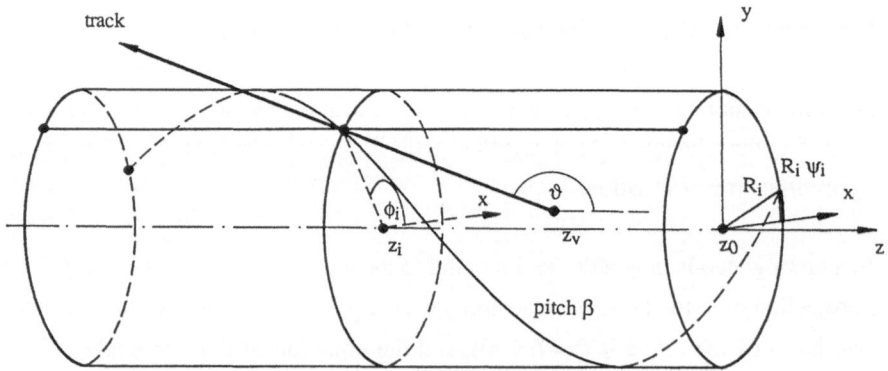

Fig. 2.4. Diagrammatic representation of a cylindrical wire chamber with strips in form of a helix for the resolution of multi-track ambiguities in space.

$$x_0{}^2 + y_0{}^2 = r_0{}^2,$$

x_0, y_0 ... center of the circle, r_0 ... radius of the circle.

This now can be transformed into a linear equation by a conformal transformation (Fig. 1.5, Fig. 2.5): If the i'th pad row with radius R_i measures the coordinates (x_i, y_i), the quantities

$$x_i{}' = \frac{x_i}{R_i{}^2} \, , \, y_i{}' = - \frac{y_i}{R_i{}^2}$$

follow the linear relation (x_0, y_0 are of course unknown):

$$1 - 2x_i{}' \, x_0 + 2y_i{}' y_0 = 0.$$

In practice the transformed lines are not always exactly straight, due to several contributions:

- Spread of the beam position
- Multiple scattering in the beam tube and the inner chamber wall
- Energy loss
- Measurement errors including digitizing effects

If the particle does not come at least from the beam region, standard roads with circles have to be applied. Furthermore it is important to note: if no space points are available, the detector must be designed such that one projection is normal to the main component of the magnetic field. This is also true if the field is *perpendicular to the beam direction.*

Another interesting transformation was applied in an early fixed target counter experiment for helical strips (for resolving ambiguities) around the target region without magnetic field (de Bourd and Regler 1968, private communication to the "Argonne Conference" 1968). If β is the pitch of the helix ($\beta = 2\pi\Delta z/\Delta\phi$), ϕ the azimuth angle of the trajectory, and z_0 the end of the chamber, the following relation holds (Fig. 2.4):

$$\psi_i = \phi_i + (2\pi/\beta) \, (z_0 - z_i),$$

with:

ψ_i .... azimuth angle at $z = z_0$ of the helix strip hit by the particle in cylinder i,
ϕ_i .... azimuth angle of the intersection point of the trajectory with cylinder i.

If the vertex is close to the z-axis ($\sqrt{x_v{}^2 + y_v{}^2} \ll R_1, \ldots, R_n$), ϕ_i is constant versus i, and it follows:

$$\psi_i = C_1 + C_2 \, R_i,$$

with

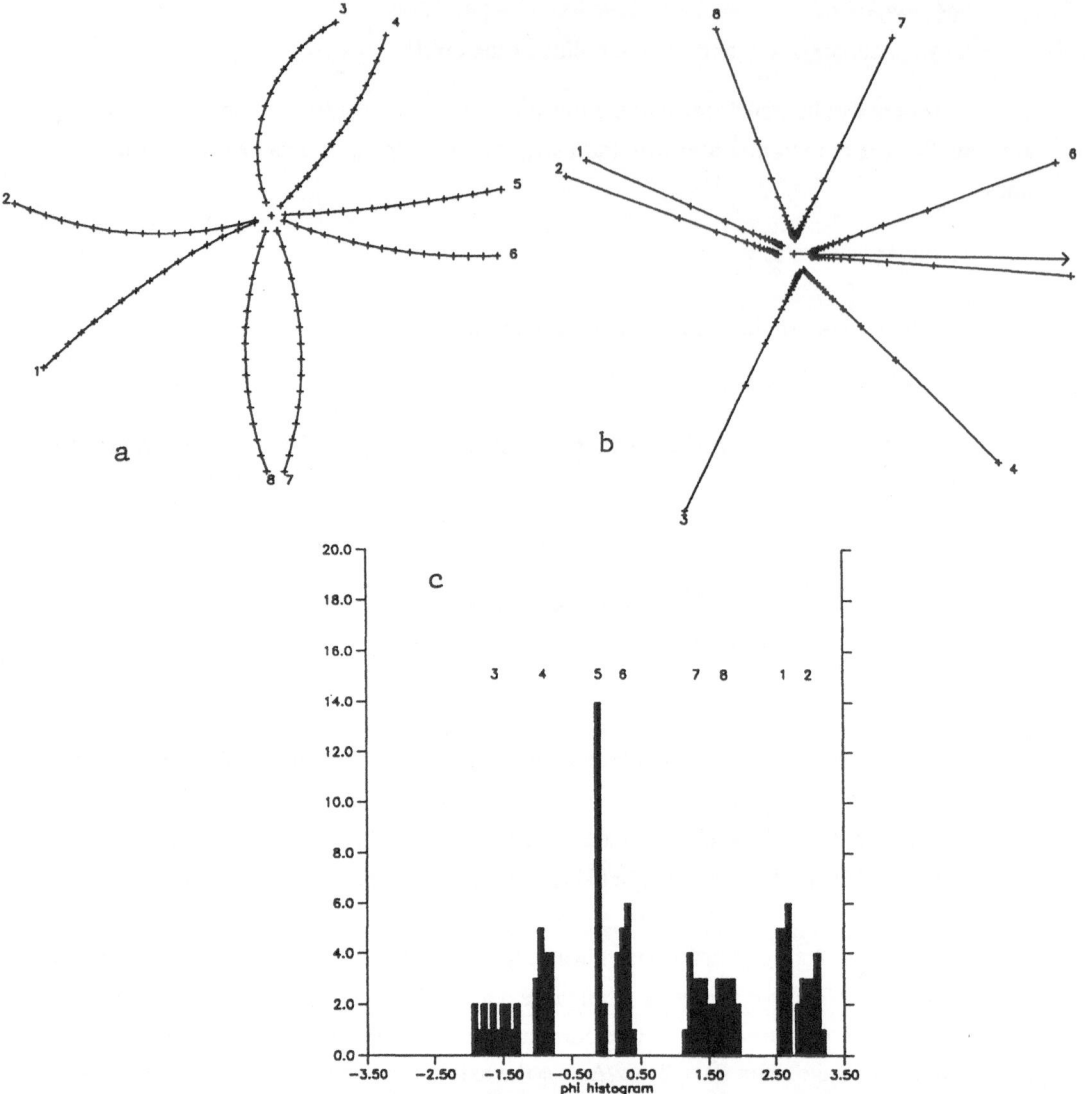

Fig. 2.5. a) Arcs of circles through the origin. The arcs could be caused by charged particles inside a homogeneous magnetic field, e.g. in the detector of fig. 2.3.

b) Arcs of circles in a) after a conformal transformation. The nearest distance of
any line to the origin is proportional to the curvature of the original circle arc.
This offset makes angular clusters of low momentum tracks in a histogram lower
and wider than those of high momentum tracks (see fig. 2.5c).

c) φ histogram of tracks in fig. 2.5b. The angle with the (arbitrary) x axis in fig.
2.5b is entered for all points there. The tracks form clusters which can be found
rather easily. The third cluster from the right contains two tracks, easily recog-
nizeable from the number of entries in it. A broad cluster (if caused by one track)
arises from a large minimum distance of the straight line in fig. 2.5b.

(From: Data Analysis Techniques in High Energy Physics Experiments,
Cambridge University Press 1989).

$$R_i / (z_i - z_v) = \tan \vartheta_i = \text{const},$$
$$C_1 = \phi_{(i)} + (2\pi/\beta)(z_0 - z_v),$$
$$C_2 = -2\pi(\beta \tan \vartheta_{(i)}).$$

If the cylinders are constructed such that the pitch of each helix is proportional to the radius R_i so that $\beta_i = kR_i$, then:

$$\psi_i = C_1' + C_2'/R_i.$$

Not only have we now obtained a linear relation between ψ_i and $1/R_i$, but the tracks coming from the same vertex are also *parallel* in the diagram ψ_i vs $(1/R_i)$.

A special problem is the reduction of the combinatorial aspect by the *removal of successfully associated coordinates*. On the one side this can be an efficient help, e.g. when removing a coordinate in one projection of a detector module all possible space points created by its association with the coordinates in the other projections disappear, i.e. the number of space point candidates is reduced by a factor of the order of the track multiplicity; but the *danger of creating a bias is high*. The stability of such removal method must be carefully checked under several conditions, and any possible bias must be carefully examined (MC, rotational invariance, comparison of known reference cross sections etc.) Many tricks have been invented to reduce the risk, like starting the track search in a different direction from one event to another, etc.

2.5 Track Following

In the literature on track finding one often distinguishes between *local* and *global methods*. A local method selects one track candidate at a time, starting by an initial track segment and then looking for further points belonging to this track, e.g. by *interpolation* (see Section 2.3) or by *extrapolation*. The definition of local methods, however, is not very strict, and in order to avoid confusion with the local pattern recognition (Section 2.3), the word local method will be avoided in the following sections, where possible.

The most traditional "local method" (from bubble chamber physics) is the "*track following*" method, which is typically applied to tracks of the "perceptual type", which is also easily accessible to the human eye. It is also the most accessible method to computer supported automatic optical measurement devices. An *initial track segment* is first selected, consisting of a few points (three or four), and this segment is normally chosen as far away from the interaction region as possible, since there the tracks are (at least on the average) more widely separated than anywhere else. Then, the next possible point is chosen by extrapolation (or even by additional interpolation, if the target position or the interaction region gives additional information about where the track must end up). The essential is that coordinates are added step by step on the inner side of the track segment, and the association criteria will get more selective with increasing knowledge about the track segment. This has the advantage that when one reaches the "crowded" vertex region, the method has gained its optimal discrimination power. This method, however, requires in general a dense sequence of coordinates, and is more frequently applied in projections than in space.

Predictions can simply be computed by the tangent of the track segment found so far, or by adding the appropriate curvature. For low energy tracks energy loss can also be considered, and multiple scattering can influence the choice of the cut values.

At the level of local pattern recognition cuts are usually defined empirically, and this method works well as long as *not too many ambiguities* (alternative choices for a given step) occur. Otherwise a tree structure must be built up, in which some branches may die out, and others may join again, or merge with a branch originating from another tree. (This is less likely if space points are available.) A complementary graph theoretical approach may be required (see section 2.6); for jets or other dense bundles of particles a combination with other methods using a model for the track as a whole might be needed.

A more ambitious application of this method is follow up of muon tracks through the iron of a hadron calorimeter, a task being made difficult by multiple scattering and the presence of a large amount of background from hadron showers.

Track following is nowadays also heavily used for track fitting where it is called "Kalman filter" (P. Billoir 1984, R. Frühwirth 1987, 1988) and is extensively discussed in section 3.4.

At the end it should be said that interpolation and extrapolation are often complementary, and none can be completely replaced by the other. They can cooperate efficiently in one and the same reconstruction program, in order to minimize the number of ambiguities and outliers from background going to the fit procedure.

2.6 Global Methods

A *global method* is defined as a method where all tracks are examined in parallel. Therefore a pure global method is *independent of the order in which coordinates enter the algorithm*. Although any method of track search must be checked against possible biases, the result of a global algorithm is at least stable with respect to the order in which the tracks are found. Global algorithms start basically with a general transformation of the totality (or, in practice, of a large suitable subset) of the event coordinates, and produce a table of tracks, or at least a table in which the final track selection can be done more easily than among the original data. The computing time is in principle proportional to n, the number of coordinates in an event, and therefore also proportional to the track multiplicity (for a more realistic estimate see below).

One of the most traditional methods is the *"histogram method"*, which can be used in the field-free telescopes of a spectrometer of a fixed target experiment (Fig. 1.2). If we assume an arrangement of m equidistant detectors (e.g. at positions $z_1, ..., z_m$), all possible slopes can be evaluated in each projection by a trivial algorithm. If we have n_i coordinates $(\alpha_i, i=1,...,n_i)$ in the i-th detector, the following quantities are evaluated and histogrammed:

$$s^x_{i,\,\alpha_i;\,j,\,\alpha_j} = \frac{x_{j,\,\alpha_j} - x_{i,\,\alpha_i}}{j - i} \,,\quad s^y_{i,\,\alpha_i;\,j,\,\alpha_j} = \frac{y_{j,\,\alpha_j} - y_{i,\,\alpha_i}}{j - i}\,.$$

A track follows the equation

$$x = t_x z + d_x, \quad y = t_y z + d_y.$$

If the slopes of all possible pairs (α_i, α_j) (with $i \neq j$) are entered into the histogram the number of entries in each projection is equal to

$$\sum_{i<j} n_i n_j$$

or roughly $<n^2> \bullet m(m-1)/2$ per projection. This number can be relatively large, but each entry requires only one subtraction and one division. These operations can even be executed on-line, and this method is therefore suitable for a higher level trigger. For circle arcs as caused by charged particles inside a homogeneous magnetic field (e.g. fig. 2.3), a conformal transformation (polar inversion) can re-establish the linear case (Fig. 2.5 a,b,c). This is, however, more time consuming. The method for straight lines can still be extended to a four-dimensional parameter vector, with

$$d_x = x_i - t_x z_i, \quad d_y = y_i - t_y z_i,$$
$$p^x_{i,\,\alpha_i;\,j,\,\alpha_j} = x_{i,\,\alpha_i} - (x_{j,\,\alpha_j} - x_{i,\,\alpha_i}) \bullet \frac{i}{j-i}\,,$$
and similar $p^y_{i,\,\alpha_i;\,j,\,\alpha_j}$.

All quadruples $\mathbf{p(m)} = (s^x, s^y, p^x, p^y)$ coming from the same genuine track form a cluster, and the problem of *track finding in the pattern space* of the measurements has thus become the problem of *cluster finding in the parameter space*. In the simple case of straight lines simple algorithms like some minimum content in a certain interval of the histogram above the "combinatorial background" can be successful, and the knowledge of the correct assignment of the coordinates in space (e.g. from cathode pad rows) will significantly reduce the pick up of wrong coordinates from combinatorial background. In more complicated cases, graph theoretical approaches like a "minimum spanning tree" (see below), linking sets of points of nearest distance in the parameter space, should be applied. In practice the histogram method is efficient if "natural projections" exist, and if appropriate parameters can by found. Otherwise it is not competitive with the usual heuristic extrapolation and interpolation methods.

Generally, cluster analysis attempts to group points in a multi-dimensional space in such a way that all points in a single group have a natural relation (*"similarity relation"*) to each other, and points not in the same group are somehow different. As discussed above, the space can be either the measurement space or a pattern space of reduced dimensionality defined by some significant parameters, including appropriate *scaling*. In track search the objects of a cluster follow a deterministic relationship as defined by the equations of motion which are randomized by several effects:

- the measurement errors,
- the multiple scattering,
- the phase space population.

Clustering techniques (i.e. a general strategy for forming clusters) in track reconstruction are still of intuitive nature with some attempt to fit them into an useful mathematical frame work, and the final performance will only be fixed after some supervised learning from the real data.

The histogram method described above is an example of the most common approach, i.e. the assumption that one can describe the similarity of pairs of objects by a numerical value (e.g. the difference in slope) called the *"similarity measure"* of a pair of objects. The objects in the parameter space were themselves again functions of pairs of coordinates.

The clustering problem in track finding is (besides the problems of ambiguities to be left for the track fit) a *disjoint clustering* ("D-clustering"); this means that we are looking for a partition $G = \{G_1, ..., G_k\}$ of our basic set, which divides it into k subsets, by optimizing a criterion function c(G) or an algorithm A(**M**), $\mathbf{M} = \{m_{i,j}; i=1, ..., k, j=n_1, ..., n_j\}$ (D-clustering algorithm). The number of tracks k can be known or not, and the number of coordinates of the j-th track, n_j, can be constant or can vary with j.

But even if the track multiplicity k is known from the final state of a reaction one is looking for, inefficiences and background will spoil the situation. On the other hand, the track search is facilitated by the fact that coordinates from the same detector do not belong to the same track (except double hits on large scintillators or cathode strips); particles can of course be bent back by the magnetic field, but in a first approach the forward and backward segments can be regarded as different tracks. In other words, there exists a partition into subsets of "units" defined by the individual detector modules, $D = \{D_1, ..., D_m\}$ such that a cluster contains at most one element from each D_i. This fact makes a *graph theoretical approach* attractive: only coordinates ("nodes") of different detectors can be associated by lines ("edges"), and this only in a well defined hierarchy given by the detector sequence. Furthermore the *compatibility graph technique* is used which provides the solution to the following problem: find the subset of a maximum number of nodes which are not linked with each other, and therefore are said to be "compatible" with the hypotheses of different tracks.

In track finding a positive weight is assigned to each line ("edge-weighted graph"), in order to link points of "nearest distance"; as this distance may be almost any positive definite function of the coordinates, it allows for considerable variations. In such a graph a *minimum spanning tree* can be found for which the sum of the edge weights has a minimum.

A dense and regular structure of the detectors, a limited amount of missing coordinates only, and a quite fixed number of coordinates per track make this minimum spanning tree method efficient, and in general it works well in projections; it was successfully applied in central detectors of storage ring experiments, while more general methods of cluster analysis are used in the analysis of particle showers in calorimeters.

Another global method for track classification is the method of "template matching". Assuming that objects are only "white or black" (two "grey levels" only), they only assume the values zero and one.

```
0 0 0 0 0 0 0 0 1
0 0 0 0 0 0 1 1 0 0
0 0 0 0 0 1 1 0 0 0
0 0 0 0 1 1 0 0 0 0
0 0 0 1 1 0 0 0 0 0
0 0 1 1 0 0 0 0 0 0
0 1 1 0 0 0 0 0 0 0
0 1 0 0 0 0 0 0 0 0
1 0 0 0 0 0 0 0 0 0
1 1 0 0 0 0 0 0 0 0
```

Each track is defined by being similar to one out of a set of prototypes $\{y_1, ..., y_m\}$. A track candidate x is then compared with all prototypes:

$$n_i = y_i^T x, \quad i=1, ..., m$$

or in matrix notation

$$n = Y^T x,$$

where the maximum component of n defines a match if it is above a certain threshold.

An example is to divide the azimuthal angle Φ measured by a detector of coaxial cylinders (Fig. 2.3) into segments, with a suitable transformation $\Phi'_i = \Phi_i - <\Phi_i>$. If a coordinate Φ_{ij} is found in sector j of detector i, the element is black, otherwise it remains white. The templates can be calculated by MC; they must consider multiple scattering and measurement errors. The sample used to define the templates is called a "training sample". Then a second independent sample must be generated to check the algorithm, the so-called "test sample". The final tuning must be performed in any case with real data.

In practice the number of possible candidates **x** is often too high for immediate template matching: for n tracks and m detectors it is n^m. A two-level approach must be chosen:

- On the one side, a "pre-selection" will be performed on the candidates by simpler methods like sectors in Φ or very loose road-width cuts.
- On the other side, the templates can be grouped into "super classes". Once the super class has been chosen only a small directory of templates has to be processed.

Although template matching is not yet very heavily used in high energy physics experiments, *it is clearly very well suited* for programs running on machines with *vector processors*, and will therefore influence the design of future detectors, also because of its practicability in fast dedicated track finders for triggering, and even in more elaborate real time systems. For instance, a pattern recognition algorithm using bit manipulation is employed for the Mark III detector at SPEAR (Stanford).

2.7 Feature Extraction and Principal Component Analysis (PCA)

In track reconstruction the objects (coordinates) are described by a small number of parameters, plus a stochastic noise contribution. The objects are *points in a pattern space* $\boldsymbol{P}$, the dimension of which can be as large as 100. In practice, however, there is usually a trivial decomposition into subspaces, e.g. the forward and the target spectrometer in fixed target experiments, or the forward, central and backward region in storage ring experiments. These subspaces need not be disjoint. This trivial decomposition will not be mentioned anymore in what follows. Note that $\boldsymbol{P}$ is not a vector space, but for convenience we use the vector notation; although the sum of two "vectors" makes no sense, adding a small difference as obtained from linear expansion does.

An undisturbed track gives rise to a point in the pattern space which lies on a five-dimensional hypersurface, while measurement errors and multiple scattering scatter the track point slightly out of the hypersurface (Fig. 2.6). However, a "track point" remains characterized by some *significant features* which should be extracted by a fast algorithm.

When speaking about the pattern space something should be said about the metric: For a stochastic approach to track finding the metric will be the inverse of the covariance matrix which is the sum of two contributions:

- The covariance matrix of the measurement errors, which is, at least blockwise, diagonal (Section 1.2, Section 3.3);
- The covariance due to multiple scattering, which is not diagonal (Section 3.3).

This has to be compared with the empirical covariance matrix C due to the population of the phase space. If the hypersurface is a plane, the latter covariance matrix is of rank five. For the

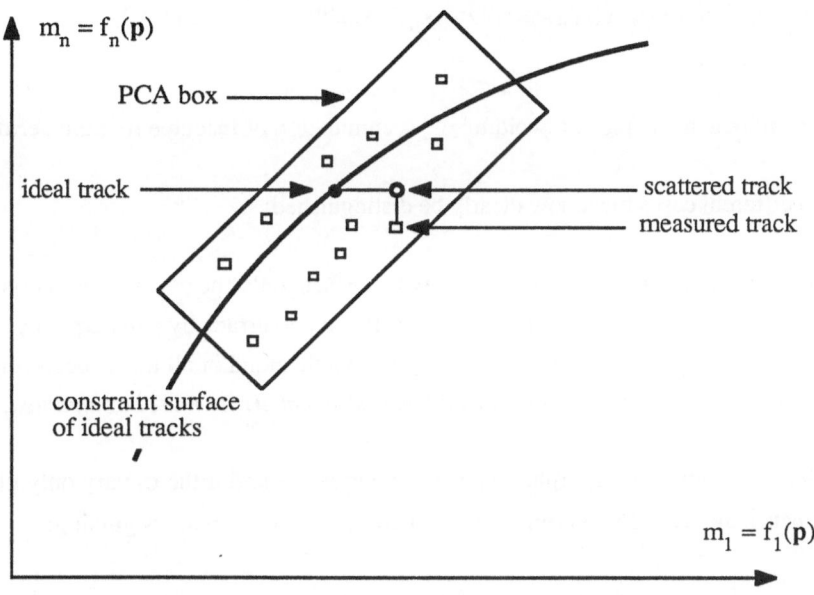

Fig. 2.6.   Track points in the n-dimensional pattern space. Defining new variables can reduce the region of interest. If n is large this becomes important. The main problem is to keep n fixed.

most general case the rank will be of the dimension of {P}. The elements of the covariance matrix C are defined as:

$$c_{ij} = (C)_{ij} = <(m_i - <m_i>)\,(m_j - <m_j>)>.$$

In practice, however, c_{ij} is replaced by its estimate s_{ij} as obtained by the mean over the "training sample". As several thousands of tracks are often used to evaluate this covariance matrix (roughly 10 to the power of the dimension of the track parameter, which is five in the most general case), *updating formulae* should be used:

$$<m_i^{(n)}> = <m_i^{(n-1)}> + \frac{1}{n}\,(x_{i,n} - <x_i^{(n-1)}>),$$

$$c_{ij}^{(n)} = c_{ij}^{(n-1)} + \frac{1}{n-1}\,((x_{i,n} - <x_i^{(n)}>)(x_{j,n} - <x_j^{(n)}>) - \frac{n-1}{n}\,c_{ij}^{(n-1)}),$$

with $c_{ij}^{(1)} = 0$.

It has the additional advantage of avoiding the accumulation of machine rounding errors.

Two different cases must now clearly be distinguished:

- If one is studying the behaviour of the hypersurface, only the covariance matrix C must be considered, e.g. for an approximation of the hypersurface by a hyperplane.
- If one is looking for a "cut box" including the whole cloud of all tracks occurring in this experiment, also *multiple scattering and measurement errors must be simulated.*

If the contribution of multiple scattering is negligible and if the σ_i vary only little with the coordinates, the reduced coordinates should be used right from the beginning:

$$m'_i = m_i/\sigma_i.$$

It is important to note that even if the track impacts in the detectors are highly non-linear functions of the five track parameters (unavoidably or by the wrong choice of track parameters), the populated *"feature space"* can still be close to a five-dimensional (linear) hyperplane. This approximate linearity can be drastically improved by correct design of the shape of the magnetic field, of the detectors and their positions. A suitable mapping like the polar inversion is often the appropriate approach.

Attempts were made to establish simple explicit expressions of low order (second or third) for a non-linear constraint surface; but it may happen that the algorithm finally obtained after a lot of work takes as much computing time as methods of fitting tracks directly as described in chapter 3, while being less transparent and less precise, and requiring an additional step for the parameter estimation. If the feature space is very non-linear, it might be convenient to subdivide it, and to "or" the different regions when looking for a track candidate.

The more successful application of feature extraction is to define a "cut box" of dimension five or more in the pattern space, according to the strength of correlation, in which all tracks must lie. In linear feature extraction one now calculates the principal components by reordering the eigenvectors of $\mathbf{C}$ according to their significance as indicated by the size of the corresponding eigenvalues.

1) First a *suitable* training sample is chosen.

It is important to notice that the choice of the training sample is essential, and *influences the performance* of the PCA. If the aim is to find tracks down to several hundred MeV/c inside a magnetic field, the training sample must cover the full momentum range. If a uniform population of the phase space in the parameter space is chosen, the appropriate "linear parameters" (e.g. 1/P instead of the momentum P) must be chosen.

Sometimes the interesting physics is hidden in a small and sparsely populated region of the total phase space, and although the total reconstruction efficiency may be 99.9%, the probability of reconstructing one of the events really looked for could be close to zero. A way out consists of generating the parameters of the training sample according to a cosine distribution, i.e.

$$p_i' = \cos (\pi \cdot (p_i - p_{i,min})/(p_{i,max} - p_{i,min})).$$

However sometimes the edges of the "phase space box" can not be realised by all five parameters due to the geometrical limitations of the detectors, and the magnetic field, and the pattern space may again be poorly populated in a boundary region.

2) Having generated a suitable training sample, the eigenvalues and eigenvectors of the matrix $\mathbf{C}$ must be computed.

The matrix $\mathbf{C}$ being positive definite, it is convenient to order the (non-negative) eigenvalues by decreasing magnitude, after having defined a new orthogonal basis of the original pattern space with the set of eigenvectors t_i, thus setting up a transformation matrix $\mathbf{T}$ with the property $\mathbf{T}^T = \mathbf{T}^{-1}$. The coordinates x_i along the feature axis t_i are obtained for each track with coordinates $\mathbf{m}$ by

$$x_i = \sum_{j=1}^{n} (T)_{ij} (m_j - <m_j>).$$

After the translation to the center of the cloud this transformation is just a rotation which has the property of minimizing the mean square sum of residuals introduced when the region populated by the training sample is approximated by the space spanned by the eigenvectors of the l largest eigenvalues (l ≤ n); in other words, this transformation has the property of minimizing the mean square (with M training track points)

$$\sum_{k=1}^{M} d^2_k(l) = minimum$$

of the Euclidean distance d(l) to the l-dimensional hyperplane spanned by the selected eigenvectors.

Now suitable cuts can be defined in this rotated space (or subspace) for $|x_i|$, i=1, ..., l, *reducing the possible combinations of coordinates to track candidates by several orders of magnitude.*

The application of the PCA-method has an essential draw back: The transformation T is defined in a pattern space $\mathcal{P}$ of fixed dimension n. If only *a single coordinate* is missing, a new pattern space $\mathcal{P}'$ must be defined. However, only a limited number of coordinate combinations can be admitted if the method should remain practicable.

To keep this method efficient, a set of n "master points" must be available for each track. To ensure their availability a redundant set of detectors like multiwire chambers are grouped together as close as possible, such that the probability to lose a space point is only of higher order in the (hopefully small) detector inefficiency. However, corrections for the distances between the individual detectors have to be applied, requiring some rough hypothesis on the track direction; due to these corrections the selectivity of this method can be far below the detector resolution. An additional difficulty arises from the ambiguities, which can be solved by detectors with two-dimensional readout (where unfortunately one coordinate is often much less precise than the other). Many algorithms were invented for fast master point evaluation and it must be strongly underlined that the efficiency of these algorithms is of vital importance for avoiding a breakdown of the feature selection method.

Several methods have been proposed to "complement" the coordinate vector, but in most of the cases the control over the PCA-method and its discrimination power gets lost even if the formal behaviour is unchanged. Designing the detector very carefully in view of the analysis algorithm, including already in the detector planning appropriate linearization and the reinstallation of the natural decomposition into projections, can make the PCA-method very powerful.

Sometimes a two level approach is advisable, e.g. in a two lever arm spectrometer: First a PCA in the two detector telescopes is performed by two different PCA models. In a second step, a PCA in the parameter space spanned by the two parameter vectors of the two track segments will link the two pieces to one track (see also next section).

2.8 PCA and Vertex Cluster Finding

An appropriate case to use the PCA-method (Section 2.7) is the search for a common vertex for tracks emerging from a region outside the active detector volume. For short lived particles in a storage ring experiment such a region would be the volume inside the beam tube.

Although the number of tracks belonging to the (primary) vertex can be variable, a simple parameter space of fixed dimension 10 can be constructed in which the similarity relation is that for each *pair of tracks* originating from a common vertex the two tracks must cross each other very closely. In order to obtain the parameter vector ($\mathbf{p_i}$ ($\mathbf{m_i}$), $\mathbf{p_j}$ ($\mathbf{m_j}$)) the beam tube is chosen as reference surface for track finding and track fitting (Chapter 3).

The next step is the linearization of the feature pattern in the projection normal to the magnetic field. The transformation of x and y is trivial (x^2 and y^2 at the beam tube being constant), but the transformation on φ and the curvature is a more complicated procedure.

The polar inversion (a conformal transformation) will produce straight lines in the (x',y') plane (Section 2.4), their nearest distance from the origin being (similar to fig. 2.5)

$$d = 1/(2r_0),$$

with r_0 being the radius of the projected helix (Fig. 2.3). The origin (0,0) is transformed into infinity. Tracks of particles emerging from secondary vertices are not transformed into straight lines, but into continuous curves which intersect in a finite point which is the transform of the secondary vertex. Due to the conformity of the tramsformation the *angles between secondary tracks are invariant.*

Now the PCA-transformation $\mathbf{T_v}$ (of dimension 10) is constructed from the appropriate training sample, which has to pass first the full single track fit procedure. The constraint hypersurface is of dimension 9, and in the example of magnetic field parallel to the detector axis the effective dimension is 8, due to the translation invariance in z (Fig. 2.3). Sometimes an appropriate choice of the track parameters on the beam tube will also reduce the measurement space to its effective dimension 8: $\Phi_j-\Phi_i$, z_j-z_i, and ($\varphi-\Phi$, $1/P_T$, ctgϑ) for i and j. This parametrization is sufficiently linear to avoid a PCA or the more complicated case of a polar inversion including the directions.

It is often convenient to perform even a fast linear fit. To make the "power" of this reasonably good one must however consider the metric: First, the error on the curvature varies considerably with the momentum, and secondly, direction and curvature are strongly correlated in the projection normal to the field.

The transformation of the five times five covariance matrix of each fitted single track can be approximated by transforming only the three times three block in the projection mentioned above, and by neglecting the correlation with the other projection. Although it needs to be done only once per track, it remains a lengthy procedure. The reward is a reasonable "χ^2-test quantity". Now from the pairwise test quantity an overall similarity measure must be constructed.

How far this method should be pushed must be decided from practical experience.

3. TRACK AND VERTEX FITTING

3.1 Introduction

Track fitting has a long tradition in high resolution *cosmic-ray* and *bubble-chamber experiments*. The importance and the feasability of track (and vertex) fitting in the more complex environment of *experiments with electronic detectors* was recognized when detectors with good and well-defined resolution came into operation. At the CERN Intersecting Storage Ring track and vertex fitting with rigorous treatment of multiple scattering was successfully applied in the early seventies (e.g. Metcalf, Regler & Broll 1973). Since then track fitting has been applied to several fixed target experiments as well as to experiments at e^+e^- and $p\bar{p}$ storage rings, where the selection of rare reactions in the presence of high multiplicities requires ultimate track separation and exact track and vertex fitting. On the other hand, the increased computing time required for complex events has sometimes discouraged the application of such rigorous reconstruction methods.

Modern experiments have features which make track fitting a difficult, but highly important task:

- The energy of secondary particles covers a large range, from particles nearly at rest up to a few hundred GeV. Hence the reconstruction precision is limited on the low energy side by multiple scattering and energy loss, and on the high energy side by the small deflection in a magnetic field.
- For fixed target experiments very long spectrometers have to be built, and the spatial separation of tracks in the vertex region is relatively small, making the final decision on ambiguities in the track search difficult.
- In storage ring detectors complex arrangements for track reconstruction and particle identification have to be compressed in a relatively small volume. Beam tube, vessels and cables are serious obstacles for the particles, causing multiple scattering, energy loss and even the creation of additional particles.
- Different techniques applied in different parts of the detector system, with resolutions ranging from a few μm in high precision vertex detectors to a few mm in tracking calorimeters, have to be combined. In addition the resolution varies often quite strongly as a function of the hit position.
- Finally, high event rates lead to a large amount of data, requiring that optimal extraction of information is performed by fast algorithms.

But also from the physics point of view optimal resolution is important, and no information produced by the detectors – which are often very expensive – should be lost:

- Invariant masses must be determined with optimal precision.
- Secondary vertices must be fully reconstructed.
- Kinks must be recognized with high probability and located as precisely as possible.

- For muon identification an enormous amount of multiple scattering in the filter material must be efficiently treated.

In addition to the choice of the proper fitting procedure, some technical problems have to be solved:
- The detector resolution must be understood.
- A precise track model is needed, allowing the fast evaluation of a charged particle's path in a magnetic field, and the values of the magnetic field at any point must be easily accessible.
- The amount of material traversed by the particle must be well known, for the evaluation of both energy loss and multiple scattering; energy loss modifies the track model, and multiple scattering must be considered as an additional error contribution in the track fit, or as a noise contribution to the trajectory.

A detailed track fit is not always performed for all kinds of final states, but the selection of rare final states embedded in millions of similar topologies may require the precise reconstruction of a high number of events if the rare event type has no clear signature.

In this chapter, first the basic concepts of track fitting will be presented, and the use of the least-squares method will be justified (Section 3.2). Section 3.3 discusses the global estimation of track parameters by the least-squares method (track model, weight matrix and minimization). Section 3.4 describes a recursive formulation of the least-squares estimation procedure, the Kalman filter and smoother. Section 3.5 presents methods of handling anomalies which may arise in track fitting, as outliers, kinks and long-tailed measurements errors. In section 3.6 the geometrical reconstruction of interaction vertices is discussed. Finally, in section 3.7 the problem of associating tracks to a secondary vertex is treated.

3.2 Basic Concepts of Track Fitting

Track reconstruction serves a two-fold purpose: first, the *parameters describing the track geometrically,* i.e. position, direction and curvature, are estimated by a procedure, which should be optimal in the statistical sense; this is called *track fit* . Secondly, the results of the track search have to be confirmed or rejected, and ambiguities have to be solved. This requires *statistical test quantities* which describe the quality of the track candidate, i.e. whether it has high or low probability to correspond to a real physical track. In addition to the estimated track parameters their *covariance matrix* should be available, so that the estimate can be fed into the next step of analysis with the proper weights.

In a discussion of track fitting one should *distinguish clearly* between the "ingredients" and the "recipe" (Fig. 3.1). The ingredients of the track fit are:
- A track candidate **m** found by the pattern recognition;

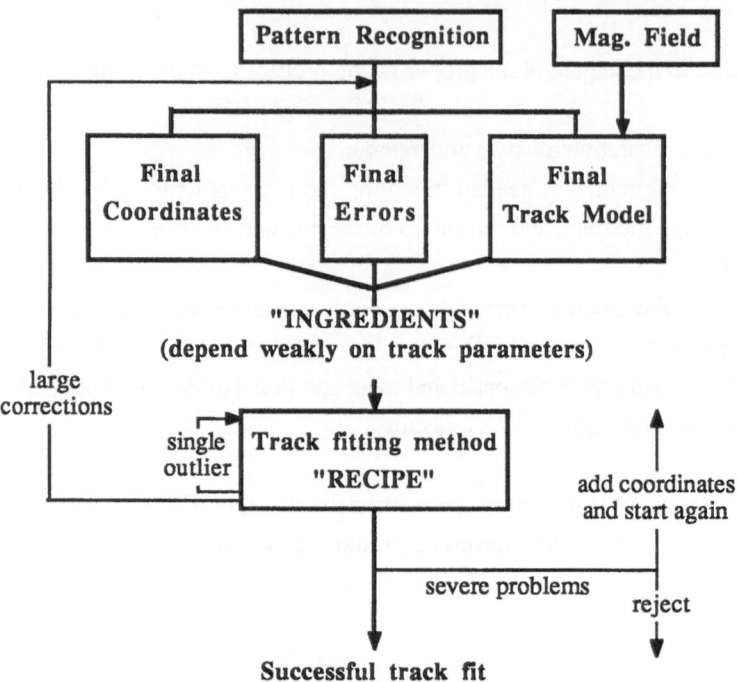

Fig. 3.1. "Ingredients" and "recipe" of track fit. Data structures and steering should not impede the choice of the appropriate algorithm.

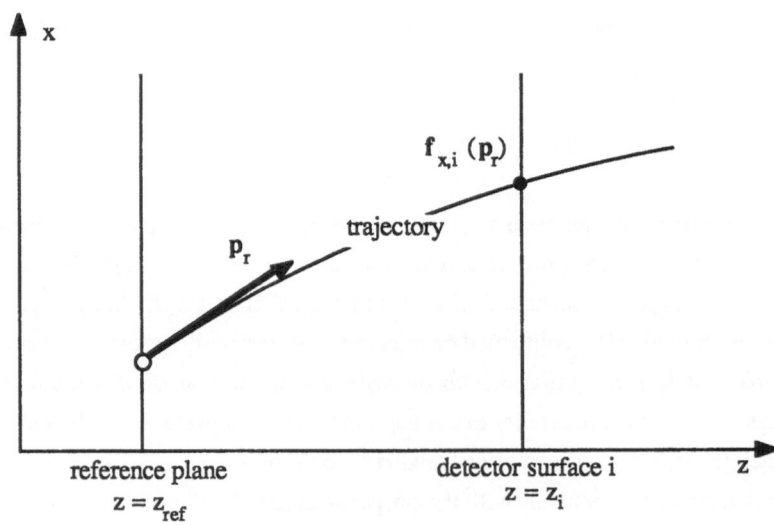

Fig. 3.2. The track model as a mapping from parameter space into measurement space. (From: Data Analysis Techniques in High Energy Physics Experiments, Cambridge University Press 1989).

- The errors on the coordinates in the track candidate (the "detector resolution");
- A mathematical model of the trajectory of a particle in a magnetic field, the "track model";
- A statistical model of the processes which disturb the trajectory when the particle crosses material ("multiple scattering"). (For energetic heavy particles the energy loss can be described with sufficient accuracy by the "deterministic" track model by modifying the momentum as a function of the path length.)

The detector resolution can be estimated from theoretical detector studies (or Monte Carlo simulations), from measurements in a calibration experiment, or from the actually recorded tracks in case they have a sufficient number of redundant measurements.

The track model is the set of solutions of the equation of motion, which describes the influence of the Lorentz force which is exerted by a magnetic field on a moving charge:

$$d^2\mathbf{x}/ds^2 = (kq/P) \cdot d\mathbf{x}/ds \otimes \mathbf{B}(\mathbf{x}(s)),$$

with:

s .... path length,
k .... constant of proportionality,
q .... charge of the particle (signed),
P .... absolute value of the momentum,
$\mathbf{B}(\mathbf{x})$ .... static magnetic field.

A track in vacuum is then represented by the five independent initial (boundary) conditions, i.e. by a point on a five dimensional hyperplane ("constraint surface"), according to the constraints given by the equation of motion.

The track candidate $\mathbf{m}$ is a sample from a random vector $\mathbf{c}$ and deviates from the constraint surface due to (randomly distributed) measurement errors and due to (random) multiple scattering:

$$\mathbf{c} = \mathbf{f}(\mathbf{p}) + \varepsilon,$$
$$\varepsilon = \delta + \gamma,$$

with:

$\mathbf{p}$ ..... 5-vector of track parameters (e.g. x,y at a reference surface z_{ref} = const, and the momentum 3-vector),
$\mathbf{f}$ ..... track model, i.e. a mapping from the parameter space into the measurement space (Fig. 3.2),
$\mathbf{c}$ ..... random measurement vector,
δ ..... random vector of the measurement errors,
γ ..... random vector of deviations due to multiple scattering.

The statistical model of multiple scattering is provided by the theory (Gluckstern 1963, Regler 1977). It will be shown below how it is incorporated into the track fit. The choice of the track model, the understanding of the detector resolution and the treatment of multiple scattering must be checked by the inspection of suitable test quantities. The most prominent of these are the chi-square statistic and the studentized residuals of the fit (pulls). They will be described in more detail in section 3.3.

It is assumed that all ingredients depend only weakly on the track parameters. If the track fit, however, introduces a large correction of the parameters, it may be necessary to go back and recompute some of the ingredients, e.g. the errors on the coordinates. It also happens that the track fit is unsuccessful and that the track candidate has to be broken up again. This implies that the coordinates have to be passed back to the pattern recognition.

The recipe, on the other hand, is a *method of computing the estimate*, its covariance matrix and suitable test statistics, given all the ingredients. It should be independent of the particularities of the ingredients. For instance, it should be possible to replace the track model by a better or a faster one, without interfering with the method of track fitting.

A method of track fitting can be regarded as a projection $\mathbf{F}$ of the coordinate vectors, $\{\mathbf{c}\}$, onto the 5-dimensional surface of the $\{\mathbf{p}\}$, without bias and with minimum variance of the fitted parameters ($\overset{t}{\mathbf{p}}$ being the true value of $\mathbf{p}$, and $\tilde{\mathbf{p}}$ the fitted value; fig. 3.3):

$$\tilde{\mathbf{p}} = \mathbf{F}\,(\mathbf{c}),$$
$$<\tilde{\mathbf{p}}> = \overset{t}{\mathbf{p}},$$
$$\sigma^2\,(\mathbf{p}_i) \equiv <\,(\tilde{\mathbf{p}}_i - \overset{t}{\mathbf{p}}_i)^2\,> \;\rightarrow\; \text{Minimum},$$

where $<\,>$ denotes an expectation value, or the average over many tracks with the same initial conditions.

For an individual track measured, $\overset{t}{\mathbf{p}}$ is a fixed but unknown quantity, while $\tilde{\mathbf{p}}$ is a function of the random vector $\mathbf{c}$ and therefore is also a random vector. The variance of $\tilde{\mathbf{p}}$ has to be considered as describing the experimental errors for repeated measurements of the same track, and it may well differ for different $\overset{t}{\mathbf{p}}$. The fact that $\overset{t}{\mathbf{p}}$ has usually a random distribution itself from one track to another does not affect what follows.

It turns out that the "*least-squares method*" (LSM) meets most of the requirements of track fitting, being simple, rather fast, and familiar to experimentalists. Its important statistical properties, together with its numerical simplicity, form the basis of the wide range of its application. If suitable parameters can be chosen such that the track model can be approximated in the neighbourhood of the measurement vector $\mathbf{m}$ by a linear model (which would coincide with a hyperplane tangent to the hypersurface at some expansion point $\overset{0}{\mathbf{p}}$, fig. 3.4), and if the errors depend only weakly on the track parameters such that they can be kept constant in the

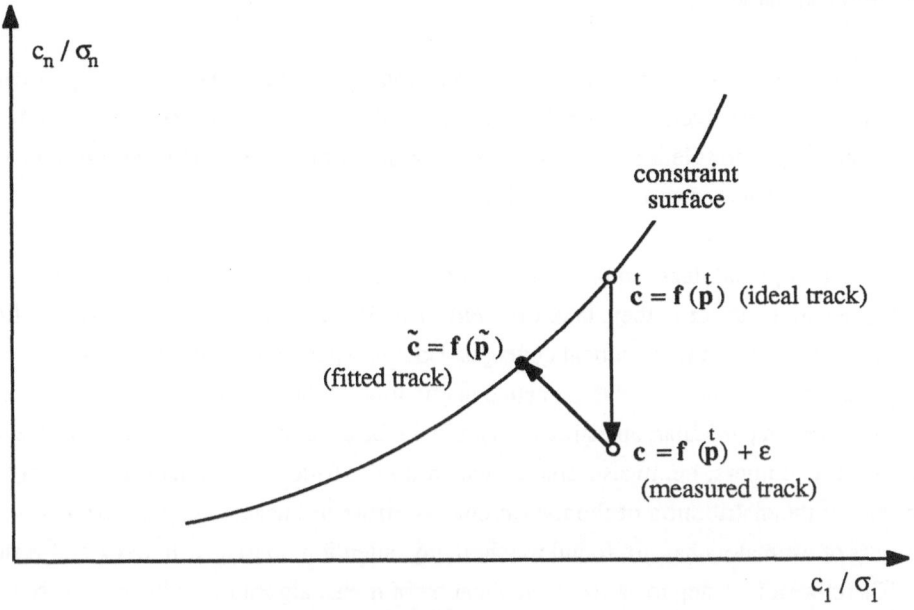

Fig. 3.3.   The track fit as a projection of the measured track onto the constraint surface.
(From: Data Analysis Techniques in High Energy Physics Experiments,
Cambridge University Press 1989)·

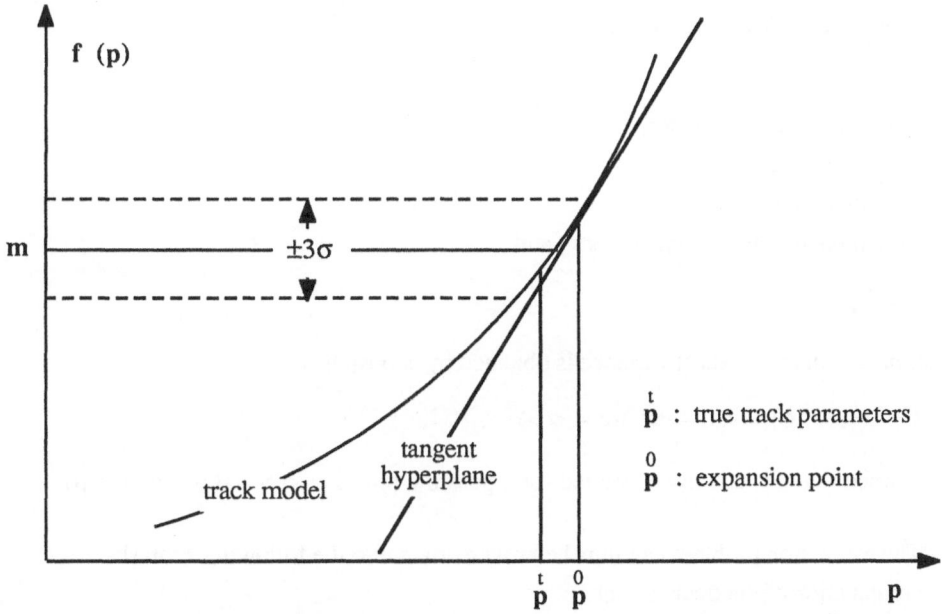

Fig. 3.4.   Linear expansion of the track model in a neighbourhood of the measured track.

neighbourhood of **m**, then the least-squares method has minimum variance among all un-biased linear estimators.

The linear estimate is then suitable for error propagation, yielding suitable ingredients for subsequent fit levels (vertex fitting, kinematics). In the asymptotic limit of a large number of measurements, or for Gaussian erors, the method is efficient, and the square root of the variance can be considered as a confidence interval.

On the other hand, least-squares methods suffer from some shortcomings. Their most important negative aspect is their lack of *robustness*. Robustness is a rather broad notion which embodies several mathematical concepts. Loosely speaking, a robust estimator is one whose performance remains good if the actual distribution of the data deviates somehow from the assumed one. In particular, an estimator should not be excessively biased by the influence of one or a few outliers, i.e. measurements which do not follow the general pattern, either due to tails in the distribution of the measurements errors or because they are background. Robustness of estimators has not found much, if any, attention up to now in track and vertex fitting. Therefore a first step towards the application of robust algorithms will be suggested in section 3.5.

3.3 Global Least-squares Estimation of Track Parameters and Related Tests

The least-squares principle requires that the track parameters **p** are estimated by mini-mizing the objective function:

$$Q(\mathbf{p}) = (\mathbf{m} - \mathbf{f}(\mathbf{p}))^T \, G \, (\mathbf{m} - \mathbf{f}(\mathbf{p})),$$

with:

m .... measurement vector,

f ..... track model,

G .... weight matrix.

If **m** is a sample of the random vector **c**, and

$$\mathbf{c} = \mathbf{f}(\mathbf{p}) + \varepsilon,$$

the estimate with the smallest variance is obtained by choosing

$$G^{-1} = cov\,\{\varepsilon\} = <(\mathbf{c} - <\mathbf{c}>)(\mathbf{c} - <\mathbf{c}>)^T>.$$

It is assumed that **c** is already corrected for a possible bias, i.e. $<\varepsilon> = \mathbf{0}$ or $<\mathbf{c}> = \mathbf{f}(\overset{t}{\mathbf{p}})$.

The estimation problem can thus be broken down into the following subtasks:

- computation of the track model
- computation of the weight matrix
- minimization of the objective function

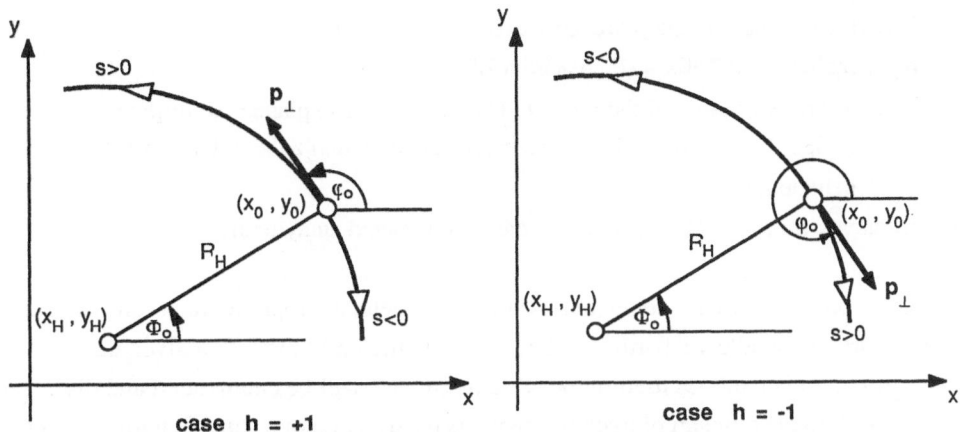

Fig. 3.5.   Parametrisation of a helix.
(From: Data Analysis Techniques in High Energy Physics Experiments,
Cambridge University Press 1989).

Since all measurements are used at once, this procedure is also called "global least-squares estimation".

The track model

In the context of global least-squares estimation the computation of the track model amounts to a solution of the equation of motion for a particular initial value **p**. Frequently also the derivatives of the track model, i.e. the partial derivatives of the function **f** with respect to **p**, are required, as will be seen in the discussion of the minimization of the objective function.

In the case of a *homogeneous magnetic field* – the z-axis is chosen parallel to **B** – the equation of motion has the following form (P is the momentum, q is the charge):

$$d^2x/ds^2 = (kq/P) * (dy/ds) * B,$$
$$d^2y/ds^2 = -(kq/P) * (dx/ds) * B,$$
$$d^2z/ds^2 = 0.$$

The general solution is a helix with the axis parallel to z (Fig. 3.5):

$$x(s) = x_0 + R_H(\cos(\Phi_0 + h*s \cos\lambda/R_H) - \cos\Phi_0),$$
$$y(s) = y_0 + R_H(\sin(\Phi_0 + h*s \cos\lambda/R_H) - \sin\Phi_0),$$
$$z(s) = z_0 + s \sin\lambda,$$

with:

   s .....path length along the helix, increasing when moving in the particle's direction,
   x_0 ...starting point at $s = s_0 = 0$,

λ.....slope ("dip") angle = arc $\sin(dz/ds)$, $-\pi/2 < \lambda \leq \pi/2$,

R_H...radius of the helix = $(P \cos\lambda)/(|kqB|)$,

h......sense of rotation of the projected helix in the (x,y) plane = $\text{sign}(qB) = \pm 1$
  ($=\text{sign}(d\varphi/ds)$, with φ being the track direction, and z the polar axis parallel to the helix axis).

The derivatives w.r.t. the initial conditions can be computed analytically.

In the presence of an *inhomogeneous* magnetic field, appropriate track following algorithms are needed to allow a particle to be followed efficiently through a given detector set-up. A few common methods used in today's experiments will be enumerated and discussed. Closely linked to the problem of track following is the question of a suitable numerical representation of the magnetic field. Numerous approaches have been tried, and the interested reader is referred to the review in (M.Regler 1989).

Helix Tracking

As mentioned above, in a homogeneous field the general solution of the equation of motion is a helix. In a *strongly inhomogeneous field*, however, stepwise tracking by small helix segments *should be avoided*, as the gradient of the field is totally neglected in one tracking step. Therefore very small steps are necessary, and tracking is very inefficient.

Numerical integration of the equation of motion

a) The fourth-order Runge-Kutta method of Nyström. The equations of motion can be rewritten more generally for a given momentum:

$$\mathbf{u}'' = \mathbf{g}(\mathbf{u}', \mathbf{B}(\mathbf{u})) = \mathbf{h}(\mathbf{u}', \mathbf{u}),$$

where ' means either derivation with respect to the path length and $\mathbf{u} = \mathbf{x}$, or derivation with respect to z and $\mathbf{u} = (x,y)^T$. For tracking, the following recursive formula is used:

$$\mathbf{u}_{n+1} = \mathbf{u}_n + h * \mathbf{u}_n' + (h^2/6) * (\mathbf{k}_1 + \mathbf{k}_2 + \mathbf{k}_3) + O(h^5),$$
$$\mathbf{u}_{n+1}' = \mathbf{u}_n' + (h/6) * (\mathbf{k}_1 + 2\mathbf{k}_2 + 2\mathbf{k}_3 + \mathbf{k}_4),$$

with:

$$\mathbf{k}_1 = \mathbf{h}(\mathbf{u}_n', \mathbf{u}_n) = \mathbf{g}(\mathbf{u}', \mathbf{B}(\mathbf{u})),$$
$$\mathbf{k}_2 = \mathbf{h}(\mathbf{u}_n' + (h/2) * \mathbf{k}_1, \mathbf{u}_n + (h/2) * \mathbf{u}_n' + (h^2/8) * \mathbf{k}_1),$$
$$\mathbf{k}_3 = \mathbf{h}(\mathbf{u}_n' + (h/2) * \mathbf{k}_2, \mathbf{u}_n + (h/2) * \mathbf{u}_n' + (h^2/8) * \mathbf{k}_2),$$
$$\mathbf{k}_4 = \mathbf{h}(\mathbf{u}_n + h*\mathbf{k}_3, \mathbf{u}_n + h*\mathbf{u}_n' + (h^2/2) * \mathbf{k}_3).$$

In practice, it is quite often sufficient to take the same field values when evaluating $\mathbf{k}_2$ and $\mathbf{k}_3$ (this is Nyström's advantage), which corresponds to replacing $\mathbf{k}_2$ by $\mathbf{k}_1$, in the second argument of $\mathbf{k}_3$. Another frequent approximation is to use the field value in $\mathbf{k}_4$ of one step to evaluate the k_1 of the subsequent one (see below). The derivatives of the track model can be computed by numerical differentation:

$$(A)_{ik} = \frac{\partial f_i}{\partial p_k} = \frac{f_i(\overset{0}{p} + \Delta_k p) - f_i(\overset{0}{p})}{\Delta p_k},$$

where $\Delta_k p$ is a vector with all components being zero but the kth component being Δp_k. Consequently, the equation of motion has to be solved not once, but six times: one "zero trajectory" and five variations. The choice of the variations is a compromise. They have to be:

- small enough to obtain a reasonable approximation of the first derivatives;
- large enough to avoid numerical problems due to the limited word length of the computer.

Typical variations are: 0.1 - 1mm for spatial coordinates, 0.1 - 1mrad for angles. The fifth parameter should be varied such that the displacement of the track at the different detectors is again 0.1 - 1.0mm. The stability of the derivatives against a change of the variations by a factor of two is a good check of the correct choice of the variations.

b) Parallel integration of the derivatives. If it is easy to obtain the field derivatives from the "field model" the derivatives of f can be integrated together with the "zero trajectory". This saves about a factor of two in computing time as compared to numerical differentation. If the gradient of the magnetic field transverse to the trajectory can be neglected, the algorithm can be simplified and made even faster. Both methods have been successfully applied in several experiments.

Parametrization

Depending on the step size and on the numerical representation of the magnetic field, track following may consume quite a lot of computing time. Therefore one may try to approximate the function f by an analytic function F of the following form:

$$F(p) = \sum_{i_1,...,i_5=0}^{m_1,...,m_5} a_{i_1,...,i_5} * \Phi_{i_1}(p_1) * ... * \Phi_{i_5}(p_5).$$

The choice of the functions Φ_i determines the properties of the approximation. For instance, the choice of Chebyshev polynomials guarantees the minimization of the maximal linear (absolute) residual, if the appropriate "training sample" of tracks is chosen $(p^{(k)}, f(p^{(k)})$; k=1...M). The coefficients a are then evaluated by minimizing the quadratic form

$$Q^2 = \sum_{k=1}^{M} [f(p^{(k)}) - F(p^{(k)})]^2.$$

This leads to a set of $m_1 \times ... \times m_5$ linear equations.

Spline interpolation

Without multiple scattering, an acceptable track model can also be obtained by spline interpolation of the measurements. Such a track model, however, is not a solution of the equation of motion. It should not be used in the presence of multiple scattering, as the statistical properties of the estimator based on it are very difficult to work out. It is equally unsuitable in a setup with a strongly varying magnetic field.

Computation of the weight matrix

As stated above, the least-squares estimator with minimum variance is obtained, if the *weight matrix is the inverse of the covariance matrix* of the random vector ε, which is defined by:

$$\mathbf{c} = \mathbf{f}(\mathbf{p}) + \varepsilon,$$

$$\varepsilon = \delta + \gamma, \ \operatorname{cov}(\varepsilon) = \operatorname{cov}(\delta) + \operatorname{cov}(\gamma),$$

where:
- δ describes the measurement errors, i.e. the difference between the true value of the measured physical quantity and the value registered by the detector, and
- γ describes a random perturbation of the track due to multiple scattering, i.e. the difference between the actual trajectory and the exact solution of the equation of motion.

As far as measurement errors are concerned, it should be noted that errors arising in different detectors are stochastically independent, except for rare cases, in which two detectors are so close to each other that correlations can be observed. This occurs only in detectors with inherent discretization, e.g. in MWPCs (discretization of space) or drift chambers (discretization of time).

Generally speaking, the distribution of the measurement error can be described by a conditional probability density function (p.d.f.) g, the so-called resolution function:

$$g_k = g_k(\varepsilon_k \mid \overset{t}{c}_k) = g_k(c_k - \overset{t}{c}_k \mid \overset{t}{c}_k),$$

with:
c_k....vector of quantities measured by detector k,
$\overset{t}{c}_k$....true values of c_k,
ε_k....vector of measurement errors of detector k.

This includes the possibility that the distribution of the measurement error depends on the impact point of the track on the detector. An eventual dependence should be sufficiently weak, so that the measurement itself or an approximate impact point obtained from the track search can be substituted for the true impact point without ill effects.

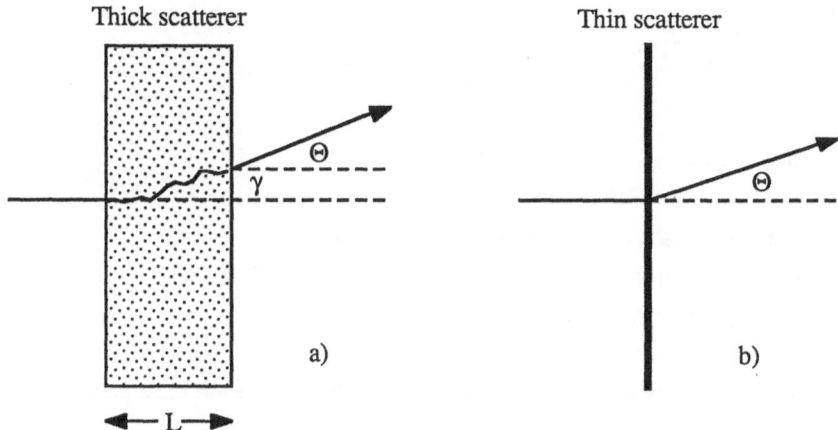

Fig. 3.6. The effects of a scatterer on a track:
a) Thick scatterer: the track emerges with offset γ and angle Θ.
b) Thin scatterer: the track emerges with the angle Θ; the offset is negligible.

For most tracking detectors in use today g_k can be approximated very well by a normal probability density function. As a normal p.d.f. is uniquely described by mean vector and covariance matrix, the problem of finding g_k is reduced in practice to the problem of determining the mean and the covariance matrix of the measurement error ε_k. For very simple detectors this can be done by theoretical considerations. In all other cases mean and covariance matrix have to be estimated from real data. Frequently a calibration experiment is carried out, in which the detector is exposed to a particle beam, a radioactive source, cosmic radiation or a laser beam. Once a sufficient number of tracks has been registered, standard techniques for the estimation of location and variance components can be applied.

Multiple Coulomb scattering acting on a particle sums up as a relatively small but random change of the direction of flight. The resulting effect is a *stochastic process*, in particular a *Markov process*, and one can only evaluate the "probable" amount of influence of a scatterer on a particle trajectory.

Within the approximation that multiple scattering can be described in two perpendicular planes by two uncorrelated random variables, the probability that a particle of momentum P travelling along the z-axis emerges from a scatterer of thickness L (Fig. 3.6) in the interval of the projected angle $(\theta, \theta+d\theta)$ with a lateral displacement $(\varepsilon, \varepsilon+d\varepsilon)$, can be approximately described by the distribution function:

$$dF\,(\varepsilon, \theta; L) = \frac{1}{2\pi\sqrt{L^4\theta_s^2/24}}*\exp-\left[\frac{4}{\theta_s^2}*\left(\frac{\theta^2}{L}-\frac{3*\theta*\varepsilon}{L^2}+\frac{3*\varepsilon^2}{L^3}\right)\right]* d\varepsilon\, d\theta,$$

where $\theta_s^2/2$ is the projected mean squared angle of scattering per unit length. It depends on the mass and the momentum of the particle, and on the kind of material traversed:

$$\frac{\theta_s{}^2}{2} = \frac{k}{L_r} \frac{m^2 + p^2}{p^4},$$

with:

      m....mass in GeV,

      p.....momentum in GeV,

      L_r....radiation length of the material (material constant),

      k.....constant of proportionality, $k \approx (0.015 \text{ GeV})^2$ x unit length.

For a proof see (Gluckstern 1963, Regler 1989).

In the general case of a random vector γ one can define a generalized geometry factor

$$<\gamma_k * \gamma_l> = \Theta_s{}^2/2 * I_{kl},$$

$$I_{kl} = \int_0^{\min(s_k, s_l)} g(s) \left(\frac{\partial f_k}{\partial \Theta_1(s)} * \frac{\partial f_l}{\partial \Theta_1(s)} + \frac{\partial f_k}{\partial \Theta_2(s)} * \frac{\partial f_l}{\partial \Theta_2(s)} \right) ds,$$

with (using the autocorrelation property of a "white noise"):

      $(\partial f_k/\partial \Theta(s'))$ $(d\Theta(s')/ds')$ $(\partial f_l/\partial \Theta(s''))$ $(d\Theta(s'')/ds'')d\Theta(s')d\Theta(s'')$

      $= (\partial f_k/\partial \Theta \delta s'))$ $(\partial f_l/\partial \Theta(s'))$ $(\Theta_s{}^2/2) * \delta(s'-s'')ds,$

      f ....track model,

      Θ_1, Θ_2 ....the two orthogonal and uncorrelated scattering angles.

Note that the parameters Θ_1 and Θ_2 have to be taken at s. The function g(s) takes care of a possible change of medium and of energy loss. It is 0 in vacuum, usually 1 when entering the first medium (reference medium) and increases with energy loss according to the equation above. A change in radiation length must also be built into the evaluation of g(s). The particle's path is usually determined from pattern recognition information.

If g(s) differs only significantly from 0 when the particle passes through discrete detector layers, the integral can be approximated by a sum (Fig. 3.7):

$$<\gamma_k * \gamma_l> = \sum_i^{\min(k-1, l-1)} <\Theta_i{}^2/2> \left(\frac{\partial f_k}{\partial \Theta_{1,i}} \frac{\partial f_l}{\partial \Theta_{1,i}} + \frac{\partial f_k}{\partial \Theta_{2,i}} \frac{\partial f_l}{\partial \Theta_{2,i}} \right),$$

where $<\Theta_i{}^2/2> = (\Theta_s{}^2/2) * (\Delta L)_i / \cos \alpha$ is the variance of the projected scattering angle at detector layer i. It is proportional to the length of material along the path and therefore direction dependent. Each layer is counted once, independent of the number of coordinates measured at this layer. α is the angle between the particle direction and the normal to the layer. For an efficient evaluation of the geometry factors using the standard track parameters see (Regler 1977). For the calculation of the above sum the covariances of the track directions are

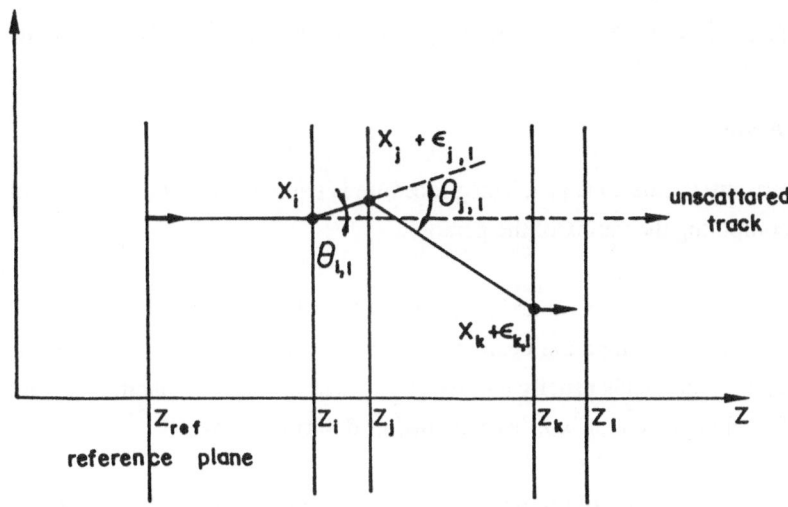

Fig. 3.7. Scattering of a track at discrete detector layers. The track is nearly parallel to the z-axis.

needed. They depend on the choice of the parameters used. For the direction cosines the following relations hold (with $p_3=dx/ds, p_4=dy/ds$):

$$< p_3 * p_3 > = (1-p_3^2) <\Theta_i^2/2 >,$$
$$< p_3 * p_4 > = -p_3*p_4 <\Theta_i^2/2 >,$$
$$< p_4 * p_4 > = (1-p_4^2) <\Theta_i^2/2 >,$$

and for the direction tangents (with $p_3=dx/dz, p_4=dy/dz$) (Eichinger & Regler 1981):

$$< p_3 * p_3 > = (1+p_3^2+p_4^2) (1+p_3^2) <\Theta_i^2/2 >,$$
$$< p_3 * p_4 > = (1+p_3^2+p_4^2) p_3*p_4 <\Theta_i^2/2 >,$$
$$< p_4 * p_4 > = (1+p_3^2+p_4^2) (1+p_4^2) <\Theta_i^2/2 >.$$

Minimization of the objective function

The least-squares estimate of **p** is obtained by minimizing the objective function:

$$Q(p) = \delta^T G \delta = [m - f(p)]^T G [m - f(p)], G = V^{-1}.$$

Differentiation of Q with respect to **p** yields:

$$\partial Q/\partial p = - 2 [m - f(p)]^T G \partial f/\partial p.$$

In order to find a zero of $\partial Q/\partial p$, one applies Newton's method and obtains the following recursion:

$$\tilde{p}_i = \tilde{p}_{i-1} + \{\ A^T\,G\,A - (\partial^2 f/\partial p_{i-1}{}^2)\,G\,[\ m - f\,(\tilde{p}_{i-1})\ ]\ \}^{-1}\,A^T\,G\,[\ m - f\,(\tilde{p}_{i-1})\ ],$$

with:

$$A = \partial f/\partial p_{i-1}.$$

The derivatives are evaluated at $\tilde{p}_{i-1}$. This procedure is equivalent to fitting a parabola to $Q(p)$ at $\tilde{p}_{i-1}$ and computing the vertex of the parabola.

The starting point of the recursion, p_0, must be sufficiently close to the true minimum of Q, in particular closer than the nearest inflection point; otherwise the iteration diverges. Since the track search yields usually a fairly good guess of the track parameters, this is rarely a problem. The derivative matrix A is calculated as described above.

The covariance matrix of multiple scattering is computed in parallel to the track following. It depends normally only weakly on the initial conditions p and thus doesn't need to be recomputed in every iteration.

If the track model is not known explicitly, the calculation of the tensor of second derivatives becomes prohibitively time consuming. In this case the term containing $\partial^2 f/\partial p^2$ is neglected. The matrix in the Newton formula remains positive definite, but the curvature of Q is clearly no longer approximated correctly. The resulting formula is the same as the one which is obtained by approximating f by a linear function:

$$f(p) \approx f(\tilde{p}_{i-1}) + A_{i-1}\,(\,p\, - \tilde{p}_{i-1}\,); \quad A_{i-1} = \partial f/\partial \tilde{p}_{i-1}.$$

Then the standard linear least-squares estimate is given by

$$\tilde{p}_i = \tilde{p}_{i-1} + (\,A^T_{i-1}\,G\,A_{i-1}\,)^{-1}\,A^T_{i-1}\,G\,[\ m - f\,(\tilde{p}_{i-1})\ ].$$

Practice shows that rarely more than 3 iterations are required until the minimum is attained, i.e. until

$$|Q(\tilde{p}_i) - Q(\tilde{p}_{i-1})| < \varepsilon.$$

A choice of $\varepsilon = 0.01$ is adequate.

It should be noted that the convergence may be slowed down by neglecting the second derivatives.Therefore care should be taken that f does not deviate too far from a linear function. This can be facilitated by the proper choice of the coordinate system in which the measurements and the track parameters are expressed. The right choice depends of course on the configuration of the magnetic field and of the tracking detectors. As an example, let us consider the frequently used setup of cylindrical detectors in a homogeneous magnetic field, which is parallel to z (Fig. 3.8). On each cylinder, two coordinates are measured: $(R\Phi)_i$ and z_i. The track parameters are estimated on a cylindrical reference surface. Then the following choice of track parameters will give a track model which is close to linear:

464

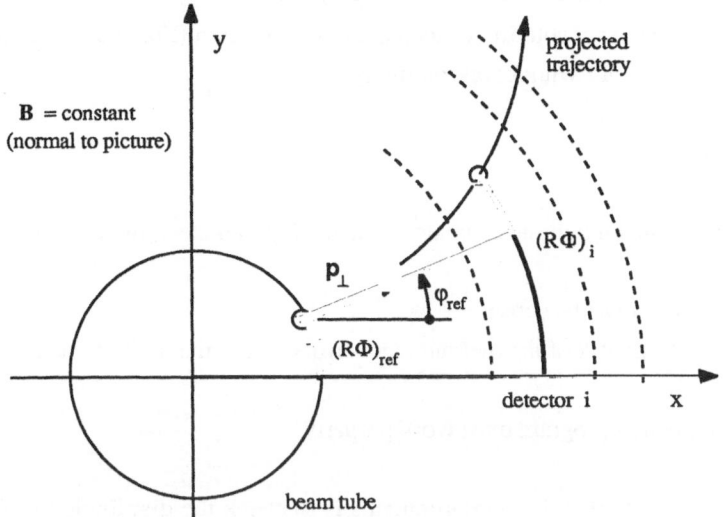

Fig. 3.8.   Cylindrical detectors in a homogeneous magnetic field. The reference surface is a
cylinder (the beam tube).
(From: Data Analysis Techniques in High Energy Physics Experiments,
Cambridge University Press 1989).

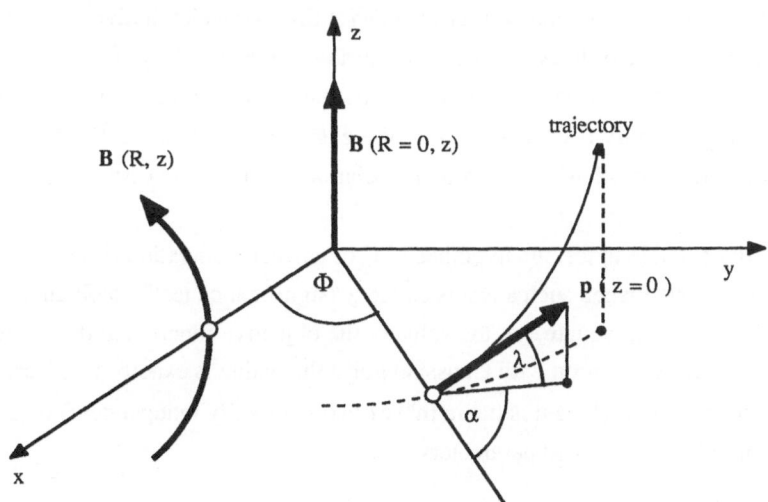

Fig. 3.9.   Choice of track parameters for a nearly linear track model.
(From: Data Analysis Techniques in High Energy Physics Experiments,
Cambridge University Press 1989).

$$p_1 = (R\Phi)_{ref}, \; p_2 = z_{ref}, \; p_3 = \varphi_{ref} - \Phi_{ref}, \; p_4 = (\tan \lambda)_{ref}, \; p_5 = (1/r)_{ref},$$

where λ is the inclination angle of the helix with respect to the x-y-plane and r is the radius of the helix, which is the exact solution of the equation of motion (Fig. 3.9). The choice of $1/r$ allows a change of sign of r without discontinuity.

Test statistics

In order to obtain an optimal estimate of the track parameters, three conditions must be fulfilled:

- The track model must be correct
- The covariance matrix of the measurement errors and of the multiple scattering must be correct
- The reconstruction program must work properly

A common way to test these requirements is to check the distribution of the *pull (or stretch) quantities*, which are the residuals divided by their standard deviation:

$$q_i = \frac{m_i - f_i(\tilde{p})}{\sqrt{var(r_i)}}, \;\; cov\{r\} = G^{-1} - A(A^T G A)^{-1} A^T.$$

All q_i have to have mean 0 and variance 1. A regular check of these quantities allows one to find out whether the detector behaviour is stable with time, or which individual track detector is giving trouble. In case of heavy multiple scattering, however, the pulls are dominated by the latter and carry little information about the behaviour of the detector. In addition, this check works properly only if the association of measured coordinates into tracks is correct. Under the alternative hypothesis of a wrong association, the pulls are distorted.

A quantity suitable to test the hypothesis H_0 of correct association (if no alternative hypothesis has been established such a test is called a "significance test") is obtained automatically by the LSM: the chi-square, i.e. the value of the objective function at the minimum. For a linear least-squares estimation with Gaussian noise this value is exactly χ^2-distributed with n–m degrees of freedom, where n is the number of (functionally independent) measurements and m is the number of estimated parameters.

As almost any kind of background has a tendency to larger chi-squares, a common way to test H_0 is to define a decision criterion making use of the χ^2 value of an individual track, e.g. to give up a certain percentage of good tracks with large χ^2, in order to eliminate the background. Therefore, one preassigns a critical value χ_c^2, dividing the interval $[0, \infty)$ into two regions: an acceptance region $[0, \chi_c^2)$ and a critical (or rejection) region $[\chi_c^2, \infty)$. The loss α caused by this "χ^2-cut" (i.e. the rejection of tracks with $\chi^2 \geq \chi_c^2$) is given by:

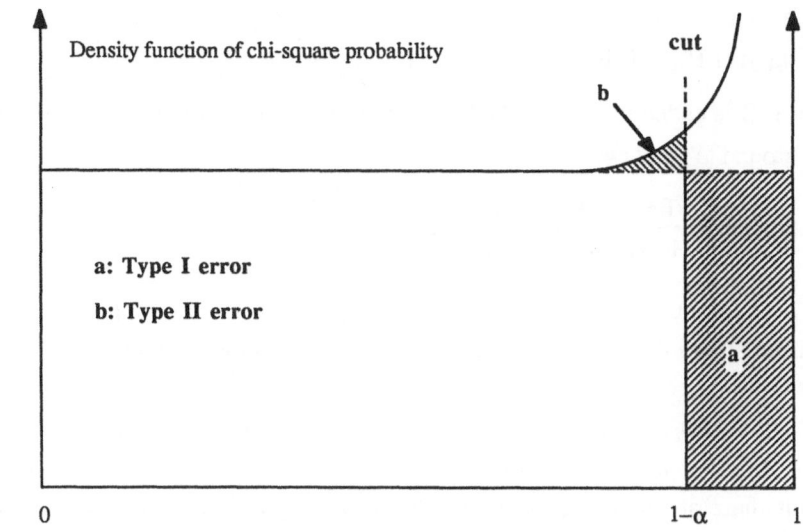

Fig. 3.10. Effects of a chi-square cut of size α.

(From: Data Analysis Techniques in High Energy Physics Experiments, Cambridge University Press 1989).

$$\alpha = \int\limits_{\chi_c^2}^{\infty} g_k(x)dx = 1-G_k(\chi^2),$$

with $G_k(\chi^2)$ being the cumulative distribution function of the χ^2-distribution.

The probability α of rejecting real tracks, for which H_0 is true, is called the "*level of significance*" or the "*size*" of the test. The choice of α is a compromise (Fig. 3.10). On the one hand only an acceptable level of background should remain, giving a lower limit for α. On the other hand, losses of good events should be kept as small as possible, giving an upper limit for α. If in the presence of high background α must be chosen large (e.g. >>1%), a bias can occur due to the nonlinearity of the track model or due to the fact that the errors are not exactly Gaussian. The quantity $1-\alpha$ is called the "*confidence level*". Rejecting the track hypothesis when it is true is called a "type I error" or an error of the first kind.

The experimental χ^2-distribution is a mixture of two probability densities: A real χ^2-distribution and a "χ^2-distribution of background", namely of wrongly associated coordinates (g_{bg}) where the χ^2 was evaluated using the LSM under the false assumption H_0 = true. Accepting the track hypothesis when it is false is called a "type II error" or an error of the second kind, and the corresponding probability is equal to

467

$$\beta = \int\limits_{0}^{\chi_c^2} g_{bg}(x| H_0 = false) \, dx.$$

The quantity $1-\beta$ is called the "*power*" of the test. The total contamination is given by the ratio of background to good events:

$$contamination = \frac{\beta * prob(H_0 = false)}{(1-\alpha) * prob(H_0 = true)} \times 100\%.$$

If the background is such that its suppression would cause too important a loss of tracks, and if no other information is obtainable, the only solution is to improve the overall detector resolution, keeping the χ^2 distribution of good events stable when the correct weights are applied for the LSM, but shifting the pseudo χ^2 obtained from background to larger values. A way to improve the background separation would also be to add additional detectors, but there are fundamental limits due to multiple scattering, energy loss and secondary processes ("*over-instrumentation*"), and also more practical ones like the space and the money available, or the manpower.

3.4 Recursive Least-squares Estimation of Track Parameters: The Kalman Filter

With the advent of more powerful accelerators and more complex detectors some shortcomings of the traditional methods of track fitting have made themselves felt. One might question, whether the traditional implementation described in the previous section is really best suited to the latest requirements. In fact, closer inspection reveals some drawbacks:

- With the global method the track parameters are estimated in a single point on the reference surface. As in the presence of multiple scattering the trajectory is not an exact solution of the equation of motion, the actual physical track may stray quite far from the "ideal" track, i.e. the extrapolation of the estimated parameters. But in a large modular detector extrapolations from the far end of the tracks are needed in order to link together the pieces of information provided from different detectors, e.g. central tracking detectors and muon chambers. Also, the points of intersection of a track with certain detectors need to be computed (e.g. with Ring Imaging Cherenkov counters). Thus it is desirable that estimates of the parameters containing the whole information are available anywhere along the track.

- A track candidate may contain wrong solutions of ambiguous measurements or measurements which are picked up from other tracks or are just noise signals. It is the duty of the track fit not to simply reject such a track candidate, but to find out which measurements might be anomalous. Attempts have been made to handle single outliers by rejecting the measurements with the largest residual whenever the latter exceeds a certain bound, and to recompute the estimate with an updated weight matrix. But in general the global method is not very well adapted to cope with multiple outliers, kinks or long tails in the distribution of the measurement errors.

- For tracks with many measurements and multiple scattering the inversion of the covariance matrix consumes too much time, and a faster track fit algorithm is needed. If there are only few scatterers, the scattering angles can be introduced as additional, directly measured parameters with measurement 0 and a variance as given by the Molière formula ("breakpoint method", e.g. Billoir et al. 1985).

In this section it will be shown how these problems can be solved by the introduction of filter techniques, in particular the Kalman filter, without leaving the realm of least-squares estimation. For additional references, see (Frühwirth 1988).

The Kalman filter is a method of analyzing a *linear discrete dynamic system*. The theory of the Kalman filter is described in many textbooks, and only a short review will be given here. A discrete dynamic system is in each point of (discrete) time, say t_k, characterized by a state vector x_k. The evolution of the state vector in time is described by a time-dependent transformation, the *system equation*:

$$x_k = f_k(x_{k-1}) + w_k,$$

where f_k is a deterministic function and w_k is a random disturbance of the system, the so-called *process noise*.

The state vector doesn't need to be observed directly. Generally a function of the state vector is observed, corrupted by *measurement noise*:

$$m_k = h_k(x_k) + \varepsilon_k,$$

where m_k is the vector of observations at time t_k. It will be assumed in the following that all w_k and ε_k are independent random vectors and have mean 0 and a finite covariance matrix. In the simplest case both f_k and h_k are linear functions for all k:

$$f_k(x_{k-1}) = F_k\, x_{k-1} + a_k,$$
$$h_k(x_k) = H_k\, x_k + b_k.$$

In the following all constants a_k and b_k are assumed to be 0, without loss of generality.

There are three types of operations to be performed in the analysis of a dynamic system (here described in terms of "time"):
- *Filtering* is the estimation of the "present" state vector, based upon all "past" measurements.
- *Prediction* is the estimation of the state vector at a "future" time.
- *Smoothing* is the estimation of the state vector at some time in the "past" based on all measurements taken up to the "present" time.

The Kalman filter is the optimum solution of these three problems in the sense that it minimizes the mean square estimation error. If the system is linear and if w_k and ε_k are gaussian random variables for all k, the Kalman filter is efficient, i.e. it is *the* optimal filter; no non-linear filter can do better. In other cases it is simply the optimal linear filter. Prediction, filtering and smoothing in a linear dynamic system are done with the following formulae:

System equation:

$$\overset{t}{x}_k = F_k \overset{t}{x}_{k-1} + w_k, \; 1 \le k \le n,$$
$$\mathcal{E}\{w_k\} = 0, \; cov\{w_k\} = Q_k, \; 1 \le k \le n.$$

Measurement equation:

$$m_k = H_k \overset{t}{x}_k + \varepsilon_k,$$
$$\mathcal{E}\{\varepsilon_k\} = 0, \; cov\{\varepsilon_k\} = V_k = G_k^{-1}, \; 1 \le k \le n.$$

with:

$\overset{t}{x}_k$ = true value of the state vector at time k,

$\tilde{x}^j_k$ = estimate of $\overset{t}{x}_k$, using measurements up to time j,

(j<k: prediction, j=k: filtered estimate, j>k: smoothed estimate; $\tilde{x}^k_k$ is written as $\tilde{x}_k$),

$C^j_k = cov\{\tilde{x}^j_k - \overset{t}{x}_k\}$,

r^j_k = residual $m_k - H_k \tilde{x}^j_k$,

$R^j_k = cov\{r^j_k\}$.

In the predicted state vector, j is usually equal to k–1; in the smoothed state vector, j is usually equal to n.

Prediction:

State vector extrapolation:
$$\tilde{x}^{k-1}_k = F_k \tilde{x}_{k-1}$$
Cov. matrix extrapolation:
$$C^{k-1}_k = F_k C_{k-1} F_k^T + Q_k$$
Residuals of predictions:
$$r^{k-1}_k = m_k - H_k \tilde{x}^{k-1}_k$$
Covariance matrix of predicted residuals:
$$R^{k-1}_k = V_k + H_k C^{k-1}_k H_k^T$$

Filter (Gain matrix formalism):

State vector update:
$$\tilde{x}_k = \tilde{x}^{k-1}_k + K_k (m_k - H_k \tilde{x}^{k-1}_k)$$
Kalman gain matrix:
$$K_k = C^{k-1}_k H_k^T (V_k + H_k C^{k-1}_k H_k^T)^{-1} = C_k H_k^T G_k$$
Covariance matrix update:
$$C_k = (I - K_k H_k) C^{k-1}_k$$
Filtered residual:

470

$$r_k = m_k - H_k \tilde{x}_k = (I - H_k K_k) r^{k-1}_k$$

Covariance matrix of filtered residuals:

$$R_k = (I - H_k K_k) V_k = V_k - H_k C_k H_k^T$$

χ^2-increment:

$$\chi^2_{k,F} = r_k^T R_k^{-1} r_k$$

χ^2-update:

$$\chi^2_k = \chi^2_{k-1} + \chi^2_{k,F}$$

Filter (Weighted means formalism):

Update of the state vector:

$$\tilde{x}_k = C_k [(C^{k-1}_k)^{-1} \tilde{x}^{k-1}_k + H_k^T G_k m_k]$$

Update of the covariance matrix:

$$C_k = [(C^{k-1}_k)^{-1} + H_k^T G_k H_k]^{-1}$$

χ^2-increment:

$$\chi^2_{k,F} = r_k^T G_k r_k + (\tilde{x}_k - \tilde{x}^{k-1}_k)^T (C^{k-1}_k)^{-1} (\tilde{x}_k - \tilde{x}^{k-1}_k)$$

Smoother:

Smoothed state vector:

$$\tilde{x}^n_k = \tilde{x}_k + A_k (\tilde{x}^n_{k+1} - \tilde{x}^k_{k+1})$$

Smoother gain matrix:

$$A_k = C_k F_{k+1}^T (C^k_{k+1})^{-1}$$

Covariance matrix update:

$$C^n_k = C_k + A_k (C^n_{k+1} - C^k_{k+1}) A_k^T$$

Smoothed residuals:

$$r^n_k = m_k - H_k \tilde{x}^n_k = r_k - H_k (\tilde{x}^n_k - \tilde{x}_k)$$

Covariance matrix of smoothed residuals:

$$R^n_k = R_k - H_k A_k (C^n_{k+1} - C^k_{k+1}) A_k^T H_k^T = V_k - H_k C^n_k H_k^T$$

An inspection of the formulae for prediction, filtering and smoothing reveals the following facts:

- The gain matrix formalism and the weighted means formalism of the filter are equivalent. The choice between the two depends on the dimensions of the state vector and the measurement vector. If the dimension of the state vector is small, the computation by weighted means is faster.
- The filtered estimate of the state vector is unbiased and has minimum variance among all linear estimates using the same set of measurements. For gaussian process noise and measurement errors it is efficient. The same is true for the smoothed estimates. Therefore the Kalman filter with a subsequent smoothing is equivalent to a global linear least-squares fit which takes into account all correlations arising from the process noise.
- The computation time of the filter is basically proportional to the number of detectors

and depends (in the weighted means formalism) very little on the number of measurements per detector.

If the intermediate results of the filter are retained the smoother consists only of a few matrix multiplications and is thus very fast, as long as the the dimension of the state vector is small.

- If there is no process noise, i.e. $Q_k = O$ for all k, smoothing is equivalent to back extrapolation, as can be verified directly from the smoother equations. In this situation a global fit is to be preferred.

- Inspection of the covariance matrix update equations gives the following results, which are intuitively obvious:

  The variance of the filtered state vector is smaller than the variance of the predicted state vector (information from the measurement m_k); the mean-squared filtered residual is smaller than the mean-squared predicted residual (the state vector is pulled towards the measurement); the variance of the smoothed state vector is smaller than the variance of the filtered state vector (information from all measurements); the mean-squared smoothed residual is larger than the mean-squared filtered residual (the state vector is pulled towards the true value).

- The filtered residual vectors (also called innovations) are uncorrelated, in the gaussian case even independent. This is a characteristic property of the Kalman filter. It also proves the χ^2 update formula.

If h_k is a non-linear function, the filter is computed by approximating h_k by a linear function:

$$h_k(x) \approx h_k(\tilde{x}^{k-1}{}_k) + H_k (x - \tilde{x}^{k-1}{}_k), \; H_k = (\partial h_k/\partial x)(\tilde{x}^{k-1}{}_k).$$

The update of the state vector now reads:

$$\tilde{x}_k = \tilde{x}^{k-1}{}_k + K_k [m_k - h_k (\tilde{x}^{k-1}{}_k)].$$

If necessary, the filter is iterated by re-expanding the function h_k in the point $\tilde{x}_k$, which is a better estimate of $\overset{t}{x}_k$ than $\tilde{x}^{k-1}{}_k$. This is called an *iterated Kalman filter*. The decision whether an iteration is required is made on the basis of the second derivative of h_k: No further iteration is required if the influence of the second order term in the expansion of h_k on the estimate is small as compared with the standard deviation of the estimate:

$$\{ K_k [\frac{1}{2}(\tilde{x}_k - \tilde{x}^{k-1}{}_k)^T \frac{\partial^2 h_k}{\partial x_k \partial x_k} (\tilde{x}_k - \tilde{x}^{k-1}{}_k)] \}_i \ll \sqrt{(C_k)_{ii}} .$$

h_k is a vector of functions, and the expression in square brackets is a vector of the dimension of h_k. The decision does not need to be made for each track, but is made once for all on the basis of a sample of typical tracks.

If f_k is non-linear, the predicted covariance matrix $C^{k-1}{}_k$ is computed from the approximation of f_k by a linear function:

472

$$\mathbf{f}_k(\mathbf{x}) \approx \mathbf{f}_k(\widetilde{\mathbf{x}}_{k-1}) + \mathbf{F}_k\ (\ \mathbf{x} - \widetilde{\mathbf{x}}_{k-1}\ ),\ \mathbf{F}_k = (\partial \mathbf{f}_k / \partial \mathbf{x})(\widetilde{\mathbf{x}}_{k-1}),$$

$$\mathbf{C}^{k-1}{}_k = \mathbf{F}_k\ \mathbf{C}_{k-1}\ \mathbf{F}_k{}^T + \mathbf{Q}_k.$$

The natural choice of the expansion point is the filtered estimate $\widetilde{\mathbf{x}}_{k-1}$. Apart from the state vector extrapolation which now reads:

$$\widetilde{\mathbf{x}}^{k-1}{}_k = \mathbf{f}_k(\widetilde{\mathbf{x}}_{k-1}),$$

the prediction, filter and smoother equations remain the same. This is called an *extended Kalman filter*. If both $\mathbf{h}_k$ and $\mathbf{f}_k$ are non-linear functions, the extended and the iterated filter have to be combined. If $\widetilde{\mathbf{x}}_{k-1}$ is far away from $\overset{t}{\mathbf{x}}_{k-1}$, the prediction $\widetilde{\mathbf{x}}^{k-1}{}_k$ may be heavily biased. In this case a special trick has to be applied (see below).

In order to apply the Kalman filter to track fitting the trajectory in space has to be interpreted as a dynamic system. This can be done quite naturally by identifying the *state vector* of the dynamic system with the *vector* $\mathbf{x}$ *of the five parameters* which describe the track uniquely at any given surface. In fact the space points (and possibly directions) measured by a tracking detector can be considered as lying on a fixed surface defined by the shape of the detector, usually either a plane or a cylinder. Consequently only two coordinates (and the directions) are affected by measurement errors, whereas the third one is known precisely. For the purpose of track fitting it is sufficient to consider the state vector in a discrete set of points, namely in the intersection points of the trajectory with those surfaces where measurements are available. The state vector of the track on detector surface k will be written as $\mathbf{x}_k$.

If the state vector, i.e. the vector of track parameters, is known on surface k–1, the trajectory can be extrapolated to surface k by means of the track model discussed in section 3.3. But in the presence of multiple scattering the extrapolated track does not coincide with the actual physical track. If the perturbation of the track by multiple scattering is interpreted as the process noise, the following system equation can be written down:

$$\mathbf{x}_k = \mathbf{f}_k(\mathbf{x}_{k-1}) + \mathbf{w}_k,\ 1 \leq k \leq n.$$

$\mathbf{f}_k$ is the *track propagator* from surface k–1 to surface k; according to the previous section it is computed either analytically or by numerical integration of the equation of motion, or else by an approximating analytical function. The random vector $\mathbf{w}_k$ describes the random deviation of the actual from the extrapolated track due to multiple scattering between surface k–1 and surface k. $\mathbf{w}_k$ is in good approximation normally distributed and satisfies

$$\mathcal{E}\{\mathbf{w}_k\} = \mathbf{0},\ cov\{\mathbf{w}_k\} = \mathbf{Q}_k,\ 1 \leq k \leq n.$$

$\mathbf{Q}_k$ is computed as indicated in section 3.3, by integrating over an infinite number of infinitesimal scattering processes.

473

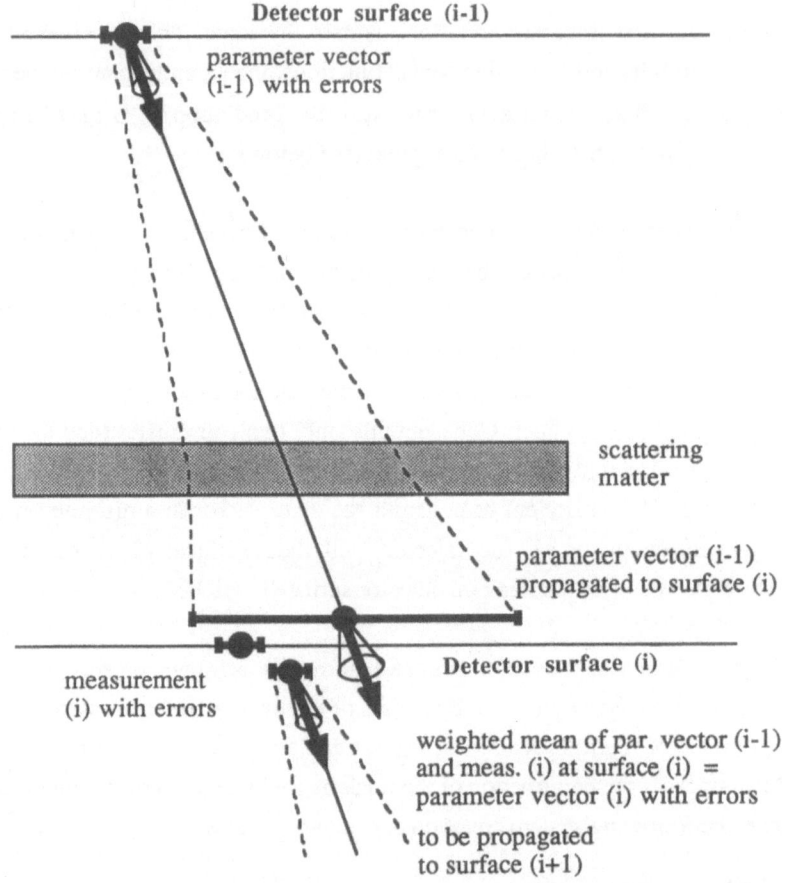

Fig. 3.11. The Kalman Filter: Prediction, error propagation and weighted mean.
(From: Data Analysis Techniques in High Energy Physics Experiments,
Cambridge University Press 1989).

The quantities measured by detector k depend in some way on the state vector of the track in detector k. This dependence yields the measurement equation, which is in general non-linear:

$$\mathbf{m}_k = \mathbf{h}_k(\mathbf{x}_k) + \boldsymbol{\varepsilon}_k, \quad \mathcal{E}\{\boldsymbol{\varepsilon}_k\} = \mathbf{0}, \quad cov\{\boldsymbol{\varepsilon}_k\} = \mathbf{V}_k = \mathbf{G}_k^{-1}, \quad 1 \leq k \leq n.$$

The measurement equation can in principle always be made linear by the appropriate choice of the coordinate system in which the state vector is expressed. It is, however, more important that the track propagator $\mathbf{f}_k$ is linear or nearly linear, so that $\tilde{\mathbf{x}}^{k-1}_k$ is (nearly) unbiased for all values of k.

The covariance matrix of the measurement errors is known from theoretical considerations or from a calibration experiment (see section 3.3). It should be noted that all $\boldsymbol{\varepsilon}_k$ and all $\mathbf{w}_k$ are actually stochastically independent.

Finally the track parameters are updated by the computation of the weighted mean of the predicted state vector and the measurement vector (Fig.3.11). Thus the track parameters are estimated by an extended Kalman filter which proceeds from one end of the track to the other one. At the end of the filter the information from all detectors is included in the estimate, with the exception of measurements which obviously do not belong to the track (see section 3.5). After a subsequent smoothing *estimates of the track parameters using the full information are available on all detector surfaces*. This procedure has the following advantages as compared with the global fit:

- Extrapolations from both ends of the track as well as intersections of the trajectory with other detectors can be computed using the full information.
- The computation time of the filter is always proportional to the number of detectors in the track fit, independent of whether there is multiple scattering or not.
- The filter can be used for track finding and track fitting at the same time, provided that the track density is not too high. If the track density is high, track finding with the filter runs into combinatorial difficulties, since at any point several measurements may be compatible with the predicted state vectors. This is particularly true at the beginning of the track, where the state vector is only poorly defined.
- The linear approximation of the track model does not need to be valid along the entire trajectory, but only from one detector to the next one.

The time spent in the calculation of the track model, of its derivatives and of the covariance matrix of multiple scattering is the same with the Kalman filter as with a global fit; what is saved is the inversion of the global covariance matrix. A comparison of computing times is given in fig. 3.12.

In order to be able to start the filter an initial value $\mathbf{x}_0$ with its covariance matrix $\mathbf{C}_0$ is needed, except in the case that $\mathbf{x}_1$ can be uniquely determined from $\mathbf{m}_1$ alone. In all other cases $\mathbf{x}_0$ is a rough estimate of the track parameters which is usually supplied by the track

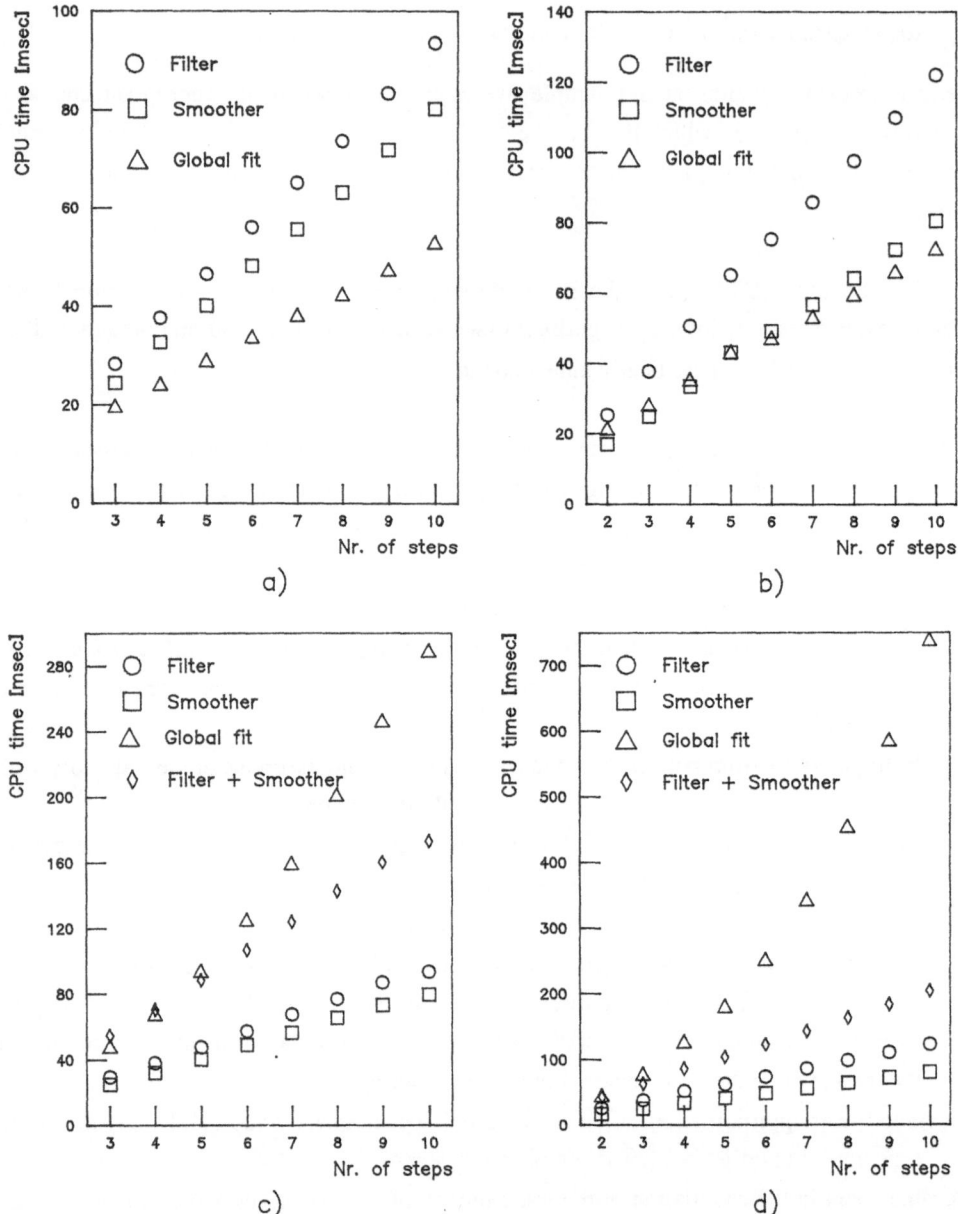

Fig. 3.12. Execution times per track of filter, smoother and global fit (µVAX II). The track model is a helix.

a) 2 measurements per step, no m.s. b) 4 measurements per step, no m.s.

c) 2 measurements per step, with m.s. d) 4 measurements per step, with m.s.

search. In order not to bias the exact filter, x_0 should be given an "infinite" covariance matrix. In practice, C_0 is a matrix with sufficiently large diagonal elements and zeroes otherwise.

In specific cases it may occur that the state vector is only poorly defined by the first measurement(s), and the extrapolation may go astray, although the track is in principle well defined by the subsequent measurements. In extreme cases the extrapolated track may even fail to intersect with the next detector. These problems are due to the fact that for a non-linear track model the prediction is not unbiased. If the estimated state vector is far away from the true value, the bias of the subsequent prediction may become arbitrarily large. In such a case the following trick can be applied: Instead of the state vector itself the *difference* between the state vector and some fixed *reference track* is filtered and smoothed, by a linear expansion of the track model around the reference track. If the reference track is close to the real track, the bias of the prediction is much reduced this way. The reference track is either supplied by the track search or is computed from a measurement which defines the track sufficiently well.

The validity of the reference track must be continuously checked. If the fitted track strays too far from the reference track, it may be necessary to revise certain ingredients, e.g. the measurement errors as computed by an earlier stage of the analysis.

Another problem may arise if the gap between two subsequent measurements is very large relative to the errors on the predicted state vector. In this case the correlation between track position and track direction becomes nearly 1, so that the covariance matrix of the prediction is numerically singular. If this occurs, error propagation and inversion have to be computed in double precision arithmetics.

In a large modular detector it is frequently the case that a track is reconstructed not as a whole, but in several pieces which have then to be put together. This can be done efficiently with the Kalman filter-smoother algorithm. For the sake of simplicity the case of two track segments will be considered; the generalization to several segments is obvious. It is assumed that two track segments have been fitted individually in two separate detector modules (Fig. 3.13) with state vector $\tilde{x}^i_j$ and $\tilde{y}^k_l$, respectively. In order to merge the information from the two track segments one proceeds as follows:

- Take the smoothed estimate $\tilde{y}^m_l$ at the beginning of track segment 2.
- Extrapolate $\tilde{y}^m_l$ to the end of track segment 1, i.e the surface of $\tilde{x}_n$, including error propagation and multiple scattering. The extrapolation is called $\tilde{y}_1{}'$.
- Compute the weighted mean of $\tilde{y}_1{}'$ and $\tilde{x}_n$ and store it in $\tilde{x}_n{}'$.
- (Re)compute the smoother in track segment 1, starting with $\tilde{x}_n{}'$ in place of $\tilde{x}_n$.

Only the smoother has to be (re)computed in all track segments, plus n–1 extrapolation steps between adjacent segments. It is essential, however, that all segments are filtered along the same direction, and that the intermediate results of the filter are retained in all segments. This

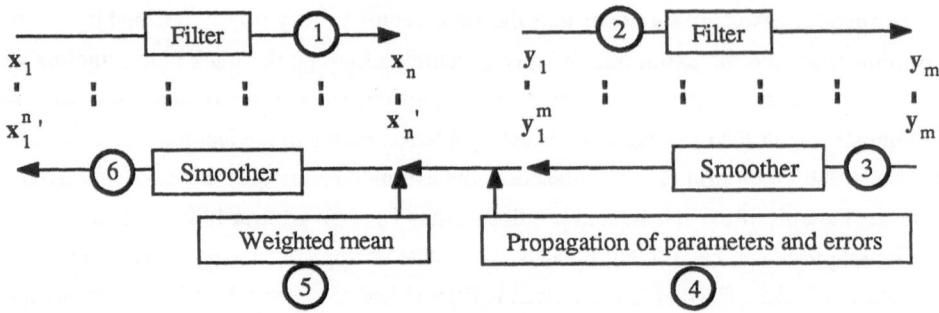

Fig. 3.13. Merging of track segments by successive filtering and smoothing.

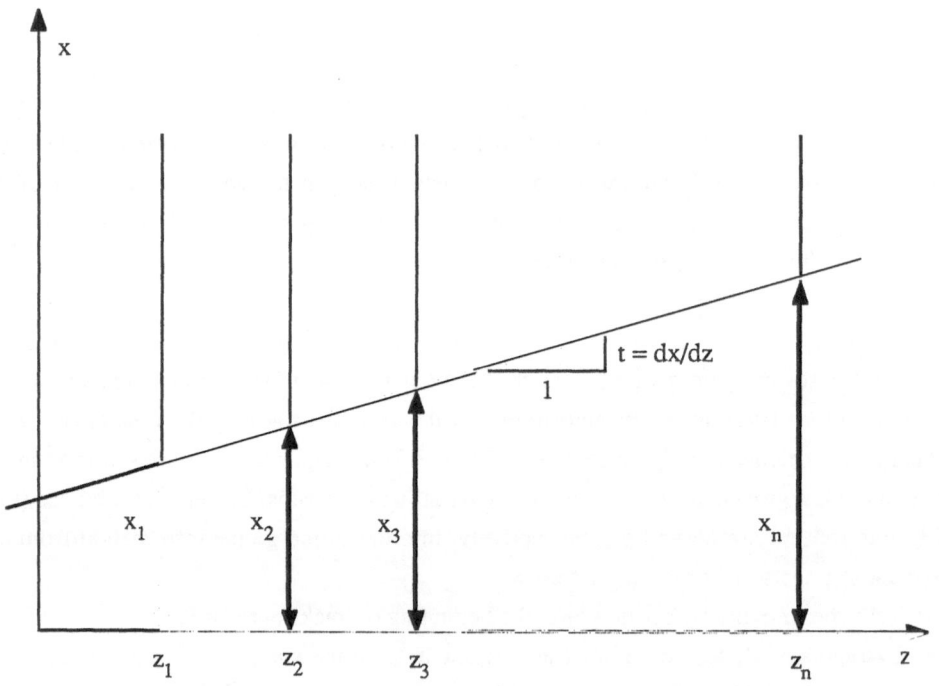

Fig. 3.14. Fit of a straight line in two dimensions. The x-position of the line is measured in n planes $z=z_i$, $1 \le i \le n$. The state vector consists of the position x and the slope t.

has to be taken into consideration during the design of the data structures of the track fit program.

The principal features of track fitting with the Kalman filter shall now be demonstrated on a simple example: the fit of a straight line in two dimensions (Fig. 3.14). The x-position of the line is measured with variance σ^2 at n planes ($z=z_1, \ldots, z=z_n$). The state vector at plane i consists of the position x_i and the slope $t_i = dx/dz$ (z_i):

$$x_i = \begin{bmatrix} x_i \\ t_i \end{bmatrix}.$$

It is assumed that there is already an estimate $\tilde{x}_{k-1}$ at $z=z_{k-1}$ with its covariance matrix C_{k-1}. Then the filter proceeds in the following way. First, $\tilde{x}_{k-1}$ is extrapolated to $z=z_k$:

$$\tilde{x}^{k-1}{}_k = F_k\,\tilde{x}_{k-1}\,, \quad F_k = \begin{bmatrix} 1 & z_k-z_{k-1} \\ 0 & 1 \end{bmatrix}.$$

Also, C_{k-1} is propagated:

$$C^{k-1}{}_k = F_k\, C_{k-1}\, F_k^T.$$

Then a new estimate in plane k is computed from $\tilde{x}^{k-1}{}_k$ and the measurement m_k:

$$\tilde{x}_k = C_k\,[\,(C^{k-1}{}_k)^{-1}\,\tilde{x}^{k-1}{}_k + H_k^T\, m_k/\sigma^2\,],$$
$$C_k = [\,(C^{k-1}{}_k)^{-1} + H_k^T\, H_k/\sigma_k^2\,]^{-1},$$
$$H_k = [\,1 \quad 0\,],$$
$$\sigma^2 = var\,\{m_k\}.$$

The filter starts at $z=z_1$ with the following state vector:

$$\tilde{x}_1 = \begin{bmatrix} m_1 \\ t_1 \end{bmatrix}, \quad C_1 = \begin{bmatrix} \sigma^2 & 0 \\ 0 & M \end{bmatrix}.$$

where t_1 is an approximate track direction and M is a large number. A step of the filter and neglecting terms in 1/M yields the following estimate at z_2 ($\Delta = z_2 - z_1$):

$$\tilde{x}_1 = \begin{bmatrix} m_2 \\ \dfrac{m_2-m_1}{\Delta} \end{bmatrix}, \quad C_1 = \sigma^2 \begin{bmatrix} 1 & 1/\Delta \\ 1/\Delta & 2/\Delta^2 \end{bmatrix}.$$

This result is rather obvious: The estimated position is equal to the measured position, and the slope is determined by the first two measurements. Note that if M is sufficiently large t_1 drops out again of the estimate at $z=z_2$.

3.5 Treatment of Outliers, Robust Track Fitting and Detection of Kinks

In the context of track reconstruction an outlier is defined as a measurement which does not follow the expected behaviour. This may be put into statistical terms by saying that a measurement is considered as an outlier whenever its distance from the true track position is large, the distance being expressed in terms of the covariance matrix attached to the measurement.

There are two classes of outliers. The first one consists of outliers which are *track correlated*, i.e. measurements which belong to the track but are incompatible with the assumed Gaussian distribution of measurement errors. This may for example be due to a delta-ray in a gaseous detector or to tails in the error distribution. Another example is a reconstructed space point in a drift chamber which is distorted by the choice of the wrong sign in one or several of the coordinates used for the space point (Fig. 2.2).

A second class comprises outliers which are *uncorrelated with the track*, i.e. signals wich are not caused by the track, but are nevertheless picked up by the track search. They may be signals from adjacent tracks or genuine noise, either in the detector or in the electronics.

The distribution of track correlated outliers around the true track position depends first of all on the properties of the detector, and also, in the case of delta-rays, on the physics of the underlying secondary process. The distribution of track uncorrelated outliers is in principle uniform in the whole detector volume; in practice their distribution is *imposed by the selection mechanism of the track search*. It is therefore very difficult, if not impossible, to specify the distribution of possible outliers a priori, only from theoretical considerations. The only practicable approach is the determination of the outlier distribution either from a detailed simulation or from the real data, i.e. from tracks which are well reconstructed in the adjacent tracking detectors.

An outlying measurement is characterized by the fact that it is further away from the true track position than expected. Consequently, if a large number of outliers is considered, their probability distribution function has *longer tails* than the one of "good measurements". This is to say that the conditional p.d.f. of outliers has a *larger variance* than usual:

$$g_k \, (\mathbf{m}_k - \mathbf{H}_k \, \overset{t}{\mathbf{x}}_k \mid \mathbf{m}_k \text{ is an outlier}) = g_k \, (\mathbf{0}, \mathbf{V}_k^{(2)}), \, \mathbf{V}_k^{(2)} > \mathbf{V}_k^{(1)},$$

where $\mathbf{V}_k^{(1)}$ is the covariance matrix of "good measurements" in detector k. (It is assumed that the outliers are not biased, i.e. that they do not prefer a particular direction.) This is called a *variance-inflation model* of outliers.

If the frequency of outliers is known, the unconditional p.d.f. of all measurements can be written down:

$$g_k = a_k^{(1)} \cdot \mathcal{N}(\mathbf{0}, \mathbf{V}_k^{(1)}) + a_k^{(2)} \cdot g_k^{(2)} \, (\mathbf{0}, \mathbf{V}_k^{(2)}), \, a_k^{(1)} + a_k^{(2)} = 1, \, \mathbf{V}_k^{(2)} > \mathbf{V}_k^{(1)},$$

where $g_k^{(2)}$ is some p.d.f. with mean $\mathbf{0}$ and covariance matrix $\mathbf{V}_k^{(2)}$, and $a_k^{(2)}$ is the probability of an outlier occuring in detector k. Due to lack of information one is usually forced to assume that $g_k^{(2)}$ is also a normal p.d.f. In this way the global behaviour of the data can be described by a *mixture model*.

It is well known that least-squares estimators are sensitive to outliers. In the global method of track fitting, the total chi-square of the track and the studentized residuals of the measurements (pull quantities) can be used to check whether there are outliers present. Although the *total chi-square is a powerful test against ghost tracks* (see section 1.7), it loses its power against single outliers with increasing number of measurements. The *maximum studentized residual* is a valid criterion for single outliers, but only if there is no or only little multiple scattering. If there is strong multiple scattering, the pulls lose their power as test criteria, since the outliers are masked by the multiple scattering. Furthermore removal of an outlier requires the weight matrix of all remaining measurements to be recomputed. One may conclude that it seems difficult to handle outliers with the global method.

With the Kalman filter the situation is much better insofar as ways can be found to treat outliers efficiently, with little additional computational effort. One solution is to try to *identify* outliers and to *remove* them from the track. An algorithm which accomplishes this for single or multiple outliers by means of chi-square tests is presented below. Another solution consists in *accommodating* the outlier(s) by *giving them a smaller weight*. This may be called a "robustification" of the Kalman filter and is discussed below, for the case of a mixture model. Obviously outliers which are uncorrelated with the track should always be removed since they bear no information pertaining to the track. Of course, one can never be sure to which type an outlier belongs, so that downweighting of outliers is a somewhat dangerous procedure – except in the case of truly non-Gaussian measurement errors, where also outlying observations carry useful information.

First it shall be demonstrated how outliers can be found and removed with the Kalman filter and smoother. Since an outlier in track fitting is defined as a measurement which is in some sense "too far away" from the true track position (and direction), it is natural to take as a decision criterion the *distance of the measurement from the estimated track position*. By using the predicted track position the following test statistic is obtained:

$$\chi^2_{k,P} = (r^{k-1}_k)^T (R^{k-1}_k)^{-1} r^{k-1}_k.$$

It can easily be shown that the "predicted chi-square" $\chi^2_{k,P}$ is equal to the "filtered chi-square" $\chi^2_{k,F}$:

$$\chi^2_{k,P} = \chi^2_{k,F} = r_k^T R_k^{-1} r_k.$$

If there is no outlier at $j<k$, the filtered state vector is unbiased and normally distributed. If m_k is not an outlier then $\chi^2_{k,F}$ is χ^2-distributed with $m_k = \dim(m_k)$ degrees of freedom. If m_k is an outlier, $\chi^2_{k,F}$ tends to larger values. If $\chi^2_{k,F}$ exceeds a given bound c, m_k is therefore rejected as an outlier. If c is chosen as the $(1-\alpha)$-quantile of the χ^2-distribution with m_k de-

grees of freedom, the size of the test is equal to α. If there is an (unrecognized) outlier at $j<k$, then $\tilde{x}_k$ is biased and $\chi^2_{k,F}$ is no longer exactly χ^2-distributed. However, a bias of the prediction $\tilde{x}^{k-1}_k$ is damped by the filter:

$$\text{bias}(\tilde{x}_k) = (I - K_k H_k) \text{bias}(\tilde{x}^{k-1}_k).$$

A second test can be constructed by using the distance of the measurement from the smoothed track position as a test statistic:

$$\chi^2_{k,S} = r^n_k{}^T (R^n_k)^{-1} r^n_k.$$

Since the smoothed track position is better defined than the filtered one, this test is more powerful than the previous one. If there is an outlier at $j \neq k$, $\tilde{x}^n_k$ is biased and $\chi^2_{k,S}$ is not exactly χ^2-distributed. This implies that the size of the test cannot be controlled precisely and has to be tuned with simulated data. The problem is aggravated if there are several outliers.

The conclusion is that a rough selection of measurements should take place during the filter, whereas the final search for possible outliers should be carried out during smoothing. If an outlier is found it can be removed from the track by means of the following algorithm:

$$\tilde{x}^n_k{}^* = C^n_k{}^* \cdot [(C^n_k)^{-1} \tilde{x}^n_k - H_k{}^T G_k m_k],$$
$$C^n_k{}^* = [(C^n_k)^{-1} - H_k{}^T G_k H_k]^{-1},$$

or, in gain matrix notation:

$$\tilde{x}^n_k{}^* = \tilde{x}^n_k + K^n_k{}^* (m_k - H_k \tilde{x}^n_k),$$
$$K^n_k{}^* = C^n_k H_k{}^T \cdot (- V_k + H_k C^n_k H_k{}^T)^{-1},$$
$$C^n_k{}^* = (I - K^n_k{}^* H_k) C^n_k.$$

$\tilde{x}^n_k{}^*$ is the estimate of x_k which contains the *full information with the exception of* m_k. If m_k is recognized as an outlier, the smoother has to be continued with $\tilde{x}^n_k{}^*$ and $C^n_k{}^*$ in place of $\tilde{x}^n_k$ and C^n_k. However, this doesn't update the smoothed estimates $\tilde{x}^n_j$ with $j>k$. If these are important the filter has to be recomputed, starting from $\tilde{x}^{k-1}_k$ and without using m_k, followed by a pass of the smoother over the entire track.

Finally a few words about the choice of α, the size of the chi-square test, should be said. If α is chosen too large, the variance of the final estimated track parameters will increase, because many "good" measurements are rejected as outliers. If α is too small, outliers will remain unidentified, and the estimated track parameters will be biased. It is impossible to say in general which of the two effects is more harmful, because that depends on the number of measurements in the track, on the probability of an outlier occuring, and on the distribution of the outliers. Therefore the optimum value of α can only be found by simulation studies, which try to mimic all processes leading to outliers as closely as possible. This requires feedback from detector experts and from real data.

If the distribution of outliers is known, at least approximately, they can be accommodated by a suitably modified filter. If the behaviour of outliers can be described by a mixture model, a robustified Kalman filter can be constructed via a Bayesian approach. It is almost as fast as the standard Kalman filter and is very easy to implement.

In the mixture model the distribution of measurements and outliers is approximated by a mixture of two normals:

$$\mathbf{m}_k = \mathbf{H}_k \, \mathbf{x}_k + \boldsymbol{\varepsilon}_k,$$
$$\boldsymbol{\varepsilon}_k \sim a_k^{(1)} \, \mathcal{N}(0, \mathbf{V}_k^{(1)}) + a_k^{(2)} \, \mathcal{N}(0, \mathbf{V}_k^{(2)}),$$
$$a_k^{(1)} + a_k^{(2)} = 1, \; \mathbf{V}_k^{(2)} > \mathbf{V}_k^{(1)}, \; a_k^{(1)} \gg a_k^{(2)}.$$

where $a_k^{(2)}$ is the probability of an outlier occuring in detector k. This model is a reasonable assumption for measurements of positions and derived quantities, like directions or curvature. It is inadequate for measurements of energy loss which involve long-tailed distributions of a quite different type. The distribution of $\mathbf{x}^{k-1}{}_k$ is assumed to be normal:

$$\mathbf{x}^{k-1}{}_k = \tilde{\mathbf{x}}^{k-1}{}_k + \boldsymbol{\delta}_k,$$
$$\boldsymbol{\delta}_k \sim \mathcal{N}(0, \mathbf{C}^{k-1}{}_k),$$

where $\tilde{\mathbf{x}}^{k-1}{}_k$ is the predicted estimate obtained with the robust filter. By means of Bayes's theorem one obtains the following posterior distribution of $\mathbf{x}_k$:

$$f(\mathbf{x}_k \mid \tilde{\mathbf{x}}^{k-1}{}_k, \mathbf{m}_k) = \sum_{i=1}^{2} b_k^{(i)} \, \varphi(\mathbf{x}_k; \tilde{\mathbf{x}}_k^{(i)}, \mathbf{C}_k^{(i)}),$$

with:

$$\tilde{\mathbf{x}}_k^{(i)} = \tilde{\mathbf{x}}^{k-1}{}_k + \mathbf{C}^{k-1}{}_k \, \mathbf{H}_k^T \, \mathbf{W}_k^{(i)} \, \mathbf{r}^{k-1}{}_k,$$
$$\mathbf{W}_k^{(i)} = (\mathbf{V}_k^{(i)} + \mathbf{H}_k \, \mathbf{C}^{k-1}{}_k \, \mathbf{H}_k^T)^{-1},$$
$$\mathbf{C}_k^{(i)} = [(\mathbf{C}^{k-1}{}_k)^{-1} + \mathbf{H}_k^T \, \mathbf{G}_k^{(i)} \, \mathbf{H}_k]^{-1},$$
$$\mathbf{r}^{k-1}{}_k = \mathbf{m}_k - \mathbf{H}_k \, \tilde{\mathbf{x}}^{k-1}{}_k,$$
$$\varphi(x; \mu, \mathbf{C}) = \text{normal p.d.f. with mean } \mu \text{ and covariance matrix } \mathbf{C}.$$

The coefficients $b_k^{(2)}$ and $b_k^{(1)}$ can be interpreted as the *posterior probability* of $\mathbf{m}_k$ being an outlier or not:

$$b_k^{(1)} = \left[1 + \frac{a_k^{(2)}}{a_k^{(1)}} \frac{|\mathbf{W}_k^{(2)}|}{|\mathbf{W}_k^{(1)}|} \exp\left(\tfrac{1}{2} \mathbf{r}^{k-1}{}_k{}^T \mathbf{D}_k \, \mathbf{r}^{k-1}{}_k\right) \right]^{-1},$$
$$b_k^{(2)} = 1 - b_k^{(1)},$$

with:

$$\mathbf{D}_k = \mathbf{W}_k^{(1)} - \mathbf{W}_k^{(2)}.$$

The final estimate $\tilde{\mathbf{x}}_k$ and its covariance matrix $\mathbf{C}_k$ are obtained as the mean and the covariance matrix of the posterior distribution of $\mathbf{x}_k$. The update of the state vector turns out to be a weighted sum of two Kalman filters, the weights being $b_k^{(1)}$ and $b_k^{(2)}$:

$$\tilde{\mathbf{x}}_k = \tilde{\mathbf{x}}^{k-1}_k + \mathbf{C}^{k-1}_k \, \mathbf{H}_k^T \, (\, b_k^{(1)} \, \mathbf{W}_k^{(1)} + b_k^{(2)} \, \mathbf{W}_k^{(2)} \,) \, \mathbf{r}^{k-1}_k,$$
$$\mathbf{C}_k = \mathbf{C}^{k-1}_k - \mathbf{C}^{k-1}_k \, \mathbf{H}_k^T \, (\, b_k^{(1)} \, \mathbf{W}_k^{(1)} + b_k^{(2)} \, \mathbf{W}_k^{(2)} - \mathbf{S}_k \,) \, \mathbf{H}_k \, \mathbf{C}^{k-1}_k,$$
$$\mathbf{S}_k = b_k^{(1)} \, b_k^{(2)} \, \mathbf{D}_k \, \mathbf{r}^{k-1}_k \, \mathbf{r}^{k-1}_k{}^T \, \mathbf{D}_k.$$

The posterior distribution of $\mathbf{x}_k$, being a mixture of two normals with *different means*, is asymmetric; therefore $\tilde{\mathbf{x}}_k$ is *not* the maximum-likelihood estimate! For $b_k^{(2)} = 0$ or, a fortiori, $a_k^{(2)} = 0$, the robust filter reduces to the standard Kalman filter. The smoother is not affected by the robustification of the filter and remains the same.

The robust filter accommodates outliers by giving them a smaller weight than to "good" measurements; this property should render a chi-square cut on the measurements unnecessary. If the selection criteria of the track search, however, are too weak, a generalized chi-square must be used to reject measurements which are incompatible with the state vector even under the assumptions of a long-tailed error distribution. This is a nuisance, since its exact distribution is not known. The robust filter has also troubles in resolving ambiguous measurements: Since the wrong solution gets a smaller weight, it does not necessarily yield a larger generalized chi-square than the right one. Therefore the robust filter is best used for track reconstruction in a homogeneous detector with non-ambiguous measurements, e.g. in a TPC.

Another situation in which the robust filter may prove useful, is an eventual non-normal behaviour of the process noise. It is true that multiple Coulomb scattering leads to a nearly normal process noise. But there are also other processes which contribute to the interaction of a particle with matter, like nuclear scattering and hard electromagnetic scattering. These occur rarely, but with large scattering angles. These effects are particularly serious for electrons, because of their low mass. For details of a robust filter handling this case, see (Frühwirth 1988).

The performance of the methods proposed above for the treatment of outliers shall now be demonstrated on a few examples. To this end 5000 tracks in the momentum range between 0.5 and 5 GeV/c were simulated in a simplified TPC. Each track consists of 16 space points $((R\Phi)_i, z_i)$ at constant R_i. The standard deviations of the measurement error are 200 μm in $R\Phi$ and 600μm in z. First all tracks were contaminated with a single variance-inflation outlier, with:

$$\mathbf{V}_k^{(2)} = 9 * \mathbf{V}_k^{(1)}.$$

Fig. 3.15 shows the fraction of the found outliers as a function of the distance d of the outlier from the true track position:

$$d = \sqrt{(\mathbf{m}_k - \mathbf{H}_k \tilde{\mathbf{x}}_k{}^t)^T \, \mathbf{G}_k^{(1)} \, (\mathbf{m}_k - \mathbf{H}_k \tilde{\mathbf{x}}_k{}^t)} \, ,$$

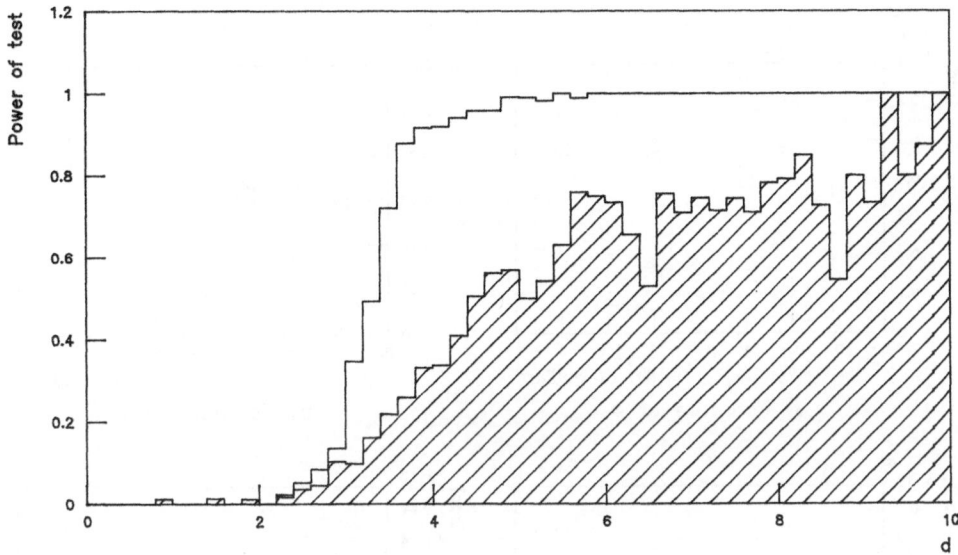

Fig. 3.15. Single variance-inflation outlier: Fraction of outliers found by the $\chi^2_{k,F}$-test (shaded) and by the $\chi^2_{k,S}$-test (blank) as a function of the distance d between the measurement and the true track position.

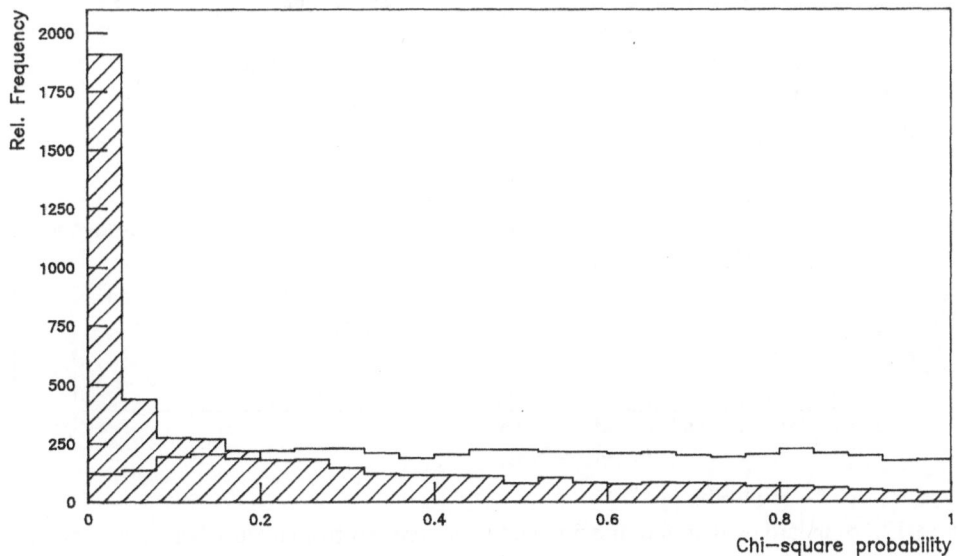

Fig. 3.16. Single-variance inflation outlier: Probability of total chi-square; without removal of outliers (shaded), and with removal of outliers (blank) found by the $\chi^2_{k,S}$-test.

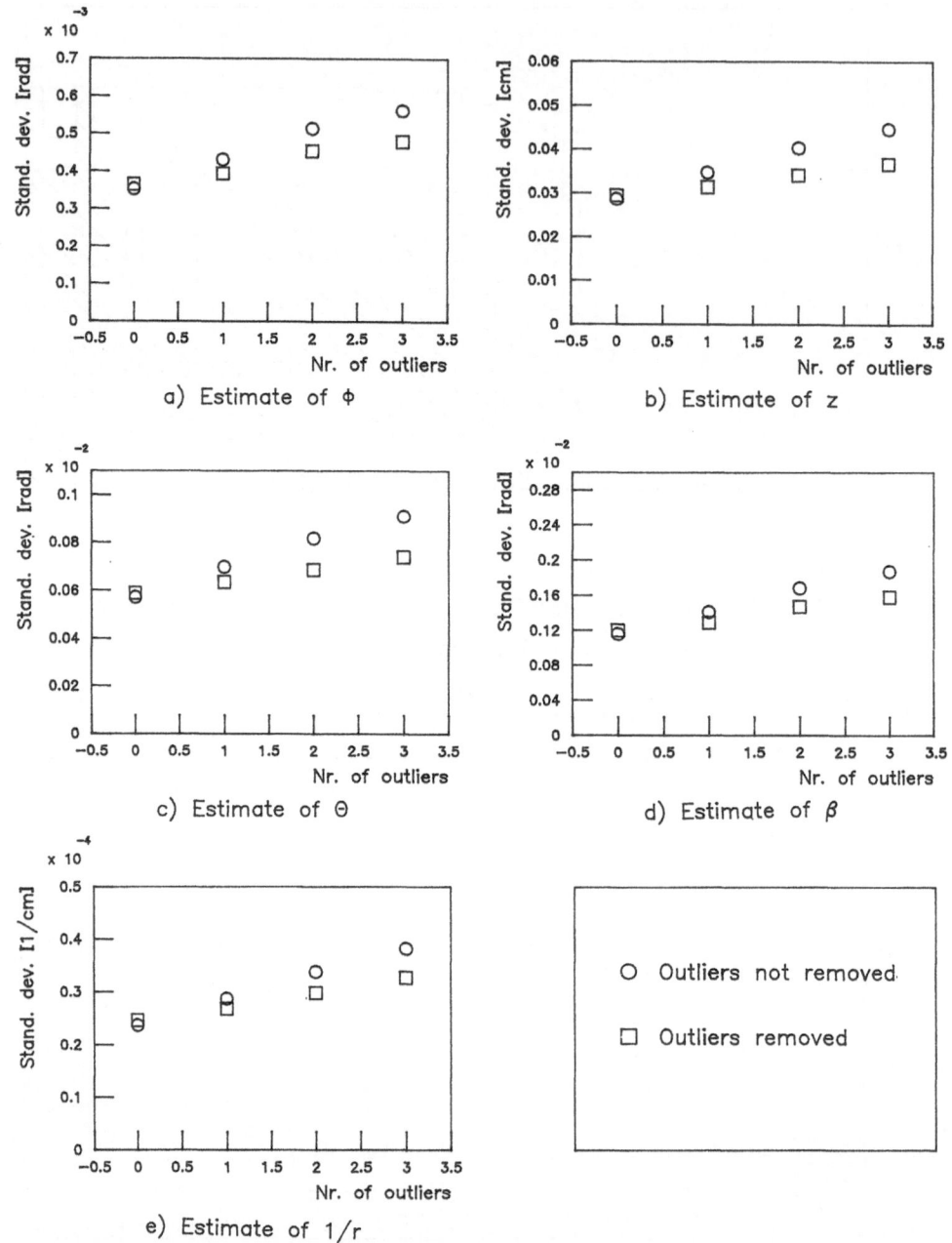

Fig. 3.17. Standard deviation of the fitted track parameters from the true value, as a function
of the number of outliers, with and without removal of outliers.
a) Φ, b) z, c) θ, d) β=φ−Φ, e) 1/r.

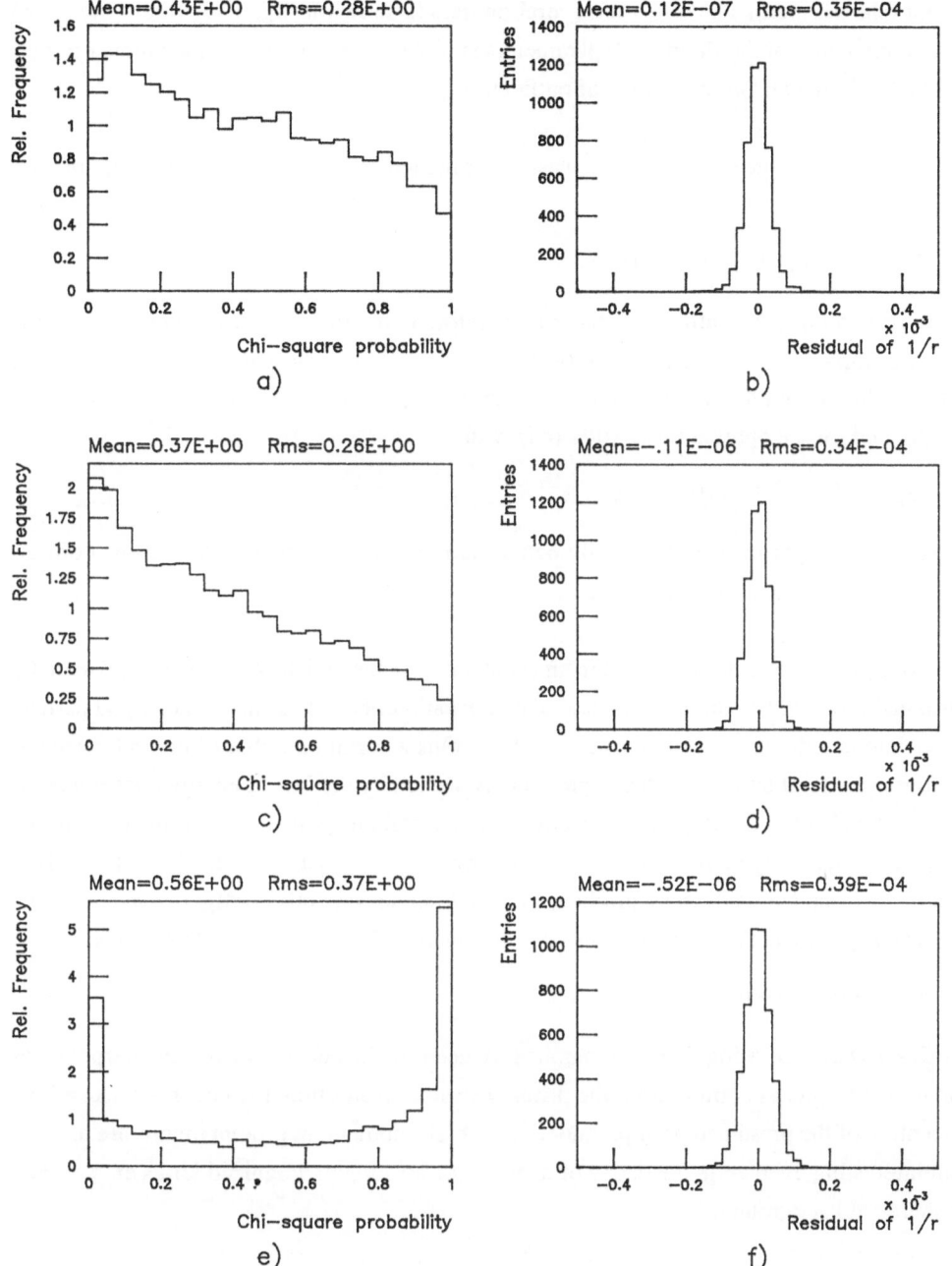

Fig. 3.18. Mixture model, average number of outliers = 3.

Probability of total chi-square and deviation of the estimated curvature from the true value.

a,b) Outliers removed by $\chi^2_{k,s}$-test; c,d) Robust filter; e,f) Optimal linear filter.

for both the $\chi^2_{k,S}$-test and the $\chi^2_{k,F}$-test (shaded area), at a confidence level of 99%. Fig. 3.16 shows the probability transform of the total chi-square, with removal of outliers and without (shaded area). The effect of several outliers is shown in fig. 3.17, where the actual standard deviation of the fitted track parameters is plotted as a function of the number of outliers, both without and with removal of outliers.

A second sample of 5000 tracks was contaminated according to the following mixture model:

$$V_k^{(2)} = 9 * V_k^{(1)}, \quad a_k^{(2)} = 3/16.$$

The average number of outliers per track is therefore equal to 3. Fig. 3.18 shows the probability transform of the total chi-square (a) and the deviation of the estimated curvature from the true value (b), with removal of outliers. The same quantities are shown for the robust filter (c,d), and for the optimal linear filter (e,f), with:

$$V_k = a_k^{(1)} V_k^{(1)} + a_k^{(2)} V_k^{(2)}.$$

The robust filter performs only slightly better than the filter/smoother with removal of outliers. The linear filter is clearly worse.

Many of the particles produced in an event are unstable and decay with an exponentially distributed lifetime. Due to the basic law of conservation of electric charge a charged particle decays into an odd number of charged particles, plus a certain number of neutral particles. (Of course the total number of decay products is limited by the law of energy conservation.) If one of the charged decay products is emitted at a small angle, its track and the track of the decaying particle may be misinterpreted by the track search as a single track, particularly if there is only one charged decay product. For obvious reasons this configuration is called a *kink*. The most frequent sources of kinks are the decays of charged π- and K-mesons:

$$\pi \to \mu\, \nu, \quad K \to \mu\, \nu.$$

If such kinks are not recognized, the muons produced by the decay may be misinterpreted as prompt muons, i.e. as coming from the primary vertex. In addition, the kink may fake a high momentum of the presumed prompt muon. Since high-momentum prompt muons are the sign of an interesting event, e.g. the decay of a W vector boson, unrecognized kinks are a potential source of background.

Actually the information provided by the filter and smoother is sufficient to allow the construction of a fast and relatively efficient kink finding algorithm. If the track is regarded as a dynamic system a kink is nothing else but a *sudden change of the state vector*. If the kink occurs somewhere between detectors k and k+1, one may expect that the state vector $\tilde{x}_k$ obtained from the first track segment $\{m_1, \ldots, m_k\}$ is significantly different from the back extrapolation $\tilde{x}^{k+1}_k{}^{(b)}$ from the second track segment $\{m_{k+1}, \ldots, m_n\}$. The back extrapolation is either computed from an actual *backward filter*, i.e. a filter starting at $\mathbf{m}_n$ and proceeding towards $\mathbf{m}_1$, or from the smoother by "removing" $\tilde{\mathbf{x}}_k$ from $\tilde{\mathbf{x}}^n_k$ (the superscript $^{(b)}$ denotes the backward filter):

$$\tilde{\mathbf{x}}^{k+1}_k{}^{(b)} = \mathbf{C}^{k+1}_k{}^{(b)} \, (\, \mathbf{C}^n_k{}^{-1} \, \tilde{\mathbf{x}}^n_k - \mathbf{C}_k{}^{-1} \, \tilde{\mathbf{x}}_k \,),$$
$$\mathbf{C}^{k+1}_k{}^{(b)} = (\, \mathbf{C}^n_k{}^{-1} - \mathbf{C}_k{}^{-1} \,)^{-1}.$$

In order to decide whether a kink has occured between detectors k and k+1, one tests whether the difference

$$\Delta_k = \tilde{\mathbf{x}}^{k+1}_k{}^{(b)} - \tilde{\mathbf{x}}_k$$

is significantly different from $\mathbf{0}$, by means of the χ^2-statistic $\chi^2_{k,\Delta}$:

$$\chi^2_{k,\Delta} = \Delta_k{}^T \cdot cov \, \{ \, \Delta_k \, \}^{-1} \, \Delta_k =$$
$$= \Delta_k{}^T \, (\, \mathbf{C}_k + \mathbf{C}^{k+1}_k{}^{(b)} \,)^{-1} \, \Delta_k =$$
$$= \Delta_k{}^T \, \mathbf{C}_k{}^{-1} \, (\, \mathbf{C}_k - \mathbf{C}^n_k \,) \, \mathbf{C}_k{}^{-1} \, \Delta_k.$$

If there is no kink, $\chi^2_{k,\Delta}$ is χ^2-distributed with 5 degrees of freedom for all k. Therefore a cut at the $(1-\alpha)$-quantile will yield a test of size α.

This test is, however, not robust against an outlier somewhere along the track: In fact, an outlier leads to a distortion of $\chi^2_{k,\Delta}$ which as a consequence may exceed the cut more often than expected, thus faking kinks where there are actually none. In addition, the test on $\chi^2_{k,\Delta}$ does not tell us where the kink is situated, as it is by no means obvious that the largest $\chi^2_{k,\Delta}$ corresponds to the position of the kink. These considerations lead to the following modified test: It is now required that

- both track segments $\{\mathbf{m}_1, \dots, \mathbf{m}_k\}$ and $\{\mathbf{m}_{k+1}, \dots, \mathbf{m}_n\}$ have small total chi-squares
- and that $\chi^2_{k,\Delta}$ is large.

These two conditions can be combined into a single test-statistic:

$$F_k = \frac{(\, \chi^2_{k,\Delta} / 5 \,)}{[\, (\, \chi^2_k + \chi^2_{k+1}{}^{(b)} \,) / n_k \,]},$$

where χ^2_k is the total χ^2 of the first track segment, $\chi^2_{k+1}{}^{(b)}$ is the total χ^2 of the second track segment, and n_k is the sum of the respective numbers of degrees of freedom. $\chi^2_{k+1}{}^{(b)}$ is either computed during the backward filter or from the following relation:

$$\chi^2_n = \chi^2_k + \chi^2_{k+1}{}^{(b)} + \chi^2_{k,\Delta}$$

which is exact in a linear model.

The case that the track contains no outliers shall be considered first. If there is no kink, F_k is F-distributed, the two degrees of freedom of the F-distribution being 5 and n_k, for all k. A test on F_k is also robust against an overall scale error of the measurement variances. If there is a kink between k and k+1, the numerator of F_k will be large and the denominator will be small, resulting in a large value of F_k. It is a slightly less powerful test than $\chi^2_{k,\Delta}$ though. If there is a kink at $j \neq k$, one of the chi-squares in the denominator will be large, thereby preventing F_k from getting too large. Therefore it is proposed to locate the kink by looking for the maximum F_k.

If there is an outlier somewhere along the track, the situation is more complicated. If there is no kink, but an outlier at j, $\chi^2_{k,\Delta}$ will be distorted, particularly if k is close to j. But also one of the chi-squares in the denominator will be large, so that a kink will be faked with much less probability than is the case with the test on $\chi^2_{k,\Delta}$. If there is a kink between k and k+1 and an outlier somewhere, both the numerator and one of the chi-squares in the denominator of F_k will be large, so that the kink may be partially masked by the outlier. Therefore the detection of kinks has to be preceded by or combined with the removal or accommodation of outliers. This problem is still under investigation.

The performance of the kink finding algorithm is demonstrated on simulated K- and π-decays in the simplified TPC (see above). Fig. 3.19 (3.20) shows the fraction of found K- (π-) decays as a function of the momentum P, for the χ^2-test (blank area) and for the F-test (shaded area), at a confidence level of 99%. In the π-sample, about 25% of the found kinks would pass a 1%-cut on the total chi-square. Therefore the kink finder proposed here is about 30% more powerful than a simple χ^2-cut for π-decays. The difference is much smaller for K-decays, where only about 8% of the detected kinks would pass a 1%-cut on the total chi-square.

3.6 Estimation of Vertex Parameters with the Kalman Filter

Throughout this section it will be assumed that n tracks are to be fitted to a common vertex. For each track participating in the vertex fit the estimated 5-vector of track parameters p_k is given on a reference surface, together with its covariance matrix V_k. If required, contributions of multiple scattering between the vertex and the reference surface are added to the matrix V_k. The vertex parameters to be estimated are (Fig. 3.21):

- the vertex position x,
- and the 3-momentum vectors q_i of the n tracks belonging to the vertex.

If the vertex is regarded as a dynamic system, the Kalman filter and smoother can be applied to the vertex fit. This results in an algorithm which allows easy addition and removal of tracks to and from the vertex. For a straightforward vertex fit its computation time is proportional to n.

Initially, the vertex is described by the prior information about the vertex position, x_0 and $C_0 = cov \{x_0\}$. x_0 may be a very crude estimate, and C_0 accordingly very large. Now one track after the other is added to the vertex. In each step of the filter the vertex position is

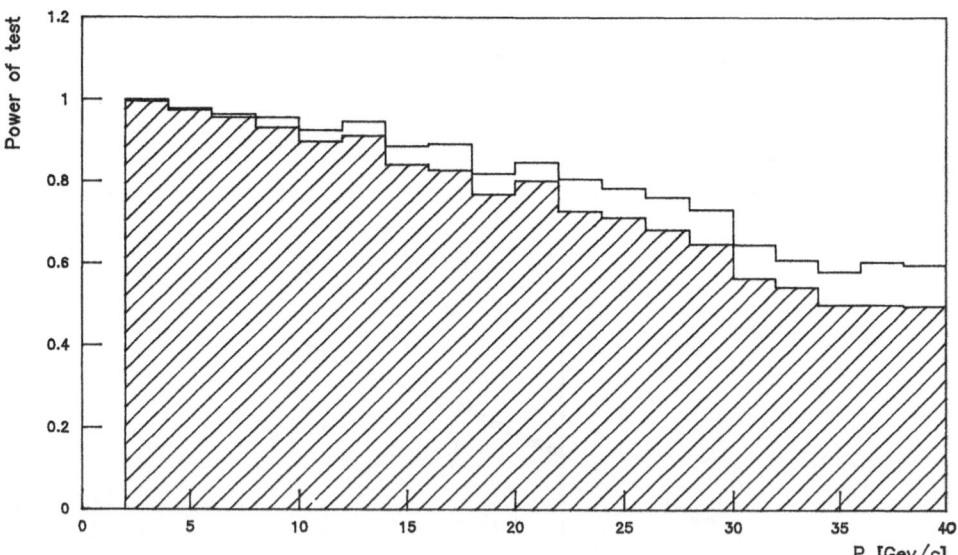

Fig. 3.19. Decay of K into μ and ν:
Fraction of kinks found by the $\chi^2_{k,\Delta}$-test (blank) and by the F_k-test (shaded) as a function of the momentum P of the K.

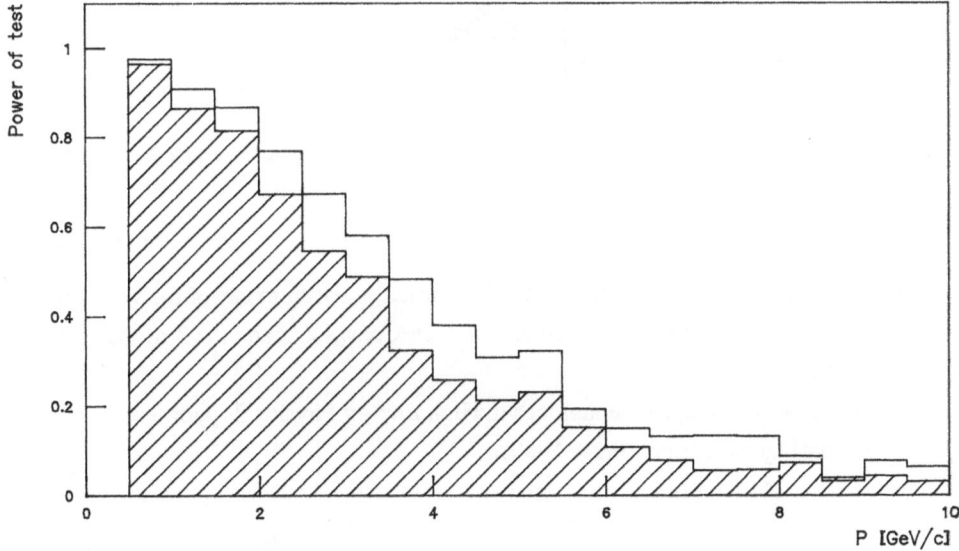

Fig. 3.20. Decay of π into μ and ν:
Fraction of kinks found by the $\chi^2_{k,\Delta}$-test (blank) and by the F_k-test (shaded) as a function of the momentum P of the π.

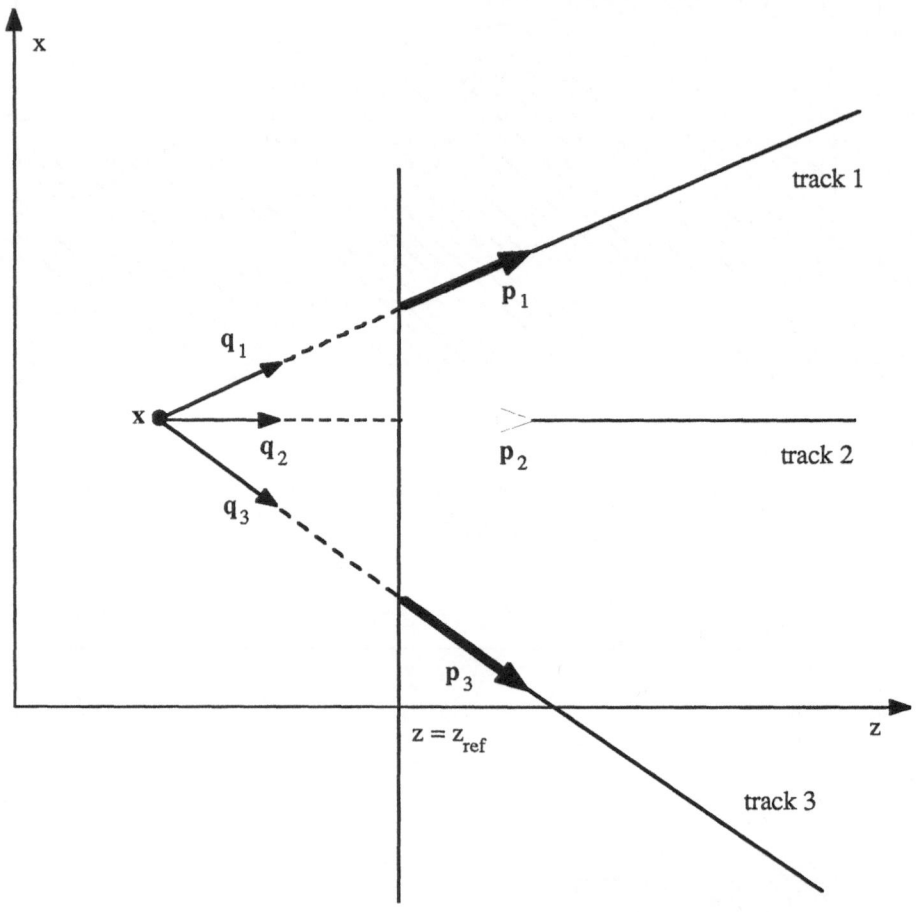

Fig. 3.21. The estimated track parameters are the (virtual) measurements of the vertex fit. The parameters of the vertex fit are the vertex position x and the 3-momentum vectors q_i.
(From: Data Analysis Techniques in High Energy Physics Experiments, Cambridge University Press 1989).

updated, and the state vector is augmented by the 3-momentum vector q_k of the track at the vertex. Thus after k steps the state vector consists of $x_k, q_1, \ldots, q_k$. The corresponding covariance matrices are:

$$C_k = cov\ \{\tilde{x}_k\},$$
$$D_i = cov\ \{\tilde{q}_i\},\ 1 \leq i \leq k,$$
$$E_i = cov\ \{\tilde{q}_i, \tilde{x}_i\},\ 1 \leq i \leq k.$$

The system equation is particularly simple, since there is no process noise:

$$x_k = x_{k-1}.$$

The measurement equation is non-linear:

$$p_k = h_k(\overset{t}{x}, \overset{t}{q_k}) + \varepsilon_k,\ cov\ \{\varepsilon_k\} = cov\ \{p_k\} = V_k = G_k^{-1}.$$

The estimated track parameters are stochastically independent.

Since the measurement equation is non-linear, the state vector is estimated by an iterated Kalman filter. In each step the function h_k is expanded into a linear function at some point $(x_k^{(0)}, q_k^{(0)})$:

$$p_k(x, q_k) \approx h_k(x_k^{(0)}, q_k^{(0)}) + A_k\ (x - x_k^{(0)}) + B_k\ (q_k - q_k^{(0)}) =$$
$$= c_k^{(0)} + A_k\ x + B_k\ q_k,$$

with:

$$A_k = (\partial p_k/\partial x)(x_k^{(0)}, q_k^{(0)}),\ B_k = (\partial p_k/\partial q_k)(x_k^{(0)}, q_k^{(0)}),$$
$$c_k^{(0)} = h_k(x_k^{(0)}, q_k^{(0)}) - A_k\ x_k^{(0)} - B_k\ q_k^{(0)}.$$

$x_k^{(0)}$ is conveniently taken equal to $\tilde{x}_{k-1}$, while $q_k^{(0)}$ is obtained by tracking back from the estimated track parameters p_k towards the approximate vertex position $x_k^{(0)}$. Since there is no prior information about q_k a zero weight matrix is assigned to the "predicted" vector q^{k-1}_k, and the prediction equations look as follows:

$$\tilde{x}^{k-1}_k = \tilde{x}_{k-1},$$
$$C^{k-1}_k = C_{k-1}.$$

Then the filter equations in the weighted means formulation look like this:

$$
\begin{bmatrix} x_k \\ q_k \end{bmatrix} =
\begin{bmatrix} C_k & E_k \\ E_k^T & D_k \end{bmatrix} \cdot
\begin{bmatrix} C_{k-1}^{-1}\ x_{k-1} + A_k^T\ G_k\ (p_k - c_k^{(0)}) \\ B_k^T\ G_k\ (p_k - c_k^{(0)}) \end{bmatrix},
$$

$$
\begin{bmatrix} C_k & E_k \\ E_k^T & D_k \end{bmatrix} =
\begin{bmatrix} C_{k-1}^{-1} + A_k^T\ G_k\ A_k & A_k^T\ G_k\ B_k \\ B_k^T\ G_k\ A_k & B_k^T\ G_k\ B_k \end{bmatrix}^{-1}.
$$

After some matrix algebra one obtains the following results:

$$\tilde{x}_k = C_k [C_{k-1}^{-1} \tilde{x}_{k-1} + A_k^T G^B_k (p_k - c_k^{(0)})],$$
$$\tilde{q}_k = W_k B_k^T G_k (p_k - c_k^{(0)} - A_k \tilde{x}_k),$$
$$C_k = (C_{k-1}^{-1} + A_k^T G^B_k A_k)^{-1},$$
$$D_k = W_k + W_k B_k^T G_k A_k C_k A_k^T G_k B_k W_k,$$
$$E_k = - W_k B_k^T G_k A_k C_k,$$

with:

$$W_k = (B_k^T G_k B_k)^{-1},$$
$$G^B_k = G_k - G_k B_k W_k B_k^T G_k,$$
$$cov \{ \tilde{x}_k \} = C_k , \; cov \{ \tilde{q}_k \} = D_k , \; cov \{ \tilde{x}_k, \tilde{q}_k \} = E_k.$$

The chi-square statistic of the filter step is given by :

$$\chi^2_{k,F} = (p_k - c_k^{(0)} - A_k \tilde{x}_k - B_k \tilde{q}_k)^T G_k (p_k - c_k^{(0)} - A_k \tilde{x}_k - B_k \tilde{q}_k) +$$
$$+ (\tilde{x}_k - \tilde{x}_{k-1})^T C_{k-1}^{-1} (\tilde{x}_k - \tilde{x}_{k-1}),$$
$$\chi^2_k = \chi^2_{k-1} + \chi^2_{k,F}.$$

If necessary, the linear expansion can now be repeated in the new point

$$x_k^{(0)} = \tilde{x}_k, \; q_k^{(0)} = \tilde{q}_k,$$

and the filter can be recomputed, until there is no significant change either in the χ^2 or in the estimate.

Since there is no process noise, the smoother is extremely simple:

$$\tilde{x}^n_k = \tilde{x}_n,$$
$$\tilde{q}^n_k = W_k B_k^T G_k (p_k - c_k^{(0)} - A_k \tilde{x}^n_k),$$
$$C^n_k = C_n,$$
$$D^n_k = W_k + W_k B_k^T G_k A_k C_n A_k^T G_k B_k W_k,$$
$$E^n_k = - W_k B_k^T G_k A_k C_{\bar{n}}.$$

If there is a significant change in the smoothed vertex position, it may be worthwhile to re-compute the derivative matrices A_k and B_k. The residuals and their covariance matrices have the following form:

$$r_k = p_k - c_k^{(0)} - A_k \tilde{x}_k - B_k \tilde{q}_k,$$
$$R_k = V_k (G^B_k - G^B_k A_k C_k A_k^T G^B_k) V_k,$$
$$r^n_k = p_k - c_k^{(0)} - A_k \tilde{x}_n - B_k \tilde{q}^n_k,$$
$$R^n_k = V_k (G^B_k - G^B_k A_k C_k A_k^T G^B_k) V_k.$$

If required, the full covariance matrix can be computed after the smoother:

$$cov \{ \tilde{q}^n_k, \tilde{q}^n_j \} = W_k B_k^T G_k A_k C_n A_j^T G_j B_j W_j.$$

3.7 Association of Tracks to Vertices

Not all tracks found in an event originate at the primary interaction vertex. Also secondary vertices are created by the decay of shortlived particles. Since certain shortlived particles are currently the focus of considerable attention, it is of great physical interest to separate secondary decay vertices from the primary interaction vertex. The shortlived particles in question have average decay lengths between a few hundred micrometers and a few millimeters. Therefore the task is not a trivial one and depends critically on the quality of the track reconstruction. It depends also very much on the properties of the particle beam in the accelerator. If the beam spot is very small and well monitored, the primary vertex position is known very accurately, and consequently secondary vertices are found much more easily.

The search for secondary vertices is done in three steps:
- First, one looks for tracks which do not originate at the primary vertex.
- Secondly, these tracks are combined to form physically possible decay vertices.
- Thirdly, the decay vertex candidates are subject to a geometrical vertex fit and a kinematics fit.

Here only the first step will be discussed: How can one find out which tracks do not come from the primary vertex? If the question is put in this way, one sees immediately that this is an outlier problem, in fact a *multiple outlier problem*, since there are normally several tracks emerging from each decay vertex, and there may be several decay vertices.

Given the high multiplicities (up to 50) and the fact that the number of "outlying" tracks is not known a priori, it is not possible – for reasons of computing time – to work through all subsets of k tracks for several values of k, say $2 \leq k \leq 8$. What is needed is a *sequential procedure*, the computing time of which is basically proportional to n, the total number of tracks. Three different approaches are proposed here. The first two have already been used in a study on vertex evaluation in the DELPHI experiment.

a) Sequential test against a reliable subset

Not all tracks have the same probability to come from a secondary vertex. It may be possible to select tracks very likely produced at the primary vertex according to physical criteria, e.g. high momentum or a certain angular region. If this can be done, a preliminary vertex can be fitted from those tracks. The remaining tracks are then subject to a χ^2-test. As a test statistic the chi-square increment of the filter, $\chi^2_{k,F}$, is used. If the track is accepted, the vertex may or may not be updated. If the beam spot is small and the beam position is precisely known, the beam can be used as a preliminary vertex.

b) Sequential test of the entire sample

In this procedure first all tracks are fitted to a common vertex. Then for each track k
($1 \le k \le n$) an updated vertex is computed, which results from the omission of track k from the
entire sample. This is done by the inverse Kalman filter:

$$\mathbf{C}^n_k{}^* = (\mathbf{C}_n^{-1} - \mathbf{A}_k^T \, \mathbf{G}^B_k \, \mathbf{A}_k)^{-1},$$
$$\tilde{\mathbf{x}}^n_k{}^* = \mathbf{C}^n_k{}^* \, [\, \mathbf{C}_n^{-1} \, \tilde{\mathbf{x}}_n - \mathbf{A}_k^T \, \mathbf{G}^B_k \, (\mathbf{p}_k - \mathbf{c}_{k0}) \,].$$

The distance of the track from the new vertex is expressed by the smoothed chi-square:

$$\chi^2_{k,S} = \mathbf{r}^n_k{}^T \, \mathbf{G}_k \, \mathbf{r}^n_k + (\tilde{\mathbf{x}}_n - \tilde{\mathbf{x}}^n_k{}^*)^T \, (\mathbf{C}^n_k{}^*)^{-1} \, (\tilde{\mathbf{x}}_n - \tilde{\mathbf{x}}^n_k{}^*).$$

$\chi^2_{k,S}$ is used as the test statistic. If it is larger than the $(1-\alpha)$-quantile of the χ^2 - distribution
with two degrees of freedom, the track is considered as an outlier, i.e. rejected from the primary vertex.

This procedure works properly only if the vertex estimated from the entire sample is not
noticeably biased by the outlying tracks. This is only true if there are only few outliers, as is
normally the case, and if the total multiplicity is large, or if the primary vertex is well defined
by several high-momentum tracks or by the beam profile. It is also assumed that obvious
outliers have been rejected already during the filter.

If there is a serious possibility that the estimated vertex is biased by outliers, one can
imagine two solutions: Either the filter is robustified (see next point) or the sequential test is
refined in the following way:
- Set i=1.
- Step 1: Look for the track with the largest $\chi^2_{k,S}$; this is track k_i.
- Step 2: If $\chi^2_{k_i,S} \le c$, stop. If $\chi^2_{k_i,S} > c$, go to Step 3.
- Step 3: Remove track k_i from the vertex and update the momenta of all remaining
  tracks. Set $i \leftarrow i+1$ and go to Step 1.

The number of operations is no longer proportional to n, but rather to m·n, where m is the
number of outliers found.

c) A robust vertex fit

If the possibility cannot be excluded that the primary vertex position is biased by
unrecognized secondary tracks, the vertex fit should be robustified (Frühwirth 1988). This
applies of course also to secondary vertices, which may be biased by a wrongly associated
(primary or secondary) track passing close to the vertex. Such a track would spoil the kinematic fit of the secondary interaction and should be detected and removed.

The final strategy for vertex association depends on the properties of the beam, on the availability of a micro-vertex detector and on the quality of the track reconstruction in general, including effects of detector alignment. It can be arrived at only by detailed simulation studies for a particular detector at a particular accelerator.

4. CONCLUSIONS

These lecture notes have first of all been written for graduate and postdoctoral students, or for new-comers in the field of track reconstruction. We hope, however, that they are also understandable for particle physicists who want to have a glance at this field for their general enlightenment.

These notes are intended to show the current state of the art; they should stimulate a desire to deepen the knowledge of this field, and serve as a guideline for further reading of the more specialized reports which are spread all over the world. They should also convince the experimentalists that efficient track reconstruction is feasible with a limited effort by using proven methods; they should also prevent them from starting to "reinvent the wheel". New effort should only be invested into new ideas, checking carefully in advance the available amount of know-how and manpower. If well planned in advance, i.e. together with the hardware and the layout of the software, the work needed for the application of efficient methods for track reconstruction is small compared to the loss of information which may occur otherwise.

The emphasis of these notes has been put on conservative and proven methods. The large variety of algorithms for pattern recognition provides the individual experimentalist with ample opportunity for picking and choosing, but it is not always easy to select the right method for a particular application. Actually it seems pretty difficult to give general criteria for deciding which method is preferable in a given case. In the field of track fitting, we have concentrated on the classical least-squares estimation of track parameters and on a more recent development, the application of the Kalman filter to track and vertex fitting. Some methods, which are still being developed, have only been hinted at, e.g. the use of vector computers for pattern recognition, or estimators and decision criteria beyond the least-squares method.

If these notes succeed in showing experimentalists how to apply known methods, and if they help them to concentrate on novel ideas, we have not spent our effort in vain.

REFERENCES

Billoir, P. (1984):

    Track Fitting with Multiple Scattering: A New Method.

    Nucl. Instr. and Meth. in Phys. Res. 225, pp. 352 - 366.

Billoir, P., Frühwirth, R. and Regler, M. (1985):

    Track Element Merging Strategy and Vertex Fitting in Complex Modular Detectors.

    Nucl. Instr. and Meth. in Phys. Res. A 241, pp. 115 - 131.

Blobel, V. (1984):

    Least Squares Methods.

    In: Formulae and Methods in Experimental Data Evaluation, Vol. 3, I1 - 33.

    European Physical Society, Geneva, 1984.

Bock, R., Grote, H., Notz, D. and Regler, M. (1989):

    Data Analysis Techniques in High Energy Physics Experiments.

    (Editor: Regler, M.)

    Cambridge University Press (in print).

Frühwirth, R. (1987):

    Application of Kalman Filtering to Track and Vertex Fitting.

    Nucl. Instr. and Meth. in Phys. Res. A 262, pp. 444 - 450.

Frühwirth, R. (1988):

    Application of Filter Methods to the Reconstruction of Tracks and Vertices in Events

    of Experimental High Energy Physics.

    Ph.D. Thesis, Technische Universität, Vienna (in print).

Gluckstern, R.L. (1963):

    Uncertainties in Track Momentum and Direction due to Multiple Scattering and

    Measurement Errors.

    Nucl. Instr. and Meth. 24, pp. 381 - 389.

Grote, H. and Zanella, P. (1980):

    Applied Software for Wire Chambers.

    Proceedings of the 2nd International Wire Chamber Conference.

    Nucl. Instr. and Meth. 176, pp. 29 - 37.

Grote, H. (1987):

    Pattern Recognition in High Energy Physics.

    Reports on Progress in Physics 50, pp. 473 - 500.

Metcalf, M., Regler, M. and Broll, C. (1973):

    A Split Field Magnet Geometry Fit Program NICOLE.

    CERN 73 - 2, CERN, Geneva.

Metcalf, M. (1986):

    Computers in High Energy Physics.

    (Editor: Yovitis, M.)

    Advances in Computers, Vol. 25, Academic Press, Inc., pp. 277 - 334.

Regler, M. (1977):

    Vielfachstreuung in der Ausgleichsrechnung.

    Acta Physica Austriaca 49, 37 - 45.

    (English translation in: Formulae and Methods in Experimental Data Evaluation,

    Vol. 2, G1 -11. European Physical Society, Geneva, 1984)

Regler, M. (1981):

    Influence of Computation Algorithms on Experimental Design.

    Proceedings of the 4th Europhysics Conference on Computational Physics.

    Computer Physics Communications 22, pp. 167 - 175.

Participants at the ASI From left to right: J. Urheim, R. Ribeiro, M. Bertani, M. Regler, L. Ibanez, S. Easo, K. Biery, N. Pastrone, G. Martinelli, J. Harvey, J. Raab, D. Jansen, S. Bianco, A. Leites, R. Mondardini, A. Bhatti, P. Weber, P. Giacomelli, C. Stoughton, P. Vaz, A. Lee, V. Kapoor, Y. Semertzidis, D. Robinson, M. Demarteau, K. Dederichs, W. Trischuk, G. Giacomelli, D. Karlen, B. Winstein, R. Van Kooten, U. Ecker, U. Das-Gupta, M. Schaefer, F. Rotondo, N. Varelas, R. Ng, M. Pia, M. Forbush, P. Kesten, U. Amaldi, S. Schaffner, B. Flaugher, S. Tkaczyk, M. Krammer, C. Bower, W. Lyle, T. Diehl, G. Alimonti, S. Lami, T. Ferbel, S. Blessing, A. Roth, G. Blair, A. Weir, C. Cochet, R. Oedingen, T. Phillips, M. Erdmann, M. Crisler, P. Fuchs, F. Perrier, G. Zioulas, A. Byon, G. Tsipolitis, P. Gouffon and B. Wyslouch. Inserted on front left: V. O'Dell, S. Peggs, J. Dorfan and A. Petradza.

501

INDEX